Advanced Textbooks in Control and Signal Processing

Other titles published in this series:

Genetic Algorithms
K.F. Man, K.S. Tang and S. Kwong

Neural Networks for Modelling and Control of Dynamic Systems
M. Nørgaard, O. Ravn, L.K. Hansen and N.K. Poulsen

Modelling and Control of Robot Manipulators (2nd Edition)
L. Sciavicco and B. Siciliano

Fault Detection and Diagnosis in Industrial Systems
L.H. Chiang, E.L. Russell and R.D. Braatz

Soft Computing
L. Fortuna, G. Rizzotto, M. Lavorgna, G. Nunnari, M.G. Xibilia and R. Caponetto

Statistical Signal Processing
T. Chonavel

Discrete-time Stochastic Processes (2nd Edition)
T. Söderström

Parallel Computing for Real-time Signal Processing and Control
M.O. Tokhi, M.A. Hossain and M.H. Shaheed

Multivariable Control Systems
P. Albertos and A. Sala

Control Systems with Input and Output Constraints
A.H. Glattfelder and W. Schaufelberger

Analysis and Control of Non-linear Process Systems
K. Hangos, J. Bokor and G. Szederkényi

Model Predictive Control (2nd Edition)
E.F. Camacho and C. Bordons

Principles of Adaptive Filters and Self-learning Systems
A. Zaknich

Digital Self-tuning Controllers
V. Bobál, J. Böhm, J. Fessl and J. Macháček

Robust Control Design with MATLAB®
D.-W. Gu, P.Hr. Petkov and M.M. Konstantinov
Publication due July 2005

Active Noise and Vibration Control
M.O. Tokhi
Publication due November 2005

R. Kelly, V. Santibáñez and A. Loría

Control of Robot Manipulators in Joint Space

With 110 Figures

Rafael Kelly, PhD
Centro de Investigación Científica y de Educación Superior de Ensenada
(CICESE), Ensenada B.C. 22800, Mexico

Victor Santibáñez Davila, PhD
Instituto Tecnologico de la Laguna, Torreón, Coahuila, 27001, Mexico

Antonio Loría, PhD
CNRS, Laboratoire des Signaux et Systèmes, Supélec, 3 rue Joliot Curie,
91192 Gif-sur-Yvette, France

British Library Cataloguing in Publication Data
Kelly, R.
Control of robot manipulators in joint space. - (Advanced
textbooks in control and signal processing)
1. Robots - Control systems 2. Manipulators (Mechanism)
3. Programmable controllers
I. Title II. Santibáñez, V. III. Loría, A.
629.8'933
ISBN-10: 1852339942

Library of Congress Control Number: 2005924306

Advanced Textbooks in Control and Signal Processing series ISSN 1439-2232
ISBN-10: 1-85233-994-2
ISBN-13: 978-1-85233-994-4
Springer Science+Business Media
springeronline.com

Typesetting: Camera ready by authors
Production: LE-TEX Jelonek, Schmidt & Vöckler GbR, Leipzig, Germany
Printed in Germany
69/3141-543210 Printed on acid-free paper SPIN 11321323

To my parents,
with everlasting love, respect and admiration.
–AL

"Attentive readers, who spread their thoughts among themselves, always go beyond the author"

—Voltaire*, 1763.

* Original citation in French: *"Des lecteurs attentifs, qui se communiquent leurs pensées, vont toujours plus loin que l'auteur"*, in *Traité sur la tolérence à l'occasion de la mort de Jean Calas*, Voltaire, 1763.

Series Editors' Foreword

The topics of control engineering and signal processing continue to flourish and develop. In common with general scientific investigation, new ideas, concepts and interpretations emerge quite spontaneously and these are then discussed, used, discarded or subsumed into the prevailing subject paradigm. Sometimes these innovative concepts coalesce into a new sub-discipline within the broad subject tapestry of control and signal processing. This preliminary battle between old and new usually takes place at conferences, through the Internet and in the journals of the discipline. After a little more maturity has been acquired by the new concepts then archival publication as a scientific or engineering monograph may occur.

A new concept in control and signal processing is known to have arrived when sufficient material has evolved for the topic to be taught as a specialized tutorial workshop or as a course to undergraduate, graduate or industrial engineers. *Advanced Textbooks in Control and Signal Processing* are designed as a vehicle for the systematic presentation of course material for both popular and innovative topics in the discipline. It is hoped that prospective authors will welcome the opportunity to publish a structured and systematic presentation of some of the newer emerging control and signal processing technologies in the textbook series.

One of our aims for the *Advanced Textbooks in Control and Signal Processing* series is to create a set of course textbooks that are comprehensive in their coverage. Even though a primary aim of the series is to service the textbook needs of various types of advanced courses we also hope that the industrial control engineer and the control academic will be able to collect the series volumes and use them as a reference library in control and signal processing.

Robotics is an area where the series has the excellent entry in the volume by L. Sciavicco and B. Siciliano entitled *Modelling and Control of Robot Manipulators*, now in its second edition. To complement our coverage in Robotics, we are pleased to welcome into the series this new volume *Control of Robot Manipulators in Joint Space* by Rafael Kelly, Víctor Santibáñez and Antonio Loría. Other topics like models, kinematics and dynamics are introduced into the narrative as and when they are needed to design and compute the robot manipulator controllers. Another novel feature of the text is the extensive use of the laboratory prototype *Pelican* robotic manipulator as the test-bed case study for the robot manipulator controllers devised. This ensures that the reader will be able to see how robot manipulator control is done in practice. Indeed, this means that the text can be closely linked to "hands on" laboratory experience. Control and mechatronics lecturers wishing to use the textbook to support their advance course on robot manipulator control will find the lecture presentation slides, and the problem solutions, which are available at `springonline.com`, an added bonus.

The style of the text is formally rigorous but avoids a lemma–theorem presentation in favour of one of thorough explanation. Chapter 2 of the text covers the main mathematical tools and introduces the concepts of the direct (or second) method of Lyapunov for system stability analysis. This is needed because the robot manipulator system is a nonlinear system. Since the coverage in this chapter includes a wide range of stability concepts, the reader will be pleased to find each new concept supported by a worked example. Robot dynamics and their implications for robot manipulator control are covered in Chapters 3 and 4 whilst Chapter 5 moves on to discuss the model details of the Pelican prototype robotic manipulator. The kinematic and dynamic models are, described and model parameter values given. This chapter shows how the Pelican prototype is "kitted out" with a set of models the properties of which are then investigated in preparation for the control studies to follow.

Parts II to IV (covering Chapters 6 to 16) are devoted to robot manipulator controller design and performance case studies. This shows just how focused the textbook is on robot manipulator *control*. This study is given in three stages: position control (Part II); motion control (Part Ill) and advanced control topics (Part IV). Remarkably, the workhorse controller type being used is from the PID family so that the control focus is close to the type of controllers used widely in industrial applications, namely from the classical Proportional, Integral, Derivative controller family. In these chapter-length controller studies, the earlier lessons in Lyapunov stability methods come to the fore, demonstrating how Lyapunov theory is used for controllers of a classical form being used with nonlinear system models to prove the necessary stability results. The advanced control topics covered in Part IV include a range of adaptive control methods. Four appendices are given with additional material on the mathematical and Lyapunov methods used and on the modelling details of direct current motors.

There is no doubt that this robot manipulator control course textbook is a challenging one but ultimately a very rewarding one. From a general viewpoint the reward of learning about how to approach classical control for systems having nonlinear models is a valuable one with potential application in other control fields. For robot manipulator control *per se*, the book is rigorous, thorough and comprehensive in its presentation and is an excellent addition to the series of advanced course textbooks in control and signal processing.

M.J. Grimble and M.A. Johnson
Glasgow, Scotland, U.K.
March 2005

Preface

The concept of *robot* has transformed from the *idea* of an artificial superhuman, materialized by the pen of science fiction writer Karel Čapek, into the *reality* of animated autonomous machines. An important class of these are the *robot manipulators*, designed to perform a wide variety of tasks in production lines of diverse industrial sectors; perhaps the most clear example is the automotive industry. *Robotics*, introduced by science fiction writer Isaac Asimov as the study of robots, has become a truly vast field of modern technology involving specialized knowledge from a range of disciplines such as electrical engineering, mechatronics, cybernetics, computer science, mechanical engineering and applied mathematics.

As a result, courses on robotics continue to gain interest and, following the demands of modern industry, every year more and more program studies, from engineering departments and faculties of universities round the globe, include robotics as a *compulsory* subject. While a complete course on robotics that is, including topics such as modeling, control, technological implementation and instrumentation, may need two terms at graduate level to be covered in fair *generality*, other more specialized courses can be studied in one senior year term. The present text addresses the subject in the second manner; it is mostly devoted to the specific but vast topic of robot *control.*

Robot control is the *spine* of robotics. It consists in studying how to make a robot manipulator do what it is desired to do *automatically*; hence, it consists in designing robot *controllers.* Typically, these take the form of an equation or an algorithm which is realized via specialized computer programs. Then, controllers form part of the so-called robot control system which is physically constituted of a computer, a data acquisition unit, *actuators* (typically electrical motors), the robot itself and some extra "electronics". Thus, the design and full implementation of a robot controller relies on every and each of the above-mentioned disciplines.

The simplest controller for industrial robot manipulators is the Proportional Integral Derivative (PID) controller. In general, this type of controller is designed on the basis that the robot model is composed of independent coupled dynamic (differential) equations. While these controllers are widely used in industrial manipulators (robotic arms), depending on the task to be carried out, they do not always result in the best performance. To improve the latter it is current practice to design so-called *model-based* controllers, which require a precise knowledge of the dynamic model including the values of the physical parameters involved. Other, non-model-based controllers, used mainly in academic applications and research prototypes include the so-called variable-structure controllers, fuzzy controllers, learning controllers, neural-net-based controllers, to mention a few.

The majority of available texts on robotics cover all of its main aspects, that is, modeling (of kinematics and dynamics), trajectory generation (that is, the mathematical setting of a *task* to be performed by the robot), robot control and some of them, instrumentation, software and other implementation issues. Because of their wide scope, texts typically broach the mentioned topics in a survey rather than a detailed manner.

Control of robot manipulators in joint space is a counter-fact to most available literature on robotics since it is mostly devoted to robot *control*, while addressing other topics, such as kinematics, mainly through case studies. Hence, we have sacrificed generality for depth and clarity of exposition by choosing to address in great detail a range of model-based controllers such as: Proportional Derivative (PD), Proportional Integral Derivative (PID), Computed torque and some variants including adaptive versions. For purely didactic reasons, we have also chosen to focus on control in joint space, totally skipping *task* space and *end-effector* space based control. These topics are addressed in a number of texts elsewhere.

The present book opens with an introductory chapter explaining, in general terms, what robot control involves. It contains a chapter on preliminaries which presents in a considerably detailed manner the main mathematical concepts and tools necessary to study robot control. In particular, this chapter introduces the student to advanced topics such as Lyapunov stability, the core of control theory and therefore, of robot control. We emphasize at this point that, while this topic is usually reserved for graduate students, we have paid special attention to include only the most basic theorems and we have reformulated the latter in simple statements. We have also included numerous examples and explanations to make this material accessible to senior year undergraduate students.

Kinematics is addressed mainly through examples of different manipulators. Dynamics is presented in two chapters but from a viewpoint that stresses the most relevant issues for robot control; *i.e.* we emphasize certain fundamental properties of the dynamic model of robots, which are commonly taken as satisfied hypotheses in control design.

We have also included a chapter entirely devoted to the detailed description of the *Pelican* prototype, a 2-degrees-of-freedom direct-drive planar articulated arm that is used throughout the book as a case study to test the performance of the studied controllers, in lab *experimentation*. Dynamic and kinematic models are derived in detail for this particular robot. The rest of the book (about 70%) is devoted to the study of a number of robot controllers, each of which is presented in a separate chapter.

The text is organized in four main parts: I) Preliminaries, which contains the two chapters on robot dynamics, the chapter on mathematical preliminaries and the chapter describing the Pelican prototype. Parts II and III contain, respectively, set-point and tracking model-based controllers. Part IV covers additional topics such as *adaptive* versions of the controllers studied in parts II and III, and a controller that does not rely on velocity measurements. Appendices containing some extra mathematical support, Lyapunov theory for the advanced reader and a short treatment on DC motors, are presented at the end of the book.

Thus, the present book is a self-contained text to serve in a course on robot control *e.g.*, within a program of Mechatronics or Electrical Engineering at senior year of BSc or first year of MSc. Chapter 1 may be covered in one or two sessions. We strongly recommend taking the time to revise thoroughly Chapter 2 which is instrumental for the remainder of the textbook. The rest of the material may be taught in different ways and depths depending on the level of students and the duration of the course. For instance, Parts I through III may be covered entirely in about 50 hours at senior year level. If the course is to be shortened, the lecturer may choose to pass over Chapters 3 and 4 faster (or even completely skip them and refer to their contents only when necessary) and to introduce the student to kinematics and dynamics using Chapter 5; then, to focus on Parts II and III. For a yet shorter but coherent basic course, the lecturer may choose to teach only Chapters 1, 2, 5 and, for the subsequent chapters of Parts II and III, concentrate on a brief study of the control laws while emphasizing the examples that concern the Pelican prototype. Further, support material for class -presentation slides for the lecturer and problems' solutions manual- are available in electronic form at `springonline.com`.

For a graduate course the lecturer may choose to cover, in addition, the three chapters on adaptive control (Chapters 14–16), or Chapter 13 on control without velocity measurements and Chapter 14, to give a short introduction to adaptive control. We remark that the advanced topics of Part IV require the material in the appendices which could be taught, for instance, at the beginning of the course or could be left as a self-study topic.

The textbook is written in a style and technical language targeted toward undergraduate senior students. Hence, we have favored a thoroughly explanatory, yet rigorous, style over a stiff mathematical (theorem-proof streamed) one. We have taken care to invoke a strictly minimum number of mathematical terms and these are mostly explained when introduced. Mathematical objects such as theorems and definitions are kept to a minimum; they are mainly present in Chapter 2 (mathematical preliminaries) and some appendices. Yet, when simplicity in the language may induce mathematical ambiguity or imprecision we have added clarifying footnotes. A large number of examples, illustrations and problems to be solved in class or as homework by the student are provided.

The precedents of the text date back to lecture notes of the first author that were printed by the National Autonomous University of Mexico (UNAM) in 1989. It has been enriched by the authors' experience of teaching the topic over more than 15 years at undergraduate (senior year) and graduate levels (first year), in several institutions in Europe and The Americas: National Autonomous Univ. of Mexico (UNAM), Mexico; Technological Institute and of High Studies of Monterrey (ITESM), Mexico; Center of Research and High Studies of Ensenada (CICESE), Mexico; Laguna Institute of Technology, Mexico; University of California at Santa Barbara, USA; National University of Science and Technology (NTNU), Norway; San Juan National University, Argentina. This has provided the text with invaluable feedback from a varied audience with different technical and cultural backgrounds. Thus, the authors are confident to say that this textbook has not been written to be *tested* but to be *used* in class.

A few final words on the nomenclature are necessary. Figures, Examples, Equations, Tables, Theorems, Lemmas, Definitions are numbered independently and carry the number of the chapter. We use the following abbreviations of Latin idioms:

i.e. –*id est*– meaning "that is";
e.g. –*exempli gratia*– meaning "for instance";
cf. –*confer*– meaning "see";
etc. –*etcetera*– meaning "and the rest".

Acknowledgments

The authors wish to thank the Mexican National Council of Science and Technology (CONACyT) whose sponsorship, to the first author, yielded an early version of this text (in Spanish). The first author also acknowledges the support of the Mexican Centre for Scientific Research and High Studies of Ensenada (CICESE). The second author acknowledges the receipt of numerous research grants from CONACyT and the Council of the National System of Technological Education (COSNET), which served in part in the elaboration of this text. Most of the writing of this textbook was realized while the third author was holding a visiting professorship at CICESE in 2002 and 2003. The third author acknowledges the grants obtained and praises the extraordinary working conditions provided by the French National Centre for Scientific Research (CNRS).

The realization of this textbook would not have been possible without the valuable feedback of numerous colleagues and students throughout the years. In particular, the first author is thanks Ricardo Carelli and Romeo Ortega, the collaboration with whom extended over almost 20 years, and which considerably improved both the contents and writing of the present book. The authors also acknowledge the numerous exchanges on the topics of the present book, with Mark Spong, Suguru Arimoto, Carlos Canudas de Wit, Jean-Jacques Slotine, John T. Wen, Roberto Horowitz, Daniel Koditschek, Claude Samson, Louis Whitcomb, Harry Berghuis, Henk Nijmeijer, Hebertt Sira-Ramírez, Juan M. Ibarra, Alfonso Pámanes, Ilse Cervantes, José Alvarez-Ramírez, Antoine Chaillet and Marco A. Arteaga.

Special words of thanks go to Ricardo Campa who actively participated in the lab experiments presented in the examples throughout the book. The authors wish to single out the invaluable comments, remarks and corrections provided by the students of the numerous institutions where this material has been taught.

The third author takes this opportunity to mention that it was with an early version of the lecture notes that evolved into this text, that he was introduced to Lyapunov theory and robotics, *by* the first author. It is a honor and a great pleasure to participate in writing this book. He also wishes to express his deep gratitude to his friend and scientific mentor Romeo Ortega for his valuable teaching, in particular, on robot control.

The authors acknowledge the valuable assistance of Oliver Jacksson, their contact editor at Springer-Verlag, London, along the publication process of this book; from the state of proposal to its realization. Last but not least, the authors acknowledge both their technical and language reviewers; it goes without saying that any error in the contents or in the typeset of the present text is the entire responsibility of the authors.

Ensenada, Mexico
Torreón, Mexico
Gif sur Yvette, France

Rafael Kelly,
Víctor Santibáñez,
Antonio Loría

May 2005

Contents

List of Figures

Part I

Preliminaries

Introduction to Part I

The high quality and rapidity requirements in production systems of our globalized contemporary world demand a wide variety of technological advancements. Moreover, the incorporation of these advancements in modern industrial plants grows rapidly. A notable example of this situation, is the privileged place that *robots* occupy in the modernization of numerous sectors of the society.

The word *robot* finds its origins in *robota* which means *work* in Czech. In particular, *robot* was introduced by the Czech science fiction writer Karel Čapek to name artificial *humanoids* – biped robots – which helped human beings in physically difficult tasks. Thus, beyond its literal definition the term robot is nowadays used to denote animated *autonomous* machines. These machines may be roughly classified as follows:

- Robot manipulators
- Mobile robots
  - Ground robots
    - Wheeled robots
    - Legged robots
  - Submarine robots
  - Aerial robots

Both, mobile robots and manipulators are key pieces of the mosaic that constitutes *robotics* nowadays. This book is exclusively devoted to robot *manipulators.*

Robotics – a term coined by the science fiction writer Isaac Asimov – is as such a rather recent field in modern technology. The good understanding and development of robotics applications are conditioned to the good knowledge of different disciplines. Among these, electrical engineering, mechanical engineering, industrial engineering, computer science and applied mathematics. Hence, robotics incorporates a variety of fields among which is automatic *control of robot manipulators.*

To date, we count several definitions of *industrial robot manipulator* not without polemic among authors. According to the definition adopted by the International Federation of Robotics under standard ISO/TR 8373, a robot manipulator is defined as follows:

> A manipulating industrial robot is an automatically controlled, reprogrammable, multipurpose manipulator programmable in three or more axes, which may be either fixed in place or mobile for use in industrial automation applications.

In spite of the above definition, we adopt the following one for the pragmatic purposes of the present textbook: a robot manipulator – or simply, manipulator – is a mechanical articulated arm that is constituted of links interconnected through hinges or joints that allow a relative movement between two consecutive links.

The movement of each joint may be prismatic, revolute or a combination of both. In this book we consider only joints which are *either* revolute or prismatic. Under reasonable considerations, the number of joints of a manipulator determines also its number of *degrees of freedom* (*DOF*). Typically, a manipulator possesses 6 DOF, among which 3 determine the position of the end of the last link in the Cartesian space and 3 more specify its orientation.

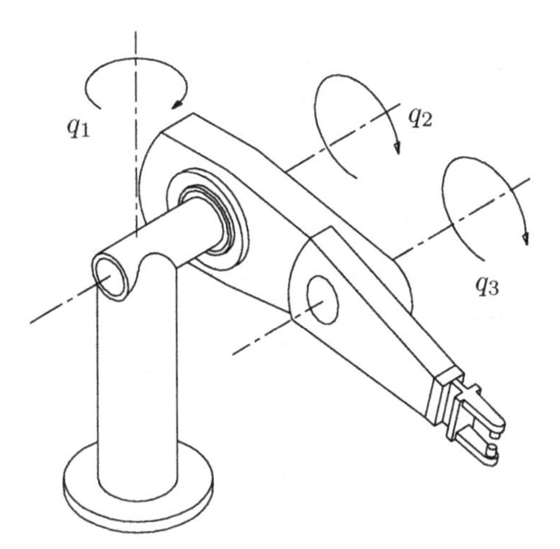

Figure I.1. Robot manipulator

Figure I.1 illustrates a robot manipulator. The variables q_1, q_2 and q_3 are referred to as the joint positions of the robot. Consequently, these positions denote under the definition of an adequate reference frame, the positions (displacements) of the robot's joints which may be linear or angular. For analytical purposes, considering an n-DOF robot manipulator, the joint positions are collected in the vector $\boldsymbol{q}$, *i.e.*[2]

$$\boldsymbol{q} := \begin{bmatrix} q_1 \\ q_2 \\ \vdots \\ q_n \end{bmatrix}.$$

Physically, the joint positions $\boldsymbol{q}$ are measured by sensors conveniently located on the robot. The corresponding *joint velocities* $\dot{\boldsymbol{q}} := \frac{d}{dt}\boldsymbol{q}$ may also be measured or estimated from joint position evolution.

To each joint corresponds an actuator which may be electromechanical, pneumatic or hydraulic. The *actuators* have as objective to generate the forces or torques which produce the movement of the links and consequently, the movement of the robot as a whole. For analytical purposes these torques and forces are collected in the vector $\boldsymbol{\tau}$, *i.e.*

$$\boldsymbol{\tau} := \begin{bmatrix} \tau_1 \\ \tau_2 \\ \vdots \\ \tau_n \end{bmatrix}.$$

In its industrial application, robot manipulators are commonly employed in repetitive tasks of precision and others, which may be hazardous for human beings. The main arguments in favor of the use of manipulators in industry is the reduction of production costs, enhancement of precision, quality and productivity while having greater flexibility than specialized machines. In addition to this, there exist applications which are monopolized by robot manipulators, as is the case of tasks in hazardous conditions such as in radioactive, toxic zones or where a risk of explosion exists, as well as spatial and submarine applications. Nonetheless, short-term projections show that assembly tasks will continue to be the main applications of robot manipulators.

[2] The symbol ":=" stands for *is defined as*.

1

What Does "Control of Robots" Involve?

The present textbook focuses on the interaction between robotics and electrical engineering and more specifically, in the area of *automatic control*. From this interaction emerges what we call *robot control*.

Loosely speaking (in this textbook), robot control consists in studying how to make a robot manipulator perform a task and in materializing the results of this study in a lab prototype.

In spite of the numerous existing commercial robots, robot control design is still a field of intensive study among robot constructors and research centers. Some specialists in automatic control might argue that today's industrial robots are already able to perform a variety of complex tasks and therefore, at first sight, the research on robot control is not justified anymore. Nevertheless, not only is research on robot control an interesting topic by itself but it also offers important theoretical challenges and more significantly, its study is indispensable in specific tasks which cannot be performed by the present commercial robots.

As a general rule, control design may be divided roughly into the following steps:

- familiarization with the physical system under consideration;
- modeling;
- control specifications.

In the sequel we develop further on these stages, emphasizing specifically their application in robot control.

1.1 Familiarization with the Physical System under Consideration

On a general basis, during this stage one must determine the physical variables of the system whose behavior is desired to control. These may be temperature, pressure, displacement, velocity, *etc.* These variables are commonly referred to as the system's *outputs*. In addition to this, we must also clearly identify those variables that are available and that have an influence on the behavior of the system and more particularly, on its outputs. These variables are referred to as *inputs* and may correspond for instance, to the opening of a valve, voltage, torque, force, *etc.*

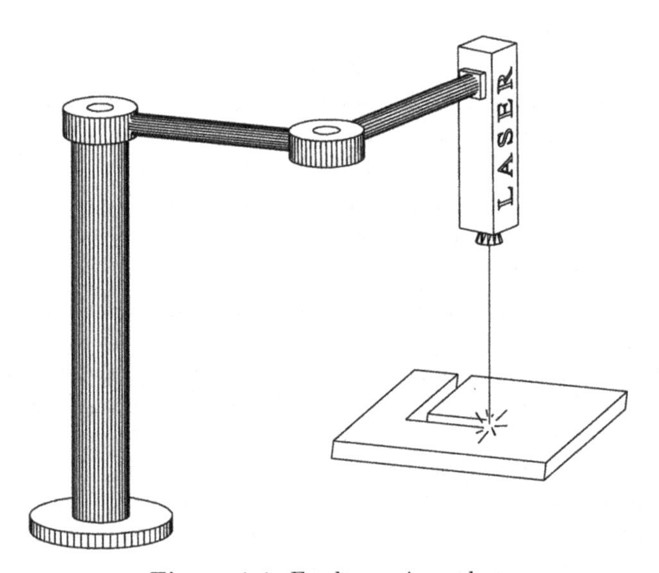

Figure 1.1. Freely moving robot

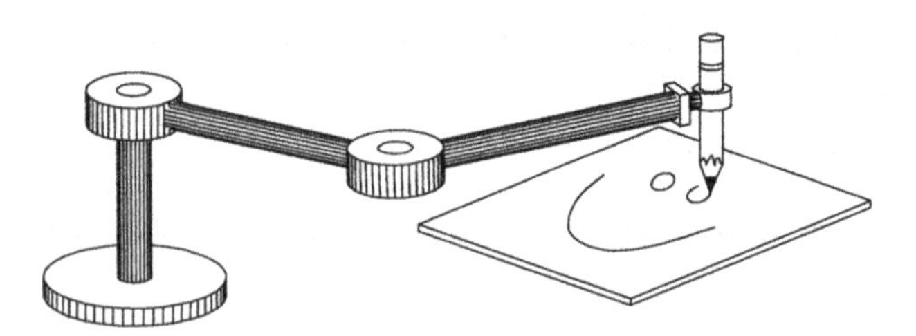

Figure 1.2. Robot interacting with its environment

In the particular case of robot manipulators, there is a wide variety of outputs – temporarily denoted by $\boldsymbol{y}$ – whose behavior one may wish to control.

For robots moving freely in their *workspace*, *i.e.* without interacting with their environment (*cf.* Figure 1.1) as for instance robots used for painting, "pick and place", laser cutting, *etc.*, the output $\boldsymbol{y}$ to be controlled, may correspond to the joint positions $\boldsymbol{q}$ and joint velocities $\dot{\boldsymbol{q}}$ or alternatively, to the position and orientation of the end-effector (also called end-tool).

For robots such as the one depicted in Figure 1.2 that have physical contact with their environment, *e.g.* to perform tasks involving polishing, deburring of materials, high quality assembling, *etc.*, the output $\boldsymbol{y}$ may include the torques and forces $\boldsymbol{f}$ exerted by the end-tool over its environment.

Figure 1.3 shows a manipulator holding a marked tray, and a camera which provides an image of the tray with marks. The output $\boldsymbol{y}$ in this system may correspond to the coordinates associated to each of the marks with reference to a screen on a monitor. Figure 1.4 depicts a manipulator whose end-effector has a camera attached to capture the scenery of its environment. In this case, the output $\boldsymbol{y}$ may correspond to the coordinates of the dots representing the marks on the screen and which represent visible objects from the environment of the robot.

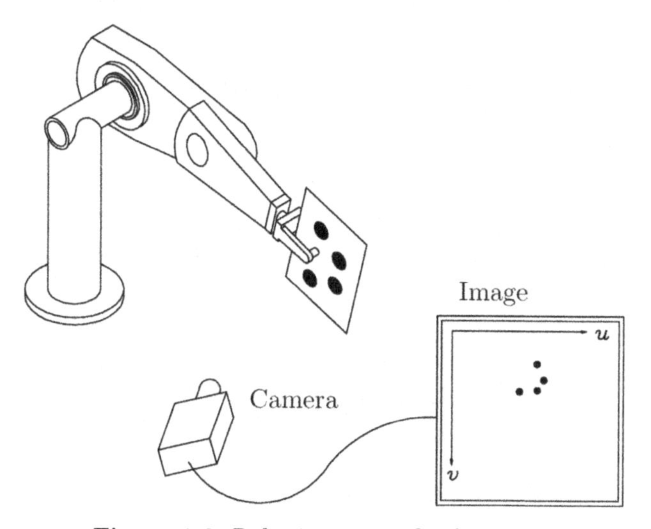

Figure 1.3. Robotic system: fixed camera

From these examples we conclude that the corresponding output $\boldsymbol{y}$ of a robot system – involved in a specific class of tasks – may in general, be of the form

$$\boldsymbol{y} = \boldsymbol{y}(\boldsymbol{q}, \dot{\boldsymbol{q}}, \boldsymbol{f}) .$$

On the other hand, the input variables, that is, those that may be modified to affect the evolution of the output, are basically the torques and forces $\boldsymbol{\tau}$ applied by the actuators over the robot's joints. In Figure 1.5 we show

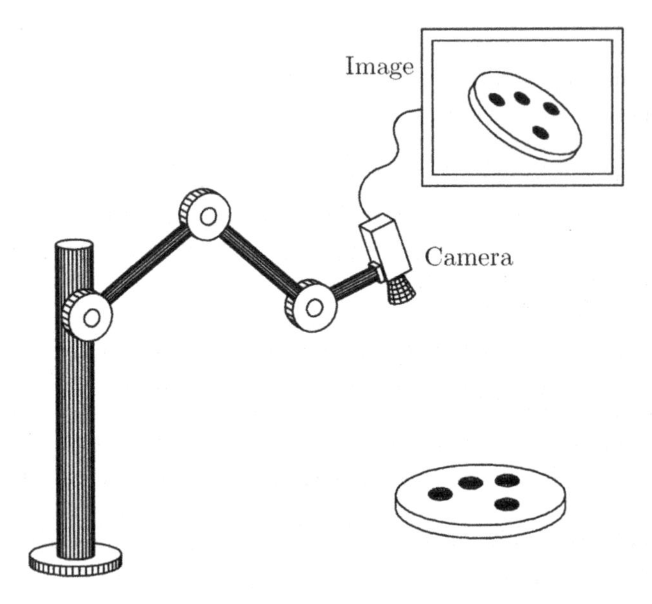

Figure 1.4. Robotic system: camera in hand

the block-diagram corresponding to the case when the outputs are the joint positions and velocities, that is,

$$\boldsymbol{y} = \boldsymbol{y}(\boldsymbol{q}, \dot{\boldsymbol{q}}, \boldsymbol{f}) = \begin{bmatrix} \boldsymbol{q} \\ \dot{\boldsymbol{q}} \end{bmatrix}$$

while $\boldsymbol{\tau}$ is the input. In this case notice that for robots with n joints one has, in general, $2n$ outputs and n inputs.

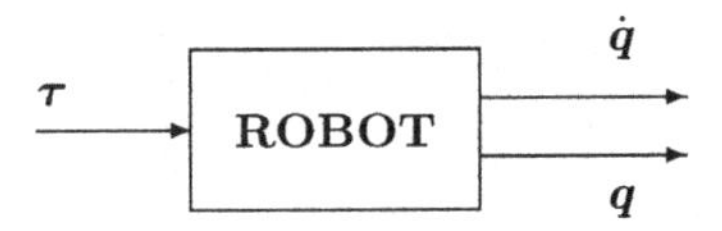

Figure 1.5. Input–output representation of a robot

1.2 Dynamic Model

At this stage, one determines the mathematical model which relates the input variables to the output variables. In general, such mathematical representation of the system is realized by ordinary differential equations. The system's mathematical model is obtained typically via one of the two following techniques.

- *Analytical*: this procedure is based on physical laws of the system's motion. This methodology has the advantage of yielding a mathematical model as precise as is wanted.
- *Experimental*: this procedure requires a certain amount of experimental data collected from the system itself. Typically one examines the system's behavior under specific input signals. The model so obtained is in general more imprecise than the analytic model since it largely depends on the inputs and the operating point[1]. However, in many cases it has the advantage of being much easier and quicker to obtain.

On certain occasions, at this stage one proceeds to a simplification of the system model to be controlled in order to design a relatively simple controller. Nevertheless, depending on the degree of simplification, this may yield malfunctioning of the overall controlled system due to potentially neglected physical phenomena. The ability of a control system to cope with errors due to neglected dynamics is commonly referred to as *robustness*. Thus, one typically is interested in designing robust controllers.

In other situations, after the modeling stage one performs the *parametric identification*. The objective of this task is to obtain the numerical values of different physical parameters or quantities involved in the dynamic model. The identification may be performed via techniques that require the measurement of inputs and outputs to the controlled system.

The dynamic model of robot manipulators is typically derived in the analytic form, that is, using the laws of physics. Due to the mechanical nature of robot manipulators, the laws of physics involved are basically the laws of mechanics.

On the other hand, from a dynamical systems viewpoint, an n-DOF system may be considered as a *multivariable nonlinear* system. The term "multivariable" denotes the fact that the system has multiple (*e.g.* n) inputs (the forces and torques $\boldsymbol{\tau}$ applied to the joints by the electromechanical, hydraulic or pneumatic actuators) and, multiple ($2n$) state variables typically associated to the n positions $\boldsymbol{q}$, and n joint velocities $\dot{\boldsymbol{q}}$. In Figure 1.5 we depict the corresponding block-diagram assuming that the state variables also correspond to the outputs. The topic of robot dynamics is presented in Chapter 3. In Chapter 5 we provide the specific dynamic model of a two-DOF prototype of a robot manipulator that we use to illustrate through examples, the performance of the controllers studied in the succeeding chapters. Readers interested in the aspects of dynamics are invited to see the references listed on page 16.

As was mentioned earlier, the dynamic models of robot manipulators are in general characterized by ordinary nonlinear and nonautonomous[2] differential equations. This fact limits considerably the use of control techniques

[1] That is the working regime.

[2] That is, they depend on the state variables and time. See Chapter 2.

tailored for linear systems, in robot control. In view of this and the present requirements of precision and rapidity of robot motion it has become necessary to use increasingly sophisticated control techniques. This class of control systems may include nonlinear and adaptive controllers.

1.3 Control Specifications

During this last stage one proceeds to dictate the desired characteristics for the control system through the definition of control objectives such as:

- stability;
- regulation (position control);
- trajectory tracking (motion control);
- optimization.

The most important property in a control system, in general, is *stability*. This fundamental concept from control theory basically consists in the property of a system to go on working at a regime or *closely* to it *for ever*.

Two techniques of analysis are typically used in the analytical study of the stability of controlled robots. The first is based on the so-called *Lyapunov* stability theory. The second is the so-called *input–output* stability theory. Both techniques are complementary in the sense that the interest in Lyapunov theory is the study of stability of the system using a *state* variables description, while in the second one, we are interested in the stability of the system from an input–output perspective. In this text we concentrate our attention on Lyapunov stability in the development and analysis of controllers. The foundations of Lyapunov theory are presented in the Chapter 2.

In accordance with the adopted definition of a robot manipulator's output $\boldsymbol{y}$, the control objectives related to *regulation* and *trajectory tracking* receive special names. In particular, in the case when the output $\boldsymbol{y}$ corresponds to the joint position $\boldsymbol{q}$ and velocity $\dot{\boldsymbol{q}}$, we refer to the control objectives as "*position* control in joint coordinates" and "*motion* control in joint coordinates" respectively. Or we may simply say "position" and "motion" control respectively. The relevance of these problems motivates a more detailed discussion which is presented next.

1.4 Motion Control of Robot Manipulators

The simplest way to specify the movement of a manipulator is the so-called "point-to-point" method. This methodology consists in determining a series of points in the manipulator's workspace, which the end-effector is required to pass through (*cf.* Figure 1.6). Thus, the position control problem consists in making the end-effector go to a specified point regardless of the trajectory followed from its initial configuration.

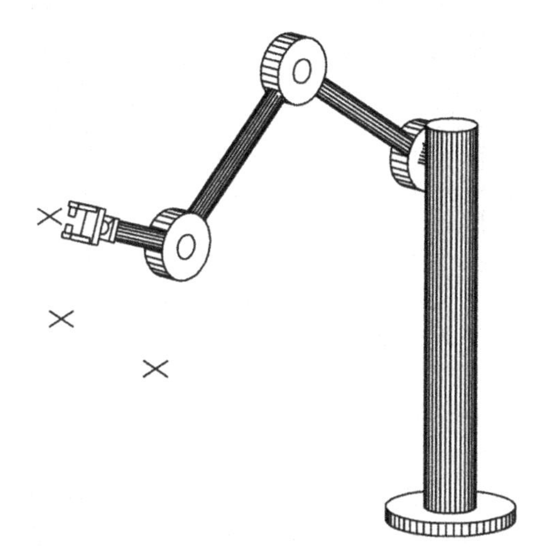

Figure 1.6. Point-to-point motion specification

A more general way to specify a robot's motion is via the so-called (continuous) trajectory. In this case, a (continuous) curve, or path in the state space and parameterized in time, is available to achieve a desired task. Then, the *motion* control problem consists in making the end-effector follow this trajectory as closely as possible (*cf.* Figure 1.7). This control problem, whose study is our central objective, is also referred to as trajectory tracking control.

Let us briefly recapitulate a simple formulation of robot control which, as a matter of fact, is a particular case of motion control; that is, the position control problem. In this formulation the specified trajectory is simply a point in the workspace (which may be translated under appropriate conditions into a point in the joint space). The position control problem consists in driving the manipulator's end-effector (resp. the joint variables) to the desired position, regardless of the initial posture.

The topic of motion control may in its turn, be fitted in the more general framework of the so-called *robot navigation*. The robot navigation problem consists in solving, in one single step, the following subproblems:

- path planning;
- trajectory generation;
- control design.

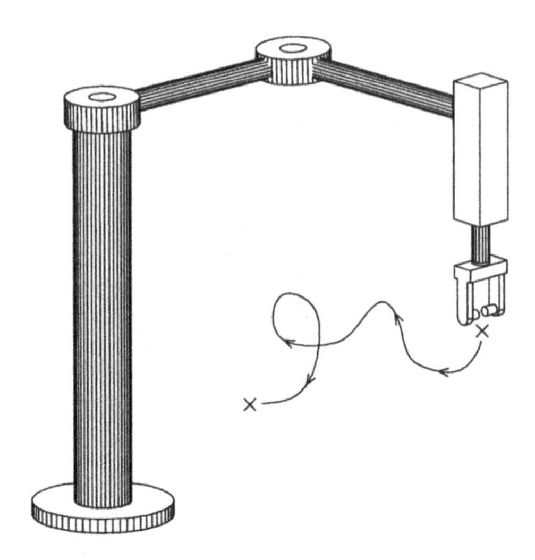

Figure 1.7. Trajectory motion specification

Path planning consists in determining a curve in the state space, connecting the initial and final desired posture of the end-effector, while avoiding any obstacle. Trajectory generation consists in parameterizing in time the so-obtained curve during the path planning. The resulting time-parameterized trajectory which is commonly called the *reference* trajectory, is obtained primarily in terms of the coordinates in the workspace. Then, following the so-called method of *inverse kinematics* one may obtain a time-parameterized trajectory for the joint coordinates. The control design consists in solving the control problem mentioned above.

The main interest of this textbook is the study of motion controllers and more particularly, the analysis of their inherent stability in the sense of Lyapunov. Therefore, we assume that the problems of path planning and trajectory generation are previously solved.

The dynamic models of robot manipulators possess parameters which depend on physical quantities such as the mass of the objects possibly held by the end-effector. This mass is typically unknown, which means that the values of these parameters are unknown. The problem of controlling systems with unknown parameters is the main objective of the *adaptive* controllers. These owe their name to the addition of an adaptation law which updates on-line, an estimate of the unknown parameters to be used in the control law. This motivates the study of adaptive control techniques applied to robot control. In the past two decades a large body of literature has been devoted to the adaptive control of manipulators. This problem is examined in Chapters 15 and 16.

We must mention that in view of the scope and audience of the present textbook, we have excluded some control techniques whose use in robot motion control is supported by a large number of publications contributing both theoretical and experimental achievements. Among such strategies we mention the so-called passivity-based control, variable-structure control, learning control, fuzzy control and neural-networks-based. These topics, which demand a deeper knowledge of control and stability theory, may make part of a second course on robot control.

Bibliography

A number of concepts and data related to robot manipulators may be found in the introductory chapters of the following textbooks.

- Paul R., 1981, *"Robot manipulators: Mathematics programming and control"*, MIT Press, Cambridge, MA.
- Asada H., Slotine J. J., 1986, *"Robot analysis and control "*, Wiley, New York.
- Fu K., Gonzalez R., Lee C., 1987, *"Robotics: Control, sensing, vision and intelligence"*, McGraw–Hill.
- Craig J., 1989, *"Introduction to robotics: Mechanics and control"*, Addison-Wesley, Reading, MA.
- Spong M., Vidyasagar M., 1989, *"Robot dynamics and control"*, Wiley, New York.
- Yoshikawa T., 1990, *"Foundations of robotics: Analysis and control"*, The MIT Press.
- Nakamura Y., 1991, *"Advanced robotics: Redundancy and optimization"*, Addison–Wesley, Reading, MA.
- Spong M., Lewis F. L., Abdallah C. T., 1993, *"Robot control: Dynamics, motion planning and analysis"*, IEEE Press, New York.
- Lewis F. L., Abdallah C. T., Dawson D. M., 1993, *"Control of robot manipulators"*, Macmillan Pub. Co.
- Murray R. M., Li Z., Sastry S., 1994, *"A mathematical introduction to robotic manipulation"*, CRC Press, Inc., Boca Raton, FL.
- Qu Z., Dawson D. M., 1996, *"Robust tracking control of robot manipulators"*, IEEE Press, New York.
- Canudas C., Siciliano B., Bastin G., (Eds), 1996, *"Theory of robot control"*, Springer-Verlag, London.
- Arimoto S., 1996, *"Control theory of non–linear mechanical systems"*, Oxford University Press, New York.
- Sciavicco L., Siciliano B., 2000, *"Modeling and control of robot manipulators"*, Second Edition, Springer-Verlag, London.

- de Queiroz M., Dawson D. M., Nagarkatti S. P., Zhang F., 2000, *"Lyapunov–based control of mechanical systems"*, Birkhäuser, Boston, MA.

Robot dynamics is thoroughly discussed in Spong, Vidyasagar (1989) and Sciavicco, Siciliano (2000).

To read more on the topics of force control, impedance control and hybrid motion/force see among others, the texts of Asada, Slotine (1986), Craig (1989), Spong, Vidyasagar (1989), and Sciavicco, Siciliano (2000), previously cited, and the book

- Natale C., 2003, *"Interaction control of robot manipulators"*, Springer, Germany.
- Siciliano B., Villani L., *"Robot force control"*, 1999, Kluwer Academic Publishers, Norwell, MA.

Aspects of stability in the input–output framework (in particular, passivity-based control) are studied in the first part of the book

- Ortega R., Loría A., Nicklasson P. J. and Sira-Ramírez H., 1998, *"Passivity-based control of Euler-Lagrange Systems Mechanical, Electrical and Electromechanical Applications"*, Springer-Verlag: London, Communications and Control Engg. Series.

In addition, we may mention the following classic texts.

- Raibert M., Craig J., 1981, *"Hybrid position/force control of manipulators"*, ASME Journal of Dynamic Systems, Measurement and Control, June.
- Hogan N., 1985, *"Impedance control: An approach to manipulation. Parts I, II, and III"*, ASME Journal of Dynamic Systems, Measurement and Control, Vol. 107, March.
- Whitney D., 1987, *" Historical perspective and state of the art in robot force control"*, The International Journal of Robotics Research, Vol. 6, No. 1, Spring.

The topic of robot navigation may be studied from

- Rimon E., Koditschek D. E., 1992, *"Exact robot navigation using artificial potential functions"*, IEEE Transactions on Robotics and Automation, Vol. 8, No. 5, October.

Several theoretical and technological aspects on the guidance of manipulators involving the use of vision sensors may be consulted in the following books.

- Hashimoto K., 1993, *"Visual servoing: Real–time control of robot manipulators based on visual sensory feedback"*, World Scientific Publishing Co., Singapore.
- Corke P.I., 1996, *"Visual control of robots: High–performance visual servoing"*, Research Studies Press Ltd., Great Britain.
- Vincze M., Hager G. D., 2000, *"Robust vision for vision-based control of motion"*, IEEE Press, Washington, USA.

The definition of robot manipulator is taken from

- United Nations/Economic Commission for Europe and International Federation of Robotics, 2001, *"World robotics 2001"*, United Nation Publication *sales No.* GV.E.01.0.16, ISBN 92–1–101043–8, ISSN 1020–1076, Printed at United Nations, Geneva, Switzerland.

We list next some of the most significant journals focused on robotics research.

- *Advanced Robotics,*
- *Autonomous Robots,*
- *IASTED International Journal of Robotics and Automation*
- *IEEE/ASME Transactions on Mechatronics,*
- *IEEE Transactions on Robotics and Automation*[3],
- *IEEE Transactions on Robotics,*
- *Journal of Intelligent and Robotic Systems,*
- *Journal of Robotic Systems,*
- *Mechatronics,*
- *The International Journal of Robotics Research,*
- *Robotica.*

Other journals, which in particular, provide a discussion forum on robot control are

- *ASME Journal of Dynamic Systems, Measurement and Control,*
- *Automatica,*
- *IEEE Transactions on Automatic Control,*
- *IEEE Transactions on Industrial Electronics,*
- *IEEE Transactions on Systems, Man, and Cybernetics,*
- *International Journal of Adaptive Control and Signal Processing,*
- *International Journal of Control,*
- *Systems and Control Letters.*

[3] Until June 2004 only.

2

Mathematical Preliminaries

In this chapter we present the foundations of Lyapunov stability theory. The definitions, lemmas and theorems are borrowed from specialized texts and, as needed, their statements are adapted for the purposes of this text. The proofs of these statements are beyond the scope of the present text hence, are omitted. The interested reader is invited to consult the list of references cited at the end of the chapter. The proofs of less common results are presented.

The chapter starts by briefly recalling basic concepts of linear algebra which, together with integral and differential undergraduate calculus, are a requirement for this book.

Basic Notation

Throughout the text we employ the following mathematical symbols:

$\forall$ meaning "for all";
$\exists$ meaning "there exists";
$\in$ meaning "belong(s) to";
$\Longrightarrow$ meaning "implies";
$\Longleftrightarrow$ meaning "is equivalent to" or "if and only if";
$\rightarrow$ meaning "tends to" or "maps onto";
$:=$ and $=:$ meaning "is defined as" and "equals by definition" respectively;
$\dot{x}$ meaning $\dfrac{dx}{dt}$.

We denote functions f with domain $\mathcal{D}$ and taking values in a set $\mathcal{R}$ by $f : \mathcal{D} \rightarrow \mathcal{R}$. With an abuse of notation we may also denote a function by $f(\boldsymbol{x})$ where $\boldsymbol{x} \in \mathcal{D}$.

2.1 Linear Algebra

Vectors

Basic notation and definitions of linear algebra are the starting point of our exposition.

The set of *real numbers* is denoted by the symbol $\mathbb{R}$. The real numbers are expressed by italic small capitalized letters and occasionally, by small Greek letters.

The set of non-negative real numbers, $\mathbb{R}_+$, is defined as

$$\mathbb{R}_+ = \{\alpha \in \mathbb{R} : \alpha \in [0, \infty)\}.$$

The absolute value of a real number $x \in \mathbb{R}$ is denoted by $|x|$.

We denote by $\mathbb{R}^n$, the real vector space of dimension n, that is, the set of all vectors $\boldsymbol{x}$ of dimension n formed by n real numbers in the column format

$$\boldsymbol{x} = \begin{bmatrix} x_1 \\ x_2 \\ \vdots \\ x_n \end{bmatrix} = [x_1 \ \ x_2 \ \ \cdots \ \ x_n]^T,$$

where $x_1, x_2, \cdots, x_n \in \mathbb{R}$ are the coordinates or components of the vector $\boldsymbol{x}$ and the super-index ^{T} denotes transpose. The associated vectors are denoted by **bold** small letters, either Latin or Greek.

Vector Product

The *inner product* of two vectors $\boldsymbol{x}, \boldsymbol{y} \in \mathbb{R}^n$ is defined as

$$\boldsymbol{x}^T\boldsymbol{y} = \sum_{i=1}^{n} x_i y_i = [x_1 \ \ x_2 \ \ \cdots \ \ x_n] \begin{bmatrix} y_1 \\ y_2 \\ \vdots \\ y_n \end{bmatrix} = \begin{bmatrix} x_1 \\ x_2 \\ \vdots \\ x_n \end{bmatrix}^T \begin{bmatrix} y_1 \\ y_2 \\ \vdots \\ y_n \end{bmatrix}.$$

It can be verified that the inner product of two vectors satisfies the following:

- $\boldsymbol{x}^T\boldsymbol{y} = \boldsymbol{y}^T\boldsymbol{x}$ for all $\boldsymbol{x}, \boldsymbol{y} \in \mathbb{R}^n$;
- $\boldsymbol{x}^T(\boldsymbol{y} + \boldsymbol{z}) = \boldsymbol{x}^T\boldsymbol{y} + \boldsymbol{x}^T\boldsymbol{z}$ for all $\boldsymbol{x}, \boldsymbol{y}, \boldsymbol{z} \in \mathbb{R}^n$.

Euclidean Norm

The *Euclidean norm* $\|\boldsymbol{x}\|$ of a vector $\boldsymbol{x} \in \mathbb{R}^n$ is defined as

$$\|\boldsymbol{x}\| := \sqrt{\sum_{i=1}^{n} x_i^2} = \sqrt{\boldsymbol{x}^T\boldsymbol{x}},$$

and satisfies the following axioms and properties:

- $\|\boldsymbol{x}\| = 0$ if and only if $\boldsymbol{x} = \boldsymbol{0} \in \mathbb{R}^n$;
- $\|\boldsymbol{x}\| > 0$ for all $\boldsymbol{x} \in \mathbb{R}^n$ with $\boldsymbol{x} \neq \boldsymbol{0} \in \mathbb{R}^n$;
- $\|\alpha\boldsymbol{x}\| = |\alpha| \, \|\boldsymbol{x}\|$ for all $\alpha \in \mathbb{R}$ and $\boldsymbol{x} \in \mathbb{R}^n$;
- $\|\boldsymbol{x}\| - \|\boldsymbol{y}\| \leq \|\boldsymbol{x} + \boldsymbol{y}\| \leq \|\boldsymbol{x}\| + \|\boldsymbol{y}\|$ for all $\boldsymbol{x}, \boldsymbol{y} \in \mathbb{R}^n$;
- $\left|\boldsymbol{x}^T\boldsymbol{y}\right| \leq \|\boldsymbol{x}\| \, \|\boldsymbol{y}\|$ for all $\boldsymbol{x}, \boldsymbol{y} \in \mathbb{R}^n$ (Schwartz inequality).

Matrices

We denote by $\mathbb{R}^{n\times m}$ the set of real *matrices* A of dimension $n \times m$ formed by arrays of real numbers ordered in n rows and m columns,

$$A = \{a_{ij}\} = \begin{bmatrix} a_{11} & a_{12} & \cdots & a_{1m} \\ a_{21} & a_{22} & \cdots & a_{2m} \\ \vdots & \vdots & \ddots & \vdots \\ a_{n1} & a_{n2} & \cdots & a_{nm} \end{bmatrix}.$$

A vector $\boldsymbol{x} \in \mathbb{R}^n$ may be interpreted as a particular matrix belonging to $\mathbb{R}^{n\times 1} = \mathbb{R}^n$. The matrices are denoted by Latin capital letters and occasionally by Greek capital letters.

The transpose matrix $A^T = \{a_{ji}\} \in \mathbb{R}^{m\times n}$ is obtained by interchanging the rows and the columns of $A = \{a_{ij}\} \in \mathbb{R}^{n\times m}$.

Matrix Product

Consider the matrices $A \in \mathbb{R}^{m\times p}$ and $B \in \mathbb{R}^{p\times n}$. The product of matrices A and B denoted by $C = AB \in \mathbb{R}^{m\times n}$ is defined as

$$\begin{aligned} C = \{c_{ij}\} &= AB \\ &= \begin{bmatrix} a_{11} & a_{12} & \cdots & a_{1p} \\ a_{21} & a_{22} & \cdots & a_{2p} \\ \vdots & \vdots & \ddots & \vdots \\ a_{m1} & a_{m2} & \cdots & a_{mp} \end{bmatrix} \begin{bmatrix} b_{11} & b_{12} & \cdots & b_{1n} \\ b_{21} & b_{22} & \cdots & b_{2n} \\ \vdots & \vdots & \ddots & \vdots \\ b_{p1} & b_{p2} & \cdots & b_{pn} \end{bmatrix} \end{aligned}$$

$$= \begin{bmatrix} \sum_{k=1}^{p} a_{1k}b_{k1} & \sum_{k=1}^{p} a_{1k}b_{k2} & \cdots & \sum_{k=1}^{p} a_{1k}b_{kn} \\ \sum_{k=1}^{p} a_{2k}b_{k1} & \sum_{k=1}^{p} a_{2k}b_{k2} & \cdots & \sum_{k=1}^{p} a_{2k}b_{kn} \\ \vdots & \vdots & \ddots & \vdots \\ \sum_{k=1}^{p} a_{mk}b_{k1} & \sum_{k=1}^{p} a_{mk}b_{k2} & \cdots & \sum_{k=1}^{p} a_{mk}b_{kn} \end{bmatrix}.$$

It may be verified, without much difficulty, that the product of matrices satisfy the following:

- $(AB)^T = B^T A^T$ for all $A \in \mathbb{R}^{m\times p}$ and $B \in \mathbb{R}^{p\times n}$;
- in general, $AB \neq BA$;
- for all $A \in \mathbb{R}^{m\times p}$, $B \in \mathbb{R}^{p\times n}$:
  $A(B+C) = AB + AC$ with $C \in \mathbb{R}^{p\times n}$;
  $ABC = A(BC) = (AB)C$ with $C \in \mathbb{R}^{n\times r}$.

In accordance with the definition of matrix product, the expression $\boldsymbol{x}^T A \boldsymbol{y}$ where $\boldsymbol{x} \in \mathbb{R}^n$, $A \in \mathbb{R}^{n\times m}$ and $\boldsymbol{y} \in \mathbb{R}^m$ is given by

$$\begin{aligned} \boldsymbol{x}^T A \boldsymbol{y} &= \begin{bmatrix} x_1 \\ x_2 \\ \vdots \\ x_n \end{bmatrix}^T \begin{bmatrix} a_{11} & a_{12} & \cdots & a_{1m} \\ a_{21} & a_{22} & \cdots & a_{2m} \\ \vdots & \vdots & \ddots & \vdots \\ a_{n1} & a_{n2} & \cdots & a_{nm} \end{bmatrix} \begin{bmatrix} y_1 \\ y_2 \\ \vdots \\ y_m \end{bmatrix} \\ &= \sum_{i=1}^{n} \sum_{j=1}^{m} a_{ij} x_i y_j. \end{aligned}$$

Particular Matrices

A matrix A is *square* if $n = m$, *i.e.* if it has as many rows as columns. A square matrix $A \in \mathbb{R}^{n\times n}$ is *symmetric* if it is equal to its transpose that is, if $A = A^T$. A is *skew-symmetric* if $A = -A^T$. By $-A$ we obviously mean $-A := \{-a_{ij}\}$. The following property of skew-symmetric matrices is particularly useful in robot control:

$$\boldsymbol{x}^T A \boldsymbol{x} = 0, \quad \text{for all } \boldsymbol{x} \in \mathbb{R}^n.$$

A square matrix $A = \{a_{ij}\} \in \mathbb{R}^{n\times n}$ is *diagonal* if $a_{ij} = 0$ for all $i \neq j$. We denote a diagonal matrix by $\operatorname{diag}\{a_{11}, a_{22}, \cdots, a_{nn}\} \in \mathbb{R}^{n\times n}$, *i.e.*

$$\operatorname{diag}\{a_{11}, a_{22}, \cdots, a_{nn}\} = \begin{bmatrix} a_{11} & 0 & \cdots & 0 \\ 0 & a_{22} & \cdots & 0 \\ \vdots & \vdots & \ddots & \vdots \\ 0 & 0 & \cdots & a_{nn} \end{bmatrix} \in \mathbb{R}^{n\times n}.$$

Obviously, any diagonal matrix is symmetric. In the particular case when $a_{11} = a_{22} = \cdots = a_{nn} = a$, the corresponding diagonal matrix is denoted by $\mathrm{diag}\{a\} \in \mathbb{R}^{n\times n}$. Two diagonal matrices of particular importance are the following. The *identity* matrix of dimension n which is defined as

$$I = \mathrm{diag}\{1\} = \begin{bmatrix} 1 & 0 & \cdots & 0 \\ 0 & 1 & \cdots & 0 \\ \vdots & \vdots & \ddots & \vdots \\ 0 & 0 & \cdots & 1 \end{bmatrix} \in \mathbb{R}^{n\times n}$$

and the *null* matrix of dimension n which is defined as $0_{n\times n} := \mathrm{diag}\{0\} \in \mathbb{R}^{n\times n}$.

A square matrix $A \in \mathbb{R}^{n\times n}$ is *singular* if its determinant is zero that is, if $\det[A]{=}0$. In the opposite case it is *nonsingular*. The inverse matrix A^{-1} exists if and only if A is nonsingular.

A square *not necessarily* symmetric matrix $A \in \mathbb{R}^{n\times n}$, is said to be *positive definite* if

$$\boldsymbol{x}^T A \boldsymbol{x} > 0, \text{ for all } \boldsymbol{x} \in \mathbb{R}^n, \text{ with } \boldsymbol{x} \neq \boldsymbol{0} \in \mathbb{R}^n.$$

It is important to remark that in contrast to the definition given above, the majority of texts define positive definiteness for *symmetric* matrices. However, for the purposes of this textbook, we use the above-cited definition. This choice is supported by the following observation: let P be a square matrix of dimension n and define

$$A = \{a_{ij}\} = \frac{P + P^T}{2}.$$

The theorem of Sylvester establishes that the matrix P is positive definite if and only if

$$\det[a_{11}] > 0, \det\begin{bmatrix} a_{11} & a_{12} \\ a_{21} & a_{22} \end{bmatrix} > 0, \cdots, \det[A] > 0.$$

We use the notation $A > 0$ to indicate that the matrix A is positive definite[1]. Any symmetric positive definite matrix $A = A^T > 0$ is nonsingular. Moreover, $A = A^T > 0$ if and only if $A^{-1} = (A^{-1})^T > 0$.

It can also be shown that the sum of two positive definite matrices yields a positive definite matrix however, the product of two symmetric positive definite matrices $A = A^T > 0$ and $B = B^T > 0$, yields in general a matrix which is neither symmetric nor positive definite. Yet the resulting matrix AB is nonsingular.

[1] It is important to remark that $A > 0$ means that the matrix A is positive definite and shall not be read as "A is greater than 0" which makes no mathematical sense.

A square not necessarily symmetric matrix $A \in \mathbb{R}^{n\times n}$, is *positive semidefinite* if

$$x^T A x \geq 0 \text{ for all } x \in \mathbb{R}^n.$$

We employ the notation $A \geq 0$ to denote that the matrix A is positive semidefinite.

A square matrix $A \in \mathbb{R}^{n\times n}$ is *negative definite* if $-A$ is positive definite and it is *negative semidefinite* if $-A$ is positive semidefinite.

Lemma 2.1. *Given a symmetric positive definite matrix A and a nonsingular matrix B, the product $B^T AB$ is a symmetric positive definite matrix.*

Proof. Notice that the matrix $B^T AB$ is symmetric. Define $\boldsymbol{y} = B\boldsymbol{x}$ which, by virtue of the hypothesis that B is nonsingular, guarantees that $\boldsymbol{y} = \mathbf{0} \in \mathbb{R}^n$ if and only if $\boldsymbol{x} = \mathbf{0} \in \mathbb{R}^n$. From this we obtain

$$\boldsymbol{x}^T \left[B^T AB\right] \boldsymbol{x} = \boldsymbol{y}^T A \boldsymbol{y} > 0$$

for all $\boldsymbol{x} \neq \mathbf{0} \in \mathbb{R}^n$, which is equivalent to having that $B^T AB$ is positive definite. ◊◊◊

Eigenvalues

For each square matrix $A \in \mathbb{R}^{n\times n}$ there exist n eigenvalues (in general, complex numbers) denoted by $\lambda_1\{A\}, \lambda_2\{A\}, \cdots, \lambda_n\{A\}$. The eigenvalues of the matrix $A \in \mathbb{R}^{n\times n}$ are numbers that satisfy

$$\det\left[\lambda_i\{A\}I - A\right] = 0, \qquad \text{for } i = 1, 2, \cdots, n$$

where $I \in \mathbb{R}^{n\times n}$ is the identity matrix of dimension n.

For the case of a symmetric matrix $A = A^T \in \mathbb{R}^{n\times n}$, its eigenvalues are such that:

- $\lambda_1\{A\}, \lambda_2\{A\}, \cdots, \lambda_n\{A\} \in \mathbb{R}$; and,
- expressing the largest and smallest eigenvalues of A by $\lambda_{\text{Max}}\{A\}$ and $\lambda_{\min}\{A\}$ respectively, the theorem of Rayleigh–Ritz establishes that for all $\boldsymbol{x} \in \mathbb{R}^n$ we have

$$\lambda_{\text{Max}}\{A\} \|\boldsymbol{x}\|^2 \geq \boldsymbol{x}^T A \boldsymbol{x} \geq \lambda_{\min}\{A\} \|\boldsymbol{x}\|^2.$$

A symmetric matrix $A = A^T \in \mathbb{R}^{n\times n}$ is positive definite if and only if its eigenvalues are positive, *i.e.* if and only if $\lambda_i\{A\} > 0$ where $i = 1, 2, \cdots, n$. Consequently, any square matrix $A \in \mathbb{R}^{n\times n}$ is positive definite if $\lambda_i\{A + A^T\} > 0$ where $i = 1, 2, \cdots, n$.

Remark 2.1. Consider a matrix function $A : \mathbb{R}^m \to \mathbb{R}^{n\times n}$ with A symmetric. We say that A is positive definite if $B := A(\boldsymbol{y})$ is positive definite for each $\boldsymbol{y} \in \mathbb{R}^m$. In other words, if *for each* $\boldsymbol{y} \in \mathbb{R}^m$ we have

$$\boldsymbol{x}^T A(\boldsymbol{y})\boldsymbol{x} > 0 \text{ for all } \boldsymbol{x} \in \mathbb{R}^n\,, \text{ with } \boldsymbol{x} \neq \boldsymbol{0}\,.$$

By an abuse of notation, we define $\lambda_{\min}\{A\}$ as the greatest lower-bound (*i.e.* the *infimum*) of $\lambda_{\min}\{A(\boldsymbol{y})\}$ for all $\boldsymbol{y} \in \mathbb{R}^m$, that is

$$\lambda_{\min}\{A\} := \inf_{\boldsymbol{y}\in\mathbb{R}^m} \lambda_{\min}\{A(\boldsymbol{y})\}\,.$$

For the purposes of this textbook most relevant positive definite matrix functions $A : \mathbb{R}^m \to \mathbb{R}^{n\times n}$ satisfy $\lambda_{\min}\{A\} > 0$.

Spectral Norm

The *spectral norm* $\|A\|$ of a matrix $A \in \mathbb{R}^{n\times m}$ is defined as[2]

$$\|A\| = \sqrt{\lambda_{\mathrm{Max}}\{A^T A\}},$$

where $\lambda_{\mathrm{Max}}\{A^T A\}$ denotes the largest eigenvalue of the symmetric matrix $A^T A \in \mathbb{R}^{m\times m}$.

In the particular case of symmetric matrices $A = A^T \in \mathbb{R}^{n\times n}$, we have

- $\|A\| = \max_i |\lambda_i\{A\}|$;
- $\|A^{-1}\| = \dfrac{1}{\min_i |\lambda_i\{A\}|}$.

In the expressions above, the absolute value is redundant if A is symmetric positive definite, *i.e.* if $A = A^T > 0$.

The spectral norm satisfies the following properties and axioms:

- $\|A\| = 0$ if and only if $A = 0 \in \mathbb{R}^{n\times m}$;
- $\|A\| > 0$ for all $A \in \mathbb{R}^{n\times m}$ where $A \neq 0 \in \mathbb{R}^{n\times m}$;
- $\|A + B\| \leq \|A\| + \|B\|$ for all $A, B \in \mathbb{R}^{n\times m}$;
- $\|\alpha A\| = |\alpha|\,\|A\|$ for all $\alpha \in \mathbb{R}$ and $A \in \mathbb{R}^{n\times m}$;
- $\|A^T B\| \leq \|A\|\,\|B\|$ for all $A, B \in \mathbb{R}^{n\times m}$.

[2] It is important to see that we employ the same symbol for the Euclidean norm of a vector and the spectral norm of a matrix. The reader should take special care in not mistaking them. The distinction can be clearly made via the fonts used for the argument of $\|\cdot\|$, *i.e.* we use small **bold** letters for vectors and capital letters for matrices.

An important result about spectral norms is the following. Consider the matrix $A \in \mathbb{R}^{n \times m}$ and the vector $\boldsymbol{x} \in \mathbb{R}^m$. Then, the norm of the vector $A\boldsymbol{x}$ satisfies

$$\|A\boldsymbol{x}\| \leq \|A\| \, \|\boldsymbol{x}\| ,$$

where $\|A\|$ denotes the spectral norm of the matrix A while $\|\boldsymbol{x}\|$ denotes the Euclidean norm of the vector $\boldsymbol{x}$. Moreover, since $\boldsymbol{y} \in \mathbb{R}^n$, the absolute value of $\boldsymbol{y}^T A\boldsymbol{x}$ satisfies

$$\left|\boldsymbol{y}^T A\boldsymbol{x}\right| \leq \|A\| \, \|\boldsymbol{y}\| \, \|\boldsymbol{x}\| .$$

2.2 Fixed Points

We start with some basic concepts on what are called *fixed points*; these are useful to establish conditions for existence and unicity of equilibria for ordinary differential equations. Such theorems are employed later to study closed-loop dynamic systems appearing in robot control problems. To start with, we present the definition of fixed point that, in spite of its simplicity, is of great importance.

Consider a continuous function $\boldsymbol{f} : \mathbb{R}^n \to \mathbb{R}^n$. The vector $\boldsymbol{x}^* \in \mathbb{R}^n$ is a *fixed point* of $\boldsymbol{f}(\boldsymbol{x})$ if

$$\boldsymbol{f}(\boldsymbol{x}^*) = \boldsymbol{x}^*.$$

According to this definition, if $\boldsymbol{x}^*$ is a fixed point of the function $\boldsymbol{f}(\boldsymbol{x})$, then $\boldsymbol{x}^*$ is a solution of $\boldsymbol{f}(\boldsymbol{x}) - \boldsymbol{x} = \boldsymbol{0}$.

Some functions have one or multiple fixed points but there also exist functions which have no fixed points. The function $f(x) = \sin(x)$ has a unique fixed point at $x^* = 0$, while the function $f(x) = x^3$ has three fixed points: $x^* = 1$, $x^* = 0$ and $x^* = -1$. However, $f(x) = e^x$ has no fixed point.

We present next a version of the contraction mapping theorem which provides a sufficient condition for existence and unicity of fixed points.

Theorem 2.1. Contraction Mapping

Consider $\Omega \subset \mathbb{R}^m$, a vector of parameters $\boldsymbol{\theta} \in \Omega$ and the continuous function $\boldsymbol{f} : \mathbb{R}^n \times \Omega \to \mathbb{R}^n$. Assume that there exists a non-negative constant k such that for all $\boldsymbol{y}, \boldsymbol{z} \in \mathbb{R}^n$ and all $\boldsymbol{\theta} \in \Omega$ we have

$$\|\boldsymbol{f}(\boldsymbol{y}, \boldsymbol{\theta}) - \boldsymbol{f}(\boldsymbol{z}, \boldsymbol{\theta})\| \leq k \, \|\boldsymbol{y} - \boldsymbol{z}\| .$$

If the constant k is strictly smaller than one, then for each $\boldsymbol{\theta}^* \in \Omega$, the function $\boldsymbol{f}(\cdot, \boldsymbol{\theta}^*)$ possesses a unique fixed point $\boldsymbol{x}^* \in \mathbb{R}^n$.
Moreover, the fixed point $\boldsymbol{x}^*$ may be determined by

$$\boldsymbol{x}^* = \lim_{n \to \infty} \boldsymbol{x}(n, \boldsymbol{\theta}^*)$$

where $\boldsymbol{x}(n, \boldsymbol{\theta}^*) = \boldsymbol{f}(\boldsymbol{x}(n-1, \boldsymbol{\theta}^*))$ and with $\boldsymbol{x}(0, \boldsymbol{\theta}^*) \in \mathbb{R}^n$ being arbitrary.

An important interpretation of the contraction mapping theorem is the following. Assume that the function $\boldsymbol{f}(\boldsymbol{x}, \boldsymbol{\theta})$ satisfies the condition of the theorem then, for each $\boldsymbol{\theta}^* \in \Omega$ the equation $\boldsymbol{f}(\boldsymbol{x}, \boldsymbol{\theta}^*) - \boldsymbol{x} = \boldsymbol{0}$ has a solution in $\boldsymbol{x}$ and moreover it is unique. To illustrate this idea consider the function $h(x, \theta)$ defined as

$$
\begin{aligned}
h(x, \theta) &= ax - b \sin(\theta - x) \\
&= -a \left[f(x, \theta) - x \right]
\end{aligned} \tag{2.1}
$$

with $a > 0$, $b > 0$, $\theta \in \mathbb{R}$ and

$$
f(x, \theta) = \frac{b}{a} \sin(\theta - x). \tag{2.2}
$$

We wish to find conditions on a and b so that $h(x, \theta) = 0$ has a unique solution in x. From (2.1) and (2.2) we see that this is equivalent to establishing conditions to guarantee that $\frac{b}{a} \sin(\theta - x) - x = 0$ has a unique solution in x. In other words, conditions to ensure that $\frac{b}{a} \sin(\theta - x)$ has a unique fixed point for each θ. To solve this problem we may employ the contraction mapping theorem. Notice that

$$
|f(y, \theta) - f(z, \theta)| = \left| \frac{b}{a} \left[\sin(\theta - y) - \sin(\theta - z) \right] \right|
$$

and, invoking the mean value theorem (*cf.* Theorem A.2 on page 384) which ensures that $|\sin(\theta - y) - \sin(\theta - z)| \leq |y - z|$, we obtain

$$
|f(y, \theta) - f(z, \theta)| \leq \frac{b}{a} |y - z|
$$

for all y, z and $\theta \in \mathbb{R}$. Hence, if $1 > b/a \geq 0$ for each θ, $f(x, \theta)$ has a unique fixed point and consequently, $h(x, \theta) = 0$ has a unique solution in x.

2.3 Lyapunov Stability

In this section we present the basic concepts and theorems related to Lyapunov stability and, in particular the so-called **second method of Lyapunov** or **direct method of Lyapunov**.

The main objective in Lyapunov stability theory is to study the behavior of dynamical systems described by ordinary differential equations of the form

$$
\dot{\boldsymbol{x}} = \boldsymbol{f}(t, \boldsymbol{x}), \qquad \boldsymbol{x} \in \mathbb{R}^n, \quad t \in \mathbb{R}_+ , \tag{2.3}
$$

where the vector $\boldsymbol{x}$ corresponds to the state of the system represented by (2.3). We denote solutions of this differential equation by $\boldsymbol{x}(t, t_\circ, \boldsymbol{x}(t_\circ))$. That is, $\boldsymbol{x}(t, t_\circ, \boldsymbol{x}(t_\circ))$ represents the value of the system's state at time t and with arbitrary initial state $\boldsymbol{x}(t_\circ) \in \mathbb{R}^n$ and initial time $t_\circ \geq 0$. However, for simplicity in the notation and since the *initial conditions* $t_\circ$ and $\boldsymbol{x}(t_\circ)$ are fixed, most often we use $\boldsymbol{x}(t)$ to denote a solution to (2.3) in place of $\boldsymbol{x}(t, t_\circ, \boldsymbol{x}(t_\circ))$.

We assume that the function $\boldsymbol{f} : \mathbb{R}_+ \times \mathbb{R}^n \to \mathbb{R}^n$ is continuous in t and $\boldsymbol{x}$ and is such that:

- Equation (2.3) has a unique solution corresponding to each initial condition $t_\circ$, $\boldsymbol{x}(t_\circ)$;
- the solution $\boldsymbol{x}(t, t_\circ, \boldsymbol{x}(t_\circ))$ of (2.3) depends continuously on the initial conditions $t_\circ$, $\boldsymbol{x}(t_\circ)$.

Generally speaking, assuming existence of the solutions for all $t \geq t_\circ \geq 0$ is restrictive and one may simply assume that they exist on a finite interval. Then, existence on the infinite interval may be concluded from the same theorems on Lyapunov stability that we present later in this chapter. However, for the purposes of this book we assume existence on the infinite interval.

If the function $\boldsymbol{f}$ does not depend explicitly on time, that is, if $\boldsymbol{f}(t, \boldsymbol{x}) = \boldsymbol{f}(\boldsymbol{x})$ then, Equation (2.3) becomes

$$\dot{\boldsymbol{x}} = \boldsymbol{f}(\boldsymbol{x}), \qquad \boldsymbol{x} \in \mathbb{R}^n \tag{2.4}$$

and it is said to be *autonomous*. In this case it makes no sense to speak of the initial time $t_\circ$ since for any given $t_\circ$ and $t'_\circ$ such that $\boldsymbol{x}(t_\circ) = \boldsymbol{x}(t'_\circ)$, we have $\boldsymbol{x}(t_\circ + T, t_\circ, \boldsymbol{x}(t_\circ)) = \boldsymbol{x}(t'_\circ + T, t'_\circ, \boldsymbol{x}(t'_\circ))$ for any $T \geq 0$. Therefore, for all autonomous differential equations we can safely consider that $t_\circ = 0$.

If $\boldsymbol{f}(t, \boldsymbol{x}) = A(t)\boldsymbol{x} + \boldsymbol{u}(t)$ with $A(t)$ being a square matrix of dimension n and $A(t)$ and vector $\boldsymbol{u}(t)$ being functions only of t – or constant – then Equation (2.3) is said to be linear. In the opposite case it is nonlinear.

2.3.1 The Concept of Equilibrium

Among the basic concepts in Lyapunov theory that we underline are: *equilibrium*, *stability*, *asymptotic* stability, *exponential* stability and *uniformity.* We develop each of these concepts below. First, we present the concept of equilibrium which plays a central role in Lyapunov theory.

Definition 2.1. Equilibrium

A constant *vector* $\boldsymbol{x}_e \in \mathbb{R}^n$ *is an* equilibrium *or equilibrium state of the system (2.3) if*

$$\boldsymbol{f}(t, \boldsymbol{x}_e) = \boldsymbol{0} \qquad \forall\, t \geq 0\,.$$

A direct consequence of the definition of equilibrium, under regularity appropriate conditions that exclude "pathological" situations, is that if the initial state $\boldsymbol{x}(t_\circ) \in \mathbb{R}^n$ is an equilibrium ($\boldsymbol{x}(t_\circ) = \boldsymbol{x}_e \in \mathbb{R}^n$) then,

- $\boldsymbol{x}(t) = \boldsymbol{x}_e \quad \forall\, t \geq t_\circ \geq 0$
- $\dot{\boldsymbol{x}}(t) = \boldsymbol{0} \quad \forall\, t \geq t_\circ \geq 0$.

This idea is illustrated in Figure 2.1, where the case $\boldsymbol{x}(t_\circ) \in \mathbb{R}^2$ is depicted. The initial state $\boldsymbol{x}(t_\circ)$ is precisely $\boldsymbol{x}_e$, so the evolution of the solution $\boldsymbol{x}(t)$ corresponds exactly to the constant value $\boldsymbol{x}_e$ for all times $t \geq t_\circ$.

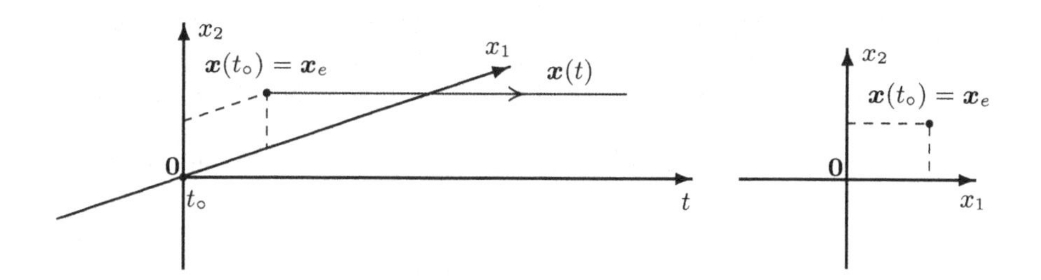

Figure 2.1. Concept of equilibrium

It is typically assumed that the origin of the state space $\mathbb{R}^n$, that is, $\boldsymbol{x} = \boldsymbol{0} \in \mathbb{R}^n$, is an equilibrium of (2.3). If this is not the case, it may be shown that by a suitable change of variable, any equilibrium of (2.3) may be translated to the origin.

In general, a differential equation may have more than one equilibrium, indeed even an infinite number of them! However, it is also possible that a differential equation does not possess any equilibrium at all. This is illustrated by the following example.

Example 2.1. Consider the following linear differential equation

$$\dot{x} = a\, x + b\, u(t),$$

with initial conditions $(t_\circ, x(t_\circ)) \in \mathbb{R}_+ \times \mathbb{R}$ and where $a \neq 0$ and $b \neq 0$ are real constants and $u : \mathbb{R}_+ \to \mathbb{R}$ is a continuous function. If $u(t) = u_0$ for all $t \geq 0$, that is, if the differential equation is autonomous, then the unique equilibrium point of this equation is $x_e = -bu_0/a$.

On the other hand, one must be careful in concluding that any autonomous system has equilibria. For instance, the autonomous nonlinear system

$$\dot{x} = e^{-x}$$

has no equilibrium point.

Consider the following nonlinear autonomous differential equation

$$\dot{x}_1 = x_2$$
$$\dot{x}_2 = \sin(x_1) .$$

The previous set of equations has an infinite number of (isolated) equilibria, which are given by $\boldsymbol{x}_e = [x_{1e} \quad x_{2e}]^T = [n\pi \quad 0]^T$ with $n = \cdots, -1, 0, 1, \cdots$. ♢

Systems with multiple equilibria are not restricted to mathematical examples but are fairly common in control practice and as a matter of fact, mechanisms are a good example of these since in general, dynamic models of robot manipulators constitute nonlinear systems. The following example shows that even for robots with a simple mathematical model, multiple equilibria may co-exist.

Example 2.2. Consider a pendulum, as depicted in Figure 2.2, of mass m, total moment of inertia about the joint axis J, and distance l from its axis of rotation to the center of mass. It is assumed that the pendulum is affected by the force of gravity induced by the gravity acceleration g.

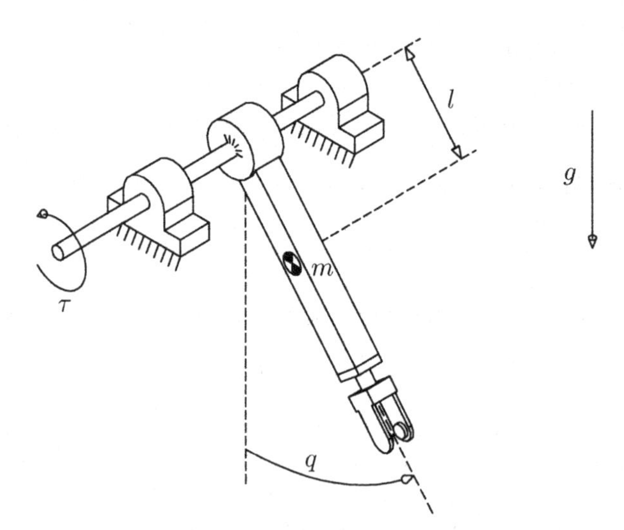

Figure 2.2. Pendulum

We assume that a torque $\tau(t)$ is applied at the axis of rotation. Then, the dynamic model which describes the motion of such a system is given by

$$J\ddot{q} + mgl\, \sin(q) = \tau(t)$$

where q is the angular position of the pendulum with respect to the vertical axis and $\dot{q}$ is the corresponding angular velocity. Or, in terms of the state $[q \;\; \dot{q}]^T$, the dynamic model is given by

$$\frac{d}{dt}\begin{bmatrix} q \\ \dot{q} \end{bmatrix} = \begin{bmatrix} \dot{q} \\ J^{-1}\left[\tau(t) - mgl\,\sin(q)\right] \end{bmatrix}.$$

If $\tau(t) = 0$, the equilibrium states are then given by $[q \;\; \dot{q}]^T = [n\pi \;\; 0]^T$ for $n = \cdots, -2, -1, 0, 1, 2, \cdots$ since $mgl\,\sin(n\pi) = 0$. Notice that if $\tau(t) = \tau^*$ and $|\tau^*| > mgl$ then, there does not exist any equilibrium since there is no $q^* \in \mathbb{R}$ such that $\tau^* = mgl\,\sin(q^*)$. ♢

2.3.2 Definitions of Stability

In this section we present the basic notions of stability of equilibria of differential equations, evoked throughout the text. We emphasize that the stability notions which are defined below are to be considered as **attributes** of the **equilibria** of the differential equations and not of the equations themselves. Without loss of generality we assume in the rest of the text that the origin of the state space, $\boldsymbol{x} = \boldsymbol{0} \in \mathbb{R}^n$, is an equilibrium of (2.3) and accordingly, we provide the definitions of stability of the origin but they can be reformulated for other equilibria by performing the appropriate changes of coordinate.

Definition 2.2. Stability

The origin is a stable equilibrium *(in the sense of Lyapunov) of Equation (2.3) if, for each pair of numbers $\varepsilon > 0$ and $t_\circ \geq 0$, there exists $\delta = \delta(t_\circ, \varepsilon) > 0$ such that*

$$\|\boldsymbol{x}(t_\circ)\| < \delta \quad \Longrightarrow \quad \|\boldsymbol{x}(t)\| < \varepsilon \qquad \forall\, t \geq t_\circ \geq 0\,. \tag{2.5}$$

Correspondingly, the origin of Equation (2.4) is said to be stable if for each $\varepsilon > 0$ there exists $\delta = \delta(\varepsilon) > 0$ such that (2.5) holds with $t_\circ = 0$.

In Definition 2.2 the constant δ (which is clearly smaller than ε) is not unique. Indeed, notice that for any given constant δ that satisfies the condition of the definition any $\delta' \leq \delta$ also satisfies it.

If one reads Definition 2.2 with appropriate care, it should be clear that the number δ depends on the number ε and in general, also on the initial time $t_\circ$. Indeed, note that the definition of stability for nonautonomous systems requires existence of a number $\delta > 0$ for each $t_\circ \geq 0$ and $\varepsilon > 0$ and not just for *some* $t_\circ \geq 0$ and $\varepsilon > 0$. Correspondingly, in the case of autonomous differential equations it is required that there exists $\delta > 0$ *for each* $\varepsilon > 0$ and not only *for some* ε.

Also, in Definition 2.2 one should not understand that the origin is Lyapunov stable if for each $\delta > 0$ one may find $\varepsilon > 0$ such that

$$\|\boldsymbol{x}(t_\circ)\| < \delta \implies \|\boldsymbol{x}(t)\| < \varepsilon \qquad \forall\, t \geq t_\circ \geq 0\,. \tag{2.6}$$

In other words, the latter statement establishes that "the origin is a stable equilibrium if for any bounded initial condition, the corresponding solution is also bounded". This is commonly known as "boundedness of solutions" or "Lagrange stability" and is a somewhat weaker property than Lyapunov stability. However, boundedness of solutions is neither a necessary nor a sufficient condition for Lyapunov stability of an equilibrium.

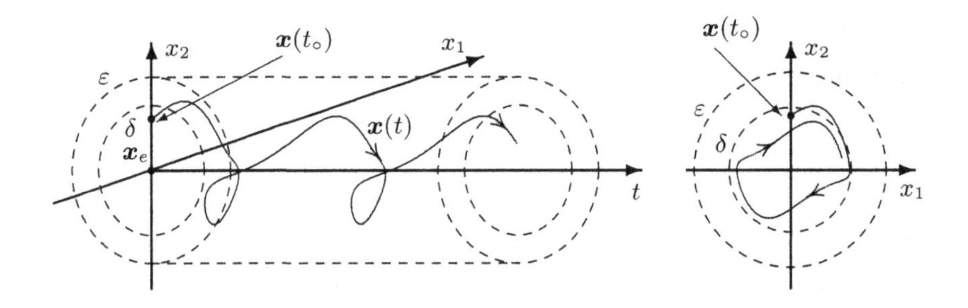

Figure 2.3. Notion of stability

To illustrate the concept of stability, Figure 2.3 shows a trajectory with an initial state $\boldsymbol{x}(t_\circ) \in \mathbb{R}^2$ such that the origin $\boldsymbol{x} = \boldsymbol{0} \in \mathbb{R}^2$ is a stable equilibrium. In Figure 2.3 we also show ε and δ which satisfy the condition from the definition of stability, that is, $\|\boldsymbol{x}(t_\circ)\| < \delta$ implies that $\|\boldsymbol{x}(t)\| < \varepsilon$ for all $t \geq t_\circ \geq 0$.

Definition 2.3. Uniform stability

The origin is a uniformly stable equilibrium *(in the sense of Lyapunov) of Equation (2.3) if for each number* $\varepsilon > 0$ *there exists* $\delta = \delta(\varepsilon) > 0$ *such that (2.5) holds.*

That is, the origin is uniformly stable if δ can be chosen independently of the initial time, $t_\circ$. For autonomous systems this is always the case, *i.e.* uniform stability and stability of the equilibrium are equivalent.

Example 2.3. Consider the system (harmonic oscillator) described by the equations:

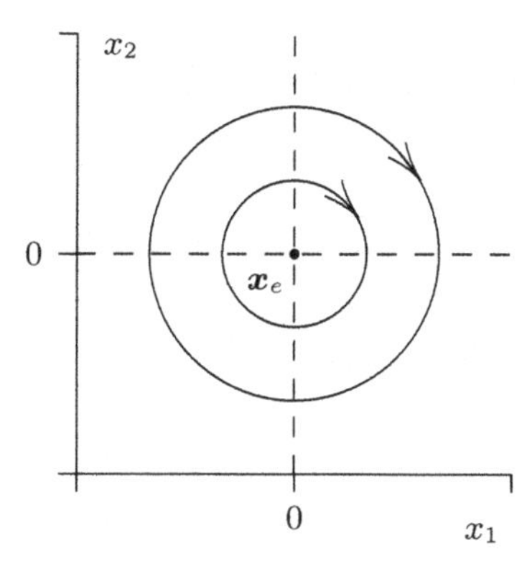

Figure 2.4. Phase plane of the harmonic oscillator

$$\dot{x}_1 = x_2 \tag{2.7}$$
$$\dot{x}_2 = -x_1 \tag{2.8}$$

and whose solution is

$$x_1(t) = x_1(0)\cos(t) + x_2(0)\sin(t)$$
$$x_2(t) = -x_1(0)\sin(t) + x_2(0)\cos(t).$$

Note that the origin is the unique equilibrium point. The graphs of some solutions of Equations (2.7)–(2.8) on the plane x_1–x_2, are depicted in Figure 2.4.

Notice that the trajectories of the system (2.7)–(2.8) describe concentric circles centered at the origin. For this example, the origin is a stable equilibrium since for any $\varepsilon > 0$ there exists $\delta > 0$ (actually any[3] $\delta \leq \varepsilon$) such that

$$\|\boldsymbol{x}(0)\| < \delta \implies \|\boldsymbol{x}(t)\| < \varepsilon \quad \forall\, t \geq 0.$$

Observe that in Example 2.3 stability is uniform since the system is autonomous (notice that the solutions do not depend on $t_\circ$). It is also important to stress that the solutions do not tend to the origin. That is, we say that the origin is stable but not *asymptotically* stable, a concept that is defined next.

Definition 2.4. Asymptotic stability

The origin is an asymptotically stable equilibrium *of Equation (2.3) if:*

1. *the origin is stable;*

[3] From this inequality the dependence of δ on ε is clear.

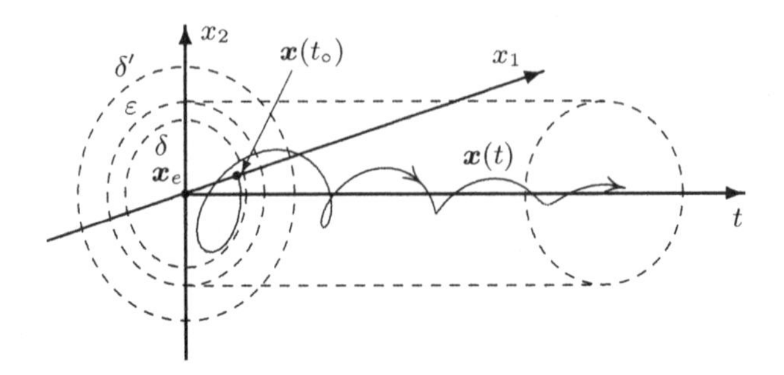

Figure 2.5. Asymptotic stability

2. the origin is attractive, i.e. for each $t_\circ \geq 0$, there exists $\delta' = \delta'(t_\circ) > 0$ such that

$$\|\boldsymbol{x}(t_\circ)\| < \delta' \quad \Longrightarrow \quad \|\boldsymbol{x}(t)\| \to 0 \quad \text{as} \quad t \to \infty\,. \tag{2.9}$$

Asymptotic stability for the origin of autonomous systems is stated by replacing above "for each $t_\circ \geq 0$, there exists $\delta' = \delta'(t_\circ) > 0$" with "there exists $\delta' > 0$".

Figure 2.5 illustrates the concept of asymptotic stability for the case of $\boldsymbol{x}(t_\circ) \in \mathbb{R}^2$.

In Definition 2.4 above, one should not read that "the origin is stable when $t \to \infty$" or that "attractivity implies stability in the sense of Definition 2.2". As a matter of fact, even though it may seem counter-intuitive to some readers, there exist systems for which all the trajectories starting close to an equilibrium tend to that equilibrium but the latter is not stable. We see in the following example that this phenomenon is not as unrealistic as one might think.

Example 2.4. Consider the autonomous system with two state variables expressed in terms of polar coordinates:

$$\begin{aligned}\dot{r} &= \frac{5}{100}r(1-r)\\ \dot{\theta} &= \sin^2(\theta/2) \quad \theta \in [0, 2\pi)\,.\end{aligned}$$

This system has an equilibrium at the origin $[r \;\; \theta]^T = [0 \;\; 0]^T$ and another one at $[r \;\; \theta]^T = [1 \;\; 0]^T$. The behavior of this system, expressed in Cartesian coordinates $x_1 = r\cos(\theta)$ and $x_2 = r\,\sin(\theta)$, is illustrated in Figure 2.6. All the solutions of the system (with the exception of those that start off at the equilibria) tend asymptotically to $[x_1 \;\; x_2]^T = [1 \;\; 0]^T$. In particular, notice from Figure 2.6

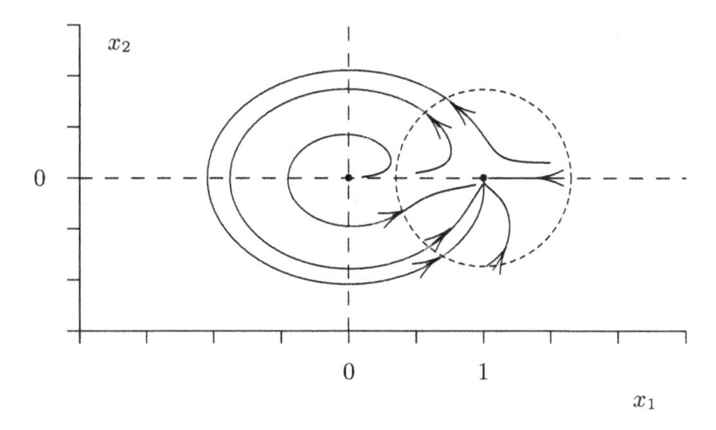

Figure 2.6. Attractive but unstable equilibrium

that for each initial condition inside the dashed disk (but excluding $[x_1 \;\; x_2]^T = [1 \;\; 0]^T$) the generated trajectory goes asymptotically to the equilibrium. That is, the equilibrium is attractive in the sense of Definition 2.4. Intuitively, it may seem reasonable that this implies that the equilibrium is also stable and therefore, asymptotically stable. However, as pointed out before, this is a fallacy since the first item of Definition 2.4 does not hold. To see this, pick ε to be the radius of the dashed disk. For this *particular* ε there does not exist a number δ such that

$$\|\boldsymbol{x}(0)\| < \delta \implies \|\boldsymbol{x}(t)\| < \varepsilon \qquad \forall\, t \geq 0,$$

because there are always solutions that leave the disk before "coming back" towards the equilibrium. Hence, the equilibrium $[x_1 \;\; x_2]^T = [1 \;\; 0]^T$ is attractive but *unstable* (*cf.* Definition 2.9). ♢

Definition 2.5. Uniform asymptotic stability

The origin is a uniformly asymptotically stable equilibrium of Equation (2.3) if:

1. *the origin is uniformly stable;*
2. *the origin is uniformly attractive, i.e. there exists a number $\delta' > 0$ such that (2.9) holds with a rate of convergence independent of $t_\circ$.*

For autonomous systems, uniform asymptotic stability and asymptotic stability are equivalent.

One may make precise the idea of "rate of convergence independent of $t_\circ$" by saying that uniform attractivity means that there exists $\delta' > 0$ and, *for each* ε' (arbitrarily small) there exists $T(\varepsilon') > 0$ such that

$$\|\boldsymbol{x}(t_\circ)\| < \delta' \quad \Longrightarrow \quad \|\boldsymbol{x}(t)\| < \varepsilon' \ \ \forall\, t \geq t_\circ + T \,. \tag{2.10}$$

Therefore, the rate of convergence is determined by the time T and we say that it is independent of $t_\circ$ if T is independent of the initial time.

The following example shows that some nonautonomous systems may be asymptotically stable but not with a uniform rate of convergence; hence, *not uniformly* asymptotically stable.

Example 2.5. Consider the system

$$\dot{x} = -\frac{x}{1+t}$$

with initial conditions $t_\circ \in \mathbb{R}_+$ and $x_\circ := x(t_\circ) \in \mathbb{R}$ and whose origin is an equilibrium point. One can solve this differential equation by simple integration of

$$\frac{dx}{x} = -\frac{dt}{1+t}\,,$$

i.e.

$$\ln\left(\frac{x(t)}{x(t_\circ)}\right) = \ln\left(\frac{1+t_\circ}{1+t}\right)\,,$$

from which we obtain

$$x(t) = \frac{1+t_\circ}{1+t} x_\circ \,. \tag{2.11}$$

From Equation (2.11) we see that $|x(t)| \to 0$ as $t \to \infty$; therefore, the origin is an attractive equilibrium. More precisely, we see that given any numbers ε' and $\delta' > 0$, if $|x_\circ| < \delta'$ then

$$|x(t)| < \frac{1+t_\circ}{1+t}\delta' \qquad \text{for all } t \geq t_\circ \,.$$

It means that for $|x(t)|$ to be smaller than ε' we need to wait at least until $t = t_\circ + T$ where

$$T > \frac{(1+t_\circ)(\delta' - \varepsilon')}{\varepsilon'}\,.$$

However, since T grows proportionally with $t_\circ$, for any fixed values of δ' and ε', the origin is not uniformly attractive. In other words, $|x(t)| \to 0$ as $t \to \infty$ but no matter what the tolerance that we impose on $|x(t)|$ to come close to zero (that is, ε') and how near the initial states are to the origin (that is, δ'), we see that the later $|x(t)|$ starts to decay the slower it tends to zero. That is, the *rate of convergence* becomes smaller as $t_\circ$ becomes larger.

On the other hand, the origin is stable and, actually, uniformly stable. To see this, observe from (2.11) that, since

$$\frac{1+t_\circ}{(1+t)} \leq 1$$

for any $t \geq t_\circ$ and any $t \geq 0$ then, $|x(t)| \leq |x_\circ|$. We conclude that for any given $\varepsilon > 0$ it holds, with $\delta = \varepsilon$, that

$$|x_\circ| < \delta \quad \Longrightarrow \quad |x(t)| < \varepsilon\,,\ \forall t \geq t_\circ\,.$$

We conclude that the origin is asymptotically stable and uniformly stable but not *uniformly asymptotically* stable. ♢

The phenomena observed in the previous examples are proper to nonautonomous systems. Nonautonomous systems appear in robot control when the desired task is to follow a time-varying trajectory, *i.e.* in motion control (*cf.* Part III) or when there is uncertainty in the physical parameters and therefore, an adaptive control approach may be used (*cf.* Part IV).

The concept of uniformity for nonautonomous systems is instrumental because uniform asymptotic stability ensures a certain *robustness* with respect to disturbances. We say that a system is robust with respect to disturbances if, in the presence of the latter, the equilibrium of the system preserves basic properties such as stability and boundedness of solutions. In the control of robot manipulators, disturbances may come from unmodeled dynamics or additive sensor noise, which are fairly common in practice. Therefore, if for instance we are able to guarantee uniform asymptotic stability for a robot control system in a motion control task, we will be sure that small measurement noise will only cause small deviations from the control objective. However, one should not understand that uniform asymptotic stability guarantees that the equilibrium remains *asymptotically* stable under disturbances or measurement noise.

In robot control we are often interested in studying the performance of controllers, considering *any* initial configuration for the robot. For this, we need to study *global* definitions of stability.

Definition 2.6. Global asymptotic stability

The origin is a globally asymptotically stable *equilibrium of Equation (2.3) if:*

1. *the origin is stable;*
2. *the origin is globally attractive, that is,*

$$\|\boldsymbol{x}(t)\| \to 0 \quad \text{as} \quad t \to \infty, \qquad \forall\, \boldsymbol{x}(t_\circ) \in \mathbb{R}^n\,,\ t_\circ \geq 0\,.$$

It should be clear from the definition above, that if the origin is globally asymptotically stable then it is also asymptotically stable, but the converse is obviously not always true.

Definition 2.7. Global uniform asymptotic stability

The origin is a globally uniformly asymptotically stable *equilibrium of Equation (2.3) if:*

1. *the origin is uniformly stable with $\delta(\varepsilon)$ in Definition 2.3 which satisfies $\delta(\varepsilon) \to \infty$ as $\varepsilon \to \infty$ (uniform boundedness) and*
2. *the origin is globally uniformly attractive, i.e. for all $\boldsymbol{x}(t_\circ) \in \mathbb{R}^n$ and all $t_\circ \geq 0$,*
$$\|\boldsymbol{x}(t)\| \to 0 \quad \text{as} \quad t \to \infty$$

with a convergence rate that is independent of $t_\circ$.

For autonomous systems, global asymptotic stability and global uniform asymptotic stability are equivalent.

As for Definition 2.5, we can make item 2 above more precise by saying that for each δ' and ε' there exists $T(\delta', \varepsilon')$ – hence, independent of $t_\circ$ – such that the implication (2.10) holds.

It is important to underline the differences between Definitions 2.5 and 2.7. First, in item 1 of Definition 2.7 one asks that δ grows unboundedly as $\varepsilon \to \infty$. In particular, this implies that the norm of *all* solutions must be bounded and, moreover, by a bound which is independent of $t_\circ$ (uniform boundedness of all solutions). Secondly, attractivity must be *global* which translates into: "for each $\delta' > 0$ and $\varepsilon' > 0$ there exists $T(\delta', \varepsilon') > 0$ such that (2.10) holds", *i.e.* in Definition 2.7 we say "*for each $\delta' > 0$*" and not "*there exists $\delta' > 0$*" as in Definition 2.5.

It is also convenient to stress that the difference between items 1 and 2 of Definitions 2.6 and 2.7 is not simply that the required properties of stability and attractivity shall be uniform but also that *all* the solutions must be uniformly bounded. The latter is ensured by the imposed condition that $\delta(\varepsilon)$ can be chosen so that $\delta(\varepsilon) \to \infty$ as $\varepsilon \to \infty$.

We finish this section on definitions with a special case of global uniform asymptotic stability.

Definition 2.8. Global exponential stability

The origin is a globally exponentially stable *equilibrium of (2.3) if there exist positive constants α and β, independent of $t_\circ$, such that*

$$\|\boldsymbol{x}(t)\| < \alpha \|\boldsymbol{x}(t_\circ)\| e^{-\beta(t-t_\circ)}, \qquad \forall\, t \geq t_\circ \geq 0, \qquad \forall\, \boldsymbol{x}(t_\circ) \in \mathbb{R}^n . \quad (2.12)$$

According to the previous definitions, if the origin is a globally exponentially stable equilibrium then it is also globally uniformly asymptotically stable. The opposite is clearly not necessarily true since the convergence might not be exponential.

We emphasize that the numbers α and β must be independent of $t_\circ$ since in some cases, one may establish that for a given system the bound (2.12) holds but with constants α and β which depend on the initial times $t_\circ$. Then, we may speak of global exponential (non-uniform) convergence and as a matter of fact, of global asymptotic stability but it would be erroneous to say that the origin is globally exponentially stable. Notice that in such case, neither the origin is uniformly attractive nor the solutions are uniformly bounded.

Definition 2.9. Instability
The origin of Equation (2.3) is an unstable *equilibrium if it is* not stable.

Mathematically speaking, the property of instability means that there exists at least one $\varepsilon > 0$ for which no $\delta > 0$ can be found such that

$$\|\boldsymbol{x}(t_\circ)\| < \delta \quad \Longrightarrow \quad \|\boldsymbol{x}(t)\| < \varepsilon \qquad \forall\, t \geq t_\circ \geq 0\,.$$

Or, in other words, that there exists at least one $\varepsilon > 0$ which is desired to be a bound on the norm of the solution $\boldsymbol{x}(t)$ but there does not exist any pair of initial conditions $t_\circ \in \mathbb{R}_+$ and $\boldsymbol{x}(t_\circ) \neq \boldsymbol{0} \in \mathbb{R}^n$ whose solution $\boldsymbol{x}(t)$ may satisfy $\|\boldsymbol{x}(t)\| < \varepsilon$ for all $t \geq t_\circ \geq 0$.

It must be clear that instability does not necessarily imply that the solution $\boldsymbol{x}(t)$ grows to infinity as $t \to \infty$. The latter is a contradiction of the weaker property of boundedness of the solutions, discussed above.

We now present an example that illustrates the concept of instability.

Example 2.6. Consider the equations that define the motion of the van der Pol system,

$$\dot{x}_1 = x_2, \tag{2.13}$$

$$\dot{x}_2 = -x_1 + (1 - x_1^2)x_2, \tag{2.14}$$

where x_1 and $x_2 \in \mathbb{R}$. Notice that the origin is an equilibrium of these equations.

The graph of some solutions generated by different initial conditions of the system (2.13)–(2.14) on the *phase plane*, are depicted in Figure 2.7. The behavior of the system can be described as follows. If the initial condition of the system is inside the closed curve Γ, and is away from zero, then the solutions approach Γ asymptotically. If the initial condition is outside of the closed curve Γ, then the solutions also approach Γ.

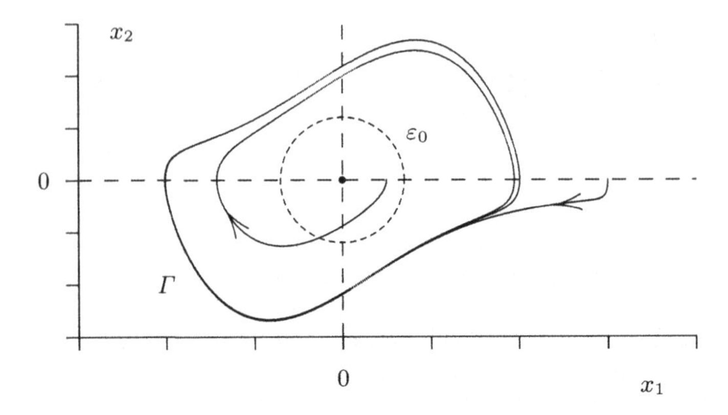

Figure 2.7. Phase plane of the van der Pol oscillator

The origin of the system (2.13)–(2.14) is an unstable equilibrium. To show this, we simply need to pick $\varepsilon := \varepsilon_0$ (*cf.* Figure 2.7) to see that there does not exist $\delta > 0$ *dependent* on ε_0 such that

$$\|\boldsymbol{x}(0)\| < \delta \quad \Longrightarrow \quad \|\boldsymbol{x}(t)\| < \varepsilon_0 \qquad \forall\, t \geq 0.$$

We close this subsection by observing that some authors speak about "stability of systems" or "stability of systems at an equilibrium" instead of "stability of equilibria (or the origin)". For instance the phrase "the system is stable at the origin" may be employed to mean that "the origin is a stable equilibrium of the system". Both ways of speaking are correct and equivalent. In this textbook we use the second one to be strict with respect to the mathematical definitions.

2.3.3 Lyapunov Functions

We present definitions that determine a particular class of functions that are fundamental in the use of Lyapunov's direct method to study the stability of equilibria of differential equations.

Definition 2.10. Locally and globally positive definite function
A continuous function $W : \mathbb{R}^n \rightarrow \mathbb{R}_+$ is said to be locally positive definite *if*

1. $W(\mathbf{0}) = 0$,
2. $W(\boldsymbol{x}) > 0 \quad$ for small $\quad \|\boldsymbol{x}\| \neq 0$.

A continuous function $W : \mathbb{R}^n \to \mathbb{R}$ *is said to be* globally positive definite *(or simply* positive definite*) if*

1. $W(\mathbf{0}) = 0$,
2. $W(\boldsymbol{x}) > 0 \qquad \forall\, \boldsymbol{x} \neq \mathbf{0}$.

According to this definition it should be clear that a positive definite function is also locally positive definite. Also, according to the definition of positive definite matrix, a *quadratic* function $f : \mathbb{R}^n \to \mathbb{R}$, *i.e.* of the form

$$f(\boldsymbol{x}) = \boldsymbol{x}^T P \boldsymbol{x}\,, \qquad P \in \mathbb{R}^{n \times n}$$

is positive definite if and only if $P > 0$.

The function $W(\boldsymbol{x})$ is said to be (locally) negative definite if $-W(\boldsymbol{x})$ is (locally) positive definite.

For a continuous function $V : \mathbb{R}_+ \times \mathbb{R}^n \to \mathbb{R}_+$, *i.e.* which also depends on time, we say that $V(t, \boldsymbol{x})$ is (resp. locally) positive definite if:

1. $V(t, \mathbf{0}) = 0 \;\; \forall\, t \geq 0$;
2. $V(t, \boldsymbol{x}) \geq W(\boldsymbol{x}), \qquad \forall\, t \geq 0, \qquad \forall\, \boldsymbol{x} \in \mathbb{R}^n$ (resp. for small $\|\boldsymbol{x}\|$)

where $W(\boldsymbol{x})$ is a (resp. locally) positive definite function.

Definition 2.11. Radially unbounded function and decrescent function

A continuous function $W : \mathbb{R}^n \to \mathbb{R}$ *is said to be* radially unbounded *if*

$$W(\boldsymbol{x}) \to \infty \qquad \text{as} \qquad \|\boldsymbol{x}\| \to \infty\,.$$

Correspondingly, we say that $V(t, \boldsymbol{x})$ *is radially unbounded if* $V(t, \boldsymbol{x}) \geq W(\boldsymbol{x})$ *for all* $t \geq 0$.

A continuous function $V : \mathbb{R}_+ \times \mathbb{R}^n \to \mathbb{R}$ *is (locally)* decrescent *if there exists a (locally) positive definite function* $W : \mathbb{R}^n \to \mathbb{R}_+$ *such that*

$$V(t, \boldsymbol{x}) \leq W(\boldsymbol{x}) \qquad \forall\, t \geq 0 \qquad \forall\, \boldsymbol{x} \in \mathbb{R}^n \;\; (\textit{for small } \|\boldsymbol{x}\|).$$

If $V(t, \boldsymbol{x})$ is independent of t, *i.e.* if $V(t, \boldsymbol{x}) = V(\boldsymbol{x})$ then $V(\boldsymbol{x})$ is decrescent.

The following examples illustrate the concepts presented above.

Example 2.7. Consider the graphs of the functions $V_i(x)$ with $i = 1, \ldots, 4$ as depicted in Figure 2.8. It is apparent from these graphs that:

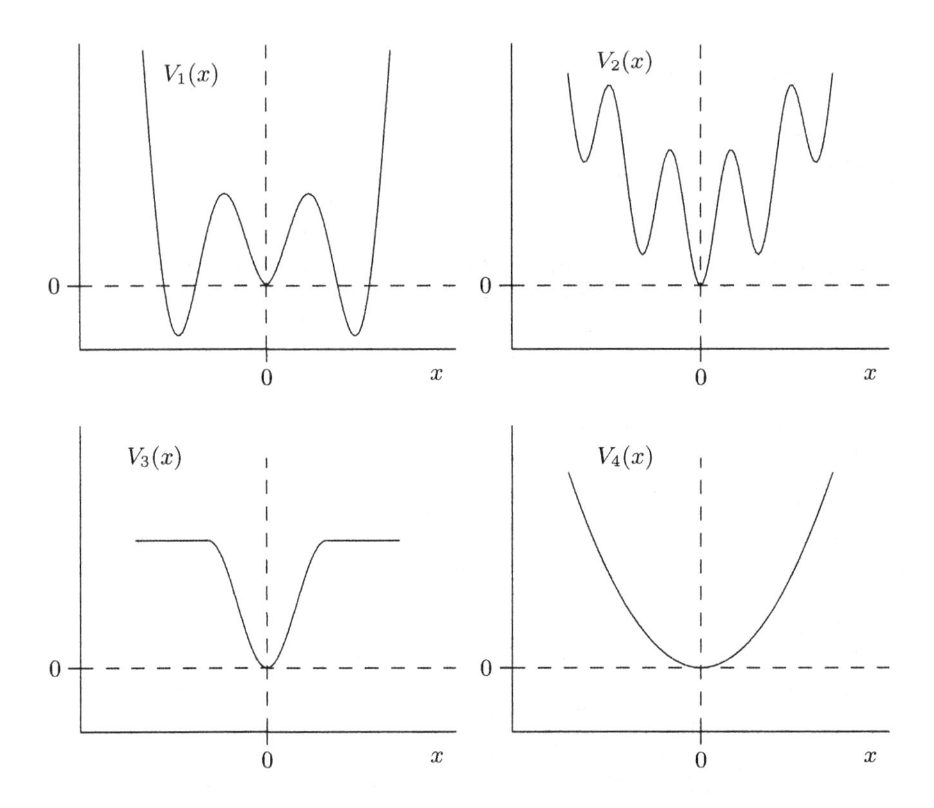

Figure 2.8. Examples

- $V_1(x)$ is locally positive definite but is not (globally) positive definite;
- $V_2(x)$ is locally positive definite, (globally) positive definite and radially unbounded;
- $V_3(x)$ is locally positive definite and (globally) positive definite but it is not radially unbounded;
- $V_4(x)$ is positive definite and radially unbounded.

◇

Example 2.8. The function $W(x_1, x_2) = x_1^2 + x_2^2$ is positive definite and radially unbounded. Since W is independent of t, it follows immediately that it is also decrescent. ◇

Example 2.9. The function $V(t, x_1, x_2) = (t+1)(x_1^2 + x_2^2)$ is positive definite since $V(t, x_1, x_2) \geq x_1^2 + x_2^2$ and $V(t, 0, 0) = 0$ for all $t \geq 0$. However, $V(t, x_1, x_2)$ is not decrescent. ♢

Example 2.10. The function $W(x_1, x_2) = (x_1 + x_2)^2$ is not positive definite since it does not satisfy $W(\boldsymbol{x}) > 0$ for all $\boldsymbol{x} \neq \boldsymbol{0}$ such that $x_1 = -x_2$. However, the function $W(x_1, x_2) = (x_1 + x_2)^2 + \alpha x_1^2$, and the function $W(x_1, x_2) = (x_1 + x_2)^2 + \alpha x_2^2$, with $\alpha > 0$ are positive definite. ♢

In order to prepare the reader for the following subsection, where we present Lyapunov's direct method for the study of stability of equilibria, we present below a series of concepts related to the notion of Lyapunov function *candidate.*

Definition 2.12. Lyapunov function candidate

A continuous and differentiable[4] *function* $V : \mathbb{R}_+ \times \mathbb{R}^n \to \mathbb{R}_+$ *is said to be a* Lyapunov function candidate *for the equilibrium* $\boldsymbol{x} = \boldsymbol{0} \in \mathbb{R}^n$ *of the equation* $\dot{\boldsymbol{x}} = \boldsymbol{f}(t, \boldsymbol{x})$ *if:*

1. $V(t, \boldsymbol{x})$ *is locally positive definite;*
2. $\dfrac{\partial V(t, \boldsymbol{x})}{\partial t}$ *is continuous with respect to t and* $\boldsymbol{x}$ *;*
3. $\dfrac{\partial V(t, \boldsymbol{x})}{\partial \boldsymbol{x}}$ *is continuous with respect to t and* $\boldsymbol{x}$ *.*

Correspondingly, a continuous and differentiable function $V : \mathbb{R}^n \to \mathbb{R}_+$ is said to be a Lyapunov function candidate for the equilibrium $\boldsymbol{x} = \boldsymbol{0} \in \mathbb{R}^n$ of Equation (2.4), *i.e.* $\dot{\boldsymbol{x}} = \boldsymbol{f}(\boldsymbol{x})$, if $V(\boldsymbol{x})$ is locally positive definite and

$$\frac{dV(\boldsymbol{x})}{d\boldsymbol{x}}$$

is continuous.

In other words, a Lyapunov function candidate for the equilibrium $\boldsymbol{x} = \boldsymbol{0} \in \mathbb{R}^n$ of Equations (2.3) or (2.4) is any locally positive definite and continuously differentiable function; that is, with continuous partial derivatives.

The time derivative of a Lyapunov function candidate plays a key role in drawing conclusions about the stability attributes of equilibria of differential equations. For this reason, we present the following definition.

[4] In some of the specialized literature authors do not assume differentiability. We shall not deal with that here.

Definition 2.13. Time derivative of a Lyapunov function candidate
Let $V(t,\boldsymbol{x})$ be a Lyapunov function candidate for the equation (2.3). The total time derivative of $V(t,\boldsymbol{x})$ along the trajectories of (2.3), denoted by $\dot{V}(t,\boldsymbol{x})$, is given by

$$\dot{V}(t,\boldsymbol{x}) := \frac{d}{dt}\{V(t,\boldsymbol{x})\} = \frac{\partial V(t,\boldsymbol{x})}{\partial t} + \frac{\partial V(t,\boldsymbol{x})}{\partial \boldsymbol{x}}^T \boldsymbol{f}(t,\boldsymbol{x}).$$

From the previous definition we observe that if $V(\boldsymbol{x})$ does not depend explicitly on time and Equation (2.3) is autonomous then,

$$\dot{V}(\boldsymbol{x}) = \frac{dV(\boldsymbol{x})}{d\boldsymbol{x}}^T \boldsymbol{f}(\boldsymbol{x})$$

which does not depend explicitly on time either.

Definition 2.14. Lyapunov function
A Lyapunov function candidate $V(t,\boldsymbol{x})$ for Equation (2.3) is a Lyapunov function *for (2.3) if its total time derivative along the trajectories of (2.3) satisfies*

$$\dot{V}(t,\boldsymbol{x}) \leq 0 \qquad \forall\, t \geq 0$$

and for small $\|\boldsymbol{x}\|$.

Correspondingly, a Lyapunov function candidate $V(\boldsymbol{x})$ for Equation (2.4) is a Lyapunov function if $\dot{V}(\boldsymbol{x}) \leq 0$ for small $\|\boldsymbol{x}\|$.

2.3.4 Lyapunov's Direct Method

With the above preliminaries we are now ready to present the basic results of Lyapunov stability theory. Indeed, the theory of Lyapunov is the product of more than a hundred years of intense study and there are numerous specialized texts. The avid reader is invited to see the texts cited at the end of the chapter. However, the list that we provide is by no means exhaustive; the cited texts have been chosen specially for the potential reader of this book.

Theorem 2.2. Stability and uniform stability
The origin is a stable *equilibrium of Equation (2.3), if there exists a Lyapunov function candidate $V(t,\boldsymbol{x})$ (i.e.* a locally *positive definite function with continuous partial derivatives with respect to t and $\boldsymbol{x}$) such that its total time derivative satisfies*

$$\dot{V}(t,\boldsymbol{x}) \leq 0, \qquad \forall\, t \geq 0 \qquad \text{for small } \|\boldsymbol{x}\|.$$

If moreover $V(t,\boldsymbol{x})$ is decrescent for small $\|\boldsymbol{x}\|$ then the origin is uniformly stable.

This theorem establishes sufficient conditions for stability of the equilibrium in the sense of Lyapunov. It is worth remarking that the conclusion of the theorem holds also if $\dot{V}(t,\boldsymbol{x}) \leq 0$ for all $t \geq 0$ and for all $\boldsymbol{x} \in \mathbb{R}^n$, or if the Lyapunov function candidate $V(t,\boldsymbol{x})$ is globally positive definite instead of being only locally positive definite. The following theorem allows us to establish some results on stability of the equilibrium and on boundedness of solutions.

Theorem 2.3. (Uniform) boundedness of solutions plus uniform stability

The origin is a uniformly stable equilibrium of Equation (2.3) and *the solutions $\boldsymbol{x}(t)$ are* uniformly *bounded for all initial conditions $(t_\circ, \boldsymbol{x}(t_\circ)) \in \mathbb{R}_+ \times \mathbb{R}^n$ if there exists a radially unbounded, globally positive definite, decrescent Lyapunov function candidate $V(t,\boldsymbol{x})$ such that its total time derivative satisfies*

$$\dot{V}(t,\boldsymbol{x}) \leq 0 \qquad \forall\, t \geq t_\circ \geq 0 \qquad \forall\, \boldsymbol{x} \in \mathbb{R}^n .$$

In particular, item 1 of Definition 2.7 holds.

Example 2.11. Consider the dynamic model of an ideal pendulum without friction as analyzed in Example 2.2 and shown in Figure 2.2 (*cf.* page 30) for which we now assume that no torque $\tau(t)$ is applied at the axis of rotation, *i.e.* we consider the system described by the differential equation

$$J\ddot{q} + mgl\,\sin(q) = 0 \quad \text{ with } \quad q(0), \dot{q}(0) \in \mathbb{R} .$$

This equation may be rewritten in the state-space form as

$$\begin{aligned}\dot{x}_1 &= x_2\\ \dot{x}_2 &= -\frac{mgl}{J}\sin(x_1),\end{aligned}$$

where $x_1 = q$ and $x_2 = \dot{q}$. Notice that these equations are autonomous nonlinear, and the origin is an equilibrium. However, we remind the reader that from Example 2.2, we know that the pendulum has multiple equilibria, more precisely at $[q \;\; \dot{q}]^T = [n\pi \;\; 0]^T$ for $n = \cdots, -2, -1, 0, 1, 2, \cdots$; that is, the origin is the equilibrium corresponding to the case $n = 0$.

In order to analyze the stability of the origin we use Theorem 2.2 with the following locally positive definite function:

$$V(x_1, x_2) = mgl\;\left[1 - \cos(x_1)\right] + J\,\frac{x_2^2}{2}.$$

Notice that the first term on the right-hand side corresponds to the potential energy, while the second one corresponds to the kinetic energy. Observe that $V(x_1, x_2)$ is not a positive definite function since it does not fulfil $V(x_1, x_2) > 0$ for all $[x_1\ x_2]^T \neq \mathbf{0} \in \mathbb{R}^2$. However, it is locally positive definite because $V(x_1, x_2) > 0$ for $\|\boldsymbol{x}\| > 0$ *small*, in the sense that $\|\boldsymbol{x}\| < 2\pi$.

Evaluating the total time derivative of $V(x_1, x_2)$ we obtain

$$\begin{aligned}\dot{V}(x_1, x_2) &= mgl\ \sin(x_1)\dot{x}_1 + Jx_2\dot{x}_2 \\ &= 0\,.\end{aligned}$$

According to Theorem 2.2, the origin is a stable equilibrium, *i.e.* the solutions $x_1(t)$ and $x_2(t)$ remain as close to the origin as desired if the initial conditions $x_1(0)$ and $x_2(0)$ are sufficiently small.

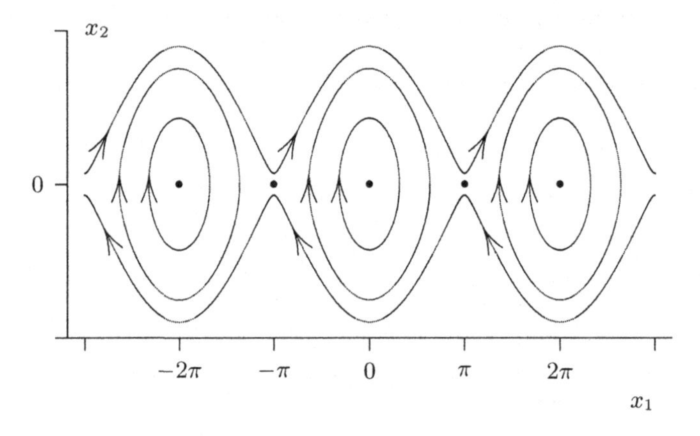

Figure 2.9. Phase plane of the pendulum

Above, we proved that the origin is a stable equilibrium point; this corresponds to $n = 0$. As a matter of fact, considering the multiple equilibria, *i.e.* $n \neq 0$, one can also show that the equilibria associated to n even are stable while those for n odd are unstable.

We stress that although the origin (and strictly speaking, infinitely many other equilibria) is Lyapunov stable, this system has unbounded solutions. Notice indeed from Figure 2.9, how the solution may grow indefinitely to the right or to the left, *i.e.* in the direction of the displacements, x_1; while the solution remains bounded in the direction of the velocities, x_2. In other words, if the pendulum starts to oscillate at a speed larger than $2\sqrt{mgl/J}$ from *any* position then it will spin for ever after! ♢

Thus, Example 2.2 shows clearly that stability of an equilibrium is not synonymous with boundedness of the solutions.

We present next, sufficient conditions for global asymptotic and exponential stability.

Theorem 2.4. Global (uniform) asymptotic stability
The origin of Equation (2.3) (respectively of Equation 2.4) is globally asymptotically stable *if there exists a radially unbounded, globally positive definite Lyapunov function candidate $V(t, \boldsymbol{x})$ (respectively $V(\boldsymbol{x})$) such that its time derivative is globally negative definite. If, moreover, the function $V(t, \boldsymbol{x})$ is decrescent, then the origin is globally uniformly asymptotically stable.*

It should be clear that the origin of the *autonomous* Equation (2.4) is globally asymptotically stable if and only if it is globally uniformly asymptotically stable.

Example 2.12. Consider the following scalar equation

$$\dot{x} = -ax^3 , \qquad x(0) \in \mathbb{R},$$

where a is a positive constant. The origin is a unique equilibrium. To analyze its stability consider the following Lyapunov function candidate which is positive definite and radially unbounded:

$$V(x) = \frac{x^2}{2} .$$

Its total time derivative is

$$\begin{aligned} \dot{V}(x) &= x\dot{x} \\ &= -ax^4 . \end{aligned}$$

So we see that from Theorem 2.4, the origin is globally asymptotically stable. ◇

In the case that Equation (2.3) is autonomous there is no difference between global asymptotic stability and global uniform asymptotic stability.

Theorem 2.5. Global exponential stability
The origin of (2.3) is globally exponentially stable *if there exists a Lyapunov function candidate $V(t, \boldsymbol{x})$ and positive constants α, β, γ and $p \geq 1$ such that:*

- $\alpha \|\boldsymbol{x}\|^p \leq V(t, \boldsymbol{x}) \leq \beta \|\boldsymbol{x}\|^p$;
- $\dot{V}(t, \boldsymbol{x}) \leq -\gamma \|\boldsymbol{x}\|^p \quad \forall\, t \geq t_\circ \geq 0 \quad \forall\, \boldsymbol{x} \in \mathbb{R}^n$.

If all the above conditions hold only for small $\|\boldsymbol{x}\|$ then we say that the origin is an exponentially stable equilibrium.

Example 2.13. Consider the following scalar equation:

$$\dot{x} = -a \left[1 - \frac{1}{2}\sin(t) \right] x \qquad x(0) \in \mathbb{R},$$

where a is a positive constant. Note that the origin is a unique equilibrium. To analyze its stability consider the following Lyapunov function candidate which is positive definite and radially unbounded:

$$V(x) = x^2 .$$

Since V does not depend on t, it is decrescent. The total time derivative of V is

$$\begin{aligned} \dot{V}(x) &= 2x\dot{x} \\ &= -2a \left[1 - \frac{1}{2}\sin(t) \right] x^2 \\ &\leq -ax^2 . \end{aligned}$$

So the conditions of Theorem 2.4 are satisfied and we conclude that the origin is globally uniformly asymptotically stable. Moreover, notice that the conditions of Theorem 2.5 also hold, with $\alpha = \beta = 1$, $\gamma = a$ and $p = 2$ so the the origin is also globally exponentially stable.

The following result establishes necessary conditions for certain global properties of stability of equilibria.

Theorem 2.6. *Consider the differential Equations (2.3) and (2.4). The unicity of an existing equilibrium point is necessary for the following properties (or, in other words, the following properties imply the unicity of an equilibrium):*

- *global asymptotic stability;*
- *global exponential stability.*

The proof of Theorem 2.6 follows straightforwardly from the observation that each of the above-cited concepts of stability implies that $\|\boldsymbol{x}(t) - \boldsymbol{x}_e\| \to \boldsymbol{0}$ when $t \to \infty$, and where $\boldsymbol{x}_e$ is the equilibrium under analysis. In view of the "globality" of the mentioned properties of stability, the convergence of the solution $\boldsymbol{x}(t)$ to the equilibrium $\boldsymbol{x}_e$, must be verified for all initial conditions. Clearly, this would not be the case if besides $\boldsymbol{x}_e$ there existed other equilibria, since in this case any solution starting off at other equilibria, by definition, would remain at that point forever after.

Notice that Theorem 2.6 does not establish as a necessary condition for stability of the equilibrium, that the equilibrium in question be unique. It is important to mention that for a given system there may coexist equilibria with local stability properties and in particular, stable equilibria with unstable equilibria.

Since for a given system the global properties of stability imply existence of a unique equilibrium, it is correct to speak of not only the properties of global stability of that equilibrium, but also of such properties for the system itself. That is, sentences such as "such a system is globally asymptotically stable" or, "the system is globally exponentially stable" are mathematically meaningful and correct.

In control theory, *i.e.* when we are required to analyze the stability of a particular system, finding a Lyapunov function with a negative definite derivative is in general very hard. Nevertheless, if in spite of painstaking efforts we are unable to find a Lyapunov function we must not conclude that the origin of the system under analysis is unstable; rather, no conclusion can be drawn. Fortunately, for autonomous systems, there are methods based on more restrictive conditions but considerably easier to verify. A notable example is the so-called La Salle's invariance principle[5] which is widely used in the analysis of robot control systems. The following theorem is a simplified version of La Salle's invariance principle that appears adequate for the purposes of this textbook.

Theorem 2.7. La Salle

Consider the autonomous differential equation

$$\dot{\boldsymbol{x}} = \boldsymbol{f}(\boldsymbol{x})$$

whose origin $\boldsymbol{x} = \boldsymbol{0} \in \mathbb{R}^n$ is an equilibrium. Assume that there exists a globally positive definite and radially unbounded Lyapunov function candidate $V(\boldsymbol{x})$, such that

$$\dot{V}(\boldsymbol{x}) \leq 0 \qquad \forall\, \boldsymbol{x} \in \mathbb{R}^n.$$

Define the set Ω as

$$\Omega = \left\{ \boldsymbol{x} \in \mathbb{R}^n : \dot{V}(\boldsymbol{x}) = 0 \right\}. \tag{2.15}$$

If $\boldsymbol{x}(0) = \boldsymbol{0}$ is the only initial state in Ω whose corresponding solution, $\boldsymbol{x}(t)$, remains forever in Ω (i.e. $\boldsymbol{x}(t) \in \Omega$ for all $t \geq 0$) then the origin $\boldsymbol{x} = \boldsymbol{0} \in \mathbb{R}^n$ is globally asymptotically stable.

[5] While in the western literature this result is mainly attributed to the French mathematician J. P. La Salle, some authors call this "Krasovskiĭ–La Salle's theorem" to give credit to the Russian mathematician N. N. Krasovskiĭ who independently reported the same theorem. See the bibliographical remarks at the end of the chapter.

We stress that the application of the theorem of La Salle to establish global asymptotic stability does not require that $\dot{V}(\boldsymbol{x})$ be a negative definite function. However, we recall that this theorem can be employed only for autonomous differential equations. A practical way to verify the condition of La Salle's theorem and which suffices for most of this textbook is given in the following statement.

Corollary 2.1. Simplified La Salle
Consider the set of autonomous differential equations

$$\dot{\boldsymbol{x}} = \boldsymbol{f}_x(\boldsymbol{x}, \boldsymbol{z}), \qquad \boldsymbol{x} \in I\!R^n \tag{2.16}$$

$$\dot{\boldsymbol{z}} = \boldsymbol{f}_z(\boldsymbol{x}, \boldsymbol{z}), \qquad \boldsymbol{z} \in I\!R^m . \tag{2.17}$$

where $\boldsymbol{f}_x(\boldsymbol{0},\boldsymbol{0}) = \boldsymbol{0}$ *and* $\boldsymbol{f}_z(\boldsymbol{0},\boldsymbol{0}) = \boldsymbol{0}$. *That is, the origin is an equilibrium point. Let* $V : I\!R^n \times I\!R^m \to I\!R_+$ *be globally positive definite and radially unbounded in both arguments. Assume that there exists a globally positive definite function* $W : I\!R^m \to I\!R_+$ *such that*

$$\dot{V}(\boldsymbol{x}, \boldsymbol{z}) = -W(\boldsymbol{z}) . \tag{2.18}$$

If $\boldsymbol{x} = \boldsymbol{0}$ *is the unique solution of* $\boldsymbol{f}_z(\boldsymbol{x}, \boldsymbol{0}) = \boldsymbol{0}$ *then the origin* $[\boldsymbol{x}^T\ \boldsymbol{z}^T]^T = \boldsymbol{0}$ *is globally asymptotically stable.*

The proof of this corollary is simple and follows by applying Theorem 2.7. Therefore, for the sake of completeness and to illustrate the use of Theorem 2.7 we present it next.

Proof of Corollary 2.1. Since $W(\boldsymbol{z})$ is globally positive definite in $\boldsymbol{z}$ we have from (2.18) that the set Ω defined in (2.15) is in this case, $\{\boldsymbol{z} = \boldsymbol{0}\} \cup \{\boldsymbol{x} \in \mathbb{R}^n\}$. This means that for the solutions of (2.16), (2.17) to be contained in Ω they must verify that for all $t \geq 0$, $\boldsymbol{z}(t) = \dot{\boldsymbol{z}}(t) = \boldsymbol{0}$ and

$$\dot{\boldsymbol{x}}(t) = \boldsymbol{f}_x(\boldsymbol{x}(t), \boldsymbol{0}) \tag{2.19}$$

$$\boldsymbol{0} = \boldsymbol{f}_z(\boldsymbol{x}(t), \boldsymbol{0}) . \tag{2.20}$$

However, by assumption the unique solution that satisfies (2.20) is the trivial solution, *i.e.* $\boldsymbol{x}(t) = \boldsymbol{0}$ for all $t \geq 0$. In its turn, the trivial solution also satisfies (2.19) since $\dot{\boldsymbol{x}}(t) = \boldsymbol{0}$ and by assumption, $\boldsymbol{f}_x(\boldsymbol{0},\boldsymbol{0}) = \boldsymbol{0}$. Thus, the only initial state for which $[\boldsymbol{x}(t)^T\ \boldsymbol{z}(t)^T]^T \in \Omega$ for all $t \geq 0$ is $[\boldsymbol{x}(0)^T\ \boldsymbol{z}(0)^T]^T = \boldsymbol{0}$. Global asymptotic stability of the origin follows from Theorem 2.7. ◊◊◊

We present next some examples to further illustrate the use of La Salle's theorem.

Example 2.14. Consider the autonomous equations

$$\dot{x} = -k\,z, \qquad x(0) \in \mathbb{R}$$
$$\dot{z} = -z^3 + k\,x, \qquad z(0) \in \mathbb{R}$$

where $k \neq 0$. Note that the origin is an equilibrium point. Consider now the following Lyapunov function candidate – positive definite and radially unbounded – to study the stability of the origin:

$$V(x,z) = \frac{1}{2}\left[x^2 + z^2\right] .$$

The total time derivative of $V(x,z)$ is given by

$$\begin{aligned}\dot{V}(x,z) &= -z^4 \\ &\leq 0 .\end{aligned}$$

Hence, according to Theorem 2.2 we may conclude stability of the origin. However, note that $\dot{V}(x,z) = 0$ for any value of x and $z = 0$, so it cannot be even locally negative definite. Therefore, since $\dot{V}(x,z)$ does not satisfy the conditions of Theorem 2.4 we may not conclude global asymptotic stability of the origin from this result.

Nevertheless, since the equations under study are autonomous, we may try to invoke La Salle's theorem (Theorem 2.7). To that end, let us follow the conditions of Corollary 2.1.

- We already verified that the origin is an equilibrium point.
- We also have verified that $V(x,z)$ is globally positive definite and radially unbounded in both arguments.
- In addition, $\dot{V}(x,z) = -z^4$ so we define $W(z) := z^4$ which is a globally positive definite function of z.
- It is only left to verify that the only solution of $f_z(x,0) = 0$ is $x = 0$ which in this case, takes the form $0 = kx$. Hence it is evident that $x = 0$ is the only solution.

We conclude from Corollary 2.1 that the origin is globally asymptotically stable. ◇

Example 2.15. Consider the following equations

$$\dot{x}_1 = -x_1 + k_1 x_2 + k_2 x_3 \qquad x_1(0) \in \mathbb{R}$$
$$\dot{x}_2 = -k_1 x_1 \qquad x_2(0) \in \mathbb{R}$$
$$\dot{x}_3 = -k_2 x_1 \qquad x_3(0) \in \mathbb{R}$$

where $k_1 \neq 0$, $k_2 \neq 0$. These equations represent an autonomous linear differential equation whose equilibria are the points

$$[x_1 \quad x_2 \quad x_3] = \left[0 \quad x_2 \quad -\frac{k_1}{k_2} x_2\right] .$$

Notice that there exist an infinite number of equilibrium points, one for each $x_2 \in \mathbb{R}$. In particular, for $x_2 = 0$, the origin is an equilibrium. To study its stability we consider the Lyapunov function candidate which is positive definite and radially unbounded,

$$V(\boldsymbol{x}) = \frac{1}{2}\left[x_1^2 + x_2^2 + x_3^2\right] ,$$

and whose total time derivative is

$$\dot{V}(\boldsymbol{x}) = -x_1^2.$$

Theorem 2.3 guarantees stability of the origin and boundedness of the solutions. Theorem 2.4 on global asymptotic stability, may not be used since $\dot{V}(\boldsymbol{x})$ is zero at $x_1 = 0$ and for any values of x_2 and x_3; that is, it does not hold that $\dot{V}(\boldsymbol{x})$ is negative definite. Even though the equations under study are autonomous they have an *infinite* number of equilibria. For this reason and according to Theorem 2.6, the origin may not be *globally* asymptotically stable. ♢

In the previous example it is not possible to conclude that the origin is an asymptotically stable equilibrium however, under the above conditions one may conclude that $\lim_{t\to\infty} x_1(t) = 0$. Indeed this can be achieved by invoking the following Lemma which guarantees boundedness of the solutions and convergence of part of the state. This is obviously a weaker property than (global) asymptotic stability but it is still a useful property to evaluate, rigorously, the performance of a controller.

Lemma 2.2. *Consider the continuously differentiable functions $\boldsymbol{x} : \mathbb{R}_+ \to \mathbb{R}^n$, $\boldsymbol{z} : \mathbb{R}_+ \to \mathbb{R}^m$, $h : \mathbb{R}_+ \to \mathbb{R}_+$ and $P : \mathbb{R}_+ \to \mathbb{R}^{(n+m)\times(n+m)}$. Assume that $P(t)$ is a symmetric positive definite matrix for each $t \in \mathbb{R}_+$ and P is continuous. Define the function $V : \mathbb{R}_+ \times \mathbb{R}^n \times \mathbb{R}^m \times \mathbb{R}_+ \to \mathbb{R}_+$ as*

$$V(t, \boldsymbol{x}, \boldsymbol{z}, h) = \begin{bmatrix} \boldsymbol{x} \\ \boldsymbol{z} \end{bmatrix}^T P(t) \begin{bmatrix} \boldsymbol{x} \\ \boldsymbol{z} \end{bmatrix} + h(t) \geq 0 .$$

If the total time derivative of $V(t, \boldsymbol{x}, \boldsymbol{z}, h)$, i.e.

$$\dot{V}(t, \boldsymbol{x}, \boldsymbol{z}, h) := \frac{\partial V(t, \boldsymbol{x}, \boldsymbol{z}, h)}{\partial t} + \frac{\partial V(t, \boldsymbol{x}, \boldsymbol{z}, h)}{\partial \boldsymbol{x}}^T \frac{d\boldsymbol{x}}{dt} + \frac{\partial V(t, \boldsymbol{x}, \boldsymbol{z}, h)}{\partial \boldsymbol{z}}^T \frac{d\boldsymbol{z}}{dt} + \frac{dh}{dt} ,$$

satisfies, for all $t \in \mathbb{R}_+$, $\boldsymbol{x} \in \mathbb{R}^n$, $\boldsymbol{z} \in \mathbb{R}^m$ *and* $h \in \mathbb{R}_+$,

$$\dot{V}(t, \boldsymbol{x}, \boldsymbol{z}, h) = -\begin{bmatrix} \boldsymbol{x} \\ \boldsymbol{z} \end{bmatrix}^T \begin{bmatrix} Q(t) & \boldsymbol{0} \\ \boldsymbol{0} & \boldsymbol{0} \end{bmatrix} \begin{bmatrix} \boldsymbol{x} \\ \boldsymbol{z} \end{bmatrix} \leq 0$$

where $Q(t) = Q(t)^T > 0$ *for all* $t \geq 0$ *then,*

1. $\boldsymbol{x}(t)$, $\boldsymbol{z}(t)$ *and* $h(t)$ *are bounded for all* $t \geq 0$ *and*
2. $\boldsymbol{x}(t)$ *is square-integrable, i.e.*

$$\int_0^\infty \|\boldsymbol{x}(t)\|^2 \, dt < \infty .$$

If, moreover, $\dot{\boldsymbol{x}}$ *is also bounded then we have*

$$\lim_{t \to \infty} \boldsymbol{x}(t) = \boldsymbol{0} .$$

The proof of this lemma is presented in Appendix A.

Bibliography

What we know nowadays as *Lyapunov theory* was launched by the Russian mathematician A. M. Lyapunov in his doctoral thesis in 1892. It is interesting to stress that his work was largely influenced by that of the French mathematicians H. Poincaré (contemporary of and personally known by A. M. Lyapunov) and Joseph La Grange, on stability of second-order differential equations such as, precisely, Lagrange's equations. The reference for the original work of A. M. Lyapunov is

- Lyapunov, A. M., 1907, *"Problème de la stabilité du mouvement"*, Annales de la faculté de sciences de Toulouse, volume 9, pp. 203-474. Translation —revised by A. M. Lyapunov— from the original published in Russian in Comm. Soc. Math., Kharkov 1892. Reprinted in Ann. Math. Studies **17**, Princeton 1949. See also the more recent edition "The general problem of stability of motion", Taylor and Francis: London, 1992.

La Salle's Theorem as adapted here for the scope of this text, is a corollary of the so-called La Salle's invariance principle which was originally and independently proposed by J. La Salle and by N. N. Krasovskiĭ and may be found in its general form in

- La Salle J., Lefschetz S., 1961, *"Stability by Lyapunov's direct method with applications"*, Academic Press, New York.

- Krasovskiĭ N. N., 1963, *"Problems of the theory of stability of motion"*, Stanford Univ. Press, 1963. Translation from the original Russian edition, Moscow, 1959.

The theorems presented in this chapter are the most commonly employed in stability analysis of control systems. The presentation that we used here has been adapted from their original statements to meet the scope of this textbook. This material is inspired from

- Vidyasagar M., 1978 and 1993, *"Nonlinear systems analysis"*, Prentice-Hall, Electrical Engineering Series.
- Khalil H. 2001, *"Nonlinear systems"*, Third Edition, Prentice-Hall.

The definition and theorems on fixed points may be found in

- Kolmogorov A. N., Fomin S. V., 1970, *"Introductory real analysis"*, Dover Pub. Inc.
- Hale J. K., 1980, *"Ordinary differential equations"*, Krieger Pub. Co.
- Khalil H., 1996, *"Nonlinear systems"*, Second Edition, Prentice-Hall.
- Sastry S., 1999, *"Nonlinear systems: analysis, stability and control"*, Springer-Verlag, New York.

Other references on differential equations and stability in the sense of Lyapunov are:

- Arnold V., 1973, *"Ordinary differential equations"*, MIT Press.
- Borrelli R., Coleman C., 1987, *"Differential equations–A modeling approach"*, Prentice-Hall.
- Slotine J. J., Li W., 1991, *"Applied nonlinear control"*, Prentice-Hall.
- Khalil H., 1996, *"Nonlinear systems"*, Second Edition, Prentice-Hall.
- Hahn W., 1967, *"Stability of motion"*, Springer-Verlag: New York.
- Rouche N., Mawhin J., 1980 *"Ordinary differential equations II: Stability and periodical solutions"*, Pitman publishing Ltd., London.

Problems

1. Consider the vectors $\boldsymbol{x} \in \mathbb{R}^n$ and $\boldsymbol{y} \in \mathbb{R}^m$. Show that

$$\left\| \begin{bmatrix} \boldsymbol{x} \\ \boldsymbol{y} \end{bmatrix} \right\| = \left\| \begin{bmatrix} \|\boldsymbol{x}\| \\ \|\boldsymbol{y}\| \end{bmatrix} \right\|.$$

2. Consider the matrix

$$P(x) = \begin{bmatrix} k & -\dfrac{\varepsilon}{1+2x^2} \\ -\dfrac{\varepsilon}{1+2x^2} & 1 \end{bmatrix}$$

where $\varepsilon > 0$. Show that if $k > \varepsilon^2$ then P is positive definite, *i.e.* $P(x) > 0$ for all $x \in \mathrm{I\!R}$.

3. Consider the differential equation that describes the behavior of a Hopfield neuron:

$$\dot{x} = -ax + w \tanh(x) + b$$

where $a > 0$, $w, b \in \mathrm{I\!R}$.

a) Show by use of the contraction mapping theorem that if $a - |w| > 0$ then the differential equation has a unique equilibrium.

b) Assume that $a = b = 1$ and $w = 1/2$. Use the contraction mapping theorem together with a numerical algorithm to obtain an approximated value of the unique equilibrium of the differential equation.

4. Consider the function

$$V(x_1, x_2) = \begin{bmatrix} x_1 & x_2 \end{bmatrix} \begin{bmatrix} 4 & 1 \\ -10 & 3 \end{bmatrix} \begin{bmatrix} x_1 \\ x_2 \end{bmatrix}.$$

Is $V(x_1, x_2)$ positive definite?

5. Consider the function

$$V(x_1, x_2) = ax_1^2 + 2bx_1x_2 + cx_2^2 .$$

Show that if $a > 0$ and $ac > b^2$ then $V(x_1, x_2)$ is positive definite.

6. Consider the linear autonomous differential equation

$$\dot{\boldsymbol{x}} = A\boldsymbol{x}, \qquad \boldsymbol{x} \in \mathrm{I\!R}^n .$$

Show that if there exists a pole of this equation at the origin of the complex plane, then the equation has an infinite number of equilibria.

Hint: Here, the "poles" are the eigenvalues of A.

7. An equilibrium $\boldsymbol{x}_e \in \mathrm{I\!R}^n$ is an *isolated equilibrium* of $\dot{\boldsymbol{x}} = \boldsymbol{f}(\boldsymbol{x})$ if there exists a real positive number $\alpha > 0$ such that there may not be any equilibrium other than $\boldsymbol{x}_e$ in Ω, where

$$\Omega = \{\boldsymbol{x} \in \mathrm{I\!R}^n : \|\boldsymbol{x} - \boldsymbol{x}_e\| < \alpha\} .$$

In the case that there does not exist any $\alpha > 0$ that satisfies the above then the equilibrium $\boldsymbol{x}_e$ is not isolated.

Assume that $\boldsymbol{x}_e$ is a non-isolated equilibrium. Answer 'true' or 'false' to the following claims:

a) the equilibrium $\boldsymbol{x}_e$ may not be asymptotically stable;
b) the equilibrium $\boldsymbol{x}_e$ is stable.

8. Consider the function $\boldsymbol{f}(x,y) : \mathbb{R}^2 \to \mathbb{R}^2$,

$$\boldsymbol{f}(x,y) = \begin{bmatrix} f_1(x,y) \\ f_2(x,y) \end{bmatrix} .$$

Assume that $\boldsymbol{f}(x,y) = \mathbf{0} \Longleftrightarrow x = 0$ and $y = 0$. Does this imply that

$$f_1(x,y) = 0 \Longleftrightarrow x = 0 \text{ and } y = 0 \, ?$$

9. Consider the following two differential equations:

$$\dot{x}_1 = \varepsilon[x_1 - \varepsilon] + x_2 - [x_1 - \varepsilon]\left[[x_1 - \varepsilon]^2 + x_2^2\right] , \qquad x_1(0) \in \mathbb{R}$$
$$\dot{x}_2 = -[x_1 - \varepsilon] + \varepsilon x_2 - x_2\left[[x_1 - \varepsilon]^2 + x_2^2\right] , \qquad x_2(0) \in \mathbb{R}$$

where $\varepsilon \in \mathbb{R}$ is constant. Determine the equilibria.

10. Consider the following second-order differential equation,

$$\ddot{y} + [y^2 - 1]\dot{y} + y^2 + 1 = 0, \qquad y(0), \dot{y}(0) \in \mathbb{R} .$$

Express this equation in the form $\dot{\boldsymbol{x}} = \boldsymbol{f}(t, \boldsymbol{x})$.
a) Is this equation linear in the state $\boldsymbol{x}$?
b) What are the equilibrium points? Discuss.

11. Consider the equation $\dot{\boldsymbol{x}} = \boldsymbol{f}(\boldsymbol{x})$. Assume that $\boldsymbol{x}_e = \mathbf{0} \in \mathbb{R}^n$ is a stable equilibrium. Does this imply that the solutions $\boldsymbol{x}(t)$ are bounded for all $t \geq 0$?

12. Consider the equations

$$\dot{x}_1 = x_2 - x_1^3$$
$$\dot{x}_2 = -x_1 - x_2^3$$

for which the origin is the unique equilibrium. Use the direct Lyapunov's method (propose a Lyapunov function) to show that the origin is stable.

13. Pick positive integer numbers m and n and appropriate constants a and b to make up a Lyapunov function of the form

$$V(x_1, x_2) = a x_1^{2m} + b x_2^{2n}$$

in order to show stability of the origin for

a)

$$\dot{x}_1 = -2x_2^3$$
$$\dot{x}_2 = 2x_1 - x_2^3$$

b)

$$\begin{aligned}\dot{x}_1 &= -x_1^3 + x_2^3 \\ \dot{x}_2 &= -x_1^3 - x_2^3 .\end{aligned}$$

14. Theorem 2.4 allows us to conclude *global* uniform asymptotic stability of an equilibrium of a differential equation. To show only uniform asymptotic stability (*i.e.* not global), the conditions of Theorem 2.4 that impose to the Lyapunov function candidate $V(t, \boldsymbol{x})$ to be:
    - (globally) positive definite;
    - radially unbounded;
    - (globally) decrescent and,
    - for $\dot{V}(t, \boldsymbol{x})$, to be (globally) negative definite;

    must be replaced by:
    - locally positive definite;
    - locally decrescent and,
    - for $\dot{V}(t, \boldsymbol{x})$, to be locally negative definite.

    If moreover, the differential equation is autonomous and the Lyapunov function candidate $V(\boldsymbol{x})$ is independent of time, then the equilibrium is (locally) asymptotically stable provided that $V(\boldsymbol{x})$ is locally positive definite and $\dot{V}(\boldsymbol{x})$ is locally negative definite.

    An application of the latter is illustrated next.

    Consider the model of a pendulum of length l and mass m concentrated at the end of the pendulum and subject to the action of gravity g, and with viscous friction at the joint (let $f > 0$ be the friction coefficient), *i.e.*

    $$ml^2\ddot{q} + f\dot{q} + mgl \sin(q) = 0,$$

    where q is the angular position with respect to the vertical. Rewrite the model in the state-space form $\dot{\boldsymbol{x}} = \boldsymbol{f}(\boldsymbol{x})$ with $\boldsymbol{x} = [q \;\; \dot{q}]^T$.

    a) Determine the equilibria of this equation.

    b) Show asymptotic stability of the origin by use of the Lyapunov function

    $$V(q, \dot{q}) = 2mgl[1 - \cos(q)] + \frac{ml^2}{2}\dot{q}^2 + \frac{1}{2}\left[\frac{f}{l\sqrt{m}}q + l\sqrt{m}\dot{q}\right]^2 .$$

    c) Is $\dot{V}(q, \dot{q})$ a negative definite function?

15. Complete the analysis of Example 2.15 by applying Lemma 2.2.

3

Robot Dynamics

Robot manipulators are articulated mechanical systems composed of links connected by joints. The joints are mainly of two types: revolute and prismatic. In this textbook we consider robot manipulators formed by an open *kinematic* chain as illustrated in Figure 3.1.

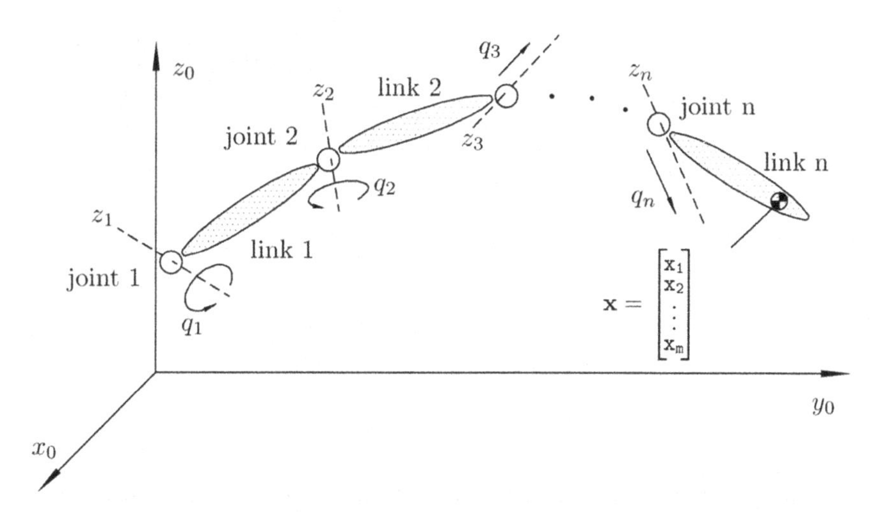

Figure 3.1. Abstract diagram of an n-DOF robot manipulator

Consider the generic configuration of an articulated arm of n links shown in Figure 3.1. In order to derive the mathematical model of a robot one typically starts by placing a 3-dimensional reference frame (*e.g.* in Cartesian coordinates) at any location in the base of the robot. Here, axes will be labeled with no distinction using $\{x \;\; y \;\; z\}$ or $\{x_0 \;\; y_0 \;\; z_0\}$ or $\{x_1 \;\; x_2 \;\; x_3\}$. The links are numbered consecutively from the base (link 0) up to the end-effector (link n). The joints correspond to the contact points between the links and are numbered in such a way that the ith joint connects the ith to the $(i-1)$th link. Each joint is independently controlled through an actuator, which is usually placed at that joint and the movement of the joints produces the relative movement of the links. We temporarily denote by z_i, the ith joint's axis of motion. The *generalized* joint coordinate denoted by q_i, corresponds to the angular displacement around z_i if the ith joint is revolute, or to the linear displacement along z_i if the ith joint is prismatic. In the typical case where the actuators are placed at the joints among the links, the generalized joint coordinates are named joint positions. Unless explicitly said otherwise, we assume that this is the case.

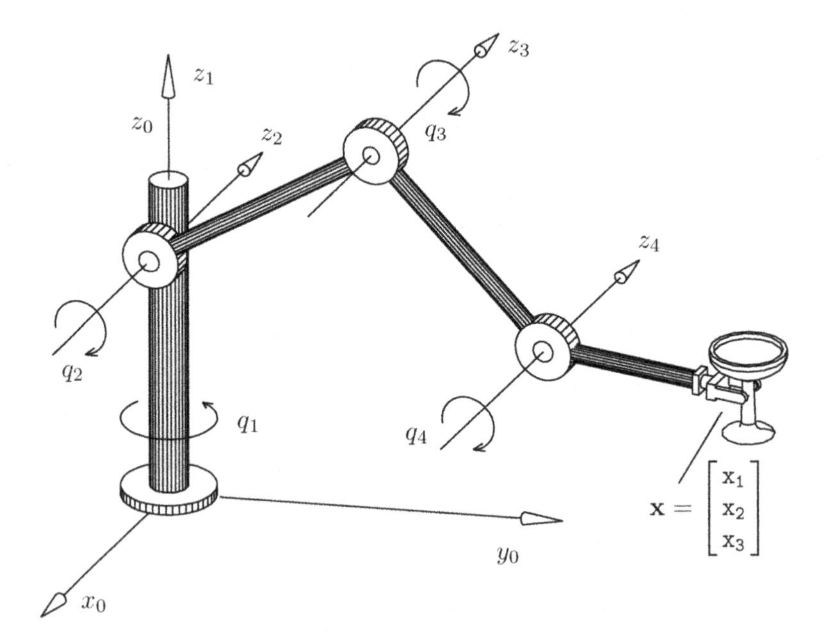

Figure 3.2. Example of a 4-DOF robot

Example 3.1. Figure 3.2 shows a 4-DOF manipulator. The placement of the axes z_i as well as the joint coordinates, are illustrated in this figure. ♢

The joint positions corresponding to each joint of the robot, and which are measured by sensors adequately placed at the actuators, that are usually located at the joints themselves, are collected for analytical purposes, in the vector of joint positions $\boldsymbol{q}$. Consequently, for a robot with n joints, that is, with n DOF (except for special cases, such as elastic-joints or flexible-link robots), the vector of joint positions $\boldsymbol{q}$ has n elements:

$$q = \begin{bmatrix} q_1 \\ q_2 \\ \vdots \\ q_n \end{bmatrix}$$

i.e. $q \in \mathbb{R}^n$. On the other hand it is also of great interest, specifically from a practical viewpoint, to determine the position and orientation (posture) of the robot's end-effector since it is the latter that actually carries out the desired task. Such position and orientation are expressed with respect to the reference frame placed at the base of the robot (*e.g.* a Cartesian frame $\{x_0, y_0, z_0\}$) and eventually in terms of the so-called Euler angles. Such coordinates (and angles) are collected in the vector $\mathbf{x}$ of *operational* positions[1]

$$\mathbf{x} = \begin{bmatrix} \mathrm{x}_1 \\ \mathrm{x}_2 \\ \vdots \\ \mathrm{x}_\mathrm{m} \end{bmatrix}$$

where $m \leq n$. In the case when the robot's end-effector can take any position and orientation in the Euclidean space of dimension 3 (*e.g.* the room where the reader is at the moment), we have $m = 6$. On the other hand, if the robot's motion is on the plane (*i.e.* in dimension 2) and only the position of the end-effector is of interest, then $m = 2$. If, however, the orientation on the plane is of concern then, $m = 3$.

The *direct kinematic model* of a robot, describes the relation between the joint position q and the position and orientation (posture) $\mathbf{x}$ of the robot's end-effector. In other words, the direct kinematic model of a robot is a function $\varphi : \mathbb{R}^n \to \mathbb{R}^m$ such that

$$\mathbf{x} = \varphi(q)\,.$$

Although lengthy, computation of the direct kinematic model, $\mathbf{x} = \varphi(q)$, is methodical and in the case of robots of only few DOF, it involves simple trigonometric expressions.

The *inverse kinematic model* consists precisely in the inverse relation of the direct kinematic model, that is, it corresponds to the relation between the operational posture $\mathbf{x}$ and the joint position q, *i.e.*

$$q = \varphi^{-1}(\mathbf{x}).$$

In contrast to the direct kinematic model, computation of the inverse kinematic model $q = \varphi^{-1}(\mathbf{x})$ may be highly complex and, as a matter of fact, may yield multiple solutions.

The *dynamic model* of a robot consists of an ordinary differential equation where the variable corresponds to the vector of positions and velocities, which

[1] These are also known as positions in the operational space or workspace.

may be in joint coordinates $\boldsymbol{q}$, $\dot{\boldsymbol{q}}$ or in operational coordinates $\mathbf{x}$, $\dot{\mathbf{x}}$. In general, these are second-order nonlinear models which can be written in the generic form

$$\boldsymbol{f}_{EL}(\boldsymbol{q}, \dot{\boldsymbol{q}}, \ddot{\boldsymbol{q}}, \boldsymbol{\tau}) = \mathbf{0}, \tag{3.1}$$

$$\boldsymbol{f}_{C}(\mathbf{x}, \dot{\mathbf{x}}, \ddot{\mathbf{x}}, \boldsymbol{\tau}) = \mathbf{0}. \tag{3.2}$$

The vector $\boldsymbol{\tau}$ stands for the forces and torques applied at the joints by the actuators. The dynamic model (3.1) is the dynamic model in joint space, while (3.2) corresponds to the dynamic model in operational space. In this text, we focus on the dynamic model in joint space and, for simplicity, we omit the words "*in joint space*".

Kinematics as much as dynamics, is fundamental to plan and carry out specific tasks for robot manipulators. Both concepts are dealt with in considerable detail in a large number of available textbooks (see the list of references at the end of this chapter). However, for the sake of completeness we present in this chapter the basic issues related to robot dynamics. In the next two chapters we present properties of the dynamic model, relevant for control. In Chapter 5 we present in detail the model of a real 2-DOF lab prototype which serves as a case study throughout the book.

Besides the unquestionable importance that robot dynamics has in control design, the dynamic models may also be used to simulate numerically (with a personal computer and specialized software), the behavior of a specific robot before actually being constructed. This simulation stage is important, since it allows us to improve the robot design and in particular, to adapt this design to the optimal execution of particular types of tasks.

One of the most common procedures followed in the computation of the dynamic model for robot manipulators, in closed form (*i.e.* not numerical), is the method which relies on the so-called Lagrange's equations of motion. The use of Lagrange's equations requires the notion of two important concepts with which we expect the reader to be familiar: *kinetic* and *potential energies*.

We describe next in some detail, how to derive the dynamic model of a robot via Lagrange's equations.

3.1 Lagrange's Equations of Motion

The dynamic equations of a robot manipulator in closed form may be obtained from Newton's equations of motion or, via Lagrange's equations. All of these are well documented in textbooks on analytical mechanics and are only briefly presented here.

The disadvantage of the first method is that the complexity of the analysis increases with the number of joints in the robot. In such cases, it is better to use Lagrange's equation of motion. The latter are named after the French mathematician Joseph Louis de La Grange (today spelled "Lagrange") which first reported them in 1788 in his celebrated work "*Mécanique analytique*"[2].

Consider the robot manipulator with n links depicted in Figure 3.1. The total energy $\mathcal{E}$ of a robot manipulator of n DOF is the sum of the kinetic and potential energy functions, $\mathcal{K}$ and $\mathcal{U}$ respectively, *i.e.*

$$\mathcal{E}(\boldsymbol{q},\dot{\boldsymbol{q}}) = \mathcal{K}(\boldsymbol{q},\dot{\boldsymbol{q}}) + \mathcal{U}(\boldsymbol{q})$$

where $\boldsymbol{q} = [q_1, \cdots, q_n]^T$.

The *Lagrangian* $\mathcal{L}(\boldsymbol{q},\dot{\boldsymbol{q}})$ of a robot manipulator of n DOF is the difference between its kinetic energy $\mathcal{K}$ and its potential energy $\mathcal{U}$, that is,

$$\mathcal{L}(\boldsymbol{q},\dot{\boldsymbol{q}}) = \mathcal{K}(\boldsymbol{q},\dot{\boldsymbol{q}}) - \mathcal{U}(\boldsymbol{q}). \tag{3.3}$$

We assume here that the potential energy $\mathcal{U}$ is due only to conservative forces such as gravitational energy and energy stored in compressed springs.

Then, the Lagrange equations of motion for a manipulator of n DOF, are given by

$$\frac{d}{dt}\left[\frac{\partial \mathcal{L}(\boldsymbol{q},\dot{\boldsymbol{q}})}{\partial \dot{\boldsymbol{q}}}\right] - \frac{\partial \mathcal{L}(\boldsymbol{q},\dot{\boldsymbol{q}})}{\partial \boldsymbol{q}} = \boldsymbol{\tau},$$

or in the equivalent form by

$$\frac{d}{dt}\left[\frac{\partial \mathcal{L}(\boldsymbol{q},\dot{\boldsymbol{q}})}{\partial \dot{q}_i}\right] - \frac{\partial \mathcal{L}(\boldsymbol{q},\dot{\boldsymbol{q}})}{\partial q_i} = \tau_i, \qquad i = 1, \cdots, n \tag{3.4}$$

where τ_i correspond to the external forces and torques (delivered by the actuators) at each joint as well as to other (nonconservative) forces. In the class of nonconservative forces we include those due to friction, the resistance to the motion of a solid in a fluid, and in general, all those that depend on time and velocity and not only on position.

Notice that one has as many scalar dynamic equations as the manipulator has degrees of freedom.

The use of Lagrange's equations in the derivation of the robot dynamics can be reduced to four main stages:

1. Computation of the kinetic energy function $\mathcal{K}(\boldsymbol{q},\dot{\boldsymbol{q}})$.
2. Computation of the potential energy function $\mathcal{U}(\boldsymbol{q})$.
3. Computation of the Lagrangian (3.3) $\mathcal{L}(\boldsymbol{q},\dot{\boldsymbol{q}})$.
4. Development of Lagrange's equations (3.4).

[2] These equations are also called Euler–Lagrange equations in honor to the Swiss scientist Leonhard Euler, contemporary of La Grange, who discovered a similar but more general set of second-order differential equations.

In the rest of this section we present some examples that illustrate the process of obtaining the robot dynamics by the use of Lagrange's equations of motion.

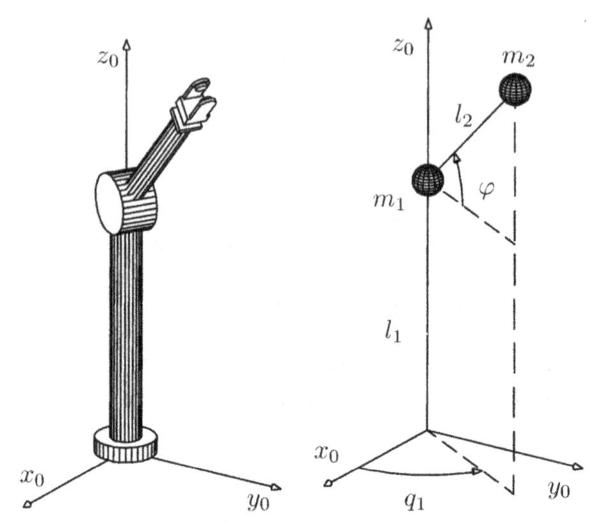

Figure 3.3. Example of a 1-DOF mechanism

Example 3.2. Consider the mechanism shown in Figure 3.3. It consists of a rigid link formed by two parts, of lengths l_1 and l_2, whose masses m_1 and m_2 are, for simplicity, considered to be concentrated at their respective centers of mass, located at the ends. The angle φ is *constant.*

The mechanism possesses only revolute motion about the z_0 axis, the angle of which is represented by q_1. For this example, the only degree-of-freedom is associated to the joint 1. Then, q is a scalar defined as $q = q_1$.

We emphasize that the dynamic model of this mechanism may be obtained using the concepts of dynamics of rigid bodies in rotation, which are subjects of study in elementary courses of physics. However, for the sake of illustration we employ Lagrange's equations of motion.

The kinetic energy function $\mathcal{K}(q, \dot{q})$ of the system is given by the product of half the moment of inertia times the angular velocity squared, *i.e.*

$$\mathcal{K}(q, \dot{q}) = \frac{1}{2} m_2 l_2^2 \cos^2(\varphi)\, \dot{q}^2$$

and the corresponding potential energy

$$\mathcal{U}(q) = m_1 l_1 g + m_2 \left(l_1 + l_2 \sin(\varphi)\right) g$$

where g is the gravity acceleration. Here, we assumed that the potential energy function is zero on the plane x_0–y_0. Actually, notice that in this example the potential energy is constant so in particular, it does not depend on the joint position q.

The Lagrangian $\mathcal{L}(q, \dot{q})$, expressed by (3.3), is in this case

$$\mathcal{L}(q, \dot{q}) = \frac{m_2}{2} l_2^2 \cos^2(\varphi)\, \dot{q}^2 - m_1 l_1 g - m_2 \left(l_1 + l_2 \sin(\varphi)\right) g\,.$$

From this, one can obtain the following equations:

$$\begin{aligned}
\frac{\partial \mathcal{L}}{\partial \dot{q}} &= m_2 l_2^2 \cos^2(\varphi)\, \dot{q} \\
\frac{d}{dt}\left[\frac{\partial \mathcal{L}}{\partial \dot{q}}\right] &= m_2 l_2^2 \cos^2(\varphi)\, \ddot{q} \\
\frac{\partial \mathcal{L}}{\partial q} &= 0\,.
\end{aligned}$$

Then, the corresponding Lagrange Equation (3.4) is

$$m_2 l_2^2 \cos^2(\varphi)\, \ddot{q} = \tau\,, \tag{3.5}$$

where τ is the torque applied at the joint 1. Equation (3.5) describes the dynamic behavior of the mechanism. Notice that the dynamic model is simply a linear second-order nonautonomous differential equation.

Equation (3.5) may be expressed in terms of the state vector $[q \;\; \dot{q}]^T$ as

$$\frac{d}{dt}\begin{bmatrix} q \\ \dot{q} \end{bmatrix} = \begin{bmatrix} \dot{q} \\ \dfrac{\tau}{m_2 l_2^2 \cos^2(\varphi)} \end{bmatrix}; \quad \begin{bmatrix} q(0) \\ \dot{q}(0) \end{bmatrix} \in \mathbb{R}^2\,.$$

The necessary and sufficient condition for the existence of equilibria is $\tau(t) = 0$ for all $t \geq 0$. In this situation, the equation has an infinite number of equilibria which are given by $[q \;\; \dot{q}]^T = [q^* \;\; 0]^T \in \mathbb{R}^2$ with $q^* \in \mathbb{R}$. The interpretation of this result is the following. If at the instant $t = 0$ the position $q(0)$ has any value $q^* \in \mathbb{R}$, the velocity $\dot{q}(0)$ is zero and moreover no torque is applied at the joint (*i.e.* $\tau(t) = 0$ for all t) then, we have $q(t) = q^*$ and $\dot{q}(t) = 0$ for all $t \geq 0$. Note that the latter is in accordance with the physical interpretation of the concept of equilibrium. ♢

The following example illustrates the derivation of the dynamic model of the 2-DOF robot with revolute joints shown in Figure 3.4 and that moves about purely on the horizontal plane. Therefore, gravity has absolutely no influence on the robot dynamics. This should not surprise the reader; note that the potential energy for this robot is constant since its mass does not move in the vertical direction.

A particularity of this example, is that the actuators that deliver the torques are not physically located at the joints themselves, but one of them transmits the motion to the link through belts. Such types of transmission are fairly common in industrial robots. The motivation for such configurations is to lighten the weight (henceforth the inertia) of the arm itself by placing the actuators as close to the base as possible.

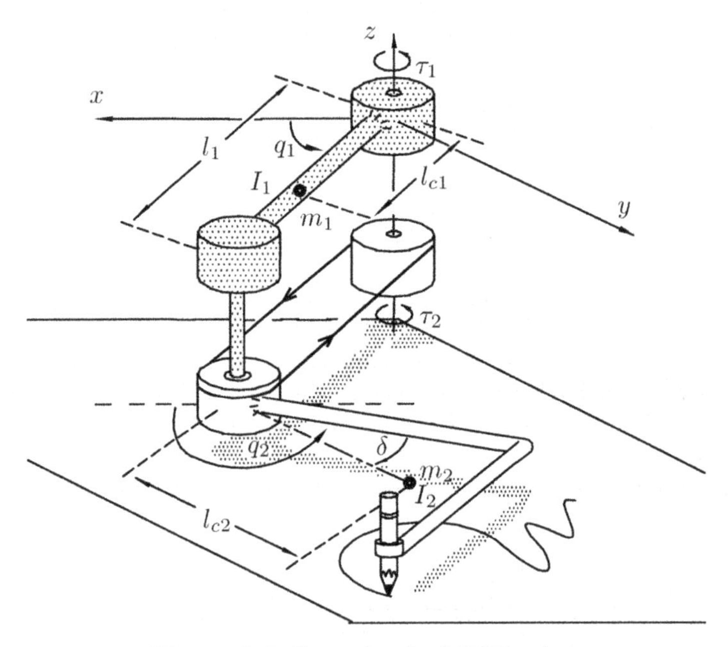

Figure 3.4. Example of a 2-DOF robot

Example 3.3. Consider the robot manipulator with 2 DOF shown in Figure 3.4. This manipulator consists of two rigid links where l_1 is the length the first link. Both joints are revolute. The robot moves about only on the *horizontal* plane x–y as is shown in Figure 3.4. The masses of the links are denoted by m_1 and m_2 respectively. Notice that the center of mass of link 2 may be physically placed "out" of the link itself! This is determined by the value of the constant angle δ. The distances of the rotating axes to the centers of mass, are denoted by l_{c1} and l_{c2} respectively. Finally, I_1 and I_2 denote the moments of inertia of the links with respect to axis that passes through their respective centers of mass and that are parallel to the z axis. The joint positions associated to the angles q_1 and q_2 are measured between each respective link and an axis which is parallel to the x axis. Both angles are taken to be positive counterclockwise. The motion is transmitted to link 2 by a belt since the corresponding actuator is placed at the base of the robot. The vector of joint positions $\boldsymbol{q}$ is defined as

$$\boldsymbol{q} = [q_1 \ \ q_2]^T .$$

The kinetic energy function $\mathcal{K}(\boldsymbol{q}, \dot{\boldsymbol{q}})$ for this arm may be decomposed into the sum of two parts: $\mathcal{K}(\boldsymbol{q}, \dot{\boldsymbol{q}}) = \mathcal{K}_1(\boldsymbol{q}, \dot{\boldsymbol{q}}) + \mathcal{K}_2(\boldsymbol{q}, \dot{\boldsymbol{q}})$, where $\mathcal{K}_1(\boldsymbol{q}, \dot{\boldsymbol{q}})$ and $\mathcal{K}_2(\boldsymbol{q}, \dot{\boldsymbol{q}})$ are the kinetic energies associated with the masses m_1 and m_2 respectively. In turn, the kinetic energy includes the linear and angular motions. Thus, we have $\mathcal{K}_1(\boldsymbol{q}, \dot{\boldsymbol{q}}) = \frac{1}{2}m_1 v_1^2 + \frac{1}{2}I_1\dot{q}_1^2$, where v_1 is the speed[3] of the center of mass of link 1. In this case,

$$\mathcal{K}_1(\boldsymbol{q}, \dot{\boldsymbol{q}}) = \frac{1}{2}m_1 l_{c1}^2 \dot{q}_1^2 + \frac{1}{2}I_1\dot{q}_1^2 . \tag{3.6}$$

On the other hand, $\mathcal{K}_2(\boldsymbol{q}, \dot{\boldsymbol{q}}) = \frac{1}{2}m_2 v_2^2 + \frac{1}{2}I_2\dot{q}_2^2$, where v_2 is the speed of the center of mass of link 2. This speed squared, *i.e.* v_2^2, is given by

$$v_2^2 = \dot{x}_2^2 + \dot{y}_2^2$$

where $\dot{x}_2$ and $\dot{y}_2$ are the components of the velocity vector of the center of mass of link 2. The latter are obtained by evaluating the time derivative of the positions x_2 and y_2 of the center of mass of link 2, *i.e.*

$$x_2 = l_1 \cos(q_1) + l_{c2} \cos(q_2 - \delta)$$

$$y_2 = l_1 \sin(q_1) + l_{c2} \sin(q_2 - \delta) .$$

Using the trigonometric identities, $\cos(\theta)^2 + \sin(\theta)^2 = 1$ and $\sin(q_1)\sin(q_2 - \delta) + \cos(q_1)\cos(q_2 - \delta) = \cos(q_1 - q_2 + \delta)$, we finally get

$$v_2^2 = l_1^2\dot{q}_1^2 + l_{c2}^2\dot{q}_2^2 + 2l_1 l_{c2} \cos(q_1 - q_2 + \delta)\dot{q}_1\dot{q}_2$$

which implies that

$$\begin{aligned}\mathcal{K}_2(\boldsymbol{q}, \dot{\boldsymbol{q}}) = {} & \frac{m_2}{2}l_1^2\dot{q}_1^2 + \frac{m_2}{2}l_{c2}^2\dot{q}_2^2 + m_2 l_1 l_{c2} \cos(q_1 - q_2 + \delta)\dot{q}_1\dot{q}_2 \\ & + \frac{1}{2}I_2\dot{q}_2^2 .\end{aligned} \tag{3.7}$$

Since the robot moves only on the horizontal plane, the potential energy is constant, *e.g.* $\mathcal{U}(\boldsymbol{q}) = 0$.

[3] Some readers may be surprised that we use the word *speed* as opposed to *velocity*. We emphasize that *velocity* is a vector quantity hence, it has a magnitude and direction. The magnitude of the velocity is called *speed*.

So from Equations (3.6) and (3.7), it is clear that the Lagrangian,

$$\mathcal{L}(\boldsymbol{q},\dot{\boldsymbol{q}}) = \mathcal{K}(\boldsymbol{q},\dot{\boldsymbol{q}}) - \mathcal{U}(\boldsymbol{q}),$$

takes the form

$$\begin{aligned}\mathcal{L}(\boldsymbol{q},\dot{\boldsymbol{q}}) &= \mathcal{K}_1(\boldsymbol{q},\dot{\boldsymbol{q}}) + \mathcal{K}_2(\boldsymbol{q},\dot{\boldsymbol{q}})\\
&= \frac{1}{2}(m_1 l_{c1}^2 + m_2 l_1^2)\dot{q}_1^2 + \frac{1}{2}m_2 l_{c2}^2 \dot{q}_2^2\\
&\quad + m_2 l_1 l_{c2}\cos(q_1 - q_2 + \delta)\dot{q}_1\dot{q}_2\\
&\quad + \frac{1}{2}I_1\dot{q}_1^2 + \frac{1}{2}I_2\dot{q}_2^2.\end{aligned}$$

From this equation we obtain the following expressions:

$$\begin{aligned}\frac{\partial \mathcal{L}}{\partial \dot{q}_1} &= (m_1 l_{c1}^2 + m_2 l_1^2 + I_1)\dot{q}_1\\
&\quad + m_2 l_1 l_{c2}\cos(q_1 - q_2 + \delta)\dot{q}_2,\end{aligned}$$

$$\begin{aligned}\frac{d}{dt}\left[\frac{\partial \mathcal{L}}{\partial \dot{q}_1}\right] &= \left(m_1 l_{c1}^2 + m_2 l_1^2 + I_1\right)\ddot{q}_1\\
&\quad + m_2 l_1 l_{c2}\cos(q_1 - q_2 + \delta)\ddot{q}_2\\
&\quad - m_2 l_1 l_{c2}\sin(q_1 - q_2 + \delta)(\dot{q}_1 - \dot{q}_2)\dot{q}_2,\end{aligned}$$

$$\frac{\partial \mathcal{L}}{\partial q_1} = -m_2 l_1 l_{c2}\sin(q_1 - q_2 + \delta)\dot{q}_1\dot{q}_2,$$

$$\frac{\partial \mathcal{L}}{\partial \dot{q}_2} = m_2 l_{c2}^2\dot{q}_2 + m_2 l_1 l_{c2}\cos(q_1 - q_2 + \delta)\dot{q}_1 + I_2\dot{q}_2,$$

$$\begin{aligned}\frac{d}{dt}\left[\frac{\partial \mathcal{L}}{\partial \dot{q}_2}\right] &= m_2 l_{c2}^2\ddot{q}_2 + m_2 l_1 l_{c2}\cos(q_1 - q_2 + \delta)\ddot{q}_1\\
&\quad - m_2 l_1 l_{c2}\sin(q_1 - q_2 + \delta)\dot{q}_1(\dot{q}_1 - \dot{q}_2) + I_2\ddot{q}_2,\end{aligned}$$

$$\frac{\partial \mathcal{L}}{\partial q_2} = m_2 l_1 l_{c2}\sin(q_1 - q_2 + \delta)\dot{q}_1\dot{q}_2.$$

Thus, the dynamic equations that model the robot manipulator may be obtained by using Lagrange's equations (3.4),

$$\begin{aligned}\tau_1 &= \left[m_1 l_{c1}^2 + m_2 l_1^2 + I_1\right]\ddot{q}_1\\
&\quad + m_2 l_1 l_{c2}\cos(q_1 - q_2 + \delta)\ddot{q}_2\\
&\quad + m_2 l_1 l_{c2}\sin(q_1 - q_2 + \delta)\dot{q}_2^2\end{aligned}$$

and,

$$\tau_2 = m_2 l_1 l_{c2} \cos(q_1 - q_2 + \delta)\ddot{q}_1 + [m_2 l_{c2}^2 + I_2]\ddot{q}_2 \\ - m_2 l_1 l_{c2} \sin(q_1 - q_2 + \delta)\dot{q}_1^2 .$$

Finally using the identities

$$\begin{aligned} \cos(q_1 - q_2 + \delta) &= \cos(\delta)\cos(q_1 - q_2) - \sin(\delta)\sin(q_1 - q_2) \\ &= \cos(\delta)\cos(q_2 - q_1) + \sin(\delta)\sin(q_2 - q_1), \\ \sin(q_1 - q_2 + \delta) &= \cos(\delta)\sin(q_1 - q_2) + \sin(\delta)\cos(q_1 - q_2) \\ &= -\cos(\delta)\sin(q_2 - q_1) + \sin(\delta)\cos(q_2 - q_1), \end{aligned}$$

and denoting $C_{21} = \cos(q_2 - q_1)$, $S_{21} = \sin(q_2 - q_1)$, one obtains

$$\begin{aligned} \tau_1 = &\left[m_1 l_{c1}^2 + m_2 l_1^2 + I_1\right] \ddot{q}_1 \\ &+ [m_2 l_1 l_{c2} \cos(\delta) C_{21} + m_2 l_1 l_{c2} \sin(\delta) S_{21}]\ddot{q}_2 \\ &+ [-m_2 l_1 l_{c2} \cos(\delta) S_{21} + m_2 l_1 l_{c2} \sin(\delta) C_{21}]\dot{q}_2^2 \end{aligned} \tag{3.8}$$

and

$$\begin{aligned} \tau_2 = &[m_2 l_1 l_{c2} \cos(\delta) C_{21} + m_2 l_1 l_{c2} \sin(\delta) S_{21}]\ddot{q}_1 \\ &+ [m_2 l_{c2}^2 + I_2]\ddot{q}_2 \\ &+ [m_2 l_1 l_{c2} \cos(\delta) S_{21} - m_2 l_1 l_{c2} \sin(\delta) C_{21}]\dot{q}_1^2 \end{aligned} \tag{3.9}$$

which are of the form (3.1) with

$$\boldsymbol{f}_{EL}(\boldsymbol{q}, \dot{\boldsymbol{q}}, \ddot{\boldsymbol{q}}, \boldsymbol{\tau}) = \begin{bmatrix} \tau_1 - \text{RHS}(3.8) \\ \tau_2 - \text{RHS}(3.9) \end{bmatrix}$$

where RHS(3.8) and RHS(3.9) denote the terms on the right-hand side of (3.8) and (3.9) respectively. ◇

In the following example we derive the dynamic model of a Cartesian robot with 3 DOF whose main feature is that it has a linear dynamic model.

Example 3.4. Consider the 3-DOF Cartesian robot manipulator shown in Figure 3.5. The manipulator consists of three rigid links mutually orthogonal. The three joints of the robot are prismatic. The robot's displacements happen in the space x_0–y_0–z_0 shown in Figure 3.5. The vector of joint positions is $\boldsymbol{q} = [q_1 \;\; q_2 \;\; q_3]^T$.

The kinetic energy function for this manipulator is given by (*cf.* Figure 3.5):

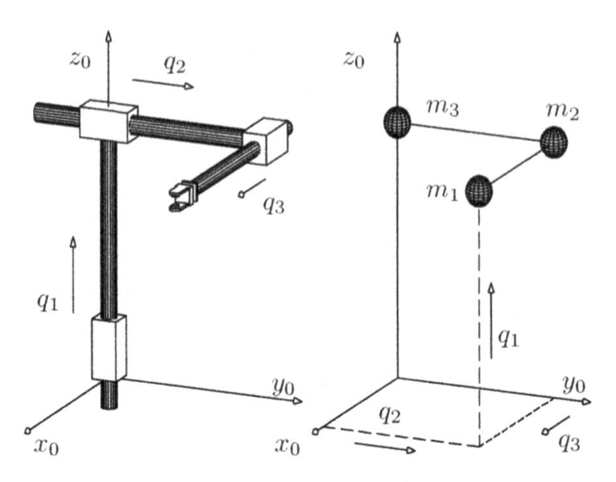

Figure 3.5. Example of a 3-DOF Cartesian robot

$$\mathcal{K}(\boldsymbol{q}, \dot{\boldsymbol{q}}) = \frac{1}{2}\left[m_1\dot{q}_3^2 + [m_1 + m_2]\dot{q}_2^2 + [m_1 + m_2 + m_3]\dot{q}_1^2\right] . \tag{3.10}$$

On the other hand, the potential energy is given by

$$\mathcal{U}(\boldsymbol{q}) = [m_1 + m_2 + m_3]gq_1 . \tag{3.11}$$

From Equations (3.10) and (3.11) we obtain the Lagrangian,

$$\begin{aligned}\mathcal{L}(\boldsymbol{q}, \dot{\boldsymbol{q}}) &= \mathcal{K}(\boldsymbol{q}, \dot{\boldsymbol{q}}) - \mathcal{U}(\boldsymbol{q}) \\ &= \frac{1}{2}\left[m_1\dot{q}_3^2 + [m_1 + m_2]\dot{q}_2^2 + [m_1 + m_2 + m_3]\dot{q}_1^2\right] \\ &\quad - [m_1 + m_2 + m_3]gq_1 .\end{aligned}$$

So we have

$$\begin{aligned}
\frac{\partial \mathcal{L}}{\partial \dot{q}_1} &= [m_1 + m_2 + m_3]\dot{q}_1 & \frac{d}{dt}\left[\frac{\partial \mathcal{L}}{\partial \dot{q}_1}\right] &= [m_1 + m_2 + m_3]\ddot{q}_1 \\
\frac{\partial \mathcal{L}}{\partial \dot{q}_2} &= [m_1 + m_2]\dot{q}_2 & \frac{d}{dt}\left[\frac{\partial \mathcal{L}}{\partial \dot{q}_2}\right] &= [m_1 + m_2]\ddot{q}_2 \\
\frac{\partial \mathcal{L}}{\partial \dot{q}_3} &= m_1\dot{q}_3 & \frac{d}{dt}\left[\frac{\partial \mathcal{L}}{\partial \dot{q}_3}\right] &= m_1\ddot{q}_3 \\
\frac{\partial \mathcal{L}}{\partial q_2} = \frac{\partial \mathcal{L}}{\partial q_3} &= 0 & \frac{\partial \mathcal{L}}{\partial q_1} &= -[m_1 + m_2 + m_3]g.
\end{aligned}$$

The dynamic equations that model this robot may be obtained by applying the Lagrange's equations (3.4) to obtain

$$[m_1 + m_2 + m_3]\ddot{q}_1 + [m_1 + m_2 + m_3]g = \tau_1 \tag{3.12}$$
$$[m_1 + m_2]\ddot{q}_2 = \tau_2 \tag{3.13}$$
$$m_1\ddot{q}_3 = \tau_3 \tag{3.14}$$

where τ_1, τ_2 and τ_3 are the external forces applied at each joint. Notice that in this example Equations (3.12)–(3.14) define a set of linear autonomous differential equations.

In terms of the state vector $[q_1 \;\; q_2 \;\; q_3 \;\; \dot{q}_1 \;\; \dot{q}_2 \;\; \dot{q}_3]$, the equations (3.12), (3.13) and (3.14) may be expressed as

$$\frac{d}{dt}\begin{bmatrix} q_1 \\ q_2 \\ q_3 \\ \dot{q}_1 \\ \dot{q}_2 \\ \dot{q}_3 \end{bmatrix} = \begin{bmatrix} \dot{q}_1 \\ \dot{q}_2 \\ \dot{q}_3 \\ \dfrac{1}{m_1 + m_2 + m_3}[\tau_1 - [m_1 + m_2 + m_3]g] \\ \dfrac{1}{m_1 + m_2}\tau_2 \\ \dfrac{1}{m_1}\tau_3 \end{bmatrix}.$$

The necessary and sufficient condition for the existence of equilibria is $\tau_1 = [m_1 + m_2 + m_3]g$, $\tau_2 = 0$ and $\tau_3 = 0$ and actually we have an infinite number of them:

$$[q_1 \;\; q_2 \;\; q_3 \;\; \dot{q}_1 \;\; \dot{q}_2 \;\; \dot{q}_3]^T = [q_1^* \;\; q_2^* \;\; q_3^* \;\; 0 \;\; 0 \;\; 0]^T$$

with q_1^*, q_2^*, $q_3^* \in \mathbb{R}$. ◊

3.2 Dynamic Model in Compact Form

In the previous section we presented some examples to illustrate the application of Lagrange's equations to obtain the dynamic equations for robots with particular geometries. This same methodology, however, may be employed in general to obtain the dynamic model of *any* robot of n DOF.

This methodology is commonly studied in classical texts on robotics and theoretical mechanics and therefore, we present it here only in compact form. The interested reader is invited to see the cited texts at the end of the chapter.

Consider a robot manipulator of n DOF composed of rigid links interconnected by frictionless joints. The kinetic energy function $\mathcal{K}(\boldsymbol{q}, \dot{\boldsymbol{q}})$ associated with such an articulated mechanism may always be expressed as

$$\mathcal{K}(\boldsymbol{q}, \dot{\boldsymbol{q}}) = \frac{1}{2}\dot{\boldsymbol{q}}^T M(\boldsymbol{q})\dot{\boldsymbol{q}} \tag{3.15}$$

where $M(\boldsymbol{q})$ is a matrix of dimension $n \times n$ referred to as the **inertia matrix.** $M(\boldsymbol{q})$ is symmetric and positive definite for all $\boldsymbol{q} \in \mathbb{R}^n$. The potential energy $\mathcal{U}(\boldsymbol{q})$ does not have a specific form as in the case of the kinetic energy but it is known that it depends on the vector of joint positions $\boldsymbol{q}$.

The Lagrangian $\mathcal{L}(\boldsymbol{q}, \dot{\boldsymbol{q}})$, given by Equation (3.3), becomes in this case

$$\mathcal{L}(\boldsymbol{q}, \dot{\boldsymbol{q}}) = \frac{1}{2}\dot{\boldsymbol{q}}^T M(\boldsymbol{q})\dot{\boldsymbol{q}} - \mathcal{U}(\boldsymbol{q}) .$$

With this Lagrangian, the Lagrange's equations of motion (3.4) may be written as

$$\frac{d}{dt}\left[\frac{\partial}{\partial \dot{\boldsymbol{q}}}\left[\frac{1}{2}\dot{\boldsymbol{q}}^T M(\boldsymbol{q})\dot{\boldsymbol{q}}\right]\right] - \frac{\partial}{\partial \boldsymbol{q}}\left[\frac{1}{2}\dot{\boldsymbol{q}}^T M(\boldsymbol{q})\dot{\boldsymbol{q}}\right] + \frac{\partial \mathcal{U}(\boldsymbol{q})}{\partial \boldsymbol{q}} = \boldsymbol{\tau} .$$

On the other hand, it holds that

$$\frac{\partial}{\partial \dot{\boldsymbol{q}}}\left[\frac{1}{2}\dot{\boldsymbol{q}}^T M(\boldsymbol{q})\dot{\boldsymbol{q}}\right] = M(\boldsymbol{q})\dot{\boldsymbol{q}} \tag{3.16}$$

$$\frac{d}{dt}\left[\frac{\partial}{\partial \dot{\boldsymbol{q}}}\left[\frac{1}{2}\dot{\boldsymbol{q}}^T M(\boldsymbol{q})\dot{\boldsymbol{q}}\right]\right] = M(\boldsymbol{q})\ddot{\boldsymbol{q}} + \dot{M}(\boldsymbol{q})\dot{\boldsymbol{q}}. \tag{3.17}$$

Considering these expressions, the equation of motion takes the form

$$M(\boldsymbol{q})\ddot{\boldsymbol{q}} + \dot{M}(\boldsymbol{q})\dot{\boldsymbol{q}} - \frac{1}{2}\frac{\partial}{\partial \boldsymbol{q}}\left[\dot{\boldsymbol{q}}^T M(\boldsymbol{q})\dot{\boldsymbol{q}}\right] + \frac{\partial \mathcal{U}(\boldsymbol{q})}{\partial \boldsymbol{q}} = \boldsymbol{\tau},$$

or, in compact form,

$$M(\boldsymbol{q})\ddot{\boldsymbol{q}} + C(\boldsymbol{q}, \dot{\boldsymbol{q}})\dot{\boldsymbol{q}} + \boldsymbol{g}(\boldsymbol{q}) = \boldsymbol{\tau} \tag{3.18}$$

where

$$C(\boldsymbol{q}, \dot{\boldsymbol{q}})\dot{\boldsymbol{q}} = \dot{M}(\boldsymbol{q})\dot{\boldsymbol{q}} - \frac{1}{2}\frac{\partial}{\partial \boldsymbol{q}}\left[\dot{\boldsymbol{q}}^T M(\boldsymbol{q})\dot{\boldsymbol{q}}\right] \tag{3.19}$$

$$\boldsymbol{g}(\boldsymbol{q}) = \frac{\partial \mathcal{U}(\boldsymbol{q})}{\partial \boldsymbol{q}} . \tag{3.20}$$

Equation (3.18) is the dynamic equation for robots of n DOF. Notice that (3.18) is a nonlinear vectorial differential equation of the state $[\boldsymbol{q}^T \;\; \dot{\boldsymbol{q}}^T]^T$. $C(\boldsymbol{q}, \dot{\boldsymbol{q}})\dot{\boldsymbol{q}}$ is a vector of dimension n called the **vector of centrifugal and Coriolis forces,** $\boldsymbol{g}(\boldsymbol{q})$ is a vector of dimension n of **gravitational forces or torques** and $\boldsymbol{\tau}$ is a vector of dimension n called the **vector of external**

forces, which in general corresponds to the torques and forces applied by the actuators at the joints.

The matrix $C(\boldsymbol{q}, \dot{\boldsymbol{q}}) \in \mathbb{R}^{n\times n}$, called the centrifugal and Coriolis forces matrix may be not unique, but the vector $C(\boldsymbol{q}, \dot{\boldsymbol{q}})\dot{\boldsymbol{q}}$ is indeed unique. One way to obtain $C(\boldsymbol{q}, \dot{\boldsymbol{q}})$ is through the coefficients or so-called Christoffel symbols of the first kind, $c_{ijk}(\boldsymbol{q})$, defined as

$$c_{ijk}(\boldsymbol{q}) = \frac{1}{2}\left[\frac{\partial M_{kj}(\boldsymbol{q})}{\partial q_i} + \frac{\partial M_{ki}(\boldsymbol{q})}{\partial q_j} - \frac{\partial M_{ij}(\boldsymbol{q})}{\partial q_k}\right]. \tag{3.21}$$

Here, $M_{ij}(\boldsymbol{q})$ denotes the ijth element of the inertia matrix $M(\boldsymbol{q})$. Indeed, the kjth element of the matrix $C(\boldsymbol{q}, \dot{\boldsymbol{q}})$, $C_{kj}(\boldsymbol{q}, \dot{\boldsymbol{q}})$, is given by (we do not show here the development of the calculations to obtain such expressions, the interested reader is invited to see the texts cited at the end of the chapter)

$$C_{kj}(\boldsymbol{q}, \dot{\boldsymbol{q}}) = \begin{bmatrix} c_{1jk}(\boldsymbol{q}) \\ c_{2jk}(\boldsymbol{q}) \\ \vdots \\ c_{njk}(\boldsymbol{q}) \end{bmatrix}^T \dot{\boldsymbol{q}}. \tag{3.22}$$

The model (3.18) may be viewed as a dynamic system with input, the vector $\boldsymbol{\tau}$, and with outputs, the vectors $\boldsymbol{q}$ and $\dot{\boldsymbol{q}}$. This is illustrated in Figure 3.6.

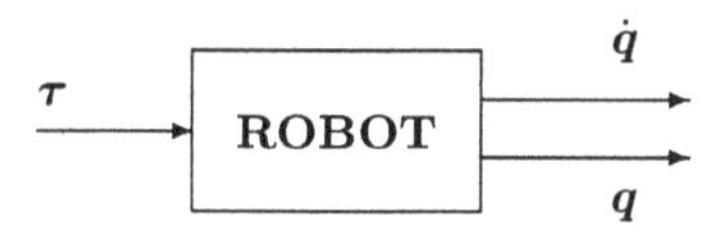

Figure 3.6. Input–output representation of a robot

Each element of $M(\boldsymbol{q})$, $C(\boldsymbol{q}, \dot{\boldsymbol{q}})$ and $\boldsymbol{g}(\boldsymbol{q})$ is in general, a relatively complex expression of the positions and velocities of all the joints, that is, of $\boldsymbol{q}$ and $\dot{\boldsymbol{q}}$. The elements of $M(\boldsymbol{q})$, $C(\boldsymbol{q}, \dot{\boldsymbol{q}})$ and $\boldsymbol{g}(\boldsymbol{q})$ depend of course, on the geometry of the robot in question. Notice that computation of the vector $\boldsymbol{g}(\boldsymbol{q})$ for a given robot may be carried out with relative ease since this is given by (3.20). In other words, the vector of gravitational torques $\boldsymbol{g}(\boldsymbol{q})$, is simply the gradient of the potential energy function $\mathcal{U}(\boldsymbol{q})$.

Example 3.5. The dynamic model of the robot from Example 3.2, that is, Equation (3.5), may be written in the generic form (3.18) by taking

$$M(q) = m_2 l_2^2 \cos^2(\varphi),$$

$$C(q,\dot{q}) = 0\,,$$
$$g(q) = 0\,.$$

Example 3.6. The Lagrangian dynamic model of the robot manipulator shown in Figure 3.4 was derived in Example 3.3. A simple inspection of Equations (3.8) and (3.9) shows that the dynamic model for this robot in compact form is

$$\underbrace{\begin{bmatrix} M_{11}(\boldsymbol{q}) & M_{12}(\boldsymbol{q}) \\ M_{21}(\boldsymbol{q}) & M_{22}(\boldsymbol{q}) \end{bmatrix}}_{M(\boldsymbol{q})} \ddot{\boldsymbol{q}} + \underbrace{\begin{bmatrix} C_{11}(\boldsymbol{q},\dot{\boldsymbol{q}}) & C_{12}(\boldsymbol{q},\dot{\boldsymbol{q}}) \\ C_{21}(\boldsymbol{q},\dot{\boldsymbol{q}}) & C_{22}(\boldsymbol{q},\dot{\boldsymbol{q}}) \end{bmatrix}}_{C(\boldsymbol{q},\dot{\boldsymbol{q}})} \dot{\boldsymbol{q}} = \boldsymbol{\tau}(t)$$

where

$$\begin{aligned} M_{11}(\boldsymbol{q}) &= \left[m_1 l_{c1}^2 + m_2 l_1^2 + I_1\right] \\ M_{12}(\boldsymbol{q}) &= \left[m_2 l_1 l_{c2} \cos(\delta) C_{21} + m_2 l_1 l_{c2} \sin(\delta) S_{21}\right] \\ M_{21}(\boldsymbol{q}) &= \left[m_2 l_1 l_{c2} \cos(\delta) C_{21} + m_2 l_1 l_{c2} \sin(\delta) S_{21}\right] \\ M_{22}(\boldsymbol{q}) &= \left[m_2 l_{c2}^2 + I_2\right] \end{aligned}$$

$$\begin{aligned} C_{11}(\boldsymbol{q},\dot{\boldsymbol{q}}) &= 0 \\ C_{12}(\boldsymbol{q},\dot{\boldsymbol{q}}) &= \left[-m_2 l_1 l_{c2} \cos(\delta) S_{21} + m_2 l_1 l_{c2} \sin(\delta) C_{21}\right]\dot{q}_2 \\ C_{21}(\boldsymbol{q},\dot{\boldsymbol{q}}) &= \left[m_2 l_1 l_{c2} \cos(\delta) S_{21} - m_2 l_1 l_{c2} \sin(\delta) C_{21}\right]\dot{q}_1 \\ C_{22}(\boldsymbol{q},\dot{\boldsymbol{q}}) &= 0 \end{aligned}$$

That is, the gravitational forces vector is zero. 

The dynamic model in compact form is important because it is the model that we use throughout the text to design controllers and to analyze the stability, in the sense of Lyapunov, of the equilibria of the closed-loop system. In anticipation of the material in later chapters of this text and in support of the material of Chapter 2 it is convenient to make some remarks at this point about the "stability of the robot system".

In the previous examples we have seen that the model in compact form may be rewritten in the state-space form. As a matter of fact, this property is not limited to particular examples but stands as a fact for robot manipulators in general. This is because the inertia matrix is positive definite and so is the matrix $M(\boldsymbol{q})^{-1}$; in particular, the latter always exists. This is what allows us to express the dynamic model (3.18) of any robot of n DOF in terms of the *state vector* $[\boldsymbol{q}^T \ \dot{\boldsymbol{q}}^T]^T$ that is, as

$$\frac{d}{dt}\begin{bmatrix} \boldsymbol{q} \\ \dot{\boldsymbol{q}} \end{bmatrix} = \begin{bmatrix} \dot{\boldsymbol{q}} \\ M(\boldsymbol{q})^{-1}\left[\boldsymbol{\tau}(t) - C(\boldsymbol{q},\dot{\boldsymbol{q}})\dot{\boldsymbol{q}} - \boldsymbol{g}(\boldsymbol{q})\right] \end{bmatrix}. \tag{3.23}$$

Note that this constitutes a set of nonlinear differential equations of the form (3.1). In view of this nonlinear nature, the concept of stability of a robot in open loop must be handled with care.

We emphasize that the definition of stability in the sense of Lyapunov, which is presented in Definition 2.2 in Chapter 2, applies to an equilibrium (typically the origin). Hence, in studying the "stability of a robot manipulator" it is indispensable to first determine the equilibria of Equation (3.23), which describes the behavior of the robot.

The necessary and sufficient condition for the existence of equilibria of Equation (3.23), is that $\boldsymbol{\tau}(t)$ be constant (say, $\boldsymbol{\tau}^*$) and that there exist a solution $\boldsymbol{q}^* \in \mathbb{R}^n$ to the algebraic possibly nonlinear equation, in $\boldsymbol{g}(\boldsymbol{q}^*) = \boldsymbol{\tau}^*$. In such a situation, the equilibria are given by $[\boldsymbol{q}^T \ \ \dot{\boldsymbol{q}}^T]^T = [\boldsymbol{q}^{*T} \ \ \boldsymbol{0}^T]^T \in \mathbb{R}^{2n}$. In the particular case of $\boldsymbol{\tau} \equiv 0$, the possible equilibria of (3.23) are given by $[\boldsymbol{q}^T \ \ \dot{\boldsymbol{q}}^T]^T = [\boldsymbol{q}^{*T} \ \ \boldsymbol{0}^T]^T$ where $\boldsymbol{q}^*$ is a solution of $\boldsymbol{g}(\boldsymbol{q}^*) = \boldsymbol{0}$. Given the definition of $\boldsymbol{g}(\boldsymbol{q})$ as the gradient of the potential energy $\mathcal{U}(\boldsymbol{q})$, we see that $\boldsymbol{q}^*$ corresponds to the vectors where the potential energy possesses extrema.

A particular case is that of robots whose workspace corresponds to the horizontal plane. In this case, $\boldsymbol{g}(\boldsymbol{q}) = \boldsymbol{0}$ and therefore it is necessary and sufficient that $\boldsymbol{\tau}(t) = \boldsymbol{0}$ for equilibria to exist. Indeed, the point $\left[\boldsymbol{q}^T \ \ \dot{\boldsymbol{q}}^T\right]^T = [\boldsymbol{q}^{*T} \ \ \boldsymbol{0}^T]^T \in \mathbb{R}^{2n}$ where $\boldsymbol{q}^*$ is any vector of dimension n is an equilibrium. This means that there exist an infinite number of equilibria. See also Example 2.2.

The development above makes it clear that if one wants to study the topic of robot stability in open loop (that is, without control) one must specify the dynamic model as well as the conditions for equilibria to exist and only then, select one among these equilibria, whose stability is of interest. Consequently, the question "is the robot stable? " is ambiguous in the present context.

3.3 Dynamic Model of Robots with Friction

It is important to notice that the generic Equation (3.18) supposes that the links are rigid, that is, they do not present any torsion or any other deformation phenomena. On the other hand, we also considered that the joints between each pair of links are stiff and frictionless. The incorporation of these phenomena in the dynamic model of robots is presented in this and the following section.

Friction effects in mechanical systems are phenomena that depend on multiple factors such as the nature of the materials in contact, lubrication of the latter, temperature, *etc.* For this reason, typically only approximate models of friction forces and torques are available. Yet, it is accepted that these forces and torques depend on the relative velocity between the bodies in contact. Thus, we distinguish two families of friction models: the static models, in which the friction force or torque depends on the *instantaneous* relative velocity between bodies and, dynamic models, which depend on the past values of the relative velocity.

Thus, in the static models, friction is modeled by a vector $\boldsymbol{f}(\dot{\boldsymbol{q}}) \in \mathbb{R}^n$ that depends only on the joint velocity $\dot{\boldsymbol{q}}$. Friction effects are local, that is, $\boldsymbol{f}(\dot{\boldsymbol{q}})$ may be written as

$$\boldsymbol{f}(\dot{\boldsymbol{q}}) = \begin{bmatrix} f_1(\dot{q}_1) \\ f_2(\dot{q}_2) \\ \vdots \\ f_n(\dot{q}_n) \end{bmatrix} .$$

An important feature of friction forces is that they dissipate energy, that is,

$$\dot{\boldsymbol{q}}^T \boldsymbol{f}(\dot{\boldsymbol{q}}) > 0 \qquad \forall\, \dot{\boldsymbol{q}} \neq \boldsymbol{0} \in \mathbb{R}^n .$$

A "classical" static friction model is one that combines the so-called viscous and Coulomb friction phenomena. This model establishes that the vector $\boldsymbol{f}(\dot{\boldsymbol{q}})$ is given by

$$\boldsymbol{f}(\dot{\boldsymbol{q}}) = F_{m1}\dot{\boldsymbol{q}} + F_{m2}\,\mathbf{sign}(\dot{\boldsymbol{q}}) \tag{3.24}$$

where F_{m1} and F_{m2} are $n \times n$ diagonal positive definite matrices. The elements of the diagonal of F_{m1} correspond to the viscous friction parameters while the elements of F_{m2} correspond to the Coulomb friction parameters. Furthermore, in the model given by (3.24)

$$\mathbf{sign}(\dot{\boldsymbol{q}}) = \begin{bmatrix} \mathrm{sign}(\dot{q}_1) \\ \mathrm{sign}(\dot{q}_2) \\ \vdots \\ \mathrm{sign}(\dot{q}_n) \end{bmatrix}$$

and $\mathrm{sign}(x)$ is the sign "function", given by

$$\mathrm{sign}(x) = \begin{cases} 1 & \text{if } x > 0 \\ -1 & \text{if } x < 0 . \end{cases}$$

However, $\mathrm{sign}(0)$ is undefined in the sense that one do not associate a particular real number to the "function" $\mathrm{sign}(x)$ when $x = 0$.

In certain applications, this fact is not of much practical relevance, as for instance, in velocity regulation, – when it is desired to maintain an operating point involving high and medium-high constant velocity, but the definition of sign(0) is crucial both from theoretical and practical viewpoints, in position control (*i.e.* when the control objective is to maintain a constant position).

For this reason, and in view of the fact that the "classical" model (3.24) describes inadequately the behavior of friction at very low velocities, that is, when bodies are at rest and start to move, this is not recommended to model friction when dealing with the position control problem (regulation). In this case it is advisable to use available dynamic models. The study of such models is beyond the scope of this textbook.

Considering friction in the joints, the general dynamic equation of the manipulator is now given by

$$M(\boldsymbol{q})\ddot{\boldsymbol{q}} + C(\boldsymbol{q}, \dot{\boldsymbol{q}})\dot{\boldsymbol{q}} + \boldsymbol{g}(\boldsymbol{q}) + \boldsymbol{f}(\dot{\boldsymbol{q}}) = \boldsymbol{\tau}\,. \tag{3.25}$$

In general, in this text we shall not assume that friction is present in the dynamic model unless it is explicitly mentioned. In such a case, we consider only viscous friction.

3.4 Dynamic Model of Elastic-joint Robots

In many industrial robots, the motion produced by the actuators is transmitted to the links via gears and belts. These, are not completely stiff but they have elasticity which can be compared to that of a spring. In the case of revolute joints, where the actuators are generally electric motors these phenomena boil down to a torsion in the axis that connects the link to the rotor of the motor. The elasticity effect in the joints is more noticeable in robots which undergo displacements with abrupt changes in velocity. A direct consequence of this effect, is the degradation of precision in the desired motion of the robot. Evidently, industrial robots are designed in a way to favor the reduction of joint elasticity, however, as mentioned above, such an effect is always present to some degree on practically any mechanical device. An exception to this rule is the case of the so-called *direct-drive* robots, in which the actuators are directly connected to the links.

Robot dynamics and control under the consideration of joint elasticity, has been an important topic of research since the mid-1980s and continues today. We present below only a brief discussion.

Consider a robot manipulator composed of rigid n links connected through revolute joints. Assume that the motion of each link is furnished by electric motors and transmitted via a set of gears. Denote by J_i, the moment of inertia of the rotors about their respective rotating axes. Let r_i be the gear reduction ratio of each rotor; *e.g.* if $r = 50$ we say that for every 50 rotor revolutions, the axis after the corresponding gear undergoes only one full turn. Joint elasticity between each link and the corresponding axis after the set of gears is modeled via a torsional spring of constant torsional 'stiffness', k_i. The larger k_i, the stiffer the joint. Figure 3.7 illustrates the case of a robot with two joints. The joint positions of each link are denoted, as usual by $\boldsymbol{q}$ while the angular positions of the axes after the set of gears are $\boldsymbol{\theta} = [\theta_1 \;\; \theta_2 \;\; \cdots \;\; \theta_n]^T$. Due to the elasticity, and while the robot is in motion we have, in general, $\boldsymbol{q} \neq \boldsymbol{\theta}$.

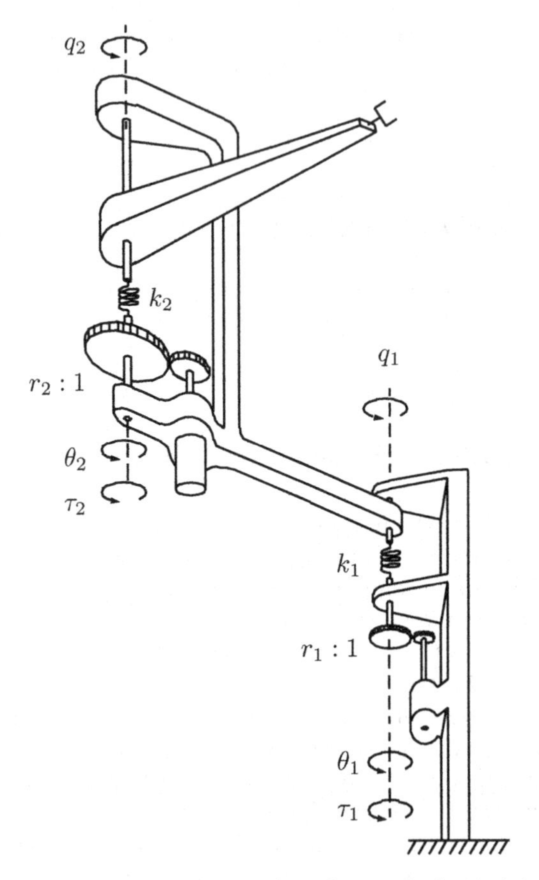

Figure 3.7. Diagram of a robot with elastic joints

Typically, the position and velocity sensors are located at the level of the rotors' axes. Thus knowing the gears reduction rate, only $\boldsymbol{\theta}$ may be determined and in particular, $\boldsymbol{q}$ is not available. This observation is of special relevance in control design since the variable to be controlled is precisely $\boldsymbol{q}$, which cannot be measured directly unless one is able to collocate appropriate sensors to measure the links' positions, giving a higher manufacturing cost.

Due to elasticity a given robot having n links has $2n$ DOF. Its generalized coordinates are $[\boldsymbol{q}^T \;\; \boldsymbol{\theta}^T]^T$. The kinetic energy function of a robot with elastic joints corresponds basically to the sum of the kinetic energies of the links and those of the rotors,[4] that is,

$$\mathcal{K}(\boldsymbol{q},\dot{\boldsymbol{q}},\dot{\boldsymbol{\theta}}) = \frac{1}{2}\dot{\boldsymbol{q}}^T M(\boldsymbol{q})\dot{\boldsymbol{q}} + \frac{1}{2}\dot{\boldsymbol{\theta}}^T J\dot{\boldsymbol{\theta}}$$

where $M(\boldsymbol{q})$ is the inertia matrix of the "rigid" (that is, assuming an infinite stiffness value of k_i for all i) robot, and J is a diagonal positive definite matrix, whose elements are the moments of inertia of the rotors, multiplied by the square of the gear reduction ratio, *i.e.*

$$J = \begin{bmatrix} J_1 r_1^2 & 0 & \cdots & 0 \\ 0 & J_2 r_2^2 & \cdots & 0 \\ \vdots & \vdots & \ddots & \vdots \\ 0 & 0 & \cdots & J_n r_n^2 \end{bmatrix}.$$

On the other hand, the potential energy is the sum of the gravitational energy plus that stored in the torsional fictitious springs[5], *i.e.*

$$\mathcal{U}(\boldsymbol{q},\boldsymbol{\theta}) = \mathcal{U}_1(\boldsymbol{q}) + \frac{1}{2}[\boldsymbol{q}-\boldsymbol{\theta}]^T K[\boldsymbol{q}-\boldsymbol{\theta}] \tag{3.26}$$

where $\mathcal{U}_1(\boldsymbol{q})$ is the potential energy due to gravity and corresponds exactly to that of the robot as if the joints were absolutely stiff. The matrix K is diagonal positive definite and its elements are the 'torsion constants', *i.e.*

$$K = \begin{bmatrix} k_1 & 0 & \cdots & 0 \\ 0 & k_2 & \cdots & 0 \\ \vdots & \vdots & \ddots & \vdots \\ 0 & 0 & \cdots & k_n \end{bmatrix}.$$

The Lagrangian is in this case

$$\mathcal{L}(\boldsymbol{q},\boldsymbol{\theta},\dot{\boldsymbol{q}},\dot{\boldsymbol{\theta}}) = \frac{1}{2}\dot{\boldsymbol{q}}^T M(\boldsymbol{q})\dot{\boldsymbol{q}} + \frac{1}{2}\dot{\boldsymbol{\theta}}^T J\dot{\boldsymbol{\theta}} - \mathcal{U}_1(\boldsymbol{q}) - \frac{1}{2}[\boldsymbol{q}-\boldsymbol{\theta}]^T K[\boldsymbol{q}-\boldsymbol{\theta}].$$

Hence, using Lagrange's motion equations (3.4) we obtain

$$\frac{d}{dt}\begin{bmatrix} \dfrac{\partial}{\partial \dot{\boldsymbol{q}}}\left[\dfrac{1}{2}\dot{\boldsymbol{q}}^T M(\boldsymbol{q})\dot{\boldsymbol{q}}\right] \\ \dfrac{\partial}{\partial \dot{\boldsymbol{\theta}}}\left[\dfrac{1}{2}\dot{\boldsymbol{\theta}}^T J\dot{\boldsymbol{\theta}}\right] \end{bmatrix} - \begin{bmatrix} \dfrac{\partial}{\partial \boldsymbol{q}}\left[\dfrac{1}{2}\dot{\boldsymbol{q}}^T M(\boldsymbol{q})\dot{\boldsymbol{q}} - \mathcal{U}_1(\boldsymbol{q}) - \dfrac{1}{2}[\boldsymbol{q}-\boldsymbol{\theta}]^T K[\boldsymbol{q}-\boldsymbol{\theta}]\right] \\ \dfrac{\partial}{\partial \boldsymbol{\theta}}\left[-\dfrac{1}{2}[\boldsymbol{q}-\boldsymbol{\theta}]^T K[\boldsymbol{q}-\boldsymbol{\theta}]\right] \end{bmatrix}$$

$$= \begin{bmatrix} \boldsymbol{0} \\ \boldsymbol{\tau} \end{bmatrix}.$$

[4] Here, we neglect the gyroscopic and other coupling effects between the rotors and the links.

[5] We assume here that the rotors constitute uniform cylinders so that they do not contribute to the total potential energy. Therefore, in (3.26) there is no term '$\mathcal{U}_2(\boldsymbol{\theta})$'.

Finally, using (3.16), (3.17), (3.19) and

$$\frac{\partial}{\partial \boldsymbol{\theta}}\left[-\frac{1}{2}[\boldsymbol{q}-\boldsymbol{\theta}]^T K[\boldsymbol{q}-\boldsymbol{\theta}]\right] = K[\boldsymbol{q}-\boldsymbol{\theta}]\,,$$

we obtain the dynamic model for elastic-joint robots as

$$M(\boldsymbol{q})\ddot{\boldsymbol{q}} + C(\boldsymbol{q},\dot{\boldsymbol{q}})\dot{\boldsymbol{q}} + \boldsymbol{g}(\boldsymbol{q}) + K(\boldsymbol{q}-\boldsymbol{\theta}) = \boldsymbol{0} \tag{3.27}$$

$$J\ddot{\boldsymbol{\theta}} - K[\boldsymbol{q}-\boldsymbol{\theta}] = \boldsymbol{\tau}\,. \tag{3.28}$$

The model above may, in turn, be written in the standard form, that is through the state vector $[\boldsymbol{q}^T \;\; \boldsymbol{\theta}^T \;\; \dot{\boldsymbol{q}}^T \;\; \dot{\boldsymbol{\theta}}^T]^T$ as

$$\frac{d}{dt}\begin{bmatrix}\boldsymbol{q}\\ \boldsymbol{\theta}\\ \dot{\boldsymbol{q}}\\ \dot{\boldsymbol{\theta}}\end{bmatrix} = \begin{bmatrix}\dot{\boldsymbol{q}}\\ \dot{\boldsymbol{\theta}}\\ M^{-1}(\boldsymbol{q})\left[-K[\boldsymbol{q}-\boldsymbol{\theta}] - C(\boldsymbol{q},\dot{\boldsymbol{q}})\dot{\boldsymbol{q}} - \boldsymbol{g}(\boldsymbol{q})\right]\\ J^{-1}\left[\boldsymbol{\tau} + K[\boldsymbol{q}-\boldsymbol{\theta}]\right]\end{bmatrix}.$$

Example 3.7. Consider the device shown in Figure 3.8, which consists of one rigid link of mass m, and whose center of mass is localized at a distance l from the rotation axis. The moment of inertia of the link with respect to the axis that passes through its center of mass is denoted by I. The joint is elastic and has a torsional constant k. The rotor's inertia is denoted by J.

The dynamic model of this device may be computed noticing that

$$\mathcal{K}(\dot{q},\dot{\theta}) = \frac{1}{2}[ml^2 + I]\dot{q}^2 + \frac{1}{2}Jr^2\dot{\theta}^2$$

$$\mathcal{U}(q,\theta) = mgl[1-\cos(q)] + \frac{1}{2}k[q-\theta]^2,$$

which, using Lagrange's equations (3.4), leads to

$$[ml^2 + I]\ddot{q} + mgl\,\sin(q) + k[q-\theta] = 0\,,$$

$$Jr^2\ddot{\theta} - k[q-\theta] = \tau\,.$$

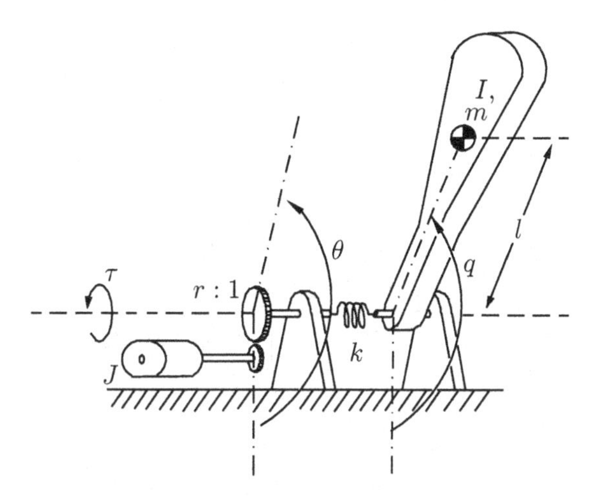

Figure 3.8. Link with an elastic joint

Unless clearly stated otherwise, in this text we consider only robots with stiff joints *i.e.* the model that we use throughout this textbook is given by (3.18).

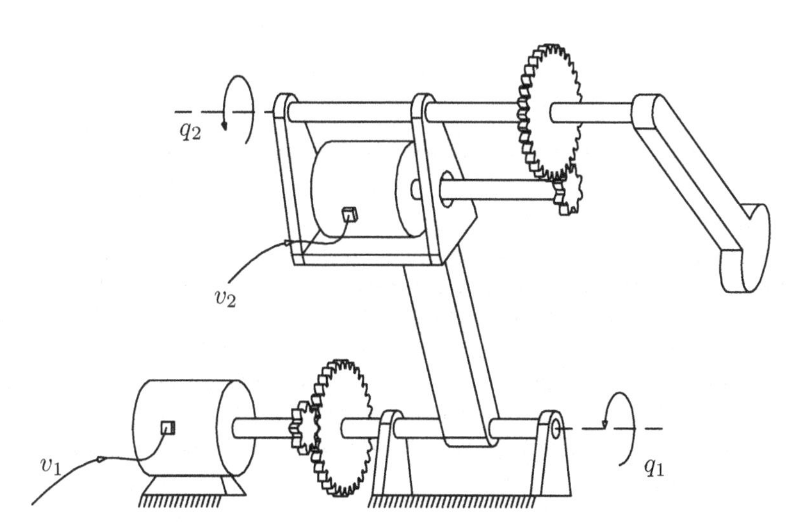

Figure 3.9. Example of a 2-DOF robot

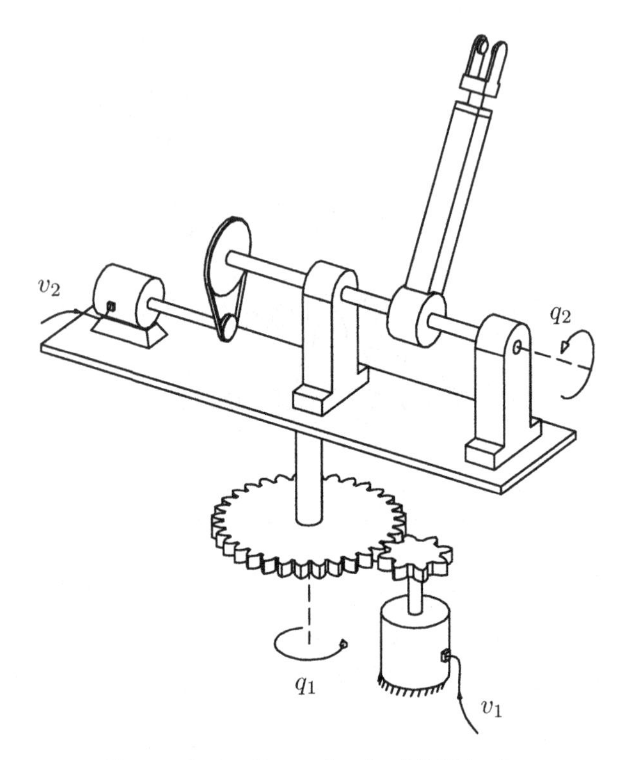

Figure 3.10. Example of a 2-DOF robot

3.5 Dynamic Model of Robots with Actuators

On a real robot manipulator the torques vector $\boldsymbol{\tau}$, is delivered by actuators that are typically electromechanical, pneumatic or hydraulic. Such actuators have their own dynamics, that is, the torque or force delivered is the product of a dynamic 'transformation' of the input to the actuator. This input may be a voltage or a current in the case of electromechanical actuators, fluid (typically oil) flux or pressure in the case of hydraulic actuators. In Figures 3.9 and 3.10 we illustrate two robotic arms with 2 DOF which have actuators transmitting the motion through gears in the first case, and through gear and belt in the second.

Actuators with Linear Dynamics

In certain situations, some types of electromechanical actuators may be modeled via second-order linear differential equations.

A common case is that of direct-current (DC) motors. The dynamic model which relates the input voltage v applied to the motor's armature, to the output torque τ delivered by the motor, is presented in some detail in Appendix

D. A simplified linear dynamic model of a DC motor with negligible armature inductance, as shown in Figure 3.11, is given by Equation (D.16) in Appendix D,

$$J_m \ddot{q} + f_m \dot{q} + \frac{K_a K_b}{R_a} \dot{q} + \frac{\tau}{r^2} = \frac{K_a}{r R_a} v \tag{3.29}$$

where:

- J_m : rotor inertia [kg m^2],
- K_a : motor-torque constant [N m/A],
- R_a : armature resistance [Ω],
- K_b : back emf [V s/rad],
- f_m : rotor friction coefficient with respect to its hinges [N m],
- τ : net applied torque after the set of gears at the load axis [N m],
- q : angular position of the load axis [rad],
- r : gear reduction ratio (in general $r \gg 1$),
- v : armature voltage [V].

Equation (3.29) relates the voltage v applied to the armature of the motor to the torque τ applied to the load, in terms of its angular position, velocity and acceleration.

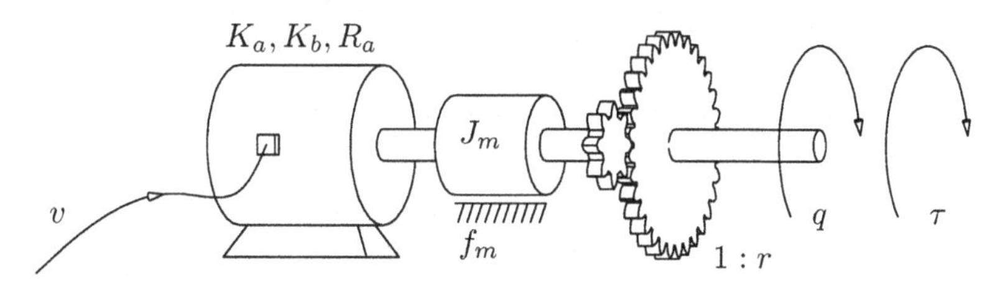

Figure 3.11. Diagram of a DC motor

Considering that each of the n joints is driven by a DC motor we obtain from Equation (3.29)

$$J\ddot{\boldsymbol{q}} + B\dot{\boldsymbol{q}} + R\boldsymbol{\tau} = K\boldsymbol{v} \tag{3.30}$$

with

$$J = \text{diag}\{J_{m_i}\}$$

$$B = \text{diag}\left\{ f_{m_i} + \left(\frac{K_a K_b}{R_a}\right)_i \right\}$$

$$R = \text{diag}\left\{\frac{1}{r_i^2}\right\} \tag{3.31}$$

$$K = \text{diag}\left\{\left(\frac{K_a}{R_a}\right)_i \frac{1}{r_i}\right\}$$

where for each motor ($i = 1, \cdots, n$), J_{m_i} corresponds to the rotor inertia, f_{m_i} to the damping coefficient, $(K_a K_b / R_a)_i$ to an electromechanical constant and r_i to the gear reduction ratio.

Thus, the complete dynamic model of a manipulator (considering friction in the joints) and having its actuators located at the joints[6] is obtained by substituting $\boldsymbol{\tau}$ from (3.30) in (3.25),

$$(R\ M(\boldsymbol{q}) + J)\,\ddot{\boldsymbol{q}} + R\ C(\boldsymbol{q},\dot{\boldsymbol{q}})\dot{\boldsymbol{q}} + R\ \boldsymbol{g}(\boldsymbol{q}) + R\ \boldsymbol{f}(\dot{\boldsymbol{q}}) + B\ \dot{\boldsymbol{q}} = K\boldsymbol{v}\,. \tag{3.32}$$

The equation above may be considered as a dynamic system whose input is $\boldsymbol{v}$ and whose outputs are $\boldsymbol{q}$ and $\dot{\boldsymbol{q}}$. A block-diagram for the model of the manipulator with actuators, given by (3.32), is depicted in Figure 3.12.

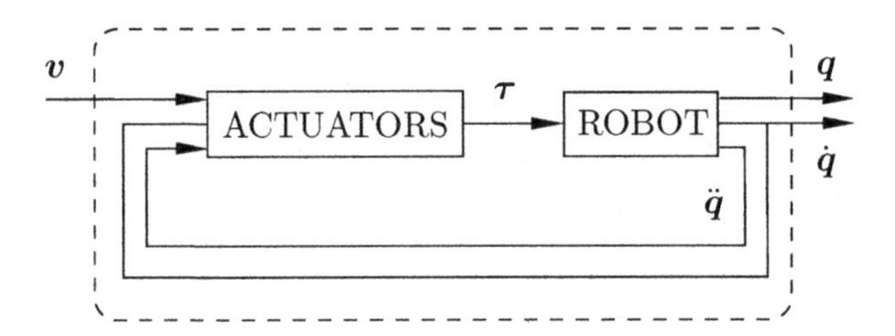

Figure 3.12. Block-diagram of a robot with its actuators

Example 3.8. Consider the pendulum depicted in Figure 3.13. The device consists of a DC motor coupled mechanically through a set of gears, to a pendular arm moving on a vertical plane under the action of gravity.

The equation of motion for this device including its load is given by

$$\left[J + ml^2\right]\ddot{q} + f_L\dot{q} + \left[m_b l_b + ml\right] g\,\sin(q) = \tau$$

where:

- J : arm inertia without load (*i.e.* with $m = 0$), with respect to the axis of rotation;
- m_b : arm mass (without load);

---

[6] Again, we neglect the gyroscopic and other coupling effects between the rotors and the links.

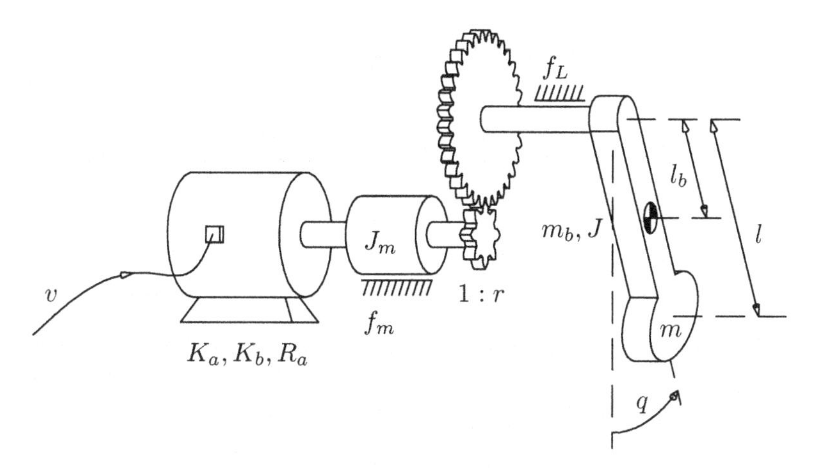

Figure 3.13. Pendular device with a DC motor

- l_b : distance from the rotating axis to the arm center of mass (without load);
- m : load mass at the tip of the arm (assumed to be punctual[7]);
- l : distance from the rotating axis to the load m;
- g : gravity acceleration;
- τ : applied torque at the rotating axis;
- f_L : friction coefficient of the arm and its load.

The equation above may also be written in the compact form

$$J_L \ddot{q} + f_L \dot{q} + k_L \sin(q) = \tau$$

where

$$J_L = J + ml^2$$

$$k_L = [m_b l_b + ml]g .$$

Hence, the complete dynamic model of the pendular device may be obtained by substituting τ from the model of the DC motor, (3.29), in the equation of the pendular arm, *i.e.*

$$\left[J_m + \frac{J_L}{r^2}\right]\ddot{q} + \left[f_m + \frac{f_L}{r^2} + \frac{K_a K_b}{R_a}\right]\dot{q} + \frac{k_L}{r^2}\sin(q) = \frac{K_a}{rR_a}v ,$$

from which, by simple inspection, comparing to (3.32), we identify

[7] That is, it is all concentrated at a point – the center of mass – and has no "size" or shape.

$$M(q) = J_L \qquad J = J_m$$
$$R = \frac{1}{r^2} \qquad K = \frac{K_a}{rR_a}$$
$$B = f_m + \frac{K_a K_b}{R_a} \qquad f(\dot{q}) = f_L \dot{q}$$
$$C(q, \dot{q}) = 0 \qquad g(q) = k_L \sin(q)\,.$$

◊

The robot-with-actuators Equation (3.32) may be simplified considerably when the gear ratios r_i are sufficiently large. In such case ($r_i \gg 1$), we have $R \approx 0$ and Equation (3.32) may be approximated by

$$J\ddot{\boldsymbol{q}} + B\dot{\boldsymbol{q}} \approx K\boldsymbol{v}\,.$$

That is, the nonlinear dynamics (3.25) of the robot may be neglected. This can be explained in the following way. If the gear reduction ratio is large enough, then the associated dynamics of the robot-with-actuators is described only by the dynamics of the actuators. This is the main argument that supports the idea that a good controller for the actuators is also appropriate to control robots having such actuators and geared transmissions with a high reduction ratio.

It is important to remark that the parameters involved in Equation (3.30) depend exclusively on the actuators, and not on the manipulator nor on its load. Therefore, it is reasonable to assume that such parameters are *constant* and known.

Since the gear ratio r_i is assumed to be nonzero, the matrix R given by (3.31) is nonsingular and therefore, R^{-1} exists and Equation (3.32) may be rewritten as

$$M'(\boldsymbol{q})\ddot{\boldsymbol{q}} + C(\boldsymbol{q}, \dot{\boldsymbol{q}})\dot{\boldsymbol{q}} + \boldsymbol{g}(\boldsymbol{q}) + \boldsymbol{f}(\dot{\boldsymbol{q}}) + R^{-1}B\dot{\boldsymbol{q}} = R^{-1}K\boldsymbol{v} \tag{3.33}$$

where $M'(\boldsymbol{q}) = M(\boldsymbol{q}) + R^{-1}J$.

The existence of the matrix $M'(\boldsymbol{q})^{-1}$ allows us to express the model (3.33) in terms of the state vector $[\boldsymbol{q} \;\; \dot{\boldsymbol{q}}]$ as

$$\frac{d}{dt}\begin{bmatrix} \boldsymbol{q} \\ \dot{\boldsymbol{q}} \end{bmatrix} = \begin{bmatrix} \dot{\boldsymbol{q}} \\ M'(\boldsymbol{q})^{-1}\left[R^{-1}\,K\boldsymbol{v} - R^{-1}\,B\dot{\boldsymbol{q}} - C(\boldsymbol{q}, \dot{\boldsymbol{q}})\dot{\boldsymbol{q}} - \boldsymbol{g}(\boldsymbol{q}) - \boldsymbol{f}(\dot{\boldsymbol{q}})\right] \end{bmatrix}.$$

That is, the input torque becomes simply a voltage scaled by $R^{-1}K$. At this point and in view of the scope of this textbook we may already formulate the problem of motion control for the system above, under the conditions previously stated. Given a set of bounded vectors $\boldsymbol{q}_d$, $\dot{\boldsymbol{q}}_d$ and $\ddot{\boldsymbol{q}}_d$, determine a vector of voltages $\boldsymbol{v}$, to be applied to the motors, in such a manner that the positions $\boldsymbol{q}$, associated to the joint positions of the robot, follow precisely $\boldsymbol{q}_d$. This is the main subject of study in the succeeding chapters.

Actuators with Nonlinear Dynamics

A dynamic model that characterizes a wide variety of actuators is described by the equations

$$\dot{\boldsymbol{x}} = \boldsymbol{m}(\boldsymbol{q}, \dot{\boldsymbol{q}}, \boldsymbol{x}) + \mathcal{G}(\boldsymbol{q}, \dot{\boldsymbol{q}}, \boldsymbol{x})\boldsymbol{u} \tag{3.34}$$

$$\boldsymbol{\tau} = \boldsymbol{l}(\boldsymbol{q}, \boldsymbol{x}) \tag{3.35}$$

where $\boldsymbol{u} \in \mathbb{R}^n$ and $\boldsymbol{x} \in \mathbb{R}^n$ are the input vectors and the state variables corresponding to the actuator and $\boldsymbol{m}$, $\mathcal{G}$ and $\boldsymbol{l}$ are nonlinear functions. In the case of DC motors, the input vector $\boldsymbol{u}$, represents the vector of voltages applied to each of the n motors. The state vector $\boldsymbol{x}$ represents, for instance, the armature current in a DC motor or the operating pressure in a hydraulic actuator. Since the torques $\boldsymbol{\tau}$ are, generally speaking, delivered by different kinds of actuators, the vector $\boldsymbol{m}(\boldsymbol{q}, \dot{\boldsymbol{q}}, \boldsymbol{x})$ and the matrix $\mathcal{G}(\boldsymbol{q}, \dot{\boldsymbol{q}}, \boldsymbol{x})$ are such that the $\partial \boldsymbol{l}(\boldsymbol{q}, \boldsymbol{x})/\partial \boldsymbol{x}^T$ and $\mathcal{G}(\boldsymbol{q}, \dot{\boldsymbol{q}}, \boldsymbol{x})$ are diagonal nonsingular matrices.

For the sake of illustration, consider the model of a DC motor with a non-negligible armature inductance ($L_a \not\approx 0$) as described in Appendix D, that is,

$$v = L_a \frac{di_a}{dt} + R_a i_a + K_b r \dot{q}$$

$$\tau = r K_a i_a$$

where v is the input (armature voltage), i_a is the direct armature current and τ is the torque applied to the load, after the set of gears. The rest of the constants are defined in Appendix D. These equations may be written in the generic form (3.34) and (3.35),

$$\frac{d \overbrace{i_a}^{x}}{dt} = \overbrace{-L_a^{-1}\left[R_a i_a + K_b r \dot{q}\right]}^{m(q,\dot{q},x)} + \overbrace{L_a^{-1}}^{\mathcal{G}(q,\dot{q},x)} \overbrace{v}^{u}$$

$$\tau = \underbrace{r K_a i_a}_{l(q,x)} .$$

Considering that the actuators' models are given by (3.34) and (3.35), the model of the robot with such actuators may be written in terms of the state vector $[\boldsymbol{q} \;\; \dot{\boldsymbol{q}} \;\; \boldsymbol{x}]$, as

$$\frac{d}{dt}\begin{bmatrix} \boldsymbol{q} \\ \dot{\boldsymbol{q}} \\ \boldsymbol{x} \end{bmatrix} = \begin{bmatrix} \dot{\boldsymbol{q}} \\ M(\boldsymbol{q})^{-1}\left[\boldsymbol{l}(\boldsymbol{q},\boldsymbol{x}) - C(\boldsymbol{q},\dot{\boldsymbol{q}})\dot{\boldsymbol{q}} - \boldsymbol{g}(\boldsymbol{q})\right] \\ \boldsymbol{m}(\boldsymbol{q},\dot{\boldsymbol{q}},\boldsymbol{x}) \end{bmatrix} + \begin{bmatrix} 0 \\ 0 \\ \mathcal{G}(\boldsymbol{q},\dot{\boldsymbol{q}},\boldsymbol{x}) \end{bmatrix}\boldsymbol{u}\,.$$

The dynamics of the actuators must be taken into account in the model of a robot, whenever these dynamics are not negligible with respect to that of the robot. Specifically for robots which are intended to perform high precision tasks.

Bibliography

Further facts and detailed developments on the kinematic and dynamic models of robot manipulators may be consulted in the following texts:

- Paul R., 1982, *"Robot manipulators: Mathematics, programming and control"*, The MIT Press, Cambridge, MA.
- Asada H., Slotine J. J., 1986, *"Robot analysis and control"*, Wiley, New York.
- Fu K., Gonzalez R., Lee C., 1987, *"Robotics: control, sensing, vision, and intelligence"*, McGraw-Hill.
- Craig J., 1989, *"Introduction to robotics: Mechanics and control"*, Addison-Wesley, Reading, MA.
- Spong M., Vidyasagar M., 1989, *"Robot dynamics and control"*, Wiley, New York.
- Yoshikawa T., 1990, *"Foundations of robotics: Analysis and control"*, The MIT Press.

The method of assigning the axis z_i as the rotation axis of the ith joint (for revolute joints) or as an axis parallel to the axis of translation at the ith joint (for prismatic joints) is taken from

- Craig J., 1989, *"Introduction to robotics: Mechanics and control"*, Addison-Wesley, Reading, MA.
- Yoshikawa T., 1990, *"Foundations of robotics: Analysis and control"*, The MIT Press.

It is worth mentioning that the notation above does not correspond to that of the so-called Denavit–Hartenberg convention, which may be familiar to some readers, but it is intuitively simpler and has several advantages.

Solution techniques to the inverse kinematics problem are detailed in

- Chiaverini S., Siciliano B., Egeland O., 1994, *"Review of the damped least square inverse kinematics with experiments on a industrial robot manipulator"*, IEEE Transactions on Control Systems Technology, Vol. 2, No. 2, June, pp. 123–134.
- Mayorga R. V., Wong A. K., Milano N., 1992, *"A fast procedure for manipulator inverse kinematics evaluation and pseudo-inverse robustness"*, IEEE Transactions on Systems, Man, and Cybernetics, Vol. 22, No. 4, July/August, pp. 790–798.

Lagrange's equations of motion are presented in some detail in the above-cited texts and also in

- Hauser W., 1966, *"Introduction to the principles of mechanics"*, Addison-Wesley, Reading MA.
- Goldstein H., 1974, *"Classical mechanics"*, Addison-Wesley, Reading MA.

A particularly simple derivation of the dynamic equations for n-DOF robots via Lagrange's equations is presented in the text by Spong and Vidyasagar (1989) previously cited.

The derivation of the dynamic model of elastic-joint robots may also be studied in the text by Spong and Vidyasagar (1989) and in

- Burkov I. V., Zaremba A. T., 1987, *"Dynamics of elastic manipulators with electric drives"*, Izv. Akad. Nauk SSSR Mekh. Tverd. Tela, Vol. 22, No. 1, pp. 57–64. English translation in *Mechanics of Solids*, Allerton Press.
- Marino R., Nicosia S., 1985, *"On the feedback control of industrial robots with elastic joints: a singular perturbation approach"*, 1st IFAC Symp. Robot Control, pp. 11–16, Barcelona, Spain.
- Spong M., 1987, *"Modeling and control of elastic joint robots"*, ASME Journal of Dynamic Systems, Measurement and Control, Vol. 109, December.

The topic of electromechanical actuator modeling and its consideration in the dynamics of manipulators is treated in the text by Spong and Vidyasagar (1989) and also in

- Luh J., 1983, *"Conventional controller design for industrial robots–A tutorial"*, IEEE Transactions on Systems, Man and Cybernetics, Vol. SMC-13, No. 3, June, pp. 298–316.
- Tourassis V., 1988, *"Principles and design of model-based robot controllers"*, International Journal of Control, Vol. 47, No. 5, pp. 1267–1275.
- Yoshikawa T., 1990, *"Foundations of robotics. Analysis and control"*, The MIT Press.

• Tarn T. J., Bejczy A. K., Yun X., Li Z., 1991, *"Effect of motor dynamics on nonlinear feedback robot arm control"*, IEEE Transactions on Robotics and Automation, Vol. 7, No. 1, February, pp. 114–122.

Problems

1. Consider the mechanical device analyzed in Example 3.2. Assume now that this device has friction on the axis of rotation, which is modeled here as a torque or force proportional to the velocity ($f > 0$ is the friction coefficient). The dynamic model in this case is

$$m_2 l_2^2 \cos^2(\varphi)\ddot{q} + f\dot{q} = \tau\,.$$

Rewrite the model in the form $\dot{\boldsymbol{x}} = \boldsymbol{f}(t, \boldsymbol{x})$ with $\boldsymbol{x} = [q \;\; \dot{q}]^T$.

a) Determine the conditions on the applied torque τ for the existence of equilibrium points.

b) Considering the condition on τ of the previous item, show by using Theorem 2.2 (see page 44), that the origin $[q \;\; \dot{q}]^T = [0 \;\; 0]^T$ is a stable equilibrium.

Hint: Use the following Lyapunov function candidate

$$V(q, \dot{q}) = \frac{1}{2}\left(q + \frac{m_2 l_2^2 \cos^2(\varphi)}{f}\dot{q}\right)^2 + \frac{1}{2}\dot{q}^2\,.$$

2. Consider the mechanical device depicted in Figure 3.14.

A simplistic model of such a device is

$$m\ddot{q} + f\dot{q} + kq + mg = \tau, \qquad q(0), \dot{q}(0) \in \mathbb{R}$$

where

- $m > 0$ is the mass
- $f > 0$ is the friction coefficient
- $k > 0$ is the stiffness coefficient of the spring
- g is the acceleration of gravity
- τ is the applied force
- q is the vertical position of the mass m with respect to origin of the plane x–y.

Write the model in the form $\dot{\boldsymbol{x}} = \boldsymbol{f}(t, \boldsymbol{x})$ where $\boldsymbol{x} = [q \;\; \dot{q}]^T$.

a) What restrictions must be imposed on τ so that there exist equilibria?

b) Is it possible to determine τ so that the only equilibrium is the origin, $\boldsymbol{x} = \boldsymbol{0} \in \mathbb{R}^2$?

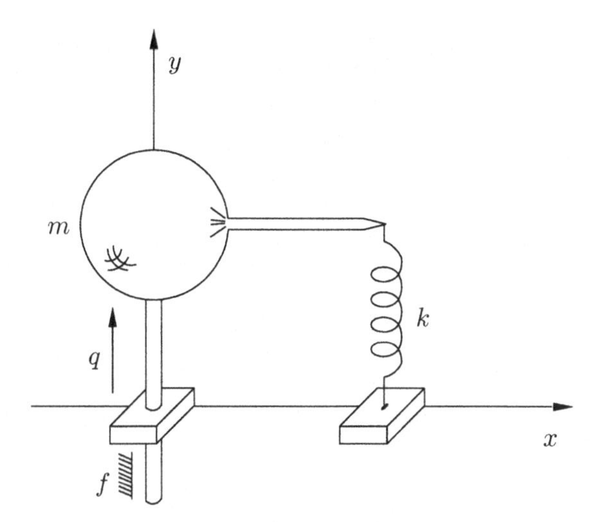

Figure 3.14. Problem 2

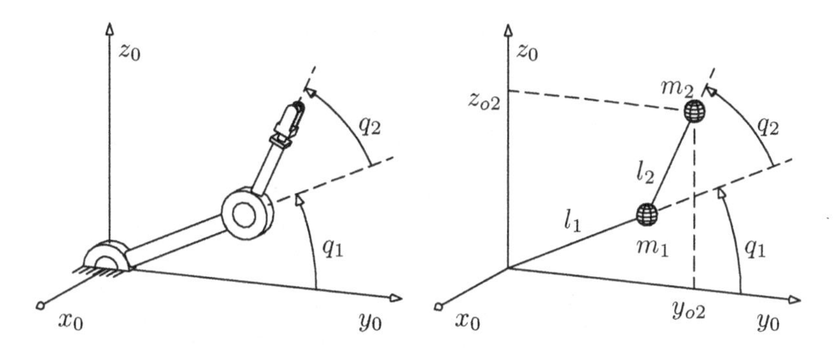

Figure 3.15. Problems 3 and 4

3. Consider the mechanical arm shown in Figure 3.15.

   Assume that the potential energy $\mathcal{U}(q_1, q_2)$ is zero when $q_1 = q_2 = 0$. Determine the vector of gravitational torques $\boldsymbol{g}(\boldsymbol{q})$,

$$\boldsymbol{g}(\boldsymbol{q}) = \begin{bmatrix} g_1(q_1, q_2) \\ g_2(q_1, q_2) \end{bmatrix}.$$

4. Consider again this mechanical device but with its simplified description as depicted in Figure 3.15.

   a) Obtain the direct kinematics model of the device, *i.e.* determine the relations

$$y_{02} = f_1(q_1, q_2)$$
$$z_{02} = f_2(q_1, q_2).$$

b) The analytical Jacobian $J(\boldsymbol{q})$ of a robot is the matrix

$$J(\boldsymbol{q}) = \frac{\partial}{\partial \boldsymbol{q}}\varphi(\boldsymbol{q}) = \begin{bmatrix} \frac{\partial}{\partial q_1}\varphi_1(\boldsymbol{q}) & \frac{\partial}{\partial q_2}\varphi_1(\boldsymbol{q}) & \cdots & \frac{\partial}{\partial q_n}\varphi_1(\boldsymbol{q}) \\ \frac{\partial}{\partial q_1}\varphi_2(\boldsymbol{q}) & \frac{\partial}{\partial q_2}\varphi_2(\boldsymbol{q}) & \cdots & \frac{\partial}{\partial q_n}\varphi_2(\boldsymbol{q}) \\ \vdots & \vdots & \ddots & \vdots \\ \frac{\partial}{\partial q_1}\varphi_m(\boldsymbol{q}) & \frac{\partial}{\partial q_2}\varphi_m(\boldsymbol{q}) & \cdots & \frac{\partial}{\partial q_n}\varphi_m(\boldsymbol{q}) \end{bmatrix}$$

where $\varphi(\boldsymbol{q})$ is the relation in the direct kinematics model ($\mathbf{x} = \varphi(\boldsymbol{q})$), n is the dimension of $\boldsymbol{q}$ and m is the dimension of $\boldsymbol{x}$. Determine the Jacobian.

5. Consider the 2-DOF robot shown in Figure 3.16, for which the meaning of the constants and variables involved is as follows:

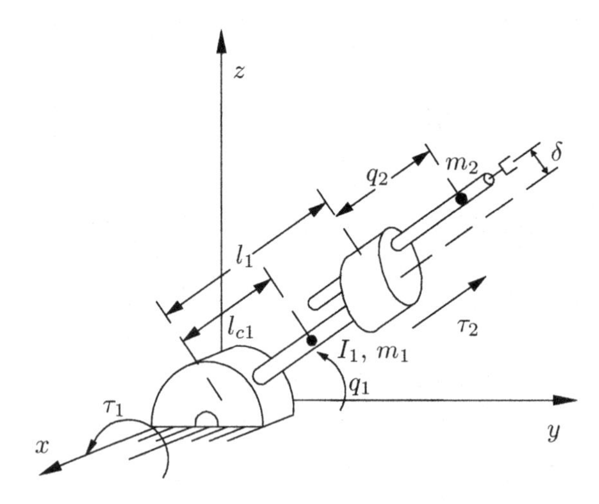

Figure 3.16. Problem 5

- m_1, m_2 are the masses of links 1 and 2 respectively;
- I_1 is the moment of inertia of link 1 with respect to the axis parallel to the axis x which passes through its center of mass; the moment of inertia of the second link is supposed negligible;
- l_1 is the length of link 1;
- l_{c1} is the distance to the center of mass of link 1 taken from its rotation axis;
- q_1 is the angular position of link 1 measured with respect to the horizontal (taken positive counterclockwise);

- q_2 is the linear position of the center of mass of link 2 measured from the edge of link 1;
- δ is negligible ($\delta = 0$).

Determine the dynamic model and write it in the form $\dot{\boldsymbol{x}} = \boldsymbol{f}(t, \boldsymbol{x})$ where $\boldsymbol{x} = [q \ \ \dot{q}]^T$.

6. Consider the 2-DOF robot depicted in Figure 3.17. Such a robot has a transmission composed of a set of bar linkage at its second joint. Assume that the mass of the lever of length l_4 associated with actuator 2 is negligible.

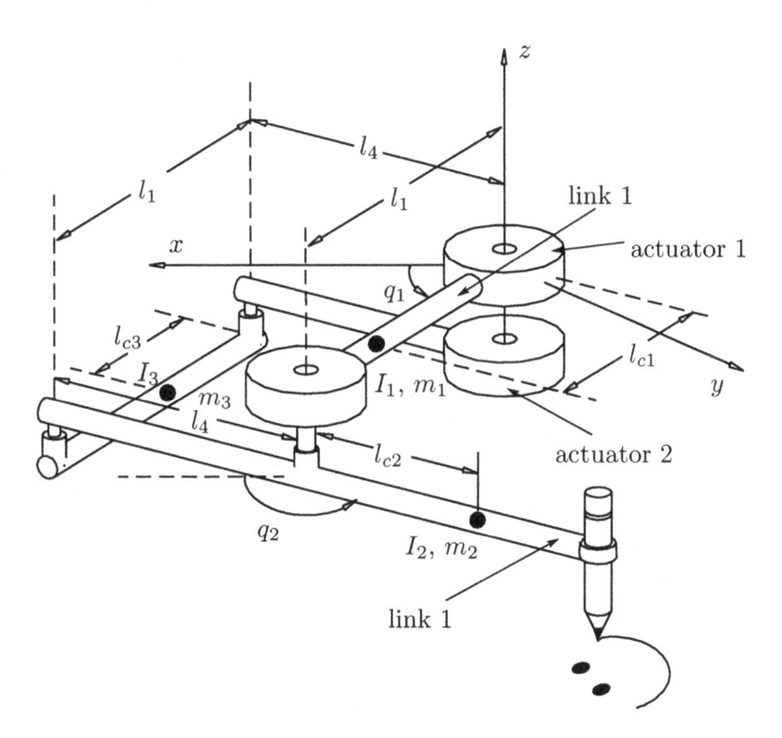

Figure 3.17. Problem 6

Determine the dynamic model. Specifically, obtain the inertia matrix $M(\boldsymbol{q})$ and the centrifugal and Coriolis matrix $C(\boldsymbol{q}, \dot{\boldsymbol{q}})$.

Hint: See the robot presented in Example 3.3. Both robots happen to be mechanically equivalent when taking $m_3 = I_3 = \delta = 0$.

4

Properties of the Dynamic Model

In this chapter we present some simple but fundamental properties of the dynamic model for n-DOF robots given by Equation (3.18), *i.e.*

$$M(\boldsymbol{q})\ddot{\boldsymbol{q}} + C(\boldsymbol{q}, \dot{\boldsymbol{q}})\dot{\boldsymbol{q}} + \boldsymbol{g}(\boldsymbol{q}) = \boldsymbol{\tau} \,. \tag{4.1}$$

In spite of the complexity of the dynamic Equation (4.1), which describes the behavior of robot manipulators, this equation and the terms which constitute it have properties that are interesting in themselves for control purposes. Besides, such properties are of particular importance in the study of control systems for robot manipulators. Only properties that are relevant to control design and stability analysis via Lyapunov's direct method (see Section 2.3.4 in Chapter 2) are presented. The reader is invited to see the references at the end of the chapter to prove further.

These properties, which we use extensively in the sequel, may be classified as follows:

- properties of the inertia matrix $M(\boldsymbol{q})$;
- properties of the centrifugal and Coriolis forces matrix $C(\boldsymbol{q}, \dot{\boldsymbol{q}})$;
- properties of the gravitational forces and torques vector $\boldsymbol{g}(\boldsymbol{q})$;
- properties of the residual dynamics.

Each of these items is treated independently and constitute the material of this chapter. Some of the proofs of the properties that are established below may be consulted in the references which are listed at the end of the chapter and others are developed in Appendix C.

4.1 The Inertia Matrix

The inertia matrix $M(\boldsymbol{q})$ plays an important role both in the robot's dynamic model as well as in control design. The properties of the inertia matrix which is closely related to the kinetic energy function $\mathcal{K} = \frac{1}{2}\dot{\boldsymbol{q}}^T M(\boldsymbol{q})\dot{\boldsymbol{q}}$, are exhaustively used in control design for robots. Among such properties we underline the following.

Property 4.1. Inertia matrix $M(\boldsymbol{q})$

The *inertia matrix* $M(\boldsymbol{q})$ is symmetric positive definite and has dimension $n \times n$. Its elements are functions only of $\boldsymbol{q}$. The inertia matrix $M(\boldsymbol{q})$ satisfies the following properties.

1. There exists a real positive number α such that

$$M(\boldsymbol{q}) \geq \alpha I \qquad \forall\, \boldsymbol{q} \in \mathbb{R}^n$$

where I denotes the identity matrix of dimension $n \times n$. The matrix $M(\boldsymbol{q})^{-1}$ exists and is positive definite.

2. For robots having only revolute joints there exists a constant $\beta > 0$ such that

$$\lambda_{\mathrm{Max}}\{M(\boldsymbol{q})\} \leq \beta \quad \forall\, \boldsymbol{q} \in \mathbb{R}^n .$$

One way of computing β is

$$\beta \geq n \left(\max_{i,j,q} |M_{ij}(\boldsymbol{q})| \right)$$

where $M_{ij}(\boldsymbol{q})$ stands for the ijth element of the matrix $M(\boldsymbol{q})$.

3. For robots having only revolute joints there exists a constant $k_M > 0$ such that

$$\|M(\boldsymbol{x})\boldsymbol{z} - M(\boldsymbol{y})\boldsymbol{z}\| \leq k_M \;\|\boldsymbol{x} - \boldsymbol{y}\|\,\|\boldsymbol{z}\| \tag{4.2}$$

for all vectors $\boldsymbol{x}, \boldsymbol{y}, \boldsymbol{z} \in \mathbb{R}^n$. One simple way to determine k_M is as follows

$$k_M \geq n^2 \left(\max_{i,j,k,q} \left| \frac{\partial M_{ij}(\boldsymbol{q})}{\partial q_k} \right| \right) . \tag{4.3}$$

4. For robots having only revolute joints there exists a number $k'_M > 0$ such that

$$\|M(\boldsymbol{x})\boldsymbol{y}\| \leq k'_M \,\|\boldsymbol{y}\|$$

for all $\boldsymbol{x}, \boldsymbol{y} \in \mathbb{R}^n$.

The reader interested in the proof of Inequality (4.2) is invited to see Appendix C.

An obvious consequence of Property 4.1 and, in particular of the fact that $M(\boldsymbol{q})$ is positive definite, is that the function $V : \mathbb{R}^n \times \mathbb{R}^n \to \mathbb{R}_+$, defined as

$$V(\boldsymbol{q}, \dot{\boldsymbol{q}}) = \dot{\boldsymbol{q}}^T M(\boldsymbol{q}) \dot{\boldsymbol{q}}$$

is positive definite in $\dot{\boldsymbol{q}}$. As a matter of fact, notice that with the previous definition we have $V(\boldsymbol{q}, \dot{\boldsymbol{q}}) = 2\mathcal{K}(\boldsymbol{q}, \dot{\boldsymbol{q}})$ where $\mathcal{K}(\boldsymbol{q}, \dot{\boldsymbol{q}})$ corresponds to the kinetic energy function of the robot, (3.15).

4.2 The Centrifugal and Coriolis Forces Matrix

The properties of the Centrifugal and Coriolis matrix $C(\boldsymbol{q}, \dot{\boldsymbol{q}})$ are important in the study of stability of control systems for robots. The main properties of such a matrix are presented below.

Property 4.2. Coriolis matrix $C(\boldsymbol{q}, \dot{\boldsymbol{q}})$

The *centrifugal and Coriolis* forces matrix $C(\boldsymbol{q}, \dot{\boldsymbol{q}})$ has dimension $n \times n$ and its elements are functions of $\boldsymbol{q}$ and $\dot{\boldsymbol{q}}$. The matrix $C(\boldsymbol{q}, \dot{\boldsymbol{q}})$ satisfies the following.

1. For a given manipulator, the matrix $C(\boldsymbol{q}, \dot{\boldsymbol{q}})$ may not be unique but the vector $C(\boldsymbol{q}, \dot{\boldsymbol{q}})\dot{\boldsymbol{q}}$ is unique.

2. $C(\boldsymbol{q}, \boldsymbol{0}) = 0$ for all vectors $\boldsymbol{q} \in \mathbb{R}^n$.

3. For all vectors $\boldsymbol{q}, \boldsymbol{x}, \boldsymbol{y}, \boldsymbol{z} \in \mathbb{R}^n$ and any scalar α we have

$$C(\boldsymbol{q}, \boldsymbol{x})\boldsymbol{y} = C(\boldsymbol{q}, \boldsymbol{y})\boldsymbol{x}$$
$$C(\boldsymbol{q}, \boldsymbol{z} + \alpha\boldsymbol{x})\boldsymbol{y} = C(\boldsymbol{q}, \boldsymbol{z})\boldsymbol{y} + \alpha C(\boldsymbol{q}, \boldsymbol{x})\boldsymbol{y}\,.$$

4. The vector $C(\boldsymbol{q}, \boldsymbol{x})\boldsymbol{y}$ may be written in the form

$$C(\boldsymbol{q}, \boldsymbol{x})\boldsymbol{y} = \begin{bmatrix} \boldsymbol{x}^T C_1(\boldsymbol{q})\boldsymbol{y} \\ \boldsymbol{x}^T C_2(\boldsymbol{q})\boldsymbol{y} \\ \vdots \\ \boldsymbol{x}^T C_n(\boldsymbol{q})\boldsymbol{y} \end{bmatrix} \tag{4.4}$$

where $C_k(\boldsymbol{q})$ are symmetric matrices of dimension $n \times n$ for all $k = 1, 2, \cdots, n$. The ij-th element $C_{k_{ij}}(\boldsymbol{q})$ of the matrix $C_k(\boldsymbol{q})$ corresponds to the so-called Christoffel symbol of the first kind $c_{jik}(\boldsymbol{q})$ and which is defined in (3.21).

5. For robots having exclusively revolute joints, there exists a number $k_{C_1} > 0$ such that

$$\|C(\boldsymbol{q}, \boldsymbol{x})\boldsymbol{y}\| \leq k_{C_1} \|\boldsymbol{x}\| \|\boldsymbol{y}\|$$

for all $\boldsymbol{q}, \boldsymbol{x}, \boldsymbol{y} \in \mathbb{R}^n$.

6. For robots having exclusively revolute joints, there exist numbers $k_{C_1} > 0$ and $k_{C_2} > 0$ such that

$$\begin{aligned}\|C(\boldsymbol{x}, \boldsymbol{z})\boldsymbol{w} - C(\boldsymbol{y}, \boldsymbol{v})\boldsymbol{w}\| \leq\; & k_{C_1} \|\boldsymbol{z} - \boldsymbol{v}\| \; \|\boldsymbol{w}\| \\ & + k_{C_2} \|\boldsymbol{x} - \boldsymbol{y}\| \|\boldsymbol{w}\| \|\boldsymbol{z}\| \end{aligned} \qquad (4.5)$$

for all vector $\boldsymbol{v}, \boldsymbol{x}, \boldsymbol{y}, \boldsymbol{z}, \boldsymbol{w} \in \mathbb{R}^n$.

7. The matrix $C(\boldsymbol{q}, \dot{\boldsymbol{q}})$, defined in (3.22) is related to the inertia matrix $M(\boldsymbol{q})$ by the expression

$$\boldsymbol{x}^T \left[\frac{1}{2}\dot{M}(\boldsymbol{q}) - C(\boldsymbol{q}, \dot{\boldsymbol{q}})\right] \boldsymbol{x} = 0 \qquad \forall\; \boldsymbol{q}, \dot{\boldsymbol{q}}, \boldsymbol{x} \in \mathbb{R}^n$$

and as a matter of fact, $\frac{1}{2}\dot{M}(\boldsymbol{q}) - C(\boldsymbol{q}, \dot{\boldsymbol{q}})$ is skew-symmetric. Equivalently, the matrix $\dot{M}(\boldsymbol{q}) - 2C(\boldsymbol{q}, \dot{\boldsymbol{q}})$ is skew-symmetric, and it is also true that

$$\dot{M}(\boldsymbol{q}) = C(\boldsymbol{q}, \dot{\boldsymbol{q}}) + C(\boldsymbol{q}, \dot{\boldsymbol{q}})^T .$$

Independently of the way in which $C(\boldsymbol{q}, \dot{\boldsymbol{q}})$ is derived, it always satisfies

$$\dot{\boldsymbol{q}}^T \left[\frac{1}{2}\dot{M}(\boldsymbol{q}) - C(\boldsymbol{q}, \dot{\boldsymbol{q}})\right] \dot{\boldsymbol{q}} = 0 \qquad \forall\; \boldsymbol{q}, \dot{\boldsymbol{q}} \in \mathbb{R}^n .$$

We present next, the proof for the existence of a positive constant k_{C_1} such that $\|C(\boldsymbol{q}, \boldsymbol{x})\boldsymbol{y}\| \leq k_{C_1} \|\boldsymbol{x}\| \|\boldsymbol{y}\|$ for all vectors $\boldsymbol{q}, \boldsymbol{x}, \boldsymbol{y} \in \mathbb{R}^n$.

Considering (4.4), the norm $\|C(\boldsymbol{q}, \boldsymbol{x})\boldsymbol{y}\|^2$ of the vector $C(\boldsymbol{q}, \boldsymbol{x})\boldsymbol{y}$ is defined in the usual way that is, as the sum of its elements squared,

$$\|C(\boldsymbol{q}, \boldsymbol{x})\boldsymbol{y}\|^2 = \sum_{k=1}^{n} \left(\boldsymbol{x}^T C_k(\boldsymbol{q})\boldsymbol{y}\right)^2 .$$

This implies that

$$\|C(\boldsymbol{q},\boldsymbol{x})\boldsymbol{y}\|^2 = \sum_{k=1}^{n} \left|\boldsymbol{x}^T C_k(\boldsymbol{q})\boldsymbol{y}\right|^2 \leq \left[\sum_{k=1}^{n} \|C_k(\boldsymbol{q})\|^2\right] \|\boldsymbol{x}\|^2 \|\boldsymbol{y}\|^2 \tag{4.6}$$

where we have used the fact that for vectors $\boldsymbol{x}$, $\boldsymbol{y}$ and a square matrix A of compatible dimensions it holds that $\left|\boldsymbol{x}^T A\boldsymbol{y}\right| \leq \|A\| \|\boldsymbol{x}\| \|\boldsymbol{y}\|$.

Taking into account that the spectral norm $\|A\|$ of a symmetric matrix $A = \{a_{ij}\}$ of dimension $n \times n$ verifies the inequality $\|A\| \leq n \ \max_{i,j} \{|a_{ij}|\}$, we have

$$\|C_k(\boldsymbol{q})\|^2 \leq n^2 \left[\max_{i,j,q} \left\{\left|C_{k_{ij}}(\boldsymbol{q})\right|\right\}\right]^2 ,$$

where $C_{k_{ij}}(\boldsymbol{q})$ stands for the ijth element of the symmetric matrix $C_k(\boldsymbol{q})$. Therefore, we obtain

$$\begin{aligned}\left[\sum_{k=1}^{n} \|C_k(\boldsymbol{q})\|^2\right] &\leq n^2 \sum_{k=1}^{n} \left[\max_{i,j,q} \left\{\left|C_{k_{ij}}(\boldsymbol{q})\right|\right\}\right]^2 \\ &\leq n^3 \left[\max_{k,i,j,q} \left\{\left|C_{k_{ij}}(\boldsymbol{q})\right|\right\}\right]^2 \\ &\leq n^4 \left[\max_{k,i,j,q} \left\{\left|C_{k_{ij}}(\boldsymbol{q})\right|\right\}\right]^2\end{aligned}$$

where we used the fact that $n \geq 1$. Though conservative, the last step above, is justified to maintain integer exponents. Using this last inequality in (4.6) we finally obtain

$$\|C(\boldsymbol{q},\boldsymbol{x})\boldsymbol{y}\| \leq n^2 \left(\max_{k,i,j,q} \left|C_{k_{ij}}(\boldsymbol{q})\right|\right) \|\boldsymbol{x}\| \ \|\boldsymbol{y}\| \tag{4.7}$$

where one clearly identifies the constant k_{C_1} as

$$k_{C_1} = n^2 \left(\max_{k,i,j,q} \left|C_{k_{ij}}(\boldsymbol{q})\right|\right) . \tag{4.8}$$

As an immediate application of the expression in (4.7), $\|C(\boldsymbol{q},\boldsymbol{x})\boldsymbol{y}\| \leq k_{C_1} \|\boldsymbol{x}\| \|\boldsymbol{y}\|$, we have

$$\|C(\boldsymbol{q},\dot{\boldsymbol{q}})\dot{\boldsymbol{q}}\| \leq \ k_{C_1}\|\dot{\boldsymbol{q}}\|^2 .$$

We present now an example with the purpose of illustrating the previous computations.

Example 4.1. Consider the centrifugal and Coriolis forces matrix

$$C(\boldsymbol{q},\dot{\boldsymbol{q}}) = \begin{bmatrix} -m_2 l_1 l_{c2}\sin(q_2)\dot{q}_2 & -m_2 l_1 l_{c2}\sin(q_2)[\dot{q}_1+\dot{q}_2] \\ \\ m_2 l_1 l_{c2}\sin(q_2)\dot{q}_1 & 0 \end{bmatrix}.$$

We wish to find a positive constant k_{C_1} such that $\|C(\boldsymbol{q},\dot{\boldsymbol{q}})\dot{\boldsymbol{q}}\| \leq k_{C_1}\|\dot{\boldsymbol{q}}\|^2$. To that end, the vector $C(\boldsymbol{q},\dot{\boldsymbol{q}})\dot{\boldsymbol{q}}$ may be rewritten as

$$C(\boldsymbol{q},\dot{\boldsymbol{q}})\dot{\boldsymbol{q}} = \begin{bmatrix} -m_2 l_1 l_{c2}\sin(q_2)\left[2\dot{q}_1\dot{q}_2+\dot{q}_2^2\right] \\ \\ m_2 l_1 l_{c2}\sin(q_2)\dot{q}_1^2 \end{bmatrix}$$

$$= \begin{bmatrix} \begin{bmatrix}\dot{q}_1\\ \dot{q}_2\end{bmatrix}^T \overbrace{\begin{bmatrix} 0 & -m_2 l_1 l_{c2}\sin(q_2) \\ -m_2 l_1 l_{c2}\sin(q_2) & -m_2 l_1 l_{c2}\sin(q_2)\end{bmatrix}}^{C_1(\boldsymbol{q})} \begin{bmatrix}\dot{q}_1\\ \dot{q}_2\end{bmatrix} \\ \\ \begin{bmatrix}\dot{q}_1\\ \dot{q}_2\end{bmatrix}^T \underbrace{\begin{bmatrix} m_2 l_1 l_{c2}\sin(q_2) & 0 \\ 0 & 0\end{bmatrix}}_{C_2(\boldsymbol{q})} \begin{bmatrix}\dot{q}_1\\ \dot{q}_2\end{bmatrix} \end{bmatrix}.$$

Using the matrices $C_1(\boldsymbol{q})$ and $C_2(\boldsymbol{q})$ one can easily verify that

$$\begin{aligned}
\max_q |C_{1_{11}}(\boldsymbol{q})| &= 0 \\
\max_q |C_{1_{12}}(\boldsymbol{q})| &= m_2 l_1 l_{c2} \\
\max_q |C_{1_{21}}(\boldsymbol{q})| &= m_2 l_1 l_{c2} \\
\max_q |C_{1_{22}}(\boldsymbol{q})| &= m_2 l_1 l_{c2} \\
\max_q |C_{2_{11}}(\boldsymbol{q})| &= m_2 l_1 l_{c2} \\
\max_q |C_{2_{12}}(\boldsymbol{q})| &= 0 \\
\max_q |C_{2_{21}}(\boldsymbol{q})| &= 0 \\
\max_q |C_{2_{22}}(\boldsymbol{q})| &= 0\,.
\end{aligned}$$

Hence, considering (4.8) we obtain

$$k_{C_1} = 4 m_2 l_1 l_{c2}\,.$$

The reader interested in the proof of Inequality (4.5) is invited to see Appendix C.

4.3 The Gravitational Torques Vector

The vector of gravitational torques, $\boldsymbol{g}(\boldsymbol{q})$, is present in robots which from a mechanical viewpoint, have not been designed with compensation of gravitational torques. For instance, without counter-weights, springs or for robots designed to move out of the horizontal plane. Some of the most relevant properties of this vector are enunciated next.

Property 4.3. Gravity vector $\boldsymbol{g}(\boldsymbol{q})$

The *gravitational torques vector* $\boldsymbol{g}(\boldsymbol{q})$, of dimension $n \times 1$, depends only on the joint positions $\boldsymbol{q}$. The vector $\boldsymbol{g}(\boldsymbol{q})$ is continuous and therefore bounded for each bounded $\boldsymbol{q}$. Moreover, $\boldsymbol{g}(\boldsymbol{q})$ also satisfies the following.

1. The vector $\boldsymbol{g}(\boldsymbol{q})$ and the velocity vector $\dot{\boldsymbol{q}}$ are correlated as

$$\int_0^T \boldsymbol{g}(\boldsymbol{q}(t))^T \dot{\boldsymbol{q}}(t)\, dt = \mathcal{U}(\boldsymbol{q}(T)) - \mathcal{U}(\boldsymbol{q}(0)) \tag{4.9}$$

for all $T \in \mathbb{R}_+$.

2. For robots having only revolute joints there exists a number $k_{\mathcal{U}}$ such that

$$\int_0^T \boldsymbol{g}(\boldsymbol{q}(t))^T \dot{\boldsymbol{q}}(t)\, dt + \mathcal{U}(\boldsymbol{q}(0)) \geq k_{\mathcal{U}}$$

for all $T \in \mathbb{R}_+$ and where $k_{\mathcal{U}} = \min_q \{\mathcal{U}(\boldsymbol{q})\}$.

3. For robots having only revolute joints, the vector $\boldsymbol{g}(\boldsymbol{q})$ is Lipschitz, that is, there exists a constant $k_g > 0$ such that

$$\|\boldsymbol{g}(\boldsymbol{x}) - \boldsymbol{g}(\boldsymbol{y})\| \leq k_g \|\boldsymbol{x} - \boldsymbol{y}\| \tag{4.10}$$

for all $\boldsymbol{x}, \boldsymbol{y} \in \mathbb{R}^n$. A simple way to compute k_g is by evaluating its partial derivative

$$k_g \geq n \left(\max_{i,j,q} \left| \frac{\partial g_i(\boldsymbol{q})}{\partial q_j} \right| \right). \tag{4.11}$$

Furthermore, k_g satisfies

$$k_g \geq \left\| \frac{\partial \boldsymbol{g}(\boldsymbol{q})}{\partial \boldsymbol{q}} \right\| \geq \lambda_{\mathrm{Max}} \left\{ \frac{\partial \boldsymbol{g}(\boldsymbol{q})}{\partial \boldsymbol{q}} \right\}.$$

4. For robots having only revolute joints there exists a constant k' such that

$$\|\boldsymbol{g}(\boldsymbol{q})\| \leq k'$$

for all $\boldsymbol{q} \in \mathbb{R}^n$.

To prove (4.9) consider the potential energy function $\mathcal{U}(\boldsymbol{q})$ for a given manipulator. The partial time derivative of $\mathcal{U}(\boldsymbol{q})$ is given by

$$\frac{d}{dt}\mathcal{U}(\boldsymbol{q}(t)) = \frac{\partial \mathcal{U}(\boldsymbol{q})}{\partial \boldsymbol{q}}^T \dot{\boldsymbol{q}}$$

hence, replacing (3.20) we obtain

$$\frac{d}{dt}\mathcal{U}(\boldsymbol{q}(t)) = \boldsymbol{g}(\boldsymbol{q})^T \dot{\boldsymbol{q}}\,.$$

To integrate the above on both sides from 0 to T, define $U_0 := \mathcal{U}(\boldsymbol{q}(0))$ and $U_T := \mathcal{U}(\boldsymbol{q}(T))$ for any $T \in \mathbb{R}_+$ then,

$$\int_{U_0}^{U_T} d\mathcal{U} = \int_0^T \boldsymbol{g}(\boldsymbol{q}(t))^T \dot{\boldsymbol{q}}(t)\, dt\,,$$

which is equivalent to

$$\mathcal{U}(\boldsymbol{q}(T)) - \mathcal{U}(\boldsymbol{q}(0)) = \int_0^T \boldsymbol{g}(\boldsymbol{q})^T \dot{\boldsymbol{q}}\, dt\,.$$

The proof of (4.10) is presented in Appendix C.

4.4 The Residual Dynamics

To each robot dynamic model there is an associated function named “residual dynamics” that is important in the study of stability of numerous controllers.

The residual dynamics $\boldsymbol{h}(t, \tilde{\boldsymbol{q}}, \dot{\tilde{\boldsymbol{q}}})$ is defined as follows[1]:

$$\begin{aligned}\boldsymbol{h}(t, \tilde{\boldsymbol{q}}, \dot{\tilde{\boldsymbol{q}}}) = {} & [M(\boldsymbol{q}_d) - M(\boldsymbol{q}_d - \tilde{\boldsymbol{q}})]\,\ddot{\boldsymbol{q}}_d \\ & + \left[C(\boldsymbol{q}_d, \dot{\boldsymbol{q}}_d) - C(\boldsymbol{q}_d - \tilde{\boldsymbol{q}}, \dot{\boldsymbol{q}}_d - \dot{\tilde{\boldsymbol{q}}})\right] \dot{\boldsymbol{q}}_d \\ & + \boldsymbol{g}(\boldsymbol{q}_d) - \boldsymbol{g}(\boldsymbol{q}_d - \tilde{\boldsymbol{q}}), \end{aligned} \tag{4.12}$$

[1] Note that, in general, because $\boldsymbol{q}_d$ depends on time so does the function of residual dynamics.

and with an abuse of notation it may be written as

$$\boldsymbol{h}(t, \tilde{\boldsymbol{q}}, \dot{\tilde{\boldsymbol{q}}}) = [M(\boldsymbol{q}_d) - M(\boldsymbol{q})]\, \ddot{\boldsymbol{q}}_d + [C(\boldsymbol{q}_d, \dot{\boldsymbol{q}}_d) - C(\boldsymbol{q}, \dot{\boldsymbol{q}})]\, \dot{\boldsymbol{q}}_d + \boldsymbol{g}(\boldsymbol{q}_d) - \boldsymbol{g}(\boldsymbol{q}).$$

This function has the characteristic that $\boldsymbol{h}(t, \boldsymbol{0}, \boldsymbol{0}) = \boldsymbol{0}$ for all t but more importantly, the residual dynamics $\boldsymbol{h}(t, \tilde{\boldsymbol{q}}, \dot{\tilde{\boldsymbol{q}}})$ has the virtue of not growing faster than $\|\dot{\tilde{\boldsymbol{q}}}\|$ and $\|\tilde{\boldsymbol{q}}\|$. Moreover it may grow arbitrarily fast only when so does $\|\dot{\tilde{\boldsymbol{q}}}\|$, independently of $\|\tilde{\boldsymbol{q}}\|$.

In order to make this statement formal we need to recall the definition and properties of a continuously differentiable monotonically increasing function: the tangent hyperbolic. As a matter of fact, the statement can be shown for a large class of monotonically increasing functions but for clarity of exposition, here we restrict our discussion to

$$\tanh(x) = \frac{e^x - e^{-x}}{e^x + e^{-x}}$$

which is illustrated in Figure 4.1.

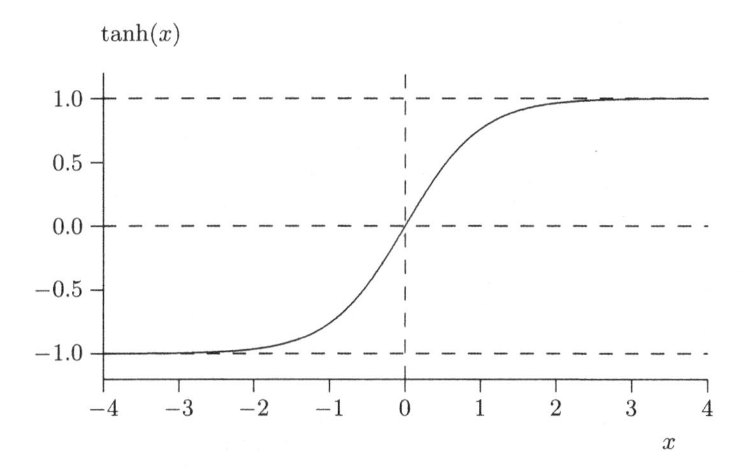

Figure 4.1. Graph of tangent hyperbolic: $\tanh(x)$

As it is clear from Figure 4.1, $\tanh(x)$ is continuous monotonically increasing. Also, it has continuous derivatives and it satisfies $|x| \geq |\tanh(x)|$ and $1 \geq |\tanh(x)|$ for all $x \in \mathbb{R}$. All these observations are stated formally below.

Definition 4.1. *Vectorial tangent hyperbolic function*

We define the vectorial tangent hyperbolic function as

$$\mathbf{tanh}(\boldsymbol{x}) := \begin{bmatrix} \tanh(x_1) \\ \vdots \\ \tanh(x_n) \end{bmatrix} \tag{4.13}$$

where $\boldsymbol{x} \in \mathbb{R}^n$. *The first partial derivative of* $\mathbf{tanh}(\boldsymbol{x})$ *is given by*

$$\frac{\partial \mathbf{tanh}}{\partial \boldsymbol{x}}(\boldsymbol{x}) =: Sech^2(\boldsymbol{x}) = diag\{sech^2(x_i)\} \tag{4.14}$$

where

$$sech(x_i) := \frac{1}{e^{x_i} - e^{-x_i}} .$$

The vectorial tangent hyperbolic function satisfies the following properties. For any $\boldsymbol{x}, \dot{\boldsymbol{x}} \in \mathbb{R}^n$

- $\|\mathbf{tanh}(\boldsymbol{x})\| \leq \alpha_1 \|\boldsymbol{x}\|$
- $\|\mathbf{tanh}(\boldsymbol{x})\| \leq \alpha_2$
- $\|\mathbf{tanh}(\boldsymbol{x})\|^2 \leq \alpha_3 \, \mathbf{tanh}(\boldsymbol{x})^T \boldsymbol{x}$
- $\left\|\mathrm{Sech}^2(\boldsymbol{x})\dot{\boldsymbol{x}}\right\| \leq \alpha_4 \|\dot{\boldsymbol{x}}\|$

where $\alpha_1, \cdots, \alpha_4 > 0$. With $\mathbf{tanh}(\boldsymbol{x})$ defined as in (4.13), the constants $\alpha_1 = 1, \alpha_2 = \sqrt{n}, \alpha_3 = 1, \alpha_4 = 1$.

Property 4.4. Residual dynamics vector $\boldsymbol{h}(t, \tilde{\boldsymbol{q}}, \dot{\tilde{\boldsymbol{q}}})$

The *vector of residual dynamics* $\boldsymbol{h}(t, \tilde{\boldsymbol{q}}, \dot{\tilde{\boldsymbol{q}}})$ of $n \times 1$ depends on the position errors $\tilde{\boldsymbol{q}}$, velocity errors $\dot{\tilde{\boldsymbol{q}}}$, and on the desired joint motion —$\boldsymbol{q}_d$, $\dot{\boldsymbol{q}}_d$, and $\ddot{\boldsymbol{q}}_d$— that is supposed to be bounded. In this respect, we denote by $\|\dot{\boldsymbol{q}}_d\|_{\mathrm{M}}$ and $\|\ddot{\boldsymbol{q}}_d\|_{\mathrm{M}}$ the supreme values over the norms of the desired velocity and acceleration. In addition, $\boldsymbol{h}(t, \tilde{\boldsymbol{q}}, \dot{\tilde{\boldsymbol{q}}})$ has the following property:

1. There exist constants $k_{h1}, k_{h2} \geq 0$ such that the norm of the residual dynamics satisfies

$$\left\|\boldsymbol{h}(t, \tilde{\boldsymbol{q}}, \dot{\tilde{\boldsymbol{q}}})\right\| \leq k_{h1}\|\dot{\tilde{\boldsymbol{q}}}\| + k_{h2}\|\mathbf{tanh}(\tilde{\boldsymbol{q}})\| \tag{4.15}$$

for all $\tilde{\boldsymbol{q}}, \dot{\tilde{\boldsymbol{q}}} \in \mathbb{R}^n$, where $\mathbf{tanh}(\tilde{\boldsymbol{q}})$ is the vectorial tangent hyperbolic function introduced in Definition 4.1.

Proof. According to the definition of the residual dynamics function (4.12), its norm satisfies

$$\begin{aligned}\left\|\boldsymbol{h}(t, \tilde{\boldsymbol{q}}, \dot{\tilde{\boldsymbol{q}}})\right\| \leq\ & \|[M(\boldsymbol{q}_d) - M(\boldsymbol{q}_d - \tilde{\boldsymbol{q}})]\,\ddot{\boldsymbol{q}}_d\| \\ & + \left\|\left[C(\boldsymbol{q}_d, \dot{\boldsymbol{q}}_d) - C(\boldsymbol{q}_d - \tilde{\boldsymbol{q}}, \dot{\boldsymbol{q}}_d - \dot{\tilde{\boldsymbol{q}}})\right]\dot{\boldsymbol{q}}_d\right\| \\ & + \|\boldsymbol{g}(\boldsymbol{q}_d) - \boldsymbol{g}(\boldsymbol{q}_d - \tilde{\boldsymbol{q}})\| \,.\end{aligned} \tag{4.16}$$

We wish to upperbound each of the three terms on the right-hand side of this inequality. We start with $\|\boldsymbol{g}(\boldsymbol{q}_d) - \boldsymbol{g}(\boldsymbol{q}_d - \tilde{\boldsymbol{q}})\|$. From Property 4.3 it follows that the vector of gravitational torques – considering robots with revolute joints – satisfies the inequalities

$$\begin{aligned}\|\boldsymbol{g}(\boldsymbol{q}_d) - \boldsymbol{g}(\boldsymbol{q}_d - \tilde{\boldsymbol{q}})\| &\leq k_g \|\tilde{\boldsymbol{q}}\| \\ \|\boldsymbol{g}(\boldsymbol{q}_d) - \boldsymbol{g}(\boldsymbol{q}_d - \tilde{\boldsymbol{q}})\| &\leq 2k'\end{aligned}$$

for all $\boldsymbol{q}_d, \tilde{\boldsymbol{q}} \in \mathbb{R}^n$ and where we have used $\|\boldsymbol{g}(\boldsymbol{q})\| \leq k'$ for the second inequality. This may be illustrated as in Figure 4.2 where $\|\boldsymbol{g}(\boldsymbol{q}_d) - \boldsymbol{g}(\boldsymbol{q}_d - \tilde{\boldsymbol{q}})\|$ is in the dotted region, for all $\boldsymbol{q}_d, \tilde{\boldsymbol{q}} \in \mathbb{R}^n$.

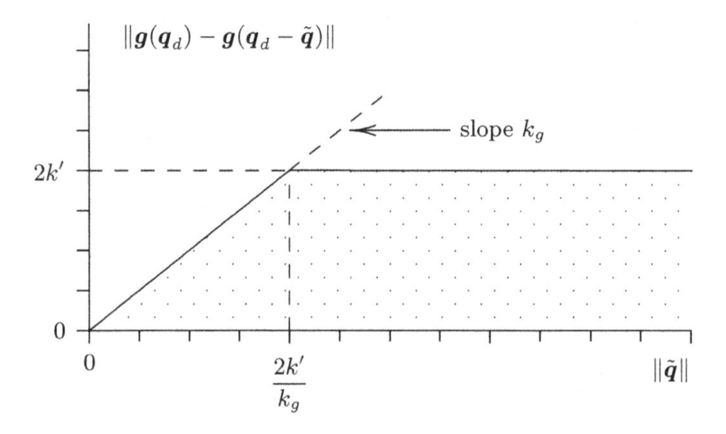

Figure 4.2. Belonging region for $\|\boldsymbol{g}(\boldsymbol{q}_d) - \boldsymbol{g}(\boldsymbol{q}_d - \tilde{\boldsymbol{q}})\|$

Regarding the first term on the right-hand side of Inequality (4.16), we have from Property 4.1 of the inertia matrix $M(\boldsymbol{q})$, that the following two inequalities hold:

$$\begin{aligned}\|[M(\boldsymbol{q}_d) - M(\boldsymbol{q}_d - \tilde{\boldsymbol{q}})]\, \ddot{\boldsymbol{q}}_d\| &\leq k_M \|\ddot{\boldsymbol{q}}_d\|_{\mathrm{M}} \|\tilde{\boldsymbol{q}}\|, \\ \|[M(\boldsymbol{q}_d) - M(\boldsymbol{q}_d - \tilde{\boldsymbol{q}})]\, \ddot{\boldsymbol{q}}_d\| &\leq 2k'_M \|\ddot{\boldsymbol{q}}_d\|_{\mathrm{M}},\end{aligned}$$

where for the second inequality, we used that $\|M(\boldsymbol{x})\ddot{\boldsymbol{q}}_d\| \leq k'_M \|\ddot{\boldsymbol{q}}_d\|$, which is valid for all $\boldsymbol{x} \in \mathbb{R}^n$.

Finally it is only left to bound the second term on the right-hand side of Inequality (4.16). This operation requires the following computations. By virtue of Property 4.2 it follows that (see (4.5))

$$\left\| \left[C(\boldsymbol{q}_d, \dot{\boldsymbol{q}}_d) - C(\boldsymbol{q}_d - \tilde{\boldsymbol{q}}, \dot{\boldsymbol{q}}_d - \dot{\tilde{\boldsymbol{q}}}) \right] \dot{\boldsymbol{q}}_d \right\| \leq k_{C_1} \|\dot{\boldsymbol{q}}_d\|_{\mathrm{M}} \left\| \dot{\tilde{\boldsymbol{q}}} \right\| + k_{C_2} \|\dot{\boldsymbol{q}}_d\|_{\mathrm{M}}^2 \|\tilde{\boldsymbol{q}}\|. \tag{4.17}$$

Also, observe that the left-hand side of (4.17) also satisfies

$$\left\| \left[C(\boldsymbol{q}_d, \dot{\boldsymbol{q}}_d) - C(\boldsymbol{q}_d - \tilde{\boldsymbol{q}}, \dot{\boldsymbol{q}}_d - \dot{\tilde{\boldsymbol{q}}}) \right] \dot{\boldsymbol{q}}_d \right\| \leq \| C(\boldsymbol{q}_d, \dot{\boldsymbol{q}}_d) \dot{\boldsymbol{q}}_d \| + \left\| C(\boldsymbol{q}_d - \tilde{\boldsymbol{q}}, \dot{\boldsymbol{q}}_d - \dot{\tilde{\boldsymbol{q}}}) \dot{\boldsymbol{q}}_d \right\| \tag{4.18}$$

but in view of the fact that $\|C(\boldsymbol{q}, \boldsymbol{x})\boldsymbol{y}\| \leq k_{C_1} \|\boldsymbol{x}\| \|\boldsymbol{y}\|$ for all $\boldsymbol{q}, \boldsymbol{x}, \boldsymbol{y} \in \mathbb{R}^n$, the terms on the right-hand side also satisfy

$$\| C(\boldsymbol{q}_d, \dot{\boldsymbol{q}}_d) \dot{\boldsymbol{q}}_d \| \leq k_{C_1} \| \dot{\boldsymbol{q}}_d \|_{\mathrm{M}}^2$$

and,

$$\begin{aligned} \left\| C(\boldsymbol{q}_d - \tilde{\boldsymbol{q}}, \dot{\boldsymbol{q}}_d - \dot{\tilde{\boldsymbol{q}}}) \dot{\boldsymbol{q}}_d \right\| &\leq k_{C_1} \| \dot{\boldsymbol{q}}_d \|_{\mathrm{M}} \left\| \dot{\boldsymbol{q}}_d - \dot{\tilde{\boldsymbol{q}}} \right\| \\ &\leq k_{C_1} \| \dot{\boldsymbol{q}}_d \|_{\mathrm{M}}^2 + k_{C_1} \| \dot{\boldsymbol{q}}_d \|_{\mathrm{M}} \left\| \dot{\tilde{\boldsymbol{q}}} \right\| . \end{aligned}$$

Using the latter in (4.18) we obtain

$$\left\| \left[C(\boldsymbol{q}_d, \dot{\boldsymbol{q}}_d) - C(\boldsymbol{q}_d - \tilde{\boldsymbol{q}}, \dot{\boldsymbol{q}}_d - \dot{\tilde{\boldsymbol{q}}}) \right] \dot{\boldsymbol{q}}_d \right\| \leq 2k_{C_1} \| \dot{\boldsymbol{q}}_d \|_{\mathrm{M}}^2 + k_{C_1} \| \dot{\boldsymbol{q}}_d \|_{\mathrm{M}} \left\| \dot{\tilde{\boldsymbol{q}}} \right\| . \tag{4.19}$$

Hence, to bound the norm of the residual dynamics (4.16) we use (4.17) and (4.19), as well as the previous bounds on the first and third terms. This yields that $\boldsymbol{h}(t, \tilde{\boldsymbol{q}}, \dot{\tilde{\boldsymbol{q}}})$ also satisfies

$$\left\| \boldsymbol{h}(t, \tilde{\boldsymbol{q}}, \dot{\tilde{\boldsymbol{q}}}) \right\| \leq k_{C_1} \| \dot{\boldsymbol{q}}_d \|_{\mathrm{M}} \left\| \dot{\tilde{\boldsymbol{q}}} \right\| + \left[k_g + k_M \| \ddot{\boldsymbol{q}}_d \|_{\mathrm{M}} + k_{C_2} \| \dot{\boldsymbol{q}}_d \|_{\mathrm{M}}^2 \right] \| \tilde{\boldsymbol{q}} \| ,$$

and

$$\left\| \boldsymbol{h}(t, \tilde{\boldsymbol{q}}, \dot{\tilde{\boldsymbol{q}}}) \right\| \leq k_{C_1} \| \dot{\boldsymbol{q}}_d \|_{\mathrm{M}} \left\| \dot{\tilde{\boldsymbol{q}}} \right\| + 2 \left[k' + k'_M \| \ddot{\boldsymbol{q}}_d \|_{\mathrm{M}} + k_{C_1} \| \dot{\boldsymbol{q}}_d \|_{\mathrm{M}}^2 \right]$$

for all $\tilde{\boldsymbol{q}} \in \mathbb{R}^n$. In other terms,

$$\left\| \boldsymbol{h}(t, \tilde{\boldsymbol{q}}, \dot{\tilde{\boldsymbol{q}}}) \right\| \leq k_{C_1} \| \dot{\boldsymbol{q}}_d \|_{\mathrm{M}} \left\| \dot{\tilde{\boldsymbol{q}}} \right\| + s(\tilde{\boldsymbol{q}}) \tag{4.20}$$

where the scalar function $s(\tilde{\boldsymbol{q}})$ is given by

$$s(\tilde{\boldsymbol{q}}) = \begin{cases} s_1 \| \tilde{\boldsymbol{q}} \| & \text{if} \quad \| \tilde{\boldsymbol{q}} \| < s_2 / s_1 \\ s_2 & \text{if} \quad \| \tilde{\boldsymbol{q}} \| \geq s_2 / s_1 \end{cases}$$

with

$$s_1 = \left[k_g + k_M \| \ddot{\boldsymbol{q}}_d \|_{\mathrm{M}} + k_{C_2} \| \dot{\boldsymbol{q}}_d \|_{\mathrm{M}}^2 \right] , \tag{4.21}$$

and

$$s_2 = 2 \left[k' + k'_M \| \ddot{\boldsymbol{q}}_d \|_{\mathrm{M}} + k_{C_1} \| \dot{\boldsymbol{q}}_d \|_{\mathrm{M}}^2 \right] . \tag{4.22}$$

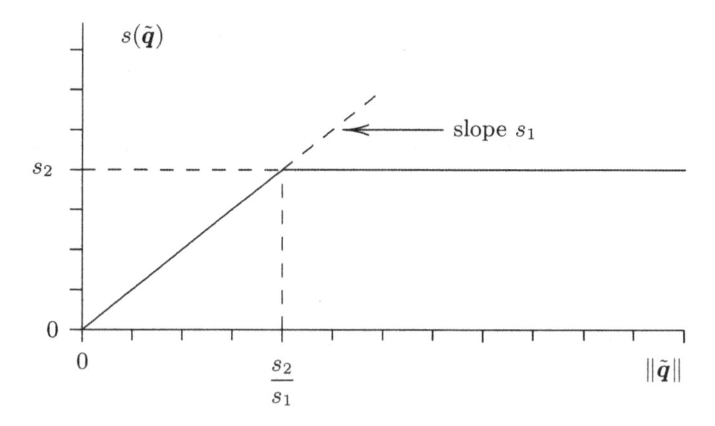

Figure 4.3. Graph of the function $s(\tilde{\boldsymbol{q}})$

The plot of $s(\tilde{\boldsymbol{q}})$ is shown in Figure 4.3. It is clear that $s(\tilde{\boldsymbol{q}})$ may be upperbounded by the tangent hyperbolic function of $\|\tilde{\boldsymbol{q}}\|$ that is,

$$|s(\tilde{\boldsymbol{q}})| \leq k_{h2} \tanh(\|\tilde{\boldsymbol{q}}\|) \tag{4.23}$$

where k_{h2} is any number that satisfies

$$k_{h2} \geq \frac{s_2}{\tanh\left(\dfrac{s_2}{s_1}\right)}. \tag{4.24}$$

Thus, we conclude that $\left\|\boldsymbol{h}(t, \tilde{\boldsymbol{q}}, \dot{\tilde{\boldsymbol{q}}})\right\|$ in (4.20) satisfies

$$\left\|\boldsymbol{h}(t, \tilde{\boldsymbol{q}}, \dot{\tilde{\boldsymbol{q}}})\right\| \leq k_{h1} \left\|\dot{\tilde{\boldsymbol{q}}}\right\| + k_{h2} \|\mathbf{tanh}(\tilde{\boldsymbol{q}})\|$$

where we have used the fact that

$$\tanh(\|\tilde{\boldsymbol{q}}\|) \leq \|\mathbf{tanh}(\tilde{\boldsymbol{q}})\|,$$

where $\mathbf{tanh}(\tilde{\boldsymbol{q}})$ is the vectorial tangent hyperbolic function (4.13) and k_{h1} is assumed to satisfy

$$k_{C_1} \|\dot{\boldsymbol{q}}_d\|_{\mathrm{M}} \leq k_{h1}. \tag{4.25}$$

Thus, Property 4.4 follows.

♢♢♢

Residual Dynamics when $\dot{\boldsymbol{q}}_d \equiv \mathbf{0}$

In the situation when $\dot{\boldsymbol{q}}_d \equiv \mathbf{0}$, and therefore $\ddot{\boldsymbol{q}}_d \equiv \mathbf{0}$, the residual dynamics (4.12) boils down to

$$\begin{aligned}\boldsymbol{h}(t, \tilde{\boldsymbol{q}}, \dot{\tilde{\boldsymbol{q}}}) &= \boldsymbol{g}(\boldsymbol{q}_d) - \boldsymbol{g}(\boldsymbol{q}_d - \tilde{\boldsymbol{q}}), \\ &= \boldsymbol{g}(\boldsymbol{q}_d) - \boldsymbol{g}(\boldsymbol{q}).\end{aligned}$$

Notice also that s_1 and s_2 in (4.21) and (4.22) respectively, become

$$\begin{aligned}s_1 &= k_g, \\ s_2 &= 2k'.\end{aligned}$$

With this information and what we know about k_{h1} from (4.25) and about k_{h2} from (4.24), we conclude that these constants

$$\begin{aligned}k_{h1} &= 0\,, \\ k_{h2} &\geq \frac{2k'}{\tanh\left(\dfrac{2k'}{k_g}\right)}\,.\end{aligned}$$

From this last inequality one can show that k_{h2} satisfies

$$k_{h2} \geq k_g\,.$$

Thus, we finally conclude from (4.20) and (4.23) that

$$\begin{aligned}\left\|\boldsymbol{h}(t, \tilde{\boldsymbol{q}}, \dot{\tilde{\boldsymbol{q}}})\right\| = \|\boldsymbol{g}(\boldsymbol{q}_d) - \boldsymbol{g}(\boldsymbol{q}_d - \tilde{\boldsymbol{q}})\| &\leq k_{h2}\tanh(\|\tilde{\boldsymbol{q}}\|), \\ &\leq k_{h2}\left\|\begin{bmatrix}\tanh(\tilde{q}_1) \\ \vdots \\ \tanh(\tilde{q}_n)\end{bmatrix}\right\|,\end{aligned}$$

for all $\boldsymbol{q}_d, \tilde{\boldsymbol{q}} \in \mathbb{R}^n$.

4.5 Conclusions

Properties 4.1, 4.2, 4.3 and 4.4 are exhaustively used in the succeeding chapters in the stability analysis of the control schemes that we present. In particular, Property 4.1 is used to construct non-negative functions and occasionally, Lyapunov functions to study stability and convergence properties for equilibria in robot control systems.

To close the chapter we summarize, in Table 4.1, the expressions involved in the computation of the main constants introduced, where s_1 and s_2 are given by Equations (4.21) and (4.22), respectively.

Table 4.1. Bounds on the matrices involved in the Lagrangian model

Bound	Definition
β	$n\ (\max_{i,j,q} \|M_{ij}(q)\|)$
k_M	$n^2 \left(\max_{i,j,k,q} \left\| \frac{\partial M_{ij}(q)}{\partial q_k} \right\| \right)$
k_{C_1}	$n^2\ \left(\max_{i,j,k,q} \left\|C_{k_{ij}}(q)\right\|\right)$
k_{C_2}	$n^3 \left(\max_{i,j,k,l,q} \left\| \frac{\partial C_{k_{ij}}(q)}{\partial q_l} \right\| \right)$
k_g	$n \left(\max_{i,j,q} \left\| \frac{\partial g_i(q)}{\partial q_j} \right\| \right)$
k_{h1}	$k_{c1}\ \|\dot{\boldsymbol{q}}_d\|_M$
k_{h2}	$\dfrac{s_2}{\tanh\left(\frac{s_2}{s_1}\right)}$

Bibliography

Properties 4.1 and 4.2 are proved in

- Spong M., Vidyasagar M., 1989, *"Robot dynamics and control"*, Wiley, New York.
- Craig J., 1988, *"Adaptive control of mechanical manipulators"*, Addison - Wesley, Reading MA.

The property of skew-symmetry of $\frac{1}{2}\dot{\overline{M(\boldsymbol{q})}} - C(\boldsymbol{q}, \dot{\boldsymbol{q}})$ was established in

- Ortega R. and Spong M., 1989, "Adaptive motion control of rigid robots: A tutorial," *Automatica*, Vol. 25-6, pp. 877–888.

The concept of residual dynamics was introduced in

- Arimoto S., 1995, *"Fundamental problems of robot control: Part I: Innovation in the realm of robot servo-loops"*, Robotica, Vol. 13, Part 1, pp. 19–27.
- Arimoto S., 1995, *"Fundamental problems of robot control: Part II: A nonlinear circuit theory towards an understanding of dexterous motions,* Robotica, Vol. 13, Part 2, pp. 111–122.

One version of the proof of Property 4.4 on the residual dynamics $\boldsymbol{h}(t, \tilde{\boldsymbol{q}}, \dot{\tilde{\boldsymbol{q}}})$ is presented in

- Santibáñez V., Kelly R., 2001, *"PD control with feedforward compensation for robot manipulators: Analysis and experimentation"*, Robotica, Vol. 19, pp. 11–19.

For other properties of robot manipulators not mentioned here and relevant to control, see

- Ortega R., Loría A., Nicklasson P. J., Sira-Ramírez H., 1998, *"Passivity-based control of Euler-Lagrange Systems Mechanical, Electrical and Electromechanical Applications"*, Springer-Verlag: London, Communications and Control Engg. Series.

Problems

1. Consider the simplified Cartesian mechanical device of Figure 4.4.

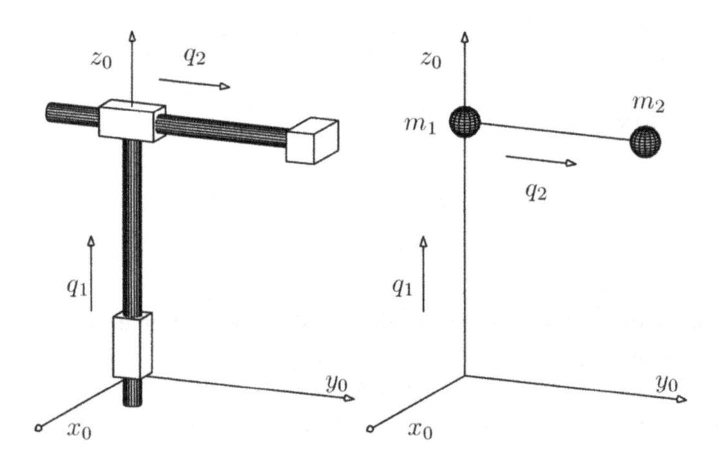

Figure 4.4. Problem 1

   a) Obtain the dynamic model using Lagrange's equations. Specifically, determine $M(\boldsymbol{q})$, $C(\boldsymbol{q}, \dot{\boldsymbol{q}})$ and $\boldsymbol{g}(\boldsymbol{q})$.
   b) Verify that the matrix $\frac{1}{2}\dot{M}(\boldsymbol{q}) - C(\boldsymbol{q}, \dot{\boldsymbol{q}})$ is skew-symmetric.
   c) Express the dynamic model in terms of the state vector $[q_1 \;\; q_2 \;\; \dot{q}_1 \;\; \dot{q}_2]^T$. Under which conditions on the external τ_1 and τ_2 do there exist equilibria?

2. Is it true that the inertia matrix $M(\boldsymbol{q})$ is constant if and only if $C(\boldsymbol{q}, \dot{\boldsymbol{q}}) = 0$? (The matrix $C(\boldsymbol{q}, \dot{\boldsymbol{q}})$ is assumed to be defined upon the Christoffel symbols of the first kind.)
3. Consider the dynamic model (3.33) of robots with (linear) actuators. Suppose that there is no friction (*i.e.* $\boldsymbol{f}(\dot{\boldsymbol{q}}) = \boldsymbol{0}$). Show that

$$\frac{1}{2}\dot{M}'(\boldsymbol{q}) - C(\boldsymbol{q}, \dot{\boldsymbol{q}}) = \frac{1}{2}\dot{M}(\boldsymbol{q}) - C(\boldsymbol{q}, \dot{\boldsymbol{q}}) .$$

4. Consider the equation that characterizes the behavior of a pendulum of length l and mass m concentrated at the edge and is submitted to the action of gravity g to which is applied a torque τ on the axis of rotation,

$$ml^2\ddot{q} + mgl \sin(q) = \tau$$

where q is the angular position of the pendulum with respect to the vertical.

Show that there exists a constant β such that

$$\int_0^T \tau(s)\dot{q}(s)\, ds \geq \beta, \qquad \forall\, T \in \mathbb{R}_+ .$$

Hint: Using Property 4.3, show that for any $T \geq 0$,

$$\int_0^T \dot{q}(s)\sin(q(s))\, ds \geq -K$$

with $K \geq 0$.

5

Case Study: The Pelican Prototype Robot

The purpose of this chapter is twofold: first, to present in detail the model of the experimental robot arm of the Robotics lab. from the CICESE Research Center, Mexico. Second, to review the topics studied in the previous chapters and to discuss, through this case study, the topics of direct kinematics and inverse kinematics, which are fundamental in determining robot models.

For the Pelican, we derive the full dynamic model of the prototype; in particular, we present the numerical values of all the parameters such as mass, inertias, lengths to centers of mass, *etc.* This is used throughout the rest of the book in numerous examples to illustrate the performance of the controllers that we study. We emphasize that *all* of these examples contain *experimentation* results.

Thus, the chapter is organized in the following sections:

- direct kinematics;
- inverse kinematics;
- dynamic model;
- properties of the dynamic model;
- reference trajectories.

For analytical purposes, further on, we refer to Figure 5.2, which represents the prototype schematically. As is obvious from this figure, the prototype is a planar arm with two links connected through revolute joints, *i.e.* it possesses 2 DOF. The links are driven by two electrical motors located at the "shoulder" (base) and at the "elbow". This is a direct-drive mechanism, *i.e.* the axes of the motors are connected directly to the links without gears or belts.

The manipulator arm consists of two rigid links of lengths l_1 and l_2, masses m_1 and m_2 respectively. The robot moves about on the plane x–y as is illustrated in Figure 5.2. The distances from the rotating axes to the centers of mass are denoted by l_{c1} and l_{c2} for links 1 and 2, respectively. Finally, I_1 and

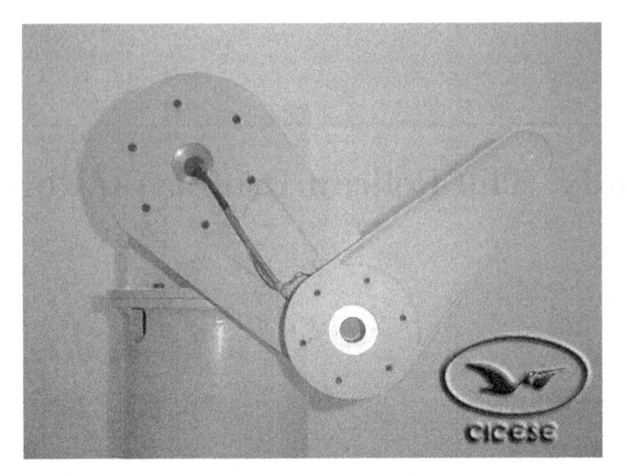

Figure 5.1. Pelican: experimental robot arm at CICESE, Robotics lab.

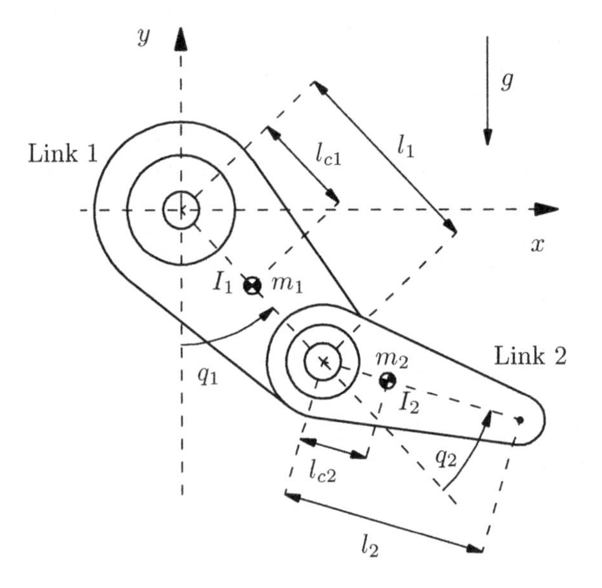

Figure 5.2. Diagram of the 2-DOF Pelican prototype robot

I_2 denote the moments of inertia of the links with respect to the axes that pass through the respective centers of mass and are parallel to the axis x. The degrees of freedom are associated with the angle q_1, which is measured from the vertical position, and q_2, which is measured relative to the extension of the first link toward the second link, both being positive counterclockwise. The vector of joint positions $\boldsymbol{q}$ is defined as

$$\boldsymbol{q} = [q_1 \;\; q_2]^T.$$

The meaning of the diverse constant parameters involved as well as their numerical values are summarized in Table 5.1.

Table 5.1. Physical parameters of Pelican robot arm

Description	Notation	Value	Units
Length of Link 1	l_1	0.26	m
Length of Link 2	l_2	0.26	m
Distance to the center of mass (Link 1)	l_{c1}	0.0983	m
Distance to the center of mass (Link 2)	l_{c2}	0.0229	m
Mass of Link 1	m_1	6.5225	kg
Mass of Link 2	m_2	2.0458	kg
Inertia rel. to center of mass (Link 1)	I_1	0.1213	kg m^2
Inertia rel. to center of mass (Link 2)	I_2	0.0116	kg m^2
Gravity acceleration	g	9.81	m/s^2

5.1 Direct Kinematics

The problem of direct kinematics for robot manipulators is formulated as follows. Consider a robot manipulator of n degrees-of-freedom placed on a fixed surface. Define a reference frame also fixed at some point on this surface. This reference frame is commonly referred to as 'base reference frame'. The problem of deriving the direct kinematic model of the robot consists in expressing the position and orientation (when the latter makes sense) of a reference frame fixed to the end of the last link of the robot, referred to the base reference frame in terms of the joint coordinates of the robot. The solution to the so-formulated problem from a mathematical viewpoint, reduces to solving a geometrical problem which always has a closed-form solution.

Regarding the Pelican robot, we start by defining the reference frame of base as a Cartesian coordinated system in two dimensions with its origin located exactly on the first joint of the robot, as is illustrated in Figure 5.2. The Cartesian coordinates x and y determine the position of the tip of the second link with respect to the base reference frame. Notice that for the present case study of a 2-DOF system, the orientation of the end-effector of the arm makes no sense. One can clearly appreciate that both Cartesian coordinates, x and

y, depend on the joint coordinates q_1 and q_2. Precisely it is this correlation that defines the direct kinematic model,

$$\begin{bmatrix} x \\ y \end{bmatrix} = \boldsymbol{\varphi}(q_1, q_2),$$

where $\boldsymbol{\varphi} : \mathbb{R}^2 \to \mathbb{R}^2$.

For the case of this robot with 2 DOF, it is immediate to verify that the direct kinematic model is given by

$$\begin{aligned} x &= l_1 \sin(q_1) + l_2 \sin(q_1 + q_2) \\ y &= -l_1 \cos(q_1) - l_2 \cos(q_1 + q_2). \end{aligned}$$

From this model is obtained: the following relation between the velocities

$$\begin{aligned} \begin{bmatrix} \dot{x} \\ \dot{y} \end{bmatrix} &= \begin{bmatrix} l_1 \cos(q_1) + l_2 \cos(q_1 + q_2) & l_2 \cos(q_1 + q_2) \\ l_1 \sin(q_1) + l_2 \sin(q_1 + q_2) & l_2 \sin(q_1 + q_2) \end{bmatrix} \begin{bmatrix} \dot{q}_1 \\ \dot{q}_2 \end{bmatrix} \\ &= J(\boldsymbol{q}) \begin{bmatrix} \dot{q}_1 \\ \dot{q}_2 \end{bmatrix} \end{aligned}$$

where $J(\boldsymbol{q}) = \dfrac{\partial \boldsymbol{\varphi}(\boldsymbol{q})}{\partial \boldsymbol{q}} \in \mathbb{R}^{2\times 2}$ is called the analytical Jacobian matrix or simply, the Jacobian of the robot. Clearly, the following relationship between accelerations also holds,

$$\begin{bmatrix} \ddot{x} \\ \ddot{y} \end{bmatrix} = \left[\frac{d}{dt} J(\boldsymbol{q})\right] \begin{bmatrix} \dot{q}_1 \\ \dot{q}_2 \end{bmatrix} + J(\boldsymbol{q}) \begin{bmatrix} \ddot{q}_1 \\ \ddot{q}_2 \end{bmatrix}.$$

The procedure by which one computes the derivatives of the Jacobian and thereby obtains expressions for the velocities in Cartesian coordinates, is called differential kinematics. This topic is not studied in more detail in this textbook since *we* do not use it for control.

5.2 Inverse Kinematics

The inverse kinematic model of robot manipulators is of great importance from a practical viewpoint. This model allows us to obtain the joint positions $\boldsymbol{q}$ in terms of the position and orientation of the end-effector of the last link referred to the base reference frame. For the case of the Pelican prototype robot, the inverse kinematic model has the form

$$\begin{bmatrix} q_1 \\ q_2 \end{bmatrix} = \boldsymbol{\varphi}^{-1}(x, y)$$

where $\varphi^{-1} : \Theta \rightarrow \mathbb{R}^2$ and $\Theta \subseteq \mathbb{R}^2$.

The derivation of the inverse kinematic model is in general rather complex and, in contrast to the direct kinematics problem, it may have multiple solutions or no solution at all! The first case is illustrated in Figure 5.3. Notice that for the same position (in Cartesian coordinates x, y) of the arm tip there exist two possible configurations of the links, *i.e.* two possible values for $\boldsymbol{q}$.

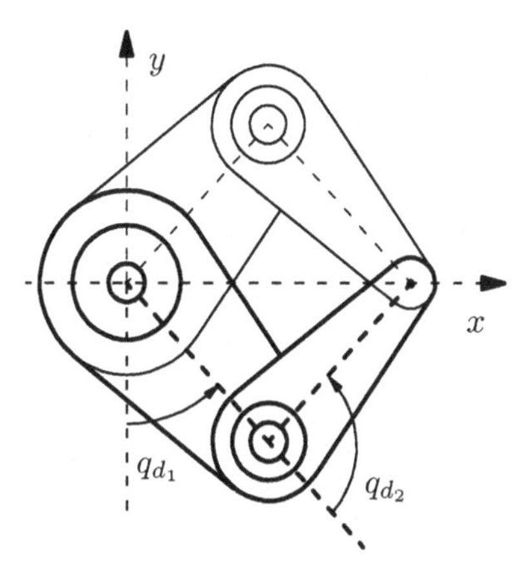

Figure 5.3. Two solutions to the inverse kinematics problem

So we see that even for this relatively simple robot configuration there exist more than one solution to the inverse kinematics problem.

The practical interest of the inverse kinematic model relies on its utility to define desired joint positions $\boldsymbol{q}_d = [q_{d_1} \;\; q_{d_2}]^T$ from specified desired positions x_d and y_d for the robot's end-effector. Indeed, note that physically, it is more intuitive to specify a task for a robot in end-effector coordinates so that interest in the inverse kinematics problem increases with the complexity of the manipulator (number of degrees of freedom).

Thus, let us now make our this discussion more precise by analytically computing the solutions $\begin{bmatrix} q_{d_1} \\ q_{d_2} \end{bmatrix} = \varphi^{-1}(x_d, y_d)$. The desired joint positions $\boldsymbol{q}_d$ can be computed using tedious but simple trigonometric manipulations to obtain

$$
\begin{aligned}
q_{d_1} &= \tan^{-1}\left(\frac{x_d}{-y_d}\right) - \tan^{-1}\left(\frac{l_2 \sin(q_{d_2})}{l_1 + l_2 \cos(q_{d_2})}\right) \\
q_{d_2} &= \cos^{-1}\left(\frac{x_d^2 + y_d^2 - l_1^2 - l_2^2}{2 l_1 l_2}\right) .
\end{aligned}
$$

The desired joint velocities and accelerations may be obtained via the *differential* kinematics[1] and its time derivative. In doing this one must keep in mind that the expressions obtained are valid only as long as the robot does not "fall" into a singular configuration, that is, as long as the Jacobian $J(\boldsymbol{q}_d)$ is square and nonsingular. These expressions are

$$\begin{bmatrix} \dot{q}_{d_1} \\ \dot{q}_{d_2} \end{bmatrix} = J^{-1}(\boldsymbol{q}_d) \begin{bmatrix} \dot{x}_d \\ \dot{y}_d \end{bmatrix}$$

$$\begin{bmatrix} \ddot{q}_{d_1} \\ \ddot{q}_{d_2} \end{bmatrix} = \underbrace{-J^{-1}(\boldsymbol{q}_d) \left[\frac{d}{dt} J(\boldsymbol{q}_d)\right] J^{-1}(\boldsymbol{q}_d)}_{\frac{d}{dt}\left[J^{-1}(\boldsymbol{q}_d)\right]} \begin{bmatrix} \dot{x}_d \\ \dot{y}_d \end{bmatrix} + J^{-1}(\boldsymbol{q}_d) \begin{bmatrix} \ddot{x}_d \\ \ddot{y}_d \end{bmatrix}$$

where $J^{-1}(\boldsymbol{q}_d)$ and $\frac{d}{dt}\left[J(\boldsymbol{q}_d)\right]$ denote the inverse of the Jacobian matrix and its time derivative respectively, evaluated at $\boldsymbol{q} = \boldsymbol{q}_d$. These are given by

$$J^{-1}(\boldsymbol{q}_d) = \begin{bmatrix} \dfrac{\mathrm{S}_{12}}{l_1 \mathrm{S}_2} & -\dfrac{\mathrm{C}_{12}}{l_1 \mathrm{S}_2} \\ \dfrac{-l_1 \mathrm{S}_1 - l_2 \mathrm{S}_{12}}{l_1 l_2 \mathrm{S}_2} & \dfrac{l_1 \mathrm{C}_1 + l_2 \mathrm{C}_{12}}{l_1 l_2 \mathrm{S}_2} \end{bmatrix},$$

and

$$\frac{d}{dt}\left[J(\boldsymbol{q}_d)\right] = \begin{bmatrix} -l_1 \mathrm{S}_1 \dot{q}_{d_1} - l_2 \mathrm{S}_{12}(\dot{q}_{d_1} + \dot{q}_{d_2}) & -l_2 \mathrm{S}_{12}(\dot{q}_{d_1} + \dot{q}_{d_2}) \\ l_1 \mathrm{C}_1 \dot{q}_{d_1} + l_2 \mathrm{C}_{12}(\dot{q}_{d_1} + \dot{q}_{d_2}) & l_2 \mathrm{C}_{12}(\dot{q}_{d_1} + \dot{q}_{d_2}) \end{bmatrix},$$

where, for simplicity, we have used the notation $\mathrm{S}_1 = \sin(q_{d_1})$, $\mathrm{S}_2 = \sin(q_{d_2})$, $\mathrm{C}_1 = \cos(q_{d_1})$, $\mathrm{S}_{12} = \sin(q_{d_1} + q_{d_2})$, $\mathrm{C}_{12} = \cos(q_{d_1} + q_{d_2})$.

Notice that the term S_2 appears in the denominator of all terms in $J(\boldsymbol{q})^{-1}$ hence, $q_{d_2} = n\pi$, with $n \in \{0, 1, 2, \ldots\}$ and any q_{d_1} also correspond to singular configurations. Physically, these configurations (for any valid n) represent the second link being completely extended or bent over the first, as is illustrated in Figure 5.4. Typically, singular configurations are those in which the end-effector of the robot is located at the physical boundary of the workspace (that is, the physical space that the end-effector can reach). For instance, the singular configuration corresponding to being stretched out corresponds to the end-effector being placed anywhere on the circumference of radius $l_1 + l_2$, which is the boundary of the robot's workspace. As for Figure 5.4 the origin of the coordinates frame constitute another point of this boundary.

Having illustrated the inverse kinematics problem through the planar manipulator of Figure 5.2 we stop our study of inverse kinematics since it is

[1] For a definition and a detailed treatment of differential kinematics see the book (Sciavicco, Siciliano 2000) —*cf.* Bibliography at the end of Chapter 1.

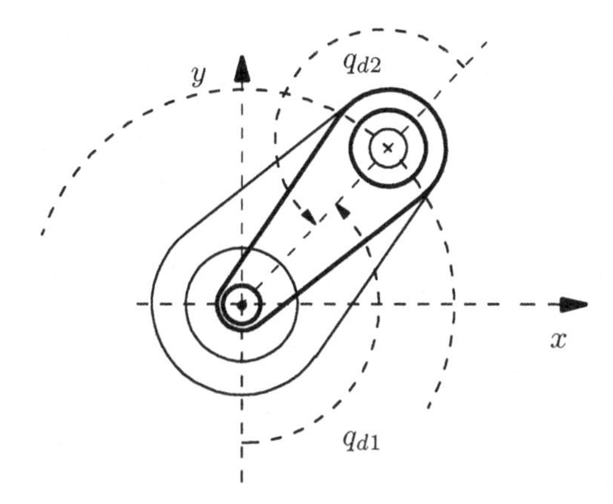

Figure 5.4. "Bent-over" singular configuration

beyond the scope of this text. However, we stress that what we have seen in the previous paragraphs extends in general.

In summary, we can say that if the control is based on the Cartesian coordinates of the end-effector, when designing the desired task for a manipulator's end-effector one must take special care that the configurations for the latter do not yield singular configurations. Concerning the controllers studied in this textbook, the reader should not worry about singular configurations since the Jacobian is not used at all: the reference trajectories are given in joint coordinates and we measure joint coordinates. This is what is called "control in joint space".

Thus, we leave the topic of kinematics to pass to the stage of modeling that is more relevant for control, *from the viewpoint of this* textbook, *i.e.* dynamics.

5.3 Dynamic Model

In this section we derive the Lagrangian equations for the CICESE prototype shown in Figure 5.1 and then we present in detail, useful bounds on the matrices of inertia, centrifugal and Coriolis forces, and on the vector of gravitational torques. Certainly, the model that we derive here applies to any planar manipulator following the same convention of coordinates as for our prototype.

5.3.1 Lagrangian Equations

Consider the 2-DOF robot manipulator shown in Figure 5.2. As we have learned from Chapter 3, to derive the Lagrangian dynamics we start by writing the kinetic energy function, $\mathcal{K}(\boldsymbol{q}, \dot{\boldsymbol{q}})$, defined in (3.15). For this manipulator, it may be decomposed into the sum of the two parts:

- the product of half the mass times the square of the speed of the center of mass; plus
- the product of half its moment of inertia (referred to the center of mass) times the square of its angular velocity (referred to the center of mass).

That is, we have $\mathcal{K}(\boldsymbol{q}, \dot{\boldsymbol{q}}) = \mathcal{K}_1(\boldsymbol{q}, \dot{\boldsymbol{q}}) + \mathcal{K}_2(\boldsymbol{q}, \dot{\boldsymbol{q}})$ where $\mathcal{K}_1(\boldsymbol{q}, \dot{\boldsymbol{q}})$ and $\mathcal{K}_2(\boldsymbol{q}, \dot{\boldsymbol{q}})$ are the kinetic energies associated with the masses m_1 and m_2 respectively. Let us now develop in more detail, the corresponding mathematical expressions. To that end, we first observe that the coordinates of the center of mass of link 1, expressed on the plane x–y, are

$$\begin{aligned} x_1 &= l_{c1} \sin(q_1) \\ y_1 &= -l_{c1} \cos(q_1) . \end{aligned}$$

The velocity vector $\boldsymbol{v}_1$ of the center of mass of such a link is then,

$$\boldsymbol{v}_1 = \begin{bmatrix} \dot{x}_1 \\ \dot{y}_1 \end{bmatrix} = \begin{bmatrix} l_{c1} \cos(q_1)\dot{q}_1 \\ l_{c1} \sin(q_1)\dot{q}_1 \end{bmatrix} .$$

Therefore, the speed squared, $\|\boldsymbol{v}_1\|^2 = \boldsymbol{v}_1^T\boldsymbol{v}_1$, of the center of mass becomes

$$\boldsymbol{v}_1^T\boldsymbol{v}_1 = l_{c1}^2 \dot{q}_1^2 .$$

Finally, the kinetic energy corresponding to the motion of link 1 can be obtained as

$$\begin{aligned} \mathcal{K}_1(\boldsymbol{q}, \dot{\boldsymbol{q}}) &= \frac{1}{2} m_1 \boldsymbol{v}_1^T \boldsymbol{v}_1 + \frac{1}{2} I_1 \dot{q}_1^2 \\ &= \frac{1}{2} m_1 l_{c1}^2 \dot{q}_1^2 + \frac{1}{2} I_1 \dot{q}_1^2 . \end{aligned} \tag{5.1}$$

On the other hand, the coordinates of the center of mass of link 2, expressed on the plane x–y are

$$\begin{aligned} x_2 &= l_1 \sin(q_1) + l_{c2} \sin(q_1 + q_2) \\ y_2 &= -l_1 \cos(q_1) - l_{c2} \cos(q_1 + q_2) . \end{aligned}$$

Consequently, the velocity vector $\boldsymbol{v}_2$ of the center of mass of such a link is

$$\begin{aligned} \boldsymbol{v}_2 &= \begin{bmatrix} \dot{x}_2 \\ \dot{y}_2 \end{bmatrix} \\ &= \begin{bmatrix} l_1 \cos(q_1)\dot{q}_1 + l_{c2} \cos(q_1 + q_2)[\dot{q}_1 + \dot{q}_2] \\ l_1 \sin(q_1)\dot{q}_1 + l_{c2} \sin(q_1 + q_2)[\dot{q}_1 + \dot{q}_2] \end{bmatrix} . \end{aligned}$$

Therefore, using the trigonometric identities $\cos(\theta)^2 + \sin(\theta)^2 = 1$ and $\sin(q_1)\sin(q_1+q_2) + \cos(q_1)\cos(q_1+q_2) = \cos(q_2)$ we conclude that the speed squared, $\|\boldsymbol{v}_2\|^2 = \boldsymbol{v}_2^T\boldsymbol{v}_2$, of the center of mass of link 2 satisfies

$$\boldsymbol{v}_2^T\boldsymbol{v}_2 = l_1^2\dot{q}_1^2 + l_{c2}^2\left[\dot{q}_1^2 + 2\dot{q}_1\dot{q}_2 + \dot{q}_2^2\right] + 2l_1l_{c2}\left[\dot{q}_1^2 + \dot{q}_1\dot{q}_2\right]\cos(q_2)$$

which implies that

$$\begin{aligned}\mathcal{K}_2(\boldsymbol{q},\dot{\boldsymbol{q}}) &= \frac{1}{2}m_2\boldsymbol{v}_2^T\boldsymbol{v}_2 + \frac{1}{2}I_2[\dot{q}_1+\dot{q}_2]^2 \\ &= \frac{m_2}{2}l_1^2\dot{q}_1^2 + \frac{m_2}{2}l_{c2}^2\left[\dot{q}_1^2 + 2\dot{q}_1\dot{q}_2 + \dot{q}_2^2\right] \\ &\quad + m_2l_1l_{c2}\left[\dot{q}_1^2 + \dot{q}_1\dot{q}_2\right]\cos(q_2) \\ &\quad + \frac{1}{2}I_2[\dot{q}_1+\dot{q}_2]^2.\end{aligned}$$

Similarly, the potential energy may be decomposed as the sum of the terms $\mathcal{U}(\boldsymbol{q}) = \mathcal{U}_1(\boldsymbol{q}) + \mathcal{U}_2(\boldsymbol{q})$, where $\mathcal{U}_1(\boldsymbol{q})$ and $\mathcal{U}_2(\boldsymbol{q})$ are the potential energies associated with the masses m_1 and m_2 respectively. Thus, assuming that the potential energy is zero at $y = 0$, we obtain

$$\mathcal{U}_1(\boldsymbol{q}) = -m_1l_{c1}g\cos(q_1)$$

and

$$\mathcal{U}_2(\boldsymbol{q}) = -m_2l_1g\cos(q_1) - m_2l_{c2}g\cos(q_1+q_2)\,. \tag{5.2}$$

From Equations (5.1)–(5.2) we obtain the Lagrangian as

$$\begin{aligned}\mathcal{L}(\boldsymbol{q},\dot{\boldsymbol{q}}) &= \mathcal{K}(\boldsymbol{q},\dot{\boldsymbol{q}}) - \mathcal{U}(\boldsymbol{q}) \\ &= \mathcal{K}_1(\boldsymbol{q},\dot{\boldsymbol{q}}) + \mathcal{K}_2(\boldsymbol{q},\dot{\boldsymbol{q}}) - \mathcal{U}_1(\boldsymbol{q}) - \mathcal{U}_2(\boldsymbol{q}) \\ &= \frac{1}{2}[m_1l_{c1}^2 + m_2l_1^2]\dot{q}_1^2 + \frac{1}{2}m_2l_{c2}^2\left[\dot{q}_1^2 + 2\dot{q}_1\dot{q}_2 + \dot{q}_2^2\right] \\ &\quad + m_2l_1l_{c2}\cos(q_2)\left[\dot{q}_1^2 + \dot{q}_1\dot{q}_2\right] \\ &\quad + [m_1l_{c1} + m_2l_1]g\cos(q_1) \\ &\quad + m_2gl_{c2}\cos(q_1+q_2) \\ &\quad + \frac{1}{2}I_1\dot{q}_1^2 + \frac{1}{2}I_2[\dot{q}_1+\dot{q}_2]^2.\end{aligned}$$

From this last equation we obtain the following expression:

$$\begin{aligned}\frac{\partial\mathcal{L}}{\partial\dot{q}_1} &= [m_1l_{c1}^2 + m_2l_1^2]\dot{q}_1 + m_2l_{c2}^2\dot{q}_1 + m_2l_{c2}^2\dot{q}_2 \\ &\quad + 2m_2l_1l_{c2}\cos(q_2)\dot{q}_1 + m_2l_1l_{c2}\cos(q_2)\dot{q}_2 \\ &\quad + I_1\dot{q}_1 + I_2[\dot{q}_1+\dot{q}_2].\end{aligned}$$

$$\begin{aligned}\frac{d}{dt}\left[\frac{\partial \mathcal{L}}{\partial \dot{q}_1}\right] &= \left[m_1 l_{c1}^2 + m_2 l_1^2 + m_2 l_{c2}^2 + 2m_2 l_1 l_{c2}\cos(q_2)\right]\ddot{q}_1 \\ &\quad + \left[m_2 l_{c2}^2 + m_2 l_1 l_{c2}\cos(q_2)\right]\ddot{q}_2 \\ &\quad - 2m_2 l_1 l_{c2}\sin(q_2)\dot{q}_1\dot{q}_2 - m_2 l_1 l_{c2}\sin(q_2)\dot{q}_2^2 \\ &\quad + I_1\ddot{q}_1 + I_2[\ddot{q}_1 + \ddot{q}_2].\end{aligned}$$

$$\frac{\partial \mathcal{L}}{\partial q_1} = -[m_1 l_{c1} + m_2 l_1]g\sin(q_1) - m_2 g l_{c2}\sin(q_1 + q_2).$$

$$\frac{\partial \mathcal{L}}{\partial \dot{q}_2} = m_2 l_{c2}^2\dot{q}_1 + m_2 l_{c2}^2\dot{q}_2 + m_2 l_1 l_{c2}\cos(q_2)\dot{q}_1 + I_2[\dot{q}_1 + \dot{q}_2].$$

$$\begin{aligned}\frac{d}{dt}\left[\frac{\partial \mathcal{L}}{\partial \dot{q}_2}\right] &= m_2 l_{c2}^2\ddot{q}_1 + m_2 l_{c2}^2\ddot{q}_2 \\ &\quad + m_2 l_1 l_{c2}\cos(q_2)\ddot{q}_1 - m_2 l_1 l_{c2}\sin(q_2)\dot{q}_1\dot{q}_2 \\ &\quad + I_2[\ddot{q}_1 + \ddot{q}_2].\end{aligned}$$

$$\frac{\partial \mathcal{L}}{\partial q_2} = -m_2 l_1 l_{c2}\sin(q_2)\left[\dot{q}_1\dot{q}_2 + \dot{q}_1^2\right] - m_2 g l_{c2}\sin(q_1 + q_2).$$

The dynamic equations that model the robot arm are obtained by applying Lagrange's Equations (3.4),

$$\frac{d}{dt}\left[\frac{\partial \mathcal{L}}{\partial \dot{q}_i}\right] - \frac{\partial \mathcal{L}}{\partial q_i} = \tau_i \qquad i = 1, 2$$

from which we finally get

$$\begin{aligned}\tau_1 &= \left[m_1 l_{c1}^2 + m_2 l_1^2 + m_2 l_{c2}^2 + 2m_2 l_1 l_{c2}\cos(q_2) + I_1 + I_2\right]\ddot{q}_1 \\ &\quad + \left[m_2 l_{c2}^2 + m_2 l_1 l_{c2}\cos(q_2) + I_2\right]\ddot{q}_2 \\ &\quad - 2m_2 l_1 l_{c2}\sin(q_2)\dot{q}_1\dot{q}_2 - m_2 l_1 l_{c2}\sin(q_2)\dot{q}_2^2 \\ &\quad + [m_1 l_{c1} + m_2 l_1]g\sin(q_1) \\ &\quad + m_2 g l_{c2}\sin(q_1 + q_2)\end{aligned} \tag{5.3}$$

and

$$\begin{aligned}\tau_2 &= \left[m_2 l_{c2}^2 + m_2 l_1 l_{c2}\cos(q_2) + I_2\right]\ddot{q}_1 + [m_2 l_{c2}^2 + I_2]\ddot{q}_2 \\ &\quad + m_2 l_1 l_{c2}\sin(q_2)\dot{q}_1^2 + m_2 g l_{c2}\sin(q_1 + q_2),\end{aligned} \tag{5.4}$$

where τ_1 and τ_2, are the external torques delivered by the actuators at joints 1 and 2.

Thus, the dynamic equations of the robot (5.3)-(5.4) constitute a set of two nonlinear differential equations of the state variables $\boldsymbol{x} = [\boldsymbol{q}^T\ \dot{\boldsymbol{q}}^T]^T$, that is, of the form (3.1) .

5.3.2 Model in Compact Form

For control purposes, it is more practical to rewrite the Lagrangian dynamic model of the robot, that is, Equations (5.3) and (5.4), in the compact form (3.18), *i.e.*

$$\underbrace{\begin{bmatrix} M_{11}(\boldsymbol{q}) & M_{12}(\boldsymbol{q}) \\ M_{21}(\boldsymbol{q}) & M_{22}(\boldsymbol{q}) \end{bmatrix}}_{M(\boldsymbol{q})} \ddot{\boldsymbol{q}} + \underbrace{\begin{bmatrix} C_{11}(\boldsymbol{q},\dot{\boldsymbol{q}}) & C_{12}(\boldsymbol{q},\dot{\boldsymbol{q}}) \\ C_{21}(\boldsymbol{q},\dot{\boldsymbol{q}}) & C_{22}(\boldsymbol{q},\dot{\boldsymbol{q}}) \end{bmatrix}}_{C(\boldsymbol{q},\dot{\boldsymbol{q}})} \dot{\boldsymbol{q}} + \underbrace{\begin{bmatrix} g_1(\boldsymbol{q}) \\ g_2(\boldsymbol{q}) \end{bmatrix}}_{\boldsymbol{g}(\boldsymbol{q})} = \boldsymbol{\tau},$$

where

$$\begin{aligned}
M_{11}(\boldsymbol{q}) &= m_1 l_{c1}^2 + m_2 \left[l_1^2 + l_{c2}^2 + 2 l_1 l_{c2} \cos(q_2) \right] + I_1 + I_2 \\
M_{12}(\boldsymbol{q}) &= m_2 \left[l_{c2}^2 + l_1 l_{c2} \cos(q_2) \right] + I_2 \\
M_{21}(\boldsymbol{q}) &= m_2 \left[l_{c2}^2 + l_1 l_{c2} \cos(q_2) \right] + I_2 \\
M_{22}(\boldsymbol{q}) &= m_2 l_{c2}^2 + I_2
\end{aligned}$$

$$\begin{aligned}
C_{11}(\boldsymbol{q},\dot{\boldsymbol{q}}) &= -m_2 l_1 l_{c2} \sin(q_2) \dot{q}_2 \\
C_{12}(\boldsymbol{q},\dot{\boldsymbol{q}}) &= -m_2 l_1 l_{c2} \sin(q_2) \left[\dot{q}_1 + \dot{q}_2 \right] \\
C_{21}(\boldsymbol{q},\dot{\boldsymbol{q}}) &= m_2 l_1 l_{c2} \sin(q_2) \dot{q}_1 \\
C_{22}(\boldsymbol{q},\dot{\boldsymbol{q}}) &= 0
\end{aligned}$$

$$\begin{aligned}
g_1(\boldsymbol{q}) &= \left[m_1 l_{c1} + m_2 l_1 \right] g \sin(q_1) + m_2 l_{c2} g \sin(q_1 + q_2) \\
g_2(\boldsymbol{q}) &= m_2 l_{c2} g \sin(q_1 + q_2) \,.
\end{aligned}$$

We emphasize that the appropriate state variables to describe the dynamic model of the robot are the positions q_1 and q_2 and the velocities $\dot{q}_1$ and $\dot{q}_2$. In terms of these state variables, the dynamic model of the robot may be written as

$$\frac{d}{dt} \begin{bmatrix} q_1 \\ q_2 \\ \dot{q}_1 \\ \dot{q}_2 \end{bmatrix} = \begin{bmatrix} \dot{q}_1 \\ \dot{q}_2 \\ M(\boldsymbol{q})^{-1} \left[\boldsymbol{\tau}(t) - C(\boldsymbol{q},\dot{\boldsymbol{q}})\dot{\boldsymbol{q}} - \boldsymbol{g}(\boldsymbol{q}) \right] \end{bmatrix} .$$

Properties of the Dynamic Model

We present now the derivation of certain bounds on the inertia matrix, the matrix of centrifugal and Coriolis forces and the vector of gravitational torques. The bounds that we derive are fundamental to properly tune the gains of the controllers studied in the succeeding chapters. We emphasize that, as studied in Chapter 4, some bounds exist for any manipulator with only revolute rigid joints. Here, we show how they can be computed for CICESE's Pelican prototype illustrated in Figure 5.2.

Derivation of $\lambda_{\min}\{M\}$

We start with the property of positive definiteness of the inertia matrix. For a symmetric 2×2 matrix

$$\begin{bmatrix} M_{11}(\boldsymbol{q}) & M_{21}(\boldsymbol{q}) \\ M_{21}(\boldsymbol{q}) & M_{22}(\boldsymbol{q}) \end{bmatrix}$$

to be positive definite for all $\boldsymbol{q} \in \mathbb{R}^n$, it is necessary and sufficient that[2] $M_{11}(\boldsymbol{q}) > 0$ and its determinant

$$M_{11}(\boldsymbol{q})M_{22}(\boldsymbol{q}) - M_{21}(\boldsymbol{q})^2$$

also be positive for all $\boldsymbol{q} \in \mathbb{R}^n$.

In the worst-case scenario $M_{11}(\boldsymbol{q}) = m_1 l_{c1}^2 + I_1 + I_2 + m_2(l_1 - l_{c2})^2 > 0$, we only need to compute the determinant of $M(\boldsymbol{q})$, that is,

$$\begin{aligned} \det[M(\boldsymbol{q})] = {} & I_1 I_2 + I_2[l_{c1}^2 m_1 + l_1^2 m_2] + l_{c2}^2 m_2 I_1 + l_{c1}^2 l_{c2}^2 m_1 m_2 \\ & + l_1^2 l_{c2}^2 m_2^2 [1 - \cos^2(q_2)] \,. \end{aligned}$$

Notice that only the last term depends on $\boldsymbol{q}$ and is positive or zero. Hence, we conclude that $M(\boldsymbol{q})$ is positive definite for all $q \in \mathbb{R}^n$, that is[3]

$$\boldsymbol{x}^T M(\boldsymbol{q})\boldsymbol{x} \geq \lambda_{\min}\{M\}\|\boldsymbol{x}\|^2 \qquad (5.5)$$

for all $\boldsymbol{q} \in \mathbb{R}^n$, where $\lambda_{\min}\{M\} > 0$.

Inequality (5.5) constitutes an important property for control purposes since for instance, it guarantees that $M(\boldsymbol{q})^{-1}$ is positive definite and bounded for all $\boldsymbol{q} \in \mathbb{R}^n$.

Let us continue with the computation of the constants β, k_M, k_{C_1}, k_{C_2} and k_g from the properties presented in Chapter 4.

Derivation of $\lambda_{\mathrm{Max}}\{M\}$

Consider the inertia matrix $M(\boldsymbol{q})$. From its components it may be verified that

[2] Consider the partitioned matrix

$$\begin{bmatrix} A & B \\ B^T & C \end{bmatrix}.$$

If $A = A^T > 0$, $C = C^T > 0$ and $C - B^T A^{-1} B \geq 0$ (resp. $C - B^T A^{-1} B > 0$), then this matrix is positive semidefinite (resp. positive definite). See Horn R. A., Johnson C. R., 1985, *Matrix analysis*, p. 473.

[3] See also Remark 2.1 on page 25.

$$\max_{i,j,\boldsymbol{q}} |M_{ij}(\boldsymbol{q})| = m_1 l_{c1}^2 + m_2 \left[l_1^2 + l_{c2}^2 + 2 l_1 l_{c2} \right] + I_1 + I_2 .$$

According to Table 4.1, the constant β may be obtained as a value larger or equal to n times the previous expression, *i.e.*

$$\beta \geq n \left[m_1 l_{c1}^2 + m_2 \left[l_1^2 + l_{c2}^2 + 2 l_1 l_{c2} \right] + I_1 + I_2 \right] .$$

Hence, defining, $\lambda_{\text{Max}}\{M\} = \beta$ we see that

$$\boldsymbol{x}^T M(\boldsymbol{q}) \boldsymbol{x} \leq \lambda_{\text{Max}}\{M\} \|\boldsymbol{x}\|^2$$

for all $\boldsymbol{q} \in \mathbb{R}^n$. Moreover, using the numerical values presented in Table 5.1, we get $\beta = 0.7193$ kg m^2, that is, $\lambda_{\text{Max}}\{M\} = 0.7193$ kg m^2.

Derivation of k_M

Consider the inertia matrix $M(\boldsymbol{q})$. From its components it may be verified that

$$\frac{\partial M_{11}(\boldsymbol{q})}{\partial q_1} = 0, \quad \frac{\partial M_{11}(\boldsymbol{q})}{\partial q_2} = -2 m_2 l_1 l_{c2} \sin(q_2)$$
$$\frac{\partial M_{12}(\boldsymbol{q})}{\partial q_1} = 0, \quad \frac{\partial M_{12}(\boldsymbol{q})}{\partial q_2} = -m_2 l_1 l_{c2} \sin(q_2)$$
$$\frac{\partial M_{21}(\boldsymbol{q})}{\partial q_1} = 0, \quad \frac{\partial M_{21}(\boldsymbol{q})}{\partial q_2} = -m_2 l_1 l_{c2} \sin(q_2)$$
$$\frac{\partial M_{22}(\boldsymbol{q})}{\partial q_1} = 0, \quad \frac{\partial M_{22}(\boldsymbol{q})}{\partial q_2} = 0 .$$

According to Table 4.1, the constant k_M may be determined as

$$k_M \geq n^2 \left(\max_{i,j,k,\boldsymbol{q}} \left| \frac{\partial M_{ij}(\boldsymbol{q})}{\partial q_k} \right| \right) ,$$

hence, this constant may be chosen to satisfy

$$k_M \geq n^2 2 m_2 l_1 l_{c2} .$$

Using the numerical values presented in Table 5.1 we get $k_M = 0.0974$ kg m^2.

Derivation of k_{C_1}

Consider the vector of centrifugal and Coriolis forces $C(\boldsymbol{q}, \dot{\boldsymbol{q}})\dot{\boldsymbol{q}}$ written as

$$C(\boldsymbol{q}, \dot{\boldsymbol{q}})\dot{\boldsymbol{q}} = \begin{bmatrix} -m_2 l_1 l_{c2} \sin(q_2) \left(2 \dot{q}_1 \dot{q}_2 + \dot{q}_2^2 \right) \\ \\ m_2 l_1 l_{c2} \sin(q_2) \dot{q}_1^2 \end{bmatrix}$$

$$= \begin{bmatrix} \begin{bmatrix} \dot{q}_1 \\ \dot{q}_2 \end{bmatrix}^T \overbrace{\begin{bmatrix} 0 & -m_2 l_1 l_{c2} \sin(q_2) \\ -m_2 l_1 l_{c2} \sin(q_2) & -m_2 l_1 l_{c2} \sin(q_2) \end{bmatrix}}^{C_1(\boldsymbol{q})} \begin{bmatrix} \dot{q}_1 \\ \dot{q}_2 \end{bmatrix} \\ \\ \begin{bmatrix} \dot{q}_1 \\ \dot{q}_2 \end{bmatrix}^T \underbrace{\begin{bmatrix} m_2 l_1 l_{c2} \sin(q_2) & 0 \\ 0 & 0 \end{bmatrix}}_{C_2(\boldsymbol{q})} \begin{bmatrix} \dot{q}_1 \\ \dot{q}_2 \end{bmatrix} \end{bmatrix} . \tag{5.6}$$

According to Table 4.1, the constant k_{C_1} may be derived as

$$k_{C_1} \geq n^2 \left(\max_{i,j,k,\boldsymbol{q}} \left| C_{k_{ij}}(\boldsymbol{q}) \right| \right)$$

hence, this constant may be chosen so that

$$k_{C_1} \geq n^2 m_2 l_1 l_{c2}$$

Consequently, in view of the numerical values from Table 5.1 we find that $k_{C_1} = 0.0487$ kg m^2.

Derivation of k_{C_2}

Consider again the vector of centrifugal and Coriolis forces $C(\boldsymbol{q}, \dot{\boldsymbol{q}})\dot{\boldsymbol{q}}$ written as in (5.6). From the matrices $C_1(\boldsymbol{q})$ and $C_2(\boldsymbol{q})$ it may easily be verified that

$$\frac{\partial C_{1_{11}}(\boldsymbol{q})}{\partial q_1} = 0, \quad \frac{\partial C_{1_{11}}(\boldsymbol{q})}{\partial q_2} = 0$$

$$\frac{\partial C_{1_{12}}(\boldsymbol{q})}{\partial q_1} = 0, \quad \frac{\partial C_{1_{12}}(\boldsymbol{q})}{\partial q_2} = -m_2 l_1 l_{c2} \cos(q_2)$$

$$\frac{\partial C_{1_{21}}(\boldsymbol{q})}{\partial q_1} = 0, \quad \frac{\partial C_{1_{21}}(\boldsymbol{q})}{\partial q_2} = -m_2 l_1 l_{c2} \cos(q_2)$$

$$\frac{\partial C_{1_{22}}(\boldsymbol{q})}{\partial q_1} = 0, \quad \frac{\partial C_{1_{22}}(\boldsymbol{q})}{\partial q_2} = -m_2 l_1 l_{c2} \cos(q_2)$$

$$\frac{\partial C_{2_{11}}(\boldsymbol{q})}{\partial q_1} = 0, \quad \frac{\partial C_{2_{11}}(\boldsymbol{q})}{\partial q_2} = m_2 l_1 l_{c2} \cos(q_2)$$

$$\frac{\partial C_{2_{12}}(\boldsymbol{q})}{\partial q_1} = 0, \quad \frac{\partial C_{2_{12}}(\boldsymbol{q})}{\partial q_2} = 0$$

$$\frac{\partial C_{2_{21}}(\boldsymbol{q})}{\partial q_1} = 0, \quad \frac{\partial C_{2_{21}}(\boldsymbol{q})}{\partial q_2} = 0$$

$$\frac{\partial C_{222}(\boldsymbol{q})}{\partial q_1} = 0, \ \frac{\partial C_{222}(\boldsymbol{q})}{\partial q_2} = 0 .$$

Furthermore, according to Table 4.1 the constant k_{C_2} may be taken to satisfy

$$k_{C_2} \geq n^3 \left(\max_{i,j,k,l,\boldsymbol{q}} \left| \frac{\partial C_{k_{ij}}(\boldsymbol{q})}{\partial q_l} \right| \right) .$$

Therefore, we may choose k_{C_2} as

$$k_{C_2} \geq n^3 m_2 l_1 l_{c2} ,$$

which, in view of the numerical values from Table 5.1, takes the numerical value $k_{C_2} = 0.0974$ kg m^2:

Derivation of k_g

According to the components of the gravitational torques vector $\boldsymbol{g}(\boldsymbol{q})$ we have

$$\begin{aligned}
\frac{\partial g_1(\boldsymbol{q})}{\partial q_1} &= (m_1 l_{c1} + m_2 l_1)\, g\, \cos(q_1) + m_2 l_{c2} g\, \cos(q_1 + q_2) \\
\frac{\partial g_1(\boldsymbol{q})}{\partial q_2} &= m_2 l_{c2} g\, \cos(q_1 + q_2) \\
\frac{\partial g_2(\boldsymbol{q})}{\partial q_1} &= m_2 l_{c2} g\, \cos(q_1 + q_2) \\
\frac{\partial g_2(\boldsymbol{q})}{\partial q_2} &= m_2 l_{c2} g\, \cos(q_1 + q_2) .
\end{aligned}$$

Notice that the Jacobian matrix $\frac{\partial \boldsymbol{g}(\boldsymbol{q})}{\partial \boldsymbol{q}}$ corresponds in fact, to the Hessian matrix (*i.e.* the second partial derivative) of the potential energy function $\mathcal{U}(\boldsymbol{q})$, and is a symmetric matrix even though not necessarily positive definite.

The positive constant k_g may be derived from the information given in Table 4.1 as

$$k_g \geq n \max_{i,j,q} \left| \frac{\partial g_i(\boldsymbol{q})}{\partial q_j} \right| .$$

That is,

$$k_g \geq n \left[m_1 l_{c1} + m_2 l_1 + m_2 l_{c2} \right] g$$

and using the numerical values from Table 5.1 may be given the numerical value $k_g = 23.94$ kg m^2/s^2.

Table 5.2. Numeric values of the parameters for the CICESE prototype

Parameter	Value	Units
$\lambda_{\text{Max}}\{M\}$	0.7193	kg m^2
k_M	0.0974	kg m^2
k_{C_1}	0.0487	kg m^2
k_{C_2}	0.0974	kg m^2
k_g	23.94	kg m^2/s^2

Summary

The numerical values of the constants $\lambda_{\text{Max}}\{M\}$, k_M, k_{C_1}, k_{C_2} and k_g obtained above are summarized in Table 5.2.

5.4 Desired Reference Trajectories

With the aim of testing in experiments the performance of the controllers presented in this book, on the Pelican robot, we have selected the following reference trajectories in *joint space*:

$$\begin{bmatrix} q_{d1} \\ \\ q_{d2} \end{bmatrix} = \begin{bmatrix} b_1[1-e^{-2.0\ t^3}] + c_1[1-e^{-2.0\ t^3}]\sin(\omega_1 t) \\ \\ b_2[1-e^{-2.0\ t^3}] + c_2[1-e^{-2.0\ t^3}]\sin(\omega_2 t) \end{bmatrix} \quad [\text{rad}] \tag{5.7}$$

where $b_1 = \pi/4$ [rad], $c_1 = \pi/9$ [rad] and $\omega_1 = 4$ [rad/s], are parameters for the desired position reference for the first joint and $b_2 = \pi/3$ [rad], $c_2 = \pi/6$ [rad] and $\omega_2 = 3$ [rad/s] correspond to parameters that determine the desired position reference for the second joint. Figure 5.5 shows graphs of these reference trajectories against time.

Note the following important features in these reference trajectories:

- the trajectory contains a sinusoidal term to evaluate the performance of the controller following relatively fast periodic motions. This test is significant since such motions excite nonlinearities in the system.
- It also contains a slowly increasing term to bring the robot to the operating point without driving the actuators into saturation.

The module and frequency of the periodic signal must be chosen with care to avoid both torque and speed saturation in the actuators. In other

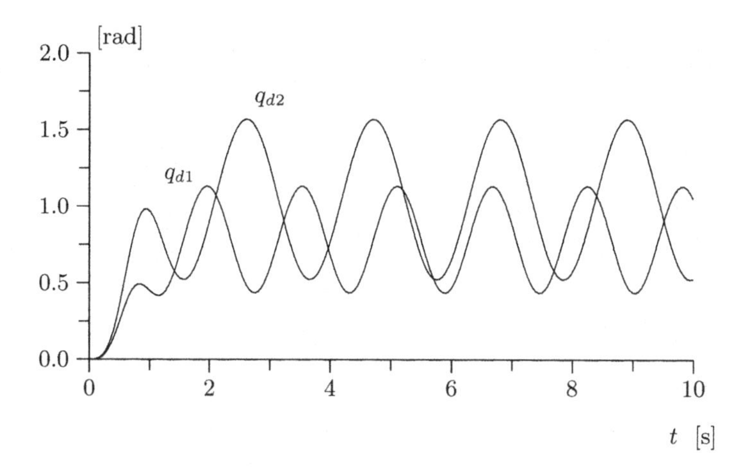

Figure 5.5. Desired reference trajectories

words, the reference trajectories must be such that the evolution of the robot dynamics along these trajectories gives admissible velocities and torques for the actuators. Otherwise, the desired reference is physically unfeasible.

Using the expressions of the desired position trajectories, (5.7), we may obtain analytically expressions for the desired velocity reference trajectories. These are obtained by direct differentiation, *i.e.*

$$\begin{aligned}
\dot{q}_{d1} &= 6b_1t^2e^{-2.0\ t^3} + 6c_1t^2e^{-2.0\ t^3}\sin(\omega_1 t) + [c_1 - c_1e^{-2.0\ t^3}]\cos(\omega_1 t)\omega_1, \\
\dot{q}_{d2} &= 6b_2t^2e^{-2.0\ t^3} + 6c_2t^2e^{-2.0\ t^3}\sin(\omega_2 t) + [c_2 - c_2e^{-2.0\ t^3}]\cos(\omega_2 t)\omega_2\,,
\end{aligned} \tag{5.8}$$

in [rad/s]. In the same way we may proceed to compute the reference accelerations to obtain

$$\begin{aligned}
\ddot{q}_{d1} &= 12b_1te^{-2.0\ t^3} - 36b_1t^4e^{-2.0\ t^3} + 12c_1te^{-2.0\ t^3}\sin(\omega_1 t) \\
&\quad - 36c_1t^4e^{-2.0\ t^3}\sin(\omega_1 t) + 12c_1t^2e^{-2.0\ t^3}\cos(\omega_1 t)\omega_1 \\
&\quad - [c_1 - c_1e^{-2.0\ t^3}]\sin(\omega_1 t)\omega_1^2 \quad [\mathrm{rad/s^2}]\,, \\[2ex]
\ddot{q}_{d2} &= 12b_2te^{-2.0\ t^3} - 36b_2t^4e^{-2.0\ t^3} + 12c_2te^{-2.0\ t^3}\sin(\omega_2 t) \\
&\quad - 36c_2t^4e^{-2.0\ t^3}\sin(\omega_2 t) + 12c_2t^2e^{-2.0\ t^3}\cos(\omega_2 t)\omega_2 \\
&\quad - [c_2 - c_2e^{-2.0\ t^3}]\sin(\omega_2 t)\omega_2^2 \quad [\mathrm{rad/s^2}]\,.
\end{aligned} \tag{5.9}$$

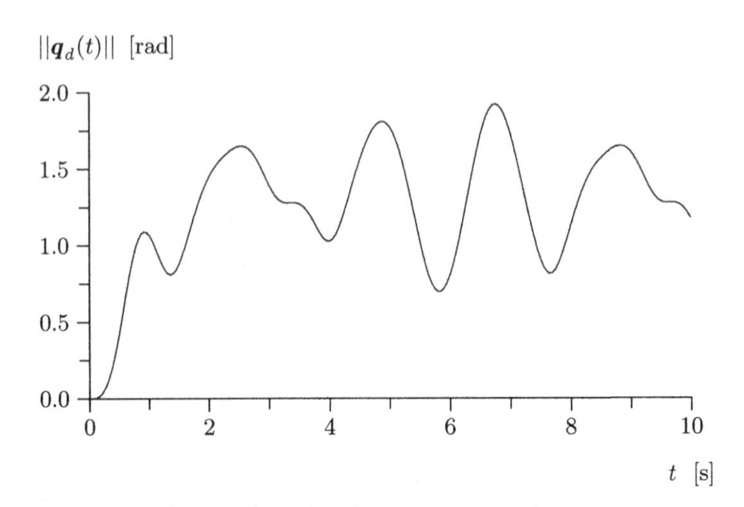

Figure 5.6. Norm of the desired positions

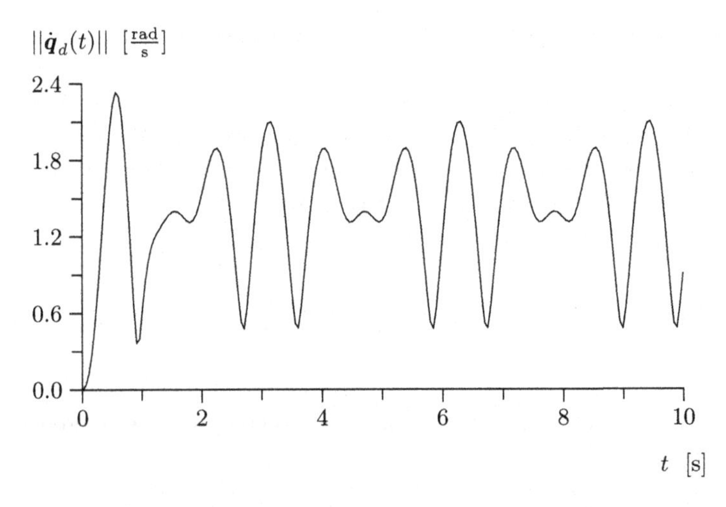

Figure 5.7. Norm of the desired velocities vector

Figures 5.6, 5.7 and 5.8 show the evolution in time of the norms corresponding to the desired joint positions, velocities and accelerations respectively. From these figures we deduce the following upper-bounds on the norms

$$\begin{aligned}
\|\mathbf{q}_d\|_{\text{Max}} &\leq 1.92 \text{ [rad]} \\
\|\dot{\mathbf{q}}_d\|_{\text{Max}} &\leq 2.33 \text{ [rad/s]} \\
\|\ddot{\mathbf{q}}_d\|_{\text{Max}} &\leq 9.52 \text{ [rad/s}^2] \,.
\end{aligned}$$

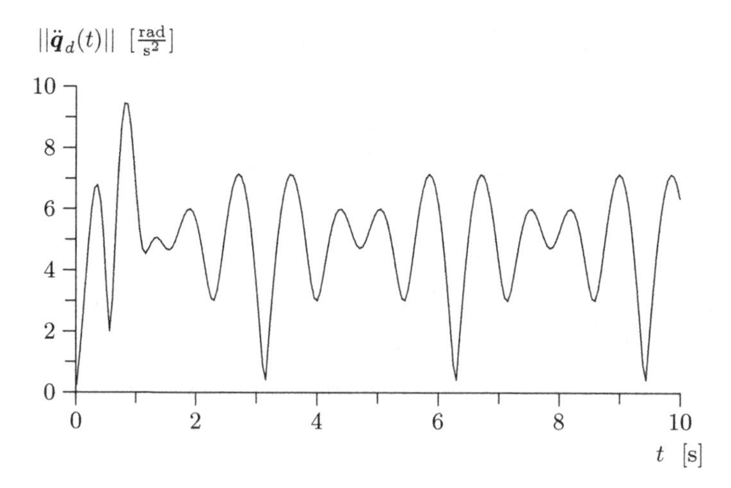

Figure 5.8. Norm of the desired accelerations vector

Bibliography

The schematic diagram of the robot depicted in Figure 5.2 and elsewhere throughout the book, corresponds to a real experimental prototype robot arm designed and constructed in the CICESE Research Center, Mexico[4].

The numerical values that appear in Table 5.1 are taken from:

- Campa R., Kelly R., Santibáñez V., 2004, *"Windows-based real-time control of direct-drive mechanisms: platform description and experiments"*, Mechatronics, Vol. 14, No. 9, pp. 1021–1036.

The constants listed in Table 5.2 may be computed based on data reported in

- Moreno J., Kelly R., Campa R., 2003, *"Manipulator velocity control using friction compensation"*, IEE Proceedings - Control Theory and Applications, Vol. 150, No. 2.

Problems

1. Consider the matrices $M(\boldsymbol{q})$ and $C(\boldsymbol{q},\dot{\boldsymbol{q}})$ from Section 5.3.2. Show that the matrix $\left[\frac{1}{2}\dot{M}(\boldsymbol{q}) - C(\boldsymbol{q},\dot{\boldsymbol{q}})\right]$ is skew-symmetric.
2. According to Property 4.2, the centrifugal and Coriolis forces matrix $C(\boldsymbol{q},\dot{\boldsymbol{q}})$, of the dynamic model of an n-DOF robot is not unique. In Section 5.3.2 we computed the elements of the matrix $C(\boldsymbol{q},\dot{\boldsymbol{q}})$ of the Pelican

[4] "Centro de Investigación Científica y de Educación Superior de Ensenada".

robot presented in this chapter. Prove also that the matrix $C(\boldsymbol{q}, \dot{\boldsymbol{q}})$ whose elements are given by

$$
\begin{aligned}
C_{11}(\boldsymbol{q}, \dot{\boldsymbol{q}}) &= -2m_2 l_1 l_{c2} \sin(q_2)\dot{q}_2 \\
C_{12}(\boldsymbol{q}, \dot{\boldsymbol{q}}) &= -m_2 l_1 l_{c2} \sin(q_2)\dot{q}_2 \\
C_{21}(\boldsymbol{q}, \dot{\boldsymbol{q}}) &= m_2 l_1 l_{c2} \sin(q_2)\dot{q}_1 \\
C_{22}(\boldsymbol{q}, \dot{\boldsymbol{q}}) &= 0
\end{aligned}
$$

characterizes the centrifugal and Coriolis forces, $C(\boldsymbol{q}, \dot{\boldsymbol{q}})\dot{\boldsymbol{q}}$. With this definition of $C(\boldsymbol{q}, \dot{\boldsymbol{q}})$, is $\frac{1}{2}\dot{M}(\boldsymbol{q}) - C(\boldsymbol{q}, \dot{\boldsymbol{q}})$ skew-symmetric?

Does it hold that $\dot{\boldsymbol{q}}^T \left[\frac{1}{2}\dot{M}(\boldsymbol{q}) - C(\boldsymbol{q}, \dot{\boldsymbol{q}})\right] \dot{\boldsymbol{q}} = 0$? Explain.

Part II

Position Control

Introduction to Part II

Depending on their application, industrial robot manipulators may be classified into two categories: the first is that of robots which move freely in their workspace (*i.e.* the physical space reachable by the end-effector) thereby undergoing movements without physical contact with their environment; tasks such as spray-painting, laser-cutting and welding may be performed by this type of manipulator. The second category encompasses robots which are designed to interact with their environment, for instance, by applying a complying force; tasks in this category include polishing and precision assembling.

In this textbook we study exclusively motion controllers for robot manipulators that move about freely in their workspace.

For clarity of exposition, we shall consider robot manipulators provided with ideal actuators, that is, actuators with negligible dynamics or in other words, that deliver torques and forces which are proportional to their inputs. This idealization is common in many theoretical works on robot control as well as in most textbooks on robotics. On the other hand, the recent technological developments in the construction of electromechanical actuators allow one to rely on direct-drive servomotors, which may be considered as ideal torque sources over a wide range of operating points. Finally, it is important to mention that even though in this textbook we assume that the actuators are ideal, most studies of controllers that we present in the sequel may be easily extended, by carrying out minor modifications, to the case of linear actuators of the second order; such is the case of DC motors.

Motion controllers that we study are classified into two main parts based on the control goal. In this second part of the book we study *position* controllers (set-point controllers) and in Part III we study *motion controllers* (tracking controllers).

Consider the dynamic model of a robot manipulator with n DOF, rigid links, no friction at the joints and with ideal actuators, (3.18), and which we recall below for convenience:

$$M(\boldsymbol{q})\ddot{\boldsymbol{q}} + C(\boldsymbol{q},\dot{\boldsymbol{q}})\dot{\boldsymbol{q}} + \boldsymbol{g}(\boldsymbol{q}) = \boldsymbol{\tau}\,. \tag{II.1}$$

where $M(\boldsymbol{q}) \in \mathbb{R}^{n\times n}$ is the inertia matrix, $C(\boldsymbol{q},\dot{\boldsymbol{q}})\dot{\boldsymbol{q}} \in \mathbb{R}^n$ is the vector of centrifugal and Coriolis forces, $\boldsymbol{g}(\boldsymbol{q}) \in \mathbb{R}^n$ is the vector of gravitational forces and torques and $\boldsymbol{\tau} \in \mathbb{R}^n$ is a vector of external forces and torques applied at the joints. The vectors $\boldsymbol{q}, \dot{\boldsymbol{q}}, \ddot{\boldsymbol{q}} \in \mathbb{R}^n$ denote the position, velocity and joint acceleration respectively.

In terms of the state vector $\begin{bmatrix}\boldsymbol{q}^T & \dot{\boldsymbol{q}}^T\end{bmatrix}^T$ these equations take the form

$$\frac{d}{dt}\begin{bmatrix}\boldsymbol{q}\\ \dot{\boldsymbol{q}}\end{bmatrix} = \begin{bmatrix}\dot{\boldsymbol{q}}\\ M(\boldsymbol{q})^{-1}\left[\boldsymbol{\tau}(t) - C(\boldsymbol{q},\dot{\boldsymbol{q}})\dot{\boldsymbol{q}} - \boldsymbol{g}(\boldsymbol{q})\right]\end{bmatrix}.$$

The problem of position control of robot manipulators may be formulated in the following terms. Consider the dynamic equation of an n-DOF robot, (II.1). Given a desired constant position (set-point reference) $\boldsymbol{q}_d$, we wish to find a vectorial function $\boldsymbol{\tau}$ such that the positions $\boldsymbol{q}$ associated with the robot's joint coordinates tend to $\boldsymbol{q}_d$ accurately.

In more formal terms, the *objective of position control* consists in finding $\boldsymbol{\tau}$ such that

$$\lim_{t\to\infty} \boldsymbol{q}(t) = \boldsymbol{q}_d$$

where $\boldsymbol{q}_d \in \mathbb{R}^n$ is a *given* constant vector which represents the desired joint positions.

The way that we evaluate whether a controller achieves the control objective is by studying the asymptotic stability of the origin of the closed-loop system in the sense of Lyapunov (*cf.* Chapter 2). For such purposes, it appears convenient to rewrite the position control objective as

$$\lim_{t\to\infty} \tilde{\boldsymbol{q}}(t) = \boldsymbol{0}$$

where $\tilde{\boldsymbol{q}} \in \mathbb{R}^n$ stands for the joint position errors vector or is simply called position error, and is defined by

$$\tilde{\boldsymbol{q}}(t) := \boldsymbol{q}_d - \boldsymbol{q}(t)\,.$$

Then, we say that the control objective is achieved, if for instance the origin of the closed-loop system (also referred to as position error dynamics) in terms of the state, *i.e.* $[\tilde{\boldsymbol{q}}^T\ \dot{\boldsymbol{q}}^T]^T = \boldsymbol{0} \in \mathbb{R}^{\mathbf{2n}}$, is asymptotically stable.

The computation of the vector $\boldsymbol{\tau}$ involves, in general, a vectorial nonlinear function of $\boldsymbol{q}$, $\dot{\boldsymbol{q}}$ and $\ddot{\boldsymbol{q}}$. This function is called the "control law" or simply, "controller". It is important to recall that robot manipulators are equipped with sensors to measure position and velocity at each joint, hence, the vectors $\boldsymbol{q}$ and $\dot{\boldsymbol{q}}$ are assumed to be measurable and may be used by the controllers. In general, a control law may be expressed as

$$\boldsymbol{\tau} = \boldsymbol{\tau}\left(\boldsymbol{q}, \dot{\boldsymbol{q}}, \ddot{\boldsymbol{q}}, \boldsymbol{q}_d, M(\boldsymbol{q}), C(\boldsymbol{q}, \dot{\boldsymbol{q}}), \boldsymbol{g}(\boldsymbol{q})\right). \tag{II.2}$$

However, for practical purposes it is desirable that the controller does not depend on the joint acceleration $\ddot{\boldsymbol{q}}$, because measurement of acceleration is unusual and accelerometers are typically highly sensitive to noise.

Figure II.1 presents the block-diagram of a robot in closed loop with a position controller.

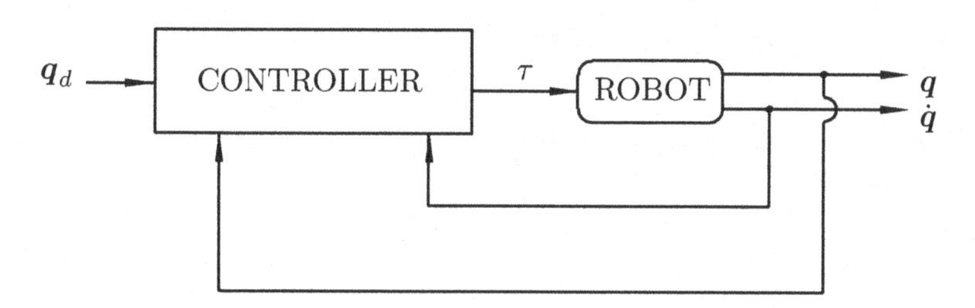

Figure II.1. Position control: closed-loop system

If the controller (II.2) does not depend explicitly on $M(\boldsymbol{q})$, $C(\boldsymbol{q}, \dot{\boldsymbol{q}})$ and $\boldsymbol{g}(\boldsymbol{q})$, it is said that the controller is not "model-based". This terminology is, however, a little misfortunate since there exist controllers, for example of the PID type (*cf.* Chapter 9), whose design parameters are computed as functions of the model of the particular robot for which the controller is designed. From this viewpoint, these controllers are model-dependent or model-based.

In this second part of the textbook we carry out stability analyses of a group of position controllers for robot manipulators. The methodology to analyze the stability may be summarized in the following steps.

1. Derivation of the closed-loop dynamic equation. This equation is obtained by replacing the control action $\boldsymbol{\tau}$ (*cf.* Equation II.2) in the dynamic model of the manipulator (*cf.* Equation II.1). In general, the closed-loop equation is a nonautonomous nonlinear ordinary differential equation.
2. Representation of the closed-loop equation in the state-space form, *i.e.*

$$\frac{d}{dt}\begin{bmatrix} \boldsymbol{q}_d - \boldsymbol{q} \\ \dot{\boldsymbol{q}} \end{bmatrix} = \boldsymbol{f}(\boldsymbol{q}, \dot{\boldsymbol{q}}, \boldsymbol{q}_d, M(\boldsymbol{q}), C(\boldsymbol{q}, \dot{\boldsymbol{q}}), \boldsymbol{g}(\boldsymbol{q})). \tag{II.3}$$

   This closed-loop equation may be regarded as a dynamic system whose inputs are $\boldsymbol{q}_d$, $\dot{\boldsymbol{q}}_d$ and $\ddot{\boldsymbol{q}}_d$, and with outputs, the state vectors $\tilde{\boldsymbol{q}} = \boldsymbol{q}_d - \boldsymbol{q}$ and $\dot{\boldsymbol{q}}$. Figure II.2 shows the corresponding block-diagram.
3. Study of the existence and possible unicity of equilibrium for the closed-loop equation. For this, we rewrite the closed-loop equation (II.3) in the state-space form choosing as the state, the position error and the velocity.

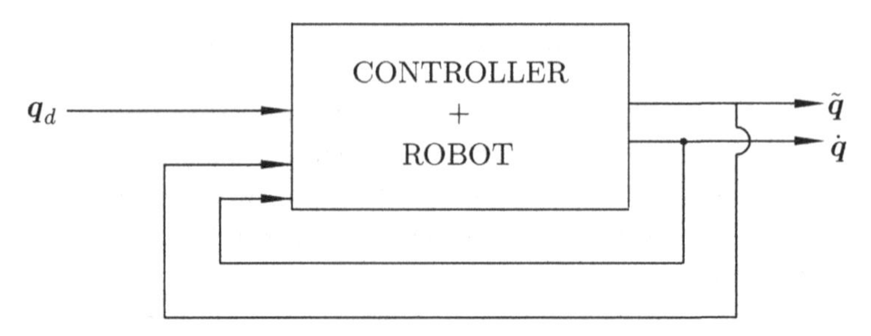

Figure II.2. Set-point control closed-loop system. Input–output representation.

That is, let $\tilde{\boldsymbol{q}} := \boldsymbol{q}_d - \boldsymbol{q}$ denote the state of the closed-loop equation. Then, (II.3) becomes

$$\frac{d}{dt}\begin{bmatrix}\tilde{\boldsymbol{q}}\\ \dot{\boldsymbol{q}}\end{bmatrix} = \tilde{\boldsymbol{f}}(\tilde{\boldsymbol{q}}, \dot{\boldsymbol{q}}) \tag{II.4}$$

where $\tilde{\boldsymbol{f}}$ is obtained by replacing $\boldsymbol{q}$ with $\boldsymbol{q}_d - \tilde{\boldsymbol{q}}$. Note that the closed-loop system equation is autonomous since $\boldsymbol{q}_d$ is constant.

Thus, for Equation (II.4) we want to verify that the origin, $[\tilde{\boldsymbol{q}}^T \, \dot{\boldsymbol{q}}^T]^T = \boldsymbol{0} \in \mathbb{R}^{\mathbf{2n}}$ is an equilibrium and whether it is unique.

4. Proposal of a Lyapunov function candidate to study the stability of the origin for the closed-loop equation, by using the Theorems 2.2, 2.3, 2.4 and 2.7. In particular, verification of the required properties, *i.e.* positivity and negativity of the time derivative.
5. Alternatively to step 4, in the case that the proposed Lyapunov function candidate appears to be inappropriate (that is, if it does not satisfy all of the required conditions) to establish the stability properties of the equilibrium under study, we may use Lemma 2.2 by proposing a positive definite function whose characteristics allow one to determine the qualitative behavior of the solutions of the closed-loop equation.

It is important to underline that if Theorems 2.2, 2.3, 2.4, 2.7 and Lemma 2.2 do not apply because one of their conditions does not hold, it does not mean that the control objective cannot be achieved with the controller under analysis but that the latter is inconclusive. In this case, one should look for other possible Lyapunov function candidates such that one of these results holds.

The rest of this second part of the textbook is divided into four chapters. The controllers that we present may be called "conventional" since they are commonly used in industrial robots. These controllers are:

- Proportional control plus velocity feedback and Proportional Derivative (PD) control;
- PD control with gravity compensation;

- PD control with desired gravity compensation;
- Proportional Integral Derivative (PID) control.

Bibliography

Among books on robotics, robot dynamics and control that include the study of tracking control systems we mention the following:

- Paul R., 1982, *"Robot manipulators: Mathematics programming and control"*, MIT Press, Cambridge, MA.
- Asada H., Slotine J. J., 1986, *"Robot analysis and control "*, Wiley, New York.
- Fu K., Gonzalez R., Lee C., 1987, *"Robotics: Control, sensing, vision and intelligence"*, McGraw–Hill.
- Craig J., 1989, *"Introduction to robotics: Mechanics and control"*, Addison-Wesley, Reading, MA.
- Spong M., Vidyasagar M., 1989, *"Robot dynamics and control"*, Wiley, New York.
- Yoshikawa T., 1990, *"Foundations of robotics: Analysis and control"*, The MIT Press.
- Spong M., Lewis F. L., Abdallah C. T., 1993, *"Robot control: Dynamics, motion planning and analysis"*, IEEE Press, New York.
- Sciavicco L., Siciliano B., 2000, *"Modeling and control of robot manipulators"*, Second Edition, Springer-Verlag, London.

Textbooks addressed to graduate students are (Sciavicco and Siciliano, 2000) and

- Lewis F. L., Abdallah C. T., Dawson D. M., 1993, *"Control of robot manipulators"*, Macmillan Pub. Co.
- Qu Z., Dawson D. M., 1996, *"Robust tracking control of robot manipulators"*, IEEE Press, New York.
- Arimoto S., 1996, *"Control theory of non–linear mechanical systems"*, Oxford University Press, New York.

More advanced monographs addressed to researchers and texts for graduate students are

- Ortega R., Loría A., Nicklasson P. J., Sira-Ramírez H., 1998, *"Passivity-based control of Euler-Lagrange Systems Mechanical, Electrical and Electromechanical Applications"*, Springer-Verlag: London, Communications and Control Engg. Series.

- Canudas C., Siciliano B., Bastin G. (Eds), 1996, *"Theory of robot control"*, Springer-Verlag: London.
- de Queiroz M., Dawson D. M., Nagarkatti S. P., Zhang F., 2000, *"Lyapunov–based control of mechanical systems"*, Birkhäuser, Boston, MA.

A particularly relevant work on robot motion control and which covers in a unified manner most of the controllers that are studied in this part of the text, is

- Wen J. T., 1990, "A unified perspective on robot control: The energy Lyapunov function approach", *International Journal of Adaptive Control and Signal Processing*, Vol. 4, pp. 487–500.

6

Proportional Control plus Velocity Feedback and PD Control

Proportional control plus velocity feedback is the simplest closed-loop controller that may be used to control robot manipulators. The conceptual application of this control strategy is common in angular position control of DC motors. In this application, the controller is also known as *proportional control with tachometric feedback*. The equation of proportional control plus velocity feedback is given by

$$\boldsymbol{\tau} = K_p\tilde{\boldsymbol{q}} - K_v\dot{\boldsymbol{q}} \tag{6.1}$$

where $K_p, K_v \in \mathbb{R}^{n\times n}$ are symmetric positive definite matrices preselected by the practitioner engineer and are commonly referred to as position gain and velocity (or derivative) gain, respectively. The vector $\boldsymbol{q}_d \in \mathbb{R}^n$ corresponds to the desired joint position, and the vector $\tilde{\boldsymbol{q}} = \boldsymbol{q}_d - \boldsymbol{q} \in \mathbb{R}^n$ is called position error. Figure 6.1 presents a block-diagram corresponding to the control system formed by the robot under proportional control plus velocity feedback.

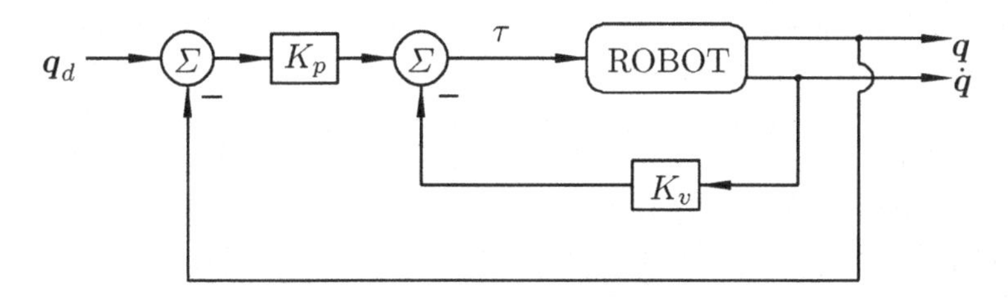

Figure 6.1. Block-diagram: Proportional control plus velocity feedback

Proportional Derivative (PD) control is an immediate extension of proportional control plus velocity feedback (6.1). As its name suggests, the control law is not only composed of a proportional term of the position error as in the case of proportional control, but also of another term which is proportional to the derivative of the position, *i.e.* to its velocity error, $\dot{\tilde{\boldsymbol{q}}}$. The PD control law is given by

$$\boldsymbol{\tau} = K_p\tilde{\boldsymbol{q}} + K_v\dot{\tilde{\boldsymbol{q}}} \tag{6.2}$$

where $K_p, K_v \in \mathbb{R}^{n\times n}$ are also symmetric positive definite and selected by the designer. In Figure 6.2 we present the block-diagram corresponding to the control system composed of a PD controller and a robot.

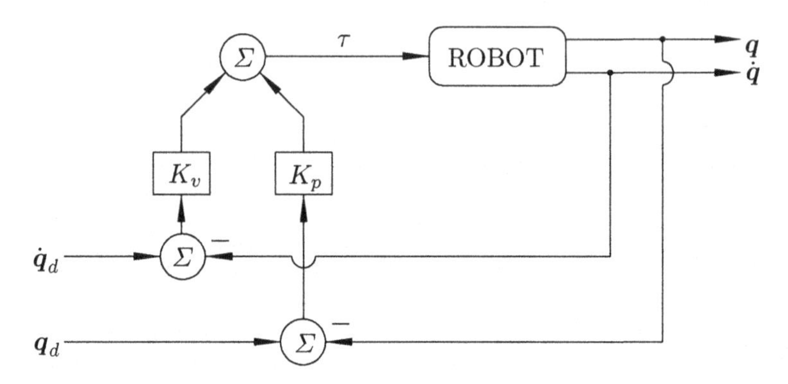

Figure 6.2. Block-diagram: PD control

So far no restriction has been imposed on the vector of desired joint positions $\boldsymbol{q}_d$ to define the proportional control law plus velocity feedback and the PD control law. This is natural, since the name that we give to a controller must characterize only its structure and should not be reference-dependent.

In spite of the veracity of the statement above, in the literature on robot control one finds that the control laws (6.1) and (6.2) are indistinctly called "PD control". The common argument in favor of this ambiguous terminology is that in the particular case when the vector of desired positions $\boldsymbol{q}_d$ is restricted to be constant, then it is clear from the definition of $\tilde{\boldsymbol{q}}$ that $\dot{\tilde{\boldsymbol{q}}} = -\dot{\boldsymbol{q}}$ and therefore, control laws (6.1) and (6.2) become identical.

With the purpose of avoiding any polemic about these observations, and to observe the use of the common nomenclature from now on, both control laws (6.1) and (6.2), are referred to in the sequel as "PD control".

In real applications, PD control is local in the sense that the torque or force determined by such a controller when applied at a particular joint, depends only on the position and velocity of the joint in question and not on those of the other joints. Mathematically, this is translated by the choice of diagonal design matrices K_p and K_v.

PD control, given by Equation (6.1), requires the measurement of positions $\boldsymbol{q}$ and velocities $\dot{\boldsymbol{q}}$ as well as specification of the desired joint position $\boldsymbol{q}_d$ (*cf.* Figure 6.1). Notice that it is not necessary to specify the desired velocity and acceleration, $\dot{\boldsymbol{q}}_d$ and $\ddot{\boldsymbol{q}}_d$.

We present next an analysis of PD control for n-DOF robot manipulators.

The behavior of an n-DOF robot in closed-loop with PD control is determined by combining the model Equation (II.1) with the control law (6.1),

$$M(\boldsymbol{q})\ddot{q} + C(\boldsymbol{q},\dot{\boldsymbol{q}})\dot{\boldsymbol{q}} + \boldsymbol{g}(\boldsymbol{q}) = K_p\tilde{\boldsymbol{q}} - K_v\dot{\boldsymbol{q}} \tag{6.3}$$

or equivalently, in terms of the state vector $\left[\tilde{\boldsymbol{q}}^T \ \dot{\tilde{\boldsymbol{q}}}^T\right]^T$

$$\frac{d}{dt}\begin{bmatrix}\tilde{\boldsymbol{q}}\\ \dot{\tilde{\boldsymbol{q}}}\end{bmatrix} = \begin{bmatrix}\dot{\tilde{\boldsymbol{q}}}\\ \ddot{\boldsymbol{q}}_d - M(\boldsymbol{q})^{-1}\left[K_p\tilde{\boldsymbol{q}} - K_v\dot{\boldsymbol{q}} - C(\boldsymbol{q},\dot{\boldsymbol{q}})\dot{\boldsymbol{q}} - \boldsymbol{g}(\boldsymbol{q})\right]\end{bmatrix}$$

which is a nonlinear nonautonomous differential equation. In the rest of this section we assume that the vector of desired joint positions, $\boldsymbol{q}_d$, is constant. Under this condition, the closed-loop equation may be rewritten in terms of the new state vector $\left[\tilde{\boldsymbol{q}}^T \ \dot{\boldsymbol{q}}^T\right]^T$, as

$$\frac{d}{dt}\begin{bmatrix}\tilde{\boldsymbol{q}}\\ \dot{\boldsymbol{q}}\end{bmatrix} = \begin{bmatrix}-\dot{\boldsymbol{q}}\\ M(\boldsymbol{q})^{-1}\left[K_p\tilde{\boldsymbol{q}} - K_v\dot{\boldsymbol{q}} - C(\boldsymbol{q},\dot{\boldsymbol{q}})\dot{\boldsymbol{q}} - \boldsymbol{g}(\boldsymbol{q})\right]\end{bmatrix}. \tag{6.4}$$

Note that the closed-loop differential equation is still nonlinear but autonomous. This is because $\boldsymbol{q}_d$ is constant. The previous equation however, may have multiple equilibria. If such is the case, they are given by $\left[\tilde{\boldsymbol{q}}^T \ \dot{\boldsymbol{q}}^T\right]^T = [\boldsymbol{s}^T \ \boldsymbol{0}^T]^T$ where $\boldsymbol{s} \in \mathbb{R}^n$ is solution of

$$K_p\boldsymbol{s} - \boldsymbol{g}(\boldsymbol{q}_d - \boldsymbol{s}) = \boldsymbol{0}\,. \tag{6.5}$$

Obviously, if the manipulator model does not include the gravitational torques term $\boldsymbol{g}(\boldsymbol{q})$, then the only equilibrium is the origin of the state space, *i.e.* $[\tilde{\boldsymbol{q}}^T \ \dot{\boldsymbol{q}}^T]^T = 0 \in \mathbb{R}^{2n}$. Also, if $\boldsymbol{g}(\boldsymbol{q})$ is independent of $\boldsymbol{q}$, *i.e.* if $\boldsymbol{g}(\boldsymbol{q}) = \boldsymbol{g}$ constant, then $\boldsymbol{s} = K_p^{-1}\boldsymbol{g}$ is the only solution.

Notice that Equation (6.5) is in general nonlinear in $\boldsymbol{s}$ due to the gravitational term $\boldsymbol{g}(\boldsymbol{q}_d - \boldsymbol{s})$. For this reason, and given the nonlinear nature of $\boldsymbol{g}(\boldsymbol{q}_d - \boldsymbol{s})$, derivation of the explicit solutions of $\boldsymbol{s}$ is in general relatively complex.

In the future sections we treat separately the cases in which the robot model contains and does not contain the vector of gravitational torques $\boldsymbol{g}(\boldsymbol{q})$.

6.1 Robots without Gravity Term

In this section we consider robots whose dynamic model does not contain the gravitational $\boldsymbol{g}(\boldsymbol{q})$, that is

$$M(\boldsymbol{q})\ddot{\boldsymbol{q}} + C(\boldsymbol{q}, \dot{\boldsymbol{q}})\dot{\boldsymbol{q}} = \boldsymbol{\tau}.$$

Robots that are described by this model are those which move only on the horizontal plane, as well as those which are mechanically designed in a specific convenient way.

Assuming that the desired joint position $\boldsymbol{q}_d$ is constant, the closed-loop Equation (6.4) becomes (with $\boldsymbol{g}(\boldsymbol{q}) = \boldsymbol{0}$),

$$\frac{d}{dt}\begin{bmatrix} \tilde{\boldsymbol{q}} \\ \dot{\boldsymbol{q}} \end{bmatrix} = \begin{bmatrix} -\dot{\boldsymbol{q}} \\ M(\boldsymbol{q}_d - \tilde{\boldsymbol{q}})^{-1}\left[K_p\tilde{\boldsymbol{q}} - K_v\dot{\boldsymbol{q}} - C(\boldsymbol{q}_d - \tilde{\boldsymbol{q}}, \dot{\boldsymbol{q}})\dot{\boldsymbol{q}}\right] \end{bmatrix} \tag{6.6}$$

which, since $\boldsymbol{q}_d$ is constant, represents an autonomous differential equation. Moreover, the origin $\begin{bmatrix}\tilde{\boldsymbol{q}}^T & \dot{\boldsymbol{q}}^T\end{bmatrix}^T = \boldsymbol{0}$ is the only equilibrium of this equation.

To study the stability of the equilibrium we appeal to Lyapunov's direct method, to which the reader has already been introduced in Section 2.3.4 of Chapter 2. Specifically, we use La Salle's Theorem 2.7 to show asymptotic stability of the equilibrium (origin).

Consider the following Lyapunov function candidate

$$\begin{aligned} V(\tilde{\boldsymbol{q}}, \dot{\boldsymbol{q}}) &= \frac{1}{2}\begin{bmatrix} \tilde{\boldsymbol{q}} \\ \dot{\boldsymbol{q}} \end{bmatrix}^T \begin{bmatrix} K_p & 0 \\ 0 & M(\boldsymbol{q}_d - \tilde{\boldsymbol{q}}) \end{bmatrix} \begin{bmatrix} \tilde{\boldsymbol{q}} \\ \dot{\boldsymbol{q}} \end{bmatrix} \\ &= \frac{1}{2}\dot{\boldsymbol{q}}^T M(\boldsymbol{q})\dot{\boldsymbol{q}} + \frac{1}{2}\tilde{\boldsymbol{q}}^T K_p \tilde{\boldsymbol{q}}. \end{aligned}$$

Notice that this function is positive definite since $M(\boldsymbol{q})$ as well as K_p are positive definite matrices.

The total derivative of $V(\tilde{\boldsymbol{q}}, \dot{\boldsymbol{q}})$ yields

$$\dot{V}(\tilde{\boldsymbol{q}}, \dot{\boldsymbol{q}}) = \dot{\boldsymbol{q}}^T M(\boldsymbol{q})\ddot{\boldsymbol{q}} + \frac{1}{2}\dot{\boldsymbol{q}}^T \dot{M}(q)\dot{\boldsymbol{q}} + \tilde{\boldsymbol{q}}^T K_p \dot{\tilde{\boldsymbol{q}}}.$$

Substituting $M(\boldsymbol{q})\ddot{\boldsymbol{q}}$ from the closed-loop Equation (6.6), we obtain

$$\begin{aligned} \dot{V}(\tilde{\boldsymbol{q}}, \dot{\boldsymbol{q}}) &= -\dot{\boldsymbol{q}}^T K_v \dot{\boldsymbol{q}} \\ &= -\begin{bmatrix} \tilde{\boldsymbol{q}} \\ \dot{\boldsymbol{q}} \end{bmatrix}^T \begin{bmatrix} 0 & 0 \\ 0 & K_v \end{bmatrix} \begin{bmatrix} \tilde{\boldsymbol{q}} \\ \dot{\boldsymbol{q}} \end{bmatrix} \leq 0, \end{aligned}$$

where we canceled the term $\dot{\boldsymbol{q}}^T\left[\frac{1}{2}\dot{M} - C\right]\dot{\boldsymbol{q}}$ by virtue of Property 4.2.7 and we used the fact that $\dot{\tilde{\boldsymbol{q}}} = -\dot{\boldsymbol{q}}$ since $\boldsymbol{q}_d$ is a constant vector.

From this and the fact that $\dot{V}(\tilde{\boldsymbol{q}}, \dot{\boldsymbol{q}}) \leq 0$ we conclude that the function $V(\tilde{\boldsymbol{q}}, \dot{\boldsymbol{q}})$ is a Lyapunov function. From Theorem 2.3 we also conclude that the origin is stable and, moreover, that the solutions $\tilde{\boldsymbol{q}}(t)$ and $\dot{\boldsymbol{q}}(t)$ are bounded.

Since the closed-loop Equation (6.6) is autonomous, we may try to apply La Salle's theorem (Theorem 2.7) to analyze the global asymptotic stability of the origin.

To that end, notice that here the set Ω is given by

$$\begin{aligned}\Omega &= \left\{\boldsymbol{x} \in \mathbb{R}^{2n} : \dot{V}(\boldsymbol{x}) = 0\right\} \\ &= \left\{\boldsymbol{x} = \begin{bmatrix}\tilde{\boldsymbol{q}} \\ \dot{\boldsymbol{q}}\end{bmatrix} \in \mathbb{R}^{2n} : \dot{V}(\tilde{\boldsymbol{q}}, \dot{\boldsymbol{q}}) = 0\right\} \\ &= \{\tilde{\boldsymbol{q}} \in \mathbb{R}^n, \dot{\boldsymbol{q}} = \boldsymbol{0} \in \mathbb{R}^n\} .\end{aligned}$$

Observe also that $\dot{V}(\tilde{\boldsymbol{q}}, \dot{\boldsymbol{q}}) = 0$ if and only if $\dot{\boldsymbol{q}} = \boldsymbol{0}$. For a solution $\boldsymbol{x}(t)$ to belong to Ω for all $t \geq 0$, it is necessary and sufficient that $\dot{\boldsymbol{q}}(t) = \boldsymbol{0}$ for all $t \geq 0$. Therefore, it must also hold that $\ddot{\boldsymbol{q}}(t) = \boldsymbol{0}$ for all $t \geq 0$. Considering all this, we conclude from the closed-loop equation (6.6), that if $\boldsymbol{x}(t) \in \Omega$ for all $t \geq 0$ then,

$$\boldsymbol{0} = M(\boldsymbol{q}_d - \tilde{\boldsymbol{q}}(t))^{-1} K_p \tilde{\boldsymbol{q}}(t) .$$

Since $M(\boldsymbol{q}_d - \tilde{\boldsymbol{q}}(t))^{-1}$ and K_p are positive definite their matrix product is nonsingular[1], this implies that $\tilde{\boldsymbol{q}}(t) = \boldsymbol{0}$ for all $t \geq 0$ and therefore, $\begin{bmatrix}\tilde{\boldsymbol{q}}(0)^T & \dot{\boldsymbol{q}}(0)^T\end{bmatrix}^T = \boldsymbol{0} \in \mathbb{R}^{2n}$ is the only initial condition in Ω for which $\boldsymbol{x}(t) \in \Omega$ for all $t \geq 0$. Thus, from La Salle's theorem (Theorem 2.7), this is enough to establish global asymptotic stability of the origin, $\begin{bmatrix}\tilde{\boldsymbol{q}}^T & \dot{\boldsymbol{q}}^T\end{bmatrix}^T = \boldsymbol{0} \in \mathbb{R}^{2n}$ and consequently,

$$\begin{aligned}\lim_{t \to \infty} \tilde{\boldsymbol{q}}(t) &= \lim_{t \to \infty} [\, \boldsymbol{q}_d - \boldsymbol{q}(t) \,] = \boldsymbol{0} \\ \lim_{t \to \infty} \dot{\boldsymbol{q}}(t) &= \boldsymbol{0} .\end{aligned}$$

In other words the position control objective is achieved.

It is interesting to emphasize at this point, that the closed-loop equation (6.6) is exactly the same as the one which will be derived for the so-called PD controller with gravity compensation and which we study in Chapter 7. In that chapter we present an alternative analysis for the asymptotic stability of the origin, by use of another Lyapunov function which does not appeal to La Salle's theorem. Certainly, this alternative analysis is also valid for the study of (6.6).

---

[1] Note that we are not claiming that the matrix product $M(\boldsymbol{q}_d - \tilde{\boldsymbol{q}}(t))^{-1} K_p$ is positive definite. This is not true in general. We are only using the fact that this matrix product is nonsingular.

6.2 Robots with Gravity Term

The behavior of the control system under PD control (*cf.* Equation 6.1) for robots whose models include explicitly the vector of gravitational torques $\boldsymbol{g}(\boldsymbol{q})$ and assuming that $\boldsymbol{q}_d$ is constant, is determined by (6.4), which we repeat below, *i.e.*

$$\frac{d}{dt}\begin{bmatrix}\tilde{\boldsymbol{q}}\\ \dot{\boldsymbol{q}}\end{bmatrix} = \begin{bmatrix}-\dot{\boldsymbol{q}}\\ M(\boldsymbol{q})^{-1}\left[K_p\tilde{\boldsymbol{q}} - K_v\dot{\boldsymbol{q}} - C(\boldsymbol{q},\dot{\boldsymbol{q}})\dot{\boldsymbol{q}} - \boldsymbol{g}(\boldsymbol{q})\right]\end{bmatrix}. \tag{6.7}$$

The study of this equation is somewhat more complex than that for the case when $\boldsymbol{g}(\boldsymbol{q}) = \boldsymbol{0}$.

In this section we analyze closed-loop Equation (6.7), and specifically, we address the following issues:

- unicity of the equilibrium;
- boundedness of solutions.

The study of this section is limited to robots having only revolute joints.

6.2.1 Unicity of the Equilibrium

In general, system (6.7) may have several equilibrium points. This is illustrated by the following example.

Example 6.1. Consider the model of an ideal pendulum, such as the one studied in Example 2.2 (*cf.* page 30)

$$J\ddot{q} + mgl\,\sin(q) = \tau\,.$$

In this case the expression (6.5) takes the form

$$k_p s - mgl\,\sin(q_d - s) = 0\,. \tag{6.8}$$

For the sake of illustration consider the following numerical values

$$J = 1 \qquad mgl = 1$$
$$k_p = 0.25\; q_d = \pi/2.$$

Either by a graphical method or using numerical algorithms, it may be verified that Equation (6.8) has exactly three solutions in s whose approximate values are: 1.25 (rad), −2.13 (rad) and −3.59 (rad). This means that the closed-loop system under PD control for the ideal pendulum, has the equilibria

$$\begin{bmatrix} \tilde{q} \\ \dot{q} \end{bmatrix} \in \left\{ \begin{bmatrix} 1.25 \\ 0 \end{bmatrix}, \begin{bmatrix} -2.13 \\ 0 \end{bmatrix}, \begin{bmatrix} -3.56 \\ 0 \end{bmatrix} \right\}.$$

Multiplicity of equilibria certainly poses a problem for the study of (global) asymptotic stability; hence, it is desirable to avoid such a situation. For the case of robots having only revolute joints we show below that, by choosing K_p sufficiently *large*, one may guarantee unicity of the equilibrium of the closed-loop Equation (6.7). To that end, we use the *contraction mapping theorem* presented in this textbook as Theorem 2.1.

The equilibria of the closed-loop Equation (6.7) satisfy

$$\left[\tilde{\boldsymbol{q}}^T \ \dot{\boldsymbol{q}}^T\right]^T = \left[\boldsymbol{s}^T \ \boldsymbol{0}^T\right]^T,$$

where $\boldsymbol{s} \in \mathrm{IR}^n$ is solution of

$$\begin{aligned} \boldsymbol{s} &= K_p^{-1}\boldsymbol{g}(\boldsymbol{q}_d - \boldsymbol{s}) \\ &= \boldsymbol{f}(\boldsymbol{s}, \boldsymbol{q}_d). \end{aligned}$$

If the function $\boldsymbol{f}(\boldsymbol{s}, \boldsymbol{q}_d)$ satisfies the condition of the contraction mapping theorem (Theorem 2.1) then the equation $\boldsymbol{s} = \boldsymbol{f}(\boldsymbol{s}, \boldsymbol{q}_d)$ has a unique solution $\boldsymbol{s}^*$ and consequently, the unique equilibrium of the closed-loop Equation (6.7) is $\left[\tilde{\boldsymbol{q}}^T \ \dot{\boldsymbol{q}}^T\right]^T = \left[\boldsymbol{s}^{*^T} \ \boldsymbol{0}^T\right]^T$.

Now, notice that for all vectors $\boldsymbol{x}, \boldsymbol{y}, \boldsymbol{q}_d \in \mathrm{IR}^n$,

$$\begin{aligned} \|\boldsymbol{f}(\boldsymbol{x}, \boldsymbol{q}_d) - \boldsymbol{f}(\boldsymbol{y}, \boldsymbol{q}_d)\| &= \left\|K_p^{-1}\boldsymbol{g}(\boldsymbol{q}_d - \boldsymbol{x}) - K_p^{-1}\boldsymbol{g}(\boldsymbol{q}_d - \boldsymbol{y})\right\| \\ &= \left\|K_p^{-1}\left\{\boldsymbol{g}(\boldsymbol{q}_d - \boldsymbol{x}) - \boldsymbol{g}(\boldsymbol{q}_d - \boldsymbol{y})\right\}\right\| \\ &\le \lambda_{\mathrm{Max}}\{K_p^{-1}\} \left\|\boldsymbol{g}(\boldsymbol{q}_d - \boldsymbol{x}) - \boldsymbol{g}(\boldsymbol{q}_d - \boldsymbol{y})\right\|. \end{aligned}$$

On the other hand, using the fact that $\lambda_{\mathrm{Max}}\{A^{-1}\} = 1/\lambda_{\min}\{A\}$ for any symmetric positive definite matrix A, and Property 4.3.3 that guarantees the existence of a positive constant k_g such that $\|\boldsymbol{g}(\boldsymbol{x}) - \boldsymbol{g}(\boldsymbol{y})\| \le k_g \|\boldsymbol{x} - \boldsymbol{y}\|$, we get

$$\|\boldsymbol{f}(\boldsymbol{x}, \boldsymbol{q}_d) - \boldsymbol{f}(\boldsymbol{y}, \boldsymbol{q}_d)\| \le \frac{k_g}{\lambda_{\min}\{K_p\}} \|\boldsymbol{x} - \boldsymbol{y}\|$$

hence, invoking the contraction mapping theorem, a sufficient condition for the unicity of the solution of $\boldsymbol{f}(\boldsymbol{s}, \boldsymbol{q}_d) - \boldsymbol{s} = K_p^{-1}\boldsymbol{g}(\boldsymbol{q}_d - \boldsymbol{s}) - \boldsymbol{s} = \boldsymbol{0}$ and consequently, for the unicity of the equilibrium of the closed-loop equation, is that K_p be selected to satisfy $\lambda_{\min}\{K_p\} > k_g$.

6.2.2 Arbitrarily Bounded Position and Velocity Error

We present next a qualitative study of the behavior of solutions of the closed-loop Equation (6.7) for the case where K_p is not restricted to satisfy $\lambda_{\min}\{K_p\} > k_g$, but it is enough that K_p be positive definite.

For the purposes of the result presented here we make use of Lemma 2.2, which, even though it does not establish any stability statement, enables one to make conclusions about the boundedness of trajectories and eventually about the convergence of some of them to zero. We assume that all joints are revolute.

Define the following non-negative function

$$V(\tilde{\boldsymbol{q}}, \dot{\boldsymbol{q}}) = \mathcal{K}(\boldsymbol{q}, \dot{\boldsymbol{q}}) + \mathcal{U}(\boldsymbol{q}) - k_U + \frac{1}{2}\tilde{\boldsymbol{q}}^T K_p \tilde{\boldsymbol{q}}$$

where $\mathcal{K}(\boldsymbol{q}, \dot{\boldsymbol{q}})$ and $\mathcal{U}(\boldsymbol{q})$ denote the kinetic and potential energy functions of the robot, and the constant k_U is defined as (*cf.* Property 4.3)

$$k_U = \min_{\boldsymbol{q}}\{\mathcal{U}(\boldsymbol{q})\}\,.$$

The function $V(\tilde{\boldsymbol{q}}, \dot{\boldsymbol{q}})$ may be expressed in the form

$$V(\tilde{\boldsymbol{q}}, \dot{\boldsymbol{q}}) = \begin{bmatrix} \tilde{\boldsymbol{q}} \\ \dot{\boldsymbol{q}} \end{bmatrix}^T \overbrace{\begin{bmatrix} \frac{1}{2}K_p & 0 \\ 0 & \frac{1}{2}M(\boldsymbol{q}_d - \tilde{\boldsymbol{q}}) \end{bmatrix}}^{P} \begin{bmatrix} \tilde{\boldsymbol{q}} \\ \dot{\boldsymbol{q}} \end{bmatrix} + \underbrace{\mathcal{U}(\boldsymbol{q}_d - \tilde{\boldsymbol{q}}) - k_U}_{h} \geq 0. \tag{6.9}$$

or equivalently, as

$$V(\tilde{\boldsymbol{q}}, \dot{\boldsymbol{q}}) = \frac{1}{2}\dot{\boldsymbol{q}}^T M(\boldsymbol{q})\dot{\boldsymbol{q}} + \frac{1}{2}\tilde{\boldsymbol{q}}^T K_p \tilde{\boldsymbol{q}} + \mathcal{U}(\boldsymbol{q}) - k_U \geq 0\,.$$

The derivative of $V(\tilde{\boldsymbol{q}}, \dot{\boldsymbol{q}})$ with respect to time yields

$$\dot{V}(\tilde{\boldsymbol{q}}, \dot{\boldsymbol{q}}) = \dot{\boldsymbol{q}}^T M(\boldsymbol{q})\ddot{\boldsymbol{q}} + \frac{1}{2}\dot{\boldsymbol{q}}^T \dot{M}(\boldsymbol{q})\dot{\boldsymbol{q}} + \tilde{\boldsymbol{q}}^T K_p \dot{\tilde{\boldsymbol{q}}} + \dot{\boldsymbol{q}}^T \boldsymbol{g}(\boldsymbol{q}) \tag{6.10}$$

where we used (3.20), *i.e.* $\boldsymbol{g}(\boldsymbol{q}) = \frac{\partial}{\partial \boldsymbol{q}}\mathcal{U}(\boldsymbol{q})$. Factoring out $M(\boldsymbol{q})\ddot{\boldsymbol{q}}$ from the closed-loop equation (6.3) and substituting in (6.10),

$$\dot{V}(\tilde{\boldsymbol{q}}, \dot{\boldsymbol{q}}) = \dot{\boldsymbol{q}}^T K_p \tilde{\boldsymbol{q}} - \dot{\boldsymbol{q}}^T K_v \dot{\boldsymbol{q}} + \tilde{\boldsymbol{q}}^T K_p \dot{\tilde{\boldsymbol{q}}}\,, \tag{6.11}$$

where the term $\dot{\boldsymbol{q}}^T\left[\frac{1}{2}\dot{M} - C\right]\dot{\boldsymbol{q}}$ has been canceled by virtue of the Property 4.2. Recalling that the vector $\boldsymbol{q}_d$ is constant and that $\tilde{\boldsymbol{q}} = \boldsymbol{q}_d - \boldsymbol{q}$, then $\dot{\tilde{\boldsymbol{q}}} = -\dot{\boldsymbol{q}}$. Taking this into account Equation (6.11) boils down to

$$\begin{aligned}\dot{V}(\tilde{\boldsymbol{q}},\dot{\boldsymbol{q}}) &= -\dot{\boldsymbol{q}}^T \overbrace{K_v}^{Q} \dot{\boldsymbol{q}} \\ &= -\begin{bmatrix}\tilde{\boldsymbol{q}}\\ \dot{\boldsymbol{q}}\end{bmatrix}^T \begin{bmatrix}0 & 0\\ 0 & K_v\end{bmatrix} \begin{bmatrix}\tilde{\boldsymbol{q}}\\ \dot{\boldsymbol{q}}\end{bmatrix} \leq 0 .\end{aligned} \tag{6.12}$$

Using $V(\tilde{\boldsymbol{q}},\dot{\boldsymbol{q}})$ and $\dot{V}(\tilde{\boldsymbol{q}},\dot{\boldsymbol{q}})$ given in (6.9) and (6.12) respectively and invoking Lemma 2.2, we conclude that $\dot{\boldsymbol{q}}(t)$ and $\tilde{\boldsymbol{q}}(t)$ are bounded for all t and moreover, the velocities vector is square integrable, that is

$$\int_0^\infty \|\dot{\boldsymbol{q}}(t)\|^2 dt < \infty . \tag{6.13}$$

Moreover, as we show next, we can determine the explicit bounds for the position and velocity errors, $\tilde{\boldsymbol{q}}$ and $\dot{\boldsymbol{q}}$. Considering that $V(\tilde{\boldsymbol{q}},\dot{\boldsymbol{q}})$ is a non-negative function and non-increasing along the trajectories ($\dot{V}(\tilde{\boldsymbol{q}},\dot{\boldsymbol{q}}) \leq 0$), we have

$$0 \leq V(\tilde{\boldsymbol{q}}(t),\dot{\boldsymbol{q}}(t)) \leq V(\tilde{\boldsymbol{q}}(0),\dot{\boldsymbol{q}}(0))$$

for all $t \geq 0$. Consequently, considering the definition of $V(\tilde{\boldsymbol{q}},\dot{\boldsymbol{q}})$ it readily follows that

$$\begin{aligned}\frac{1}{2}\tilde{\boldsymbol{q}}(t)^T K_p \tilde{\boldsymbol{q}}(t) &\leq V(\tilde{\boldsymbol{q}}(0),\dot{\boldsymbol{q}}(0)) \\ \frac{1}{2}\dot{\boldsymbol{q}}(t)^T M(\boldsymbol{q}(t))\dot{\boldsymbol{q}}(t) &\leq V(\tilde{\boldsymbol{q}}(0),\dot{\boldsymbol{q}}(0))\end{aligned}$$

for all $t \geq 0$, from which we finally conclude that the following bounds:

$$\begin{aligned}\|\tilde{\boldsymbol{q}}(t)\|^2 &\leq \frac{2V(\tilde{\boldsymbol{q}}(0),\dot{\boldsymbol{q}}(0))}{\lambda_{\min}\{K_p\}} \\ &= \frac{\dot{\boldsymbol{q}}(0)^T M(\boldsymbol{q}(0))\dot{\boldsymbol{q}}(0) + \tilde{\boldsymbol{q}}(0)^T K_p \tilde{\boldsymbol{q}}(0) + 2\mathcal{U}(\boldsymbol{q}(0)) - 2k_{\mathcal{U}}}{\lambda_{\min}\{K_p\}}\end{aligned} \tag{6.14}$$

$$\begin{aligned}\|\dot{\boldsymbol{q}}(t)\|^2 &\leq \frac{2V(\tilde{\boldsymbol{q}}(0),\dot{\boldsymbol{q}}(0))}{\lambda_{\min}\{M(\boldsymbol{q})\}} \\ &= \frac{\dot{\boldsymbol{q}}(0)^T M(\boldsymbol{q}(0))\dot{\boldsymbol{q}}(0) + \tilde{\boldsymbol{q}}(0)^T K_p \tilde{\boldsymbol{q}}(0) + 2\mathcal{U}(\boldsymbol{q}(0)) - 2k_{\mathcal{U}}}{\lambda_{\min}\{M(\boldsymbol{q})\}}\end{aligned} \tag{6.15}$$

hold for all $t \geq 0$.

We can also show that actually $\lim_{t\to\infty} \dot{\boldsymbol{q}}(t) = \mathbf{0}$. To that end, we use (6.3) to obtain

$$\ddot{\boldsymbol{q}} = M(\boldsymbol{q})^{-1}\left[K_p\tilde{\boldsymbol{q}} - K_v\dot{\boldsymbol{q}} - C(\boldsymbol{q},\dot{\boldsymbol{q}})\dot{\boldsymbol{q}} - \boldsymbol{g}(\boldsymbol{q})\right] . \tag{6.16}$$

Since $\dot{\boldsymbol{q}}(t)$ and $\tilde{\boldsymbol{q}}(t)$ are bounded functions, then $C(\boldsymbol{q}, \dot{\boldsymbol{q}})\dot{\boldsymbol{q}}$ and $\boldsymbol{g}(\boldsymbol{q})$ are also bounded, this in view of Properties 4.2 and 4.3. On the other hand, since $M(\boldsymbol{q})^{-1}$ is bounded (from Property 4.1), we conclude from (6.16) that $\ddot{\boldsymbol{q}}(t)$ is also bounded. This, and (6.13) imply in turn that (by Lemma 2.2),

$$\lim_{t\to\infty} \dot{\boldsymbol{q}}(t) = \boldsymbol{0}\,.$$

Nevertheless, it is important to underline that the limit above does not guarantee that $\boldsymbol{q}(t) \to \boldsymbol{q}_d$ as $t \to \infty$ and as a matter of fact, not even that[2] $\boldsymbol{q}(t) \to$ constant as $t \to \infty$.

Example 6.2. Consider again the ideal pendulum from Example 6.1

$$J\ddot{q} + mgl\ \sin(q) = \tau,$$

where we clearly identify $M(q) = J$ and $g(q) = mgl\ \sin(q)$. As was shown in Example 2.2 (*cf.* page 30), the potential energy function is

$$\mathcal{U}(q) = mgl[1 - \cos(q)]\,.$$

Since $\min_q\{\mathcal{U}(q)\} = 0$ the constant k_U is zero.

Consider next the numerical values from Example 6.1

$$\begin{array}{ll} J = 1 & mgl = 1 \\ k_p = 0.25 & k_v = 0.50 \\ q_d = \pi/2\,. & \end{array}$$

Assume that we apply the PD controller to drive the ideal pendulum from the initial conditions $q(0) = 0$ and $\dot{q}(0) = 0$.

According to the bounds (6.14) and (6.15) and considering the information above, we get

$$\tilde{q}^2(t) \le \tilde{q}^2(0) = 2.46\ \ \text{rad}^2 \tag{6.17}$$

$$\dot{q}^2(t) \le \frac{k_p}{J}\tilde{q}^2(0) = 0.61\ \ \left(\frac{\text{rad}}{\text{s}}\right)^2 \tag{6.18}$$

for all $t \ge 0$. Figures 6.3 and 6.4 show graphs of $\tilde{q}(t)^2$ and $\dot{q}(t)^2$ respectively, obtained in simulations. One can clearly see from these plots that both variables satisfy the inequalities (6.17) and (6.18). Finally, it is interesting to observe from these plots that $\lim_{t\to\infty} \tilde{q}^2(t) = 1.56$ and $\lim_{t\to\infty} \dot{q}^2(t) = 0$ and therefore,

$$\lim_{t\to\infty} \begin{bmatrix} \tilde{q}(t) \\ \dot{q}(t) \end{bmatrix} = \begin{bmatrix} 1.25 \\ 0 \end{bmatrix}.$$

That is, the solutions tend to one of the three equilibria determined in Example 6.1. ♢

---

[2] Counter example: For $x(t) = \ln(t+1)$ we have $\lim_{t\to\infty} \dot{x}(t) = 0$; however, $\lim_{t\to\infty} x(t) = \infty$!

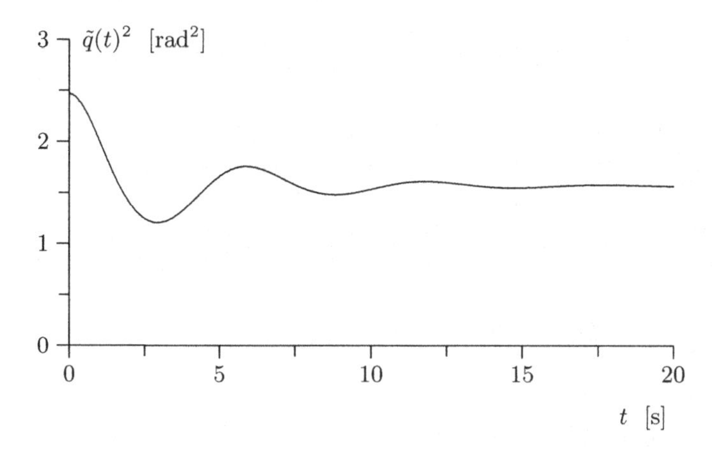

Figure 6.3. Graph of $\tilde{q}(t)^2$

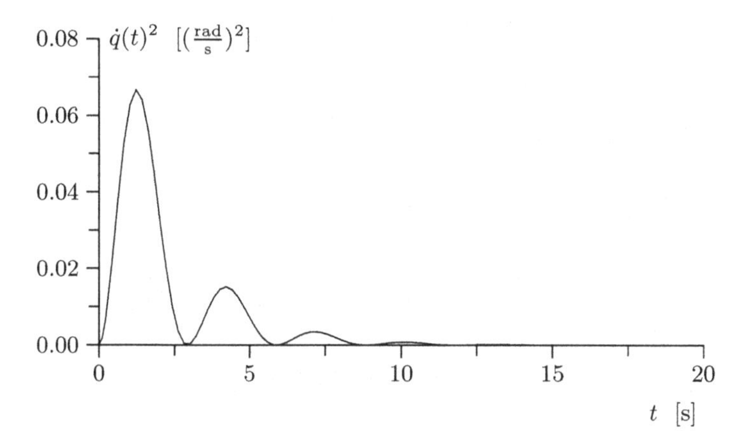

Figure 6.4. Graph of $\dot{q}(t)^2$

To close this section we present next the results we have obtained in experiments with the Pelican prototype under PD control.

Example 6.3. Consider the 2-DOF *prototype robot* studied in Chapter 5. For ease of reference, we rewrite below the vector of gravitational torques $\boldsymbol{g}(\boldsymbol{q})$ from Section 5.3.2, and its elements are

$$
\begin{aligned}
g_1(\boldsymbol{q}) &= (m_1 l_{c1} + m_2 l_1) g \sin(q_1) + m_2 g l_{c2} \sin(q_1 + q_2) \\
g_2(\boldsymbol{q}) &= m_2 g l_{c2} \sin(q_1 + q_2).
\end{aligned}
$$

The control objective consists in making

$$\lim_{t\to\infty} \boldsymbol{q}(t) = \boldsymbol{q}_d = \begin{bmatrix} \pi/10 \\ \pi/30 \end{bmatrix} \quad \text{[rad]}.$$

It may easily be verified that $\boldsymbol{g}(\boldsymbol{q}_d) \neq \boldsymbol{0}$. Therefore, the origin $\begin{bmatrix} \tilde{\boldsymbol{q}}^T & \dot{\boldsymbol{q}}^T \end{bmatrix}^T = \boldsymbol{0} \in \mathbb{R}^4$ of the closed-loop equation with the PD controller, is not an equilibrium. This means that the control objective cannot be achieved using PD control. However, with the purpose of illustrating the behavior of the system we present next some experimental results.

Consider the PD controller

$$\boldsymbol{\tau} = K_p\tilde{\boldsymbol{q}} - K_v\dot{\boldsymbol{q}}$$

with the following numerical values

$$K_p = \begin{bmatrix} 30 & 0 \\ 0 & 30 \end{bmatrix} \text{[Nm/rad]}, \qquad K_v = \begin{bmatrix} 7 & 0 \\ 0 & 3 \end{bmatrix} \text{[Nms/rad]}.$$

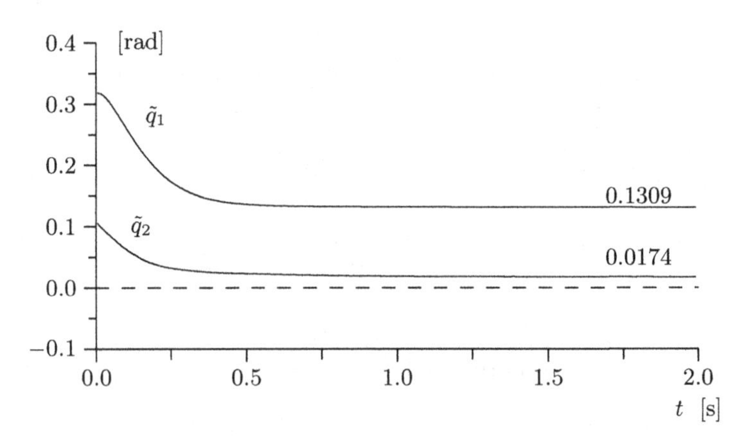

Figure 6.5. Graph of the position errors $\tilde{q}_1$ and $\tilde{q}_2$

The initial conditions are fixed at $\boldsymbol{q}(0) = \boldsymbol{0}$ and $\dot{\boldsymbol{q}}(0) = \boldsymbol{0}$. The experimental results are presented in Figure 6.5 where we show the two components of the position error, $\tilde{\boldsymbol{q}}$. One may appreciate that $\lim_{t\to\infty} \tilde{q}_1(t) = 0.1309$ and $\lim_{t\to\infty} \tilde{q}_2(t) = 0.0174$ therefore, as was expected, the control objective is not achieved. Friction at the joints may also affect the resulting position error. ♢

6.3 Conclusions

We may summarize what we have learned in this chapter, in the following ideas. Consider the PD controller of n-DOF robots. Assume that the vector of desired positions $\boldsymbol{q}_d$ is constant.

- If the vector of gravitational torques $\boldsymbol{g}(\boldsymbol{q})$ is absent in the robot model, then the origin of the closed-loop equation, expressed in terms of the state vector $\left[\tilde{\boldsymbol{q}}^T \ \dot{\boldsymbol{q}}^T\right]^T$, is globally asymptotically stable. Consequently, we have $\lim_{t\to\infty} \tilde{\boldsymbol{q}}(t) = \boldsymbol{0}$.
- For robots with only revolute joints, if the vector of gravitational torques $\boldsymbol{g}(\boldsymbol{q})$ is present in the robot model, then the origin of the closed-loop equation expressed in terms of the state vector $\left[\tilde{\boldsymbol{q}}^T \ \dot{\boldsymbol{q}}^T\right]^T$, is not necessarily an equilibrium. However, the closed-loop equation always has equilibria. In addition, if $\lambda_{\min}\{K_p\} > k_g$, then the closed-loop equation has a unique equilibrium. Finally, for any matrix $K_p = K_p^T > 0$, it is guaranteed that the position and velocity errors, $\tilde{\boldsymbol{q}}$ and $\dot{\boldsymbol{q}}$, are bounded. Moreover, the vector of joint velocities $\dot{\boldsymbol{q}}$ goes asymptotically to zero.

Bibliography

The analysis of global asymptotic stability of PD control for robots without the gravitational term (*i.e.* with $\boldsymbol{g}(\boldsymbol{q}) \equiv \boldsymbol{0}$), is identical to PD control with compensation of gravity and which was originally presented in

- Takegaki M., Arimoto S., 1981, *"A new feedback method for dynamic control of manipulators"*, Transactions ASME, Journal of Dynamic Systems, Measurement and Control, Vol. 105, p. 119–125.

Also, the same analysis for the PD control of robots without the gravitational term may be consulted in the texts

- Spong M., Vidyasagar M., 1989, *"Robot dynamics and control"*, John Wiley and Sons.
- Yoshikawa T., 1990, *"Foundations of robotics: Analysis and control"*, The MIT Press.

Problems

1. Consider the model of the ideal pendulum studied in Example 6.1

$$J\ddot{q} + mgl\,\sin(q) = \tau$$

with the numerical values

$$J = 1, \qquad mgl = 1, \qquad q_d = \pi/2$$

and under PD control. In Example 6.1 we established that the closed-loop equation possesses three equilibria for $k_p = 0.25$.

a) Determine the value of the constant k_g (*cf.* Property 4.3).
b) Determine a value of k_p for which there exists a unique equilibrium.
c) Use the value of k_p from the previous item and the contraction mapping theorem (Theorem 2.1) to obtain an approximate numerical value of the unique equilibrium.
Hint: The equilibrium is $[\tilde{q} \;\; \dot{q}]^T = [x^* \;\; 0]^T$, where $x^* = \lim_{n\to\infty} x(n)$ with
$$x(n) = \frac{mgl}{k_p}\sin(q_d - x(n-1))$$
and, for instance $x(-1) = 0$.

2. Consider the model of the ideal pendulum studied in the Example 6.1

$$J\ddot{q} + mgl\,\sin(q) = \tau$$

with the following numerical values,

$$J = 1, \qquad mgl = 1, \qquad q_d = \pi/2\,.$$

Consider the PD control with initial conditions $q(0) = 0$ and $\dot{q}(0) = 0$. From this, we have $\tilde{q}(0) = \pi/2$.

a) Obtain k_p which guarantees that
$$|\dot{q}(t)| \le c_1 \qquad \forall\, t \ge 0$$
where $c_1 > 0$. Compute a numerical value for k_p with $c_1 = 1$.
Hint: Use (6.15).

3. Consider the PD control of the 2-DOF robot studied in Example 6.3. The experimental results in this example were obtained with $K_p = \text{diag}\,\{30\}$ and the following numerical values

$$\begin{array}{lll} l_1 = 0.26 & l_{c1} = 0.0983 & l_{c2} = 0.0229 \\ m_1 = 6.5225 & m_2 = 2.0458 & g = 9.81 \\ q_{d1} = \pi/10 & q_{d2} = \pi/30 & \end{array}$$

Figure 6.5 shows that $\lim_{t\to\infty} \tilde{q}_1(t) = 0.1309$ and $\lim_{t\to\infty} \tilde{q}_2(t) = 0.0174$.

a) Show that $\left[\tilde{\boldsymbol{q}}^T \ \dot{\boldsymbol{q}}^T\right]^T = \left[\tilde{\boldsymbol{q}}^T \ \boldsymbol{0}^T\right]^T$ with

$$\tilde{\boldsymbol{q}} = \begin{bmatrix} \tilde{q}_1 \\ \tilde{q}_2 \end{bmatrix} = \begin{bmatrix} 0.1309 \\ 0.0174 \end{bmatrix}$$

is an equilibrium of the closed-loop equation. Explain.

4. Consider the 2-DOF robot from Chapter 5 and illustrated in Figure 5.2. The vector of gravitational torques $\boldsymbol{g}(\boldsymbol{q})$ for this robot is presented in Section 5.3.2, and its components are

$$g_1(\boldsymbol{q}) = (m_1 l_{c1} + m_2 l_1) g \sin(q_1) + m_2 g l_{c2} \sin(q_1 + q_2)$$
$$g_2(\boldsymbol{q}) = m_2 g l_{c2} \sin(q_1 + q_2) .$$

Consider PD control. In view of the presence of $\boldsymbol{g}(\boldsymbol{q})$, in general the origin $\left[\tilde{\boldsymbol{q}}^T \ \dot{\boldsymbol{q}}^T\right]^T = \boldsymbol{0} \in \mathbb{R}^4$ of the closed-loop equation is not an equilibrium. However, for some values of $\boldsymbol{q}_d$, the origin happens to be an equilibrium.

a) Determine all possible vectors $\boldsymbol{q}_d = [q_{d1} \ q_{d2}]^T$ for which the origin of the closed-loop equation is an equilibrium.

5. Consider the 3-DOF Cartesian robot from Example 3.4 (*cf.* page 69) illustrated in Figure 3.5. It dynamic model is given by

$$(m_1 + m_2 + m_3)\ddot{q}_1 + (m_1 + m_2 + m_3) g = \tau_1$$
$$(m_1 + m_2)\ddot{q}_2 = \tau_2$$
$$m_1 \ddot{q}_3 = \tau_3 .$$

Consider the PD control law

$$\boldsymbol{\tau} = K_p \tilde{\boldsymbol{q}} - K_v \dot{\boldsymbol{q}}$$

where $\boldsymbol{q}_d$ is constant and K_p, K_v are diagonal positive definite matrices.

a) Obtain $M(\boldsymbol{q})$, $C(\boldsymbol{q}, \dot{\boldsymbol{q}})$ and $\boldsymbol{g}(\boldsymbol{q})$. Verify that $M(\boldsymbol{q}) = M$ is a constant diagonal matrix. Verify that $\boldsymbol{g}(\boldsymbol{q}) = \boldsymbol{g}$ is a constant vector.

b) Define $\tilde{\boldsymbol{q}} = [\tilde{q}_1 \ \tilde{q}_2 \ \tilde{q}_3]^T$. Obtain the closed-loop equation.

c) Verify that the closed-loop equation has a unique equilibrium at

$$\begin{bmatrix} \tilde{\boldsymbol{q}} \\ \dot{\boldsymbol{q}} \end{bmatrix} = \begin{bmatrix} K_p^{-1} \boldsymbol{g} \\ \boldsymbol{0} \end{bmatrix} .$$

d) Define $\boldsymbol{z} = \tilde{\boldsymbol{q}} - K_p^{-1} \boldsymbol{g}$. Rewrite the closed-loop equation in terms of the new state $\left[\boldsymbol{z}^T \ \dot{\boldsymbol{q}}^T\right]^T$. Verify that the origin is the unique equilibrium. Show that the origin is a stable equilibrium.

Hint: Use the Lyapunov function,

$$V(\boldsymbol{z}, \dot{\boldsymbol{q}}) = \frac{1}{2} \dot{\boldsymbol{q}}^T \dot{\boldsymbol{q}} + \frac{1}{2} \boldsymbol{z}^T M^{-1} K_p \boldsymbol{z} .$$

e) Use La Salle's theorem (Theorem 2.7) to show that moreover the origin is globally asymptotically stable.

6. Consider the model of elastic-joint robots (3.27) and (3.28), but without the gravitational term ($\boldsymbol{g}(\boldsymbol{q}) = \boldsymbol{0}$), that is,

$$M(\boldsymbol{q})\ddot{\boldsymbol{q}} + C(\boldsymbol{q}, \dot{\boldsymbol{q}})\dot{\boldsymbol{q}} + K(\boldsymbol{q} - \boldsymbol{\theta}) = \boldsymbol{0}$$
$$J\ddot{\boldsymbol{\theta}} - K(\boldsymbol{q} - \boldsymbol{\theta}) = \boldsymbol{\tau}.$$

It is assumed that only the positions vector corresponding to the motor shafts $\boldsymbol{\theta}$, is available for measurement as well as its corresponding velocities $\dot{\boldsymbol{\theta}}$. The goal is that $\boldsymbol{q}(t) \rightarrow \boldsymbol{q}_d$ as $t \rightarrow \infty$ for any constant $\boldsymbol{q}_d$.

The PD controller is in this case,

$$\boldsymbol{\tau} = K_p\tilde{\boldsymbol{\theta}} - K_v\dot{\boldsymbol{\theta}}$$

where $\tilde{\boldsymbol{\theta}} = \boldsymbol{q}_d - \boldsymbol{\theta}$ and $K_p, K_v \in \mathbb{R}^{n \times n}$ are symmetric positive definite matrices.

a) Obtain the closed-loop equation in terms of the state vector $\left[\tilde{\boldsymbol{q}}^T \ \tilde{\boldsymbol{\theta}}^T \ \dot{\boldsymbol{q}}^T \ \dot{\boldsymbol{\theta}}^T\right]^T$ where $\tilde{\boldsymbol{q}} = \boldsymbol{q}_d - \boldsymbol{q}$. Verify that the origin is the unique equilibrium.

b) Show that the origin is a stable equilibrium.

Hint: Use the following Lyapunov function

$$V(\tilde{\boldsymbol{q}}, \tilde{\boldsymbol{\theta}}, \dot{\boldsymbol{q}}, \dot{\boldsymbol{\theta}}) = \frac{1}{2}\dot{\boldsymbol{q}}^T M(\boldsymbol{q})\dot{\boldsymbol{q}} + \frac{1}{2}\dot{\boldsymbol{\theta}}^T J\dot{\boldsymbol{\theta}} + \frac{1}{2}\left[\tilde{\boldsymbol{\theta}} - \tilde{\boldsymbol{q}}\right]^T K\left[\tilde{\boldsymbol{\theta}} - \tilde{\boldsymbol{q}}\right] + \frac{1}{2}\tilde{\boldsymbol{\theta}}^T K_p\tilde{\boldsymbol{\theta}}$$

and the skew-symmetry of $\frac{1}{2}\dot{M} - C$.

c) Use La Salle's theorem (Theorem 2.7) to show also that the origin is globally asymptotically stable.

7

PD Control with Gravity Compensation

As studied in Chapter 6, the position control objective for robot manipulators may be achieved via PD control, provided that $\boldsymbol{g}(\boldsymbol{q}) = \mathbf{0}$ or, for a suitable selection of $\boldsymbol{q}_d$. In this case, the tuning – for the purpose of stability – of this controller is trivial since it is sufficient to select the design matrices K_p and K_v as symmetric positive definite. Nevertheless, PD control does not guarantee the achievement of the position control objective for manipulators whose dynamic models contain the gravitational torques vector $\boldsymbol{g}(\boldsymbol{q})$, unless the desired position $\boldsymbol{q}_d$ is such that $\boldsymbol{g}(\boldsymbol{q}_d) = \mathbf{0}$.

In this chapter we study PD control with gravity compensation, which is able to satisfy the position control objective globally for n DOF robots; moreover, its tuning is trivial. The formal study of this controller goes back at least to 1981 and this reference is given at the end of the chapter. The previous knowledge of part of the dynamic robot model to be controlled is required in the control law, but in contrast to the PID controller which, under the tuning procedure proposed in Chapter 9, needs information on $M(\boldsymbol{q})$ and $\boldsymbol{g}(\boldsymbol{q})$, the controller studied here only uses the vector of gravitational torques $\boldsymbol{g}(\boldsymbol{q})$.

The PD control law with gravity compensation is given by

$$\boldsymbol{\tau} = K_p\tilde{\boldsymbol{q}} + K_v\dot{\tilde{\boldsymbol{q}}} + \boldsymbol{g}(\boldsymbol{q}) \tag{7.1}$$

where $K_p, K_v \in \mathbb{R}^{n\times n}$ are symmetric positive definite matrices. Notice that the only difference with respect to the PD control law (6.2) is the added term $\boldsymbol{g}(\boldsymbol{q})$. In contrast to the PD control law, which does not require any knowledge of the structure of the robot model, the controller (7.1) makes explicit use of partial knowledge of the manipulator model, specifically of $\boldsymbol{g}(\boldsymbol{q})$. However, it is important to observe that for a given robot, the vector of gravitational torques, $\boldsymbol{g}(\boldsymbol{q})$, may be obtained with relative ease since one only needs to compute the expression corresponding to the potential energy $\mathcal{U}(\boldsymbol{q})$ of the robot. The vector $\boldsymbol{g}(\boldsymbol{q})$ is obtained from (3.20) and is $\boldsymbol{g}(\boldsymbol{q}) = \partial\mathcal{U}(\boldsymbol{q})/\partial\boldsymbol{q}$.

The control law (7.1) requires information on the desired position $\boldsymbol{q}_d(t)$ and on the desired velocity $\dot{\boldsymbol{q}}_d(t)$ as well as measurement of the position $\boldsymbol{q}(t)$ and the velocity $\dot{\boldsymbol{q}}(t)$ at each instant. Figure 7.1 shows the block-diagram corresponding to the PD controller with gravity compensation.

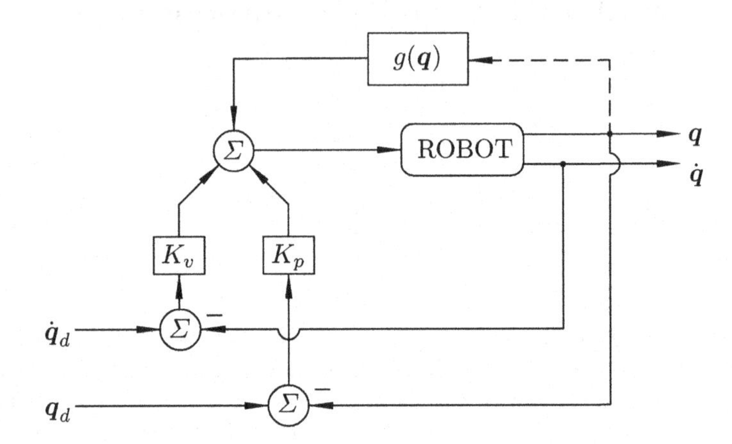

Figure 7.1. Block-diagram: PD control with gravity compensation

The equation that describes the behavior in closed loop is obtained by combining Equations (II.1) and (7.1) to obtain

$$M(\boldsymbol{q})\ddot{\boldsymbol{q}} + C(\boldsymbol{q}, \dot{\boldsymbol{q}})\dot{\boldsymbol{q}} + \boldsymbol{g}(\boldsymbol{q}) = K_p\tilde{\boldsymbol{q}} + K_v\dot{\tilde{\boldsymbol{q}}} + \boldsymbol{g}(\boldsymbol{q}) .$$

Or, in terms of the state vector $\begin{bmatrix}\tilde{\boldsymbol{q}}^T & \dot{\tilde{\boldsymbol{q}}}^T\end{bmatrix}^T$,

$$\frac{d}{dt}\begin{bmatrix}\tilde{\boldsymbol{q}} \\ \dot{\tilde{\boldsymbol{q}}}\end{bmatrix} = \begin{bmatrix}\dot{\tilde{\boldsymbol{q}}} \\ \ddot{\boldsymbol{q}}_d - M(\boldsymbol{q})^{-1}\left[K_p\tilde{\boldsymbol{q}} + K_v\dot{\tilde{\boldsymbol{q}}} - C(\boldsymbol{q}, \dot{\boldsymbol{q}})\dot{\boldsymbol{q}}\right]\end{bmatrix} .$$

A necessary and sufficient condition for the origin $\begin{bmatrix}\tilde{\boldsymbol{q}}^T & \dot{\tilde{\boldsymbol{q}}}^T\end{bmatrix}^T = \boldsymbol{0} \in \mathbb{R}^{2n}$, to be an equilibrium of the closed-loop equation is that the desired position $\boldsymbol{q}_d(t)$ satisfies

$$M(\boldsymbol{q}_d)\ddot{\boldsymbol{q}}_d + C(\boldsymbol{q}_d, \dot{\boldsymbol{q}}_d)\dot{\boldsymbol{q}}_d = \boldsymbol{0}$$

or equivalently, that $\boldsymbol{q}_d(t)$ be a solution of

$$\frac{d}{dt}\begin{bmatrix}\boldsymbol{q}_d \\ \dot{\boldsymbol{q}}_d\end{bmatrix} = \begin{bmatrix}\dot{\boldsymbol{q}}_d \\ -M(\boldsymbol{q}_d)^{-1}C(\boldsymbol{q}_d, \dot{\boldsymbol{q}}_d)\dot{\boldsymbol{q}}_d\end{bmatrix}$$

for any initial condition $\begin{bmatrix}\boldsymbol{q}_d(0)^T & \dot{\boldsymbol{q}}_d(0)^T\end{bmatrix}^T \in \mathbb{R}^{2n}$.

Obviously, in the case that the desired position $\boldsymbol{q}_d(t)$ does not satisfy the established condition, then the origin may not be an equilibrium of the closed-loop equation and, therefore, we may not expect to satisfy the control objective. That is, to drive the position error $\tilde{\boldsymbol{q}}(t)$ asymptotically to zero. Nevertheless, we may achieve the condition that the position error $\tilde{\boldsymbol{q}}(t)$ becomes, asymptotically, as small as wished if the matrices K_p and K_v are chosen sufficiently "large". For the formal proof of this claim, the reader is invited to see the corresponding cited reference at the end of the chapter.

A sufficient condition for the origin $\left[\tilde{\boldsymbol{q}}^T \ \dot{\tilde{\boldsymbol{q}}}^T\right]^T = \boldsymbol{0} \in \mathbb{R}^{2n}$ to be the unique equilibrium of the closed-loop equation is that the desired joint position $\boldsymbol{q}_d$ be a constant vector. In what is left of this chapter we assume that this is the case.

As we show next, this controller achieves the position control objective, that is,

$$\lim_{t\to\infty} \boldsymbol{q}(t) = \boldsymbol{q}_d$$

where $\boldsymbol{q}_d \in \mathbb{R}^n$ is any constant vector.

7.1 Global Asymptotic Stability by La Salle's Theorem

Considering the desired position $\boldsymbol{q}_d$ as constant, the closed-loop equation may be written in terms of the new state vector $\left[\tilde{\boldsymbol{q}}^T \ \dot{\boldsymbol{q}}^T\right]^T$ as

$$\frac{d}{dt}\begin{bmatrix}\tilde{\boldsymbol{q}}\\ \dot{\boldsymbol{q}}\end{bmatrix} = \begin{bmatrix} -\dot{\boldsymbol{q}} \\ M(\boldsymbol{q}_d-\tilde{\boldsymbol{q}})^{-1}\left[K_p\tilde{\boldsymbol{q}} - K_v\dot{\boldsymbol{q}} - C(\boldsymbol{q}_d-\tilde{\boldsymbol{q}},\dot{\boldsymbol{q}})\dot{\boldsymbol{q}}\right]\end{bmatrix} \tag{7.2}$$

which, in view of the fact that $\boldsymbol{q}_d$ is constant, is an autonomous differential equation. The origin $\left[\tilde{\boldsymbol{q}}^T \ \dot{\boldsymbol{q}}^T\right]^T = \boldsymbol{0} \in \mathbb{R}^{2n}$ is the unique equilibrium of this equation.

The stability analysis that we present next is taken from the literature. The reader may also consult the references cited at the end of the chapter.

To study the stability of the origin as an equilibrium, we use Lyapunov's direct method, which has already been presented in Chapter 2. Specifically, we use Theorem 2.2 to prove stability of the equilibrium (origin).

Consider the following Lyapunov function candidate

$$V(\tilde{\boldsymbol{q}},\dot{\boldsymbol{q}}) = \mathcal{K}(\boldsymbol{q},\dot{\boldsymbol{q}}) + \frac{1}{2}\tilde{\boldsymbol{q}}^T K_p \tilde{\boldsymbol{q}}$$

where $\mathcal{K}(\boldsymbol{q},\dot{\boldsymbol{q}})$ stands for the kinetic energy function of the robot, *i.e.* Equation (3.15). The function $V(\tilde{\boldsymbol{q}},\dot{\boldsymbol{q}})$ is globally positive definite since the kinetic energy $\mathcal{K}(\boldsymbol{q}, \dot{\boldsymbol{q}})$ is positive definite in $\dot{\boldsymbol{q}}$ and on the other hand, K_p is a positive definite matrix. Then, also the quadratic form $\tilde{\boldsymbol{q}}^T K_p \tilde{\boldsymbol{q}}$ is a positive definite function of $\tilde{\boldsymbol{q}}$.

The Lyapunov function candidate may be written as

$$
\begin{aligned}
V(\tilde{\boldsymbol{q}}, \dot{\boldsymbol{q}}) &= \frac{1}{2} \begin{bmatrix} \tilde{\boldsymbol{q}} \\ \dot{\boldsymbol{q}} \end{bmatrix}^T \begin{bmatrix} K_p & 0 \\ 0 & M(\boldsymbol{q}_d - \tilde{\boldsymbol{q}}) \end{bmatrix} \begin{bmatrix} \tilde{\boldsymbol{q}} \\ \dot{\boldsymbol{q}} \end{bmatrix} \\
&= \frac{1}{2} \dot{\boldsymbol{q}}^T M(\boldsymbol{q}) \dot{\boldsymbol{q}} + \frac{1}{2} \tilde{\boldsymbol{q}}^T K_p \tilde{\boldsymbol{q}}
\end{aligned} \tag{7.3}
$$

and its total derivative with respect to time is

$$
\dot{V}(\tilde{\boldsymbol{q}}, \dot{\boldsymbol{q}}) = \dot{\boldsymbol{q}}^T M(\boldsymbol{q}) \ddot{\boldsymbol{q}} + \frac{1}{2} \dot{\boldsymbol{q}}^T \dot{M}(\boldsymbol{q}) \dot{\boldsymbol{q}} + \tilde{\boldsymbol{q}}^T K_p \dot{\tilde{\boldsymbol{q}}} .
$$

Substituting $M(\boldsymbol{q})\ddot{\boldsymbol{q}}$ from the closed-loop Equation (7.2) we get

$$
\begin{aligned}
\dot{V}(\tilde{\boldsymbol{q}}, \dot{\boldsymbol{q}}) &= -\dot{\boldsymbol{q}}^T K_v \dot{\boldsymbol{q}} \\
&= - \begin{bmatrix} \tilde{\boldsymbol{q}} \\ \dot{\boldsymbol{q}} \end{bmatrix}^T \begin{bmatrix} 0 & 0 \\ 0 & K_v \end{bmatrix} \begin{bmatrix} \tilde{\boldsymbol{q}} \\ \dot{\boldsymbol{q}} \end{bmatrix} \leq 0
\end{aligned} \tag{7.4}
$$

where we have eliminated the term $\dot{\boldsymbol{q}}^T \left[\frac{1}{2} \dot{M} - C \right] \dot{\boldsymbol{q}}$ by virtue of Property 4.2 and used $\dot{\tilde{\boldsymbol{q}}} = -\dot{\boldsymbol{q}}$ since $\boldsymbol{q}_d$ is assumed to be a constant vector.

Therefore, the function $V(\tilde{\boldsymbol{q}}, \dot{\boldsymbol{q}})$ is a Lyapunov function since moreover $\dot{V}(\tilde{\boldsymbol{q}}, \dot{\boldsymbol{q}}) \leq 0$ for all $\tilde{\boldsymbol{q}}$ and $\dot{\boldsymbol{q}}$ and consequently, the origin is stable and all the solutions $\tilde{\boldsymbol{q}}(t)$ and $\dot{\boldsymbol{q}}(t)$ are bounded (*cf.* Theorem 2.3).

Since the closed-loop Equation (7.2) is independent of time (explicitly) we may explore the use of of La Salle's theorem (*cf.* Theorem 2.7) to analyze the global asymptotic stability of the origin.

To that end, we first remark that the set Ω is here given by

$$
\begin{aligned}
\Omega &= \left\{ \boldsymbol{x} \in \mathbb{R}^{2n} : \dot{V}(\boldsymbol{x}) = 0 \right\} \\
&= \left\{ \boldsymbol{x} = \begin{bmatrix} \tilde{\boldsymbol{q}} \\ \dot{\boldsymbol{q}} \end{bmatrix} \in \mathbb{R}^{2n} : \dot{V}(\tilde{\boldsymbol{q}}, \dot{\boldsymbol{q}}) = 0 \right\} \\
&= \{ \tilde{\boldsymbol{q}} \in \mathbb{R}^n, \dot{\boldsymbol{q}} = \boldsymbol{0} \in \mathbb{R}^n \} .
\end{aligned}
$$

Observe that $\dot{V}(\tilde{\boldsymbol{q}}, \dot{\boldsymbol{q}}) = 0$ if and only if $\dot{\boldsymbol{q}} = \boldsymbol{0}$. For a solution $\boldsymbol{x}(t)$ to belong to Ω for all $t \geq 0$, it is necessary and sufficient that $\dot{\boldsymbol{q}}(t) = \boldsymbol{0}$ for all $t \geq 0$. Therefore it must also hold that $\ddot{\boldsymbol{q}}(t) = \boldsymbol{0}$ for all $t \geq 0$. Taking this into account, we conclude from the closed-loop Equation (7.2) that if $\boldsymbol{x}(t) \in \Omega$ for all $t \geq 0$ then

$$\mathbf{0} = M(\boldsymbol{q}_d - \tilde{\boldsymbol{q}}(t))^{-1} K_p \tilde{\boldsymbol{q}}(t)$$

which means that $\tilde{\boldsymbol{q}}(t) = \mathbf{0}$ for all $t \geq 0$. Thus, $\left[\tilde{\boldsymbol{q}}(0)^T \;\; \dot{\boldsymbol{q}}(0)^T\right]^T = \mathbf{0} \in \mathbb{R}^{2n}$ is the only initial condition in Ω for which $\boldsymbol{x}(t) \in \Omega$ for all $t \geq 0$. Then, according to La Salle's theorem (*cf.* Theorem 2.7), this is enough to guarantee global asymptotic stability of the origin $\left[\tilde{\boldsymbol{q}}^T \;\; \dot{\boldsymbol{q}}^T\right]^T = \mathbf{0} \in \mathbb{R}^{2n}$.

As a result we have

$$\lim_{t\to\infty} \tilde{\boldsymbol{q}}(t) = \mathbf{0}$$
$$\lim_{t\to\infty} \dot{\boldsymbol{q}}(t) = \mathbf{0}\,,$$

that is, the position control objective is achieved.

We present next an example with the purpose of showing the performance of PD control with gravity compensation for the Pelican robot.

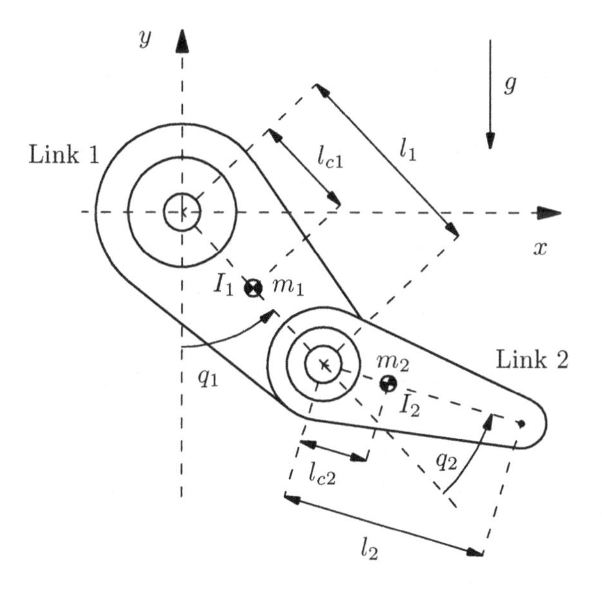

Figure 7.2. Diagram of the Pelican robot

Example 7.1. Consider the Pelican robot studied in Chapter 5, and shown in Figure 7.2.

The components of the vector of gravitational torques $\boldsymbol{g}(\boldsymbol{q})$ are given by

$$g_1(\boldsymbol{q}) = (m_1 l_{c1} + m_2 l_1) g \,\sin(q_1) + m_2 l_{c2} g \,\sin(q_1 + q_2)$$
$$g_2(\boldsymbol{q}) = m_2 l_{c2} g \,\sin(q_1 + q_2)\,.$$

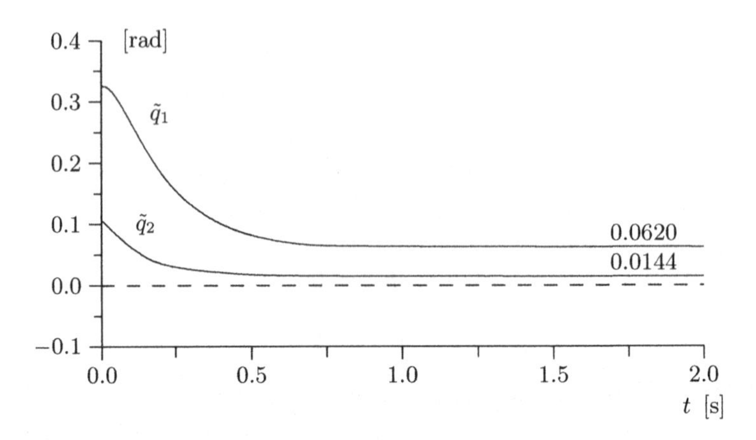

Figure 7.3. Graph of the position errors $\tilde{q}_1$ and $\tilde{q}_2$

Consider the PD control law with gravity compensation for this robot where the design matrices K_p and K_d are positive definite. In particular, let us pick (arbitrarily)

$$K_p = \text{diag}\{k_p\} = \text{diag}\{30\} \quad [\text{Nm/rad}]$$
$$K_v = \text{diag}\{k_v\} = \text{diag}\{7,\ 3\} \quad [\text{Nm s/rad}] \ .$$

The components of the control input vector $\boldsymbol{\tau}$, are given by

$$\tau_1 = k_p\tilde{q}_1 - k_v\dot{q}_1 + g_1(\boldsymbol{q})$$
$$\tau_2 = k_p\tilde{q}_2 - k_v\dot{q}_2 + g_2(\boldsymbol{q}) \, .$$

The initial conditions corresponding to the positions and velocities are chosen as

$$q_1(0) = 0,\, q_2(0) = 0$$
$$\dot{q}_1(0) = 0,\, \dot{q}_2(0) = 0 \, .$$

The desired joint positions are chosen as

$$q_{d1} = \pi/10,\, q_{d2} = \pi/30 \quad [\text{rad}] \, .$$

In terms of the state vector of the closed-loop equation, the initial state is taken to be

$$\begin{bmatrix} \tilde{\boldsymbol{q}}(0) \\ \\ \dot{\boldsymbol{q}}(0) \end{bmatrix} = \begin{bmatrix} \pi/10 \\ \pi/30 \\ 0 \\ 0 \end{bmatrix} = \begin{bmatrix} 0.3141 \\ 0.1047 \\ 0 \\ 0 \end{bmatrix} .$$

Figure 7.3 presents the components of the position error $\tilde{\boldsymbol{q}}$ obtained in the experiment. The steady state position errors shown this figure are a product of the friction phenomenon which has not been included in the robot dynamics. ♢

7.2 Lyapunov Function for Global Asymptotic Stability

In this section we present an alternative proof for global asymptotic stability without the use of La Salle's theorem. Instead, we use a strict Lyapunov function, *i.e.* a Lyapunov function whose time derivative is globally negative definite. We consider the case of robots having only revolute joints and where the proportional gain matrix K_p is diagonal instead of only symmetric, but of course, positive definite. Some readers may wish to omit this somewhat technical section and continue to Section 7.3.

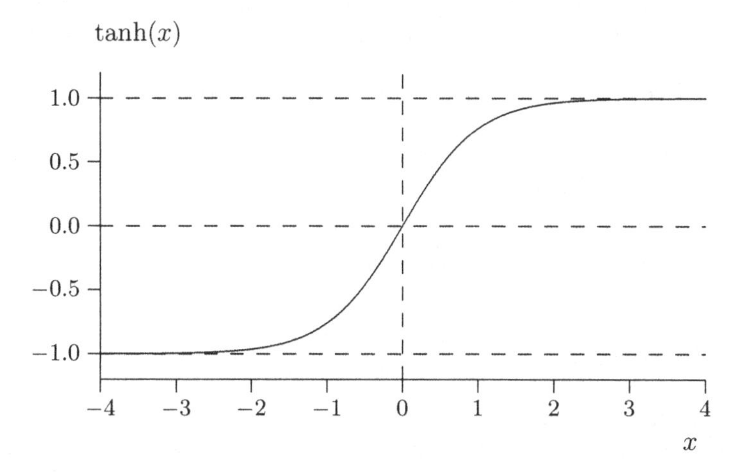

Figure 7.4. Graph of the tangent hyperbolic function: $\tanh(x)$

Before starting with the stability analysis of the equilibrium $\left[\tilde{\boldsymbol{q}}^T \ \dot{\boldsymbol{q}}^T\right]^T = \mathbf{0} \in \mathbb{R}^{2n}$ of the closed-loop Equation (7.2), it is convenient to cite some properties of the vectorial function[1]

$$\mathbf{tanh}(\boldsymbol{x}) = [\tanh(x_1) \ \tanh(x_2) \ \cdots \ \tanh(x_n)]^T \tag{7.5}$$

where $\tanh(x)$ (see Figure 7.4) denotes the hyperbolic tangent function,

$$\tanh(x) = \frac{e^x - e^{-x}}{e^x + e^{-x}} .$$

Note that this function satisfies $|x| \geq |\tanh(x)|$, and $1 \geq |\tanh(x)|$ for all $x \in \mathbb{R}$ therefore, the Euclidean norm of $\mathbf{tanh}(\boldsymbol{x})$ satisfies

$$\|\mathbf{tanh}(\boldsymbol{x})\| \leq \begin{cases} \|\boldsymbol{x}\| & \forall \ \boldsymbol{x} \in \mathbb{R}^n \\ \sqrt{n} & \forall \ \boldsymbol{x} \in \mathbb{R}^n \end{cases} \tag{7.6}$$

[1] See also Definition 4.1 on page 103.

and $\mathbf{tanh}(\boldsymbol{x}) = \mathbf{0}$ if and only if $\boldsymbol{x} = \mathbf{0}$.

One can prove without much difficulty that for a symmetric positive definite matrix A the inequality

$$\tilde{\boldsymbol{q}}^T A \tilde{\boldsymbol{q}} \geq \lambda_{\min}\{A\} \, \|\mathbf{tanh}(\tilde{\boldsymbol{q}})\|^2 \quad \forall\, \tilde{\boldsymbol{q}} \in \mathbb{R}^n$$

holds. If moreover, A is diagonal, then

$$\mathbf{tanh}(\tilde{\boldsymbol{q}})^T A \tilde{\boldsymbol{q}} \geq \lambda_{\min}\{A\} \, \|\mathbf{tanh}(\tilde{\boldsymbol{q}})\|^2 \quad \forall\, \tilde{\boldsymbol{q}} \in \mathbb{R}^n \,. \tag{7.7}$$

We present next our alternative stability analysis. To study the stability properties of the origin $\begin{bmatrix}\tilde{\boldsymbol{q}}^T & \dot{\boldsymbol{q}}^T\end{bmatrix}^T = \mathbf{0} \in \mathbb{R}^{2n}$, of the closed-loop Equation (7.2), consider the Lyapunov function (7.3) with an added term, that is,

$$V(\tilde{\boldsymbol{q}}, \dot{\boldsymbol{q}}) = \frac{1}{2}\dot{\boldsymbol{q}}^T M(\boldsymbol{q})\dot{\boldsymbol{q}} + \frac{1}{2}\tilde{\boldsymbol{q}}^T K_p \tilde{\boldsymbol{q}} - \gamma \mathbf{tanh}(\tilde{\boldsymbol{q}})^T M(\boldsymbol{q})\dot{\boldsymbol{q}} \tag{7.8}$$

where $\mathbf{tanh}(\tilde{\boldsymbol{q}})$ was defined in (7.5) and $\gamma > 0$ is a constant sufficiently small so as to satisfy simultaneously,

$$\frac{\lambda_{\min}\{K_p\}\lambda_{\min}\{M\}}{\lambda_{\mathrm{Max}}^2\{M\}} > \gamma^2 \tag{7.9}$$

and

$$\frac{4\lambda_{\min}\{K_p\}\lambda_{\min}\{K_v\}}{\lambda_{\mathrm{Max}}^2\{K_v\} + 4\lambda_{\min}\{K_p\}[\sqrt{n}\,k_{C_1} + \lambda_{\mathrm{Max}}\{M\}]} > \gamma \,. \tag{7.10}$$

Since the upper-bounds above are always strictly positive constants, there always exists $\gamma > 0$ arbitrarily small and that satisfies both inequalities.

7.2.1 Positivity of the Lyapunov Function

In order to show that the Lyapunov function candidate (7.8) is positive definite we first observe that the third term in (7.8) satisfies

$$\begin{aligned}
\gamma \mathbf{tanh}(\tilde{\boldsymbol{q}})^T M(\boldsymbol{q})\dot{\boldsymbol{q}} &\leq \gamma \, \|\mathbf{tanh}(\tilde{\boldsymbol{q}})\| \, \|M(\boldsymbol{q})\dot{\boldsymbol{q}}\| \\
&\leq \gamma \, \lambda_{\mathrm{Max}}\{M\} \, \|\mathbf{tanh}(\tilde{\boldsymbol{q}})\| \, \|\dot{\boldsymbol{q}}\| \\
&\leq \gamma \, \lambda_{\mathrm{Max}}\{M\} \, \|\tilde{\boldsymbol{q}}\| \, \|\dot{\boldsymbol{q}}\|
\end{aligned}$$

where we have used (7.6) in the last step. Therefore,

$$-\gamma \mathbf{tanh}(\tilde{\boldsymbol{q}})^T M(\boldsymbol{q})\dot{\boldsymbol{q}} \geq -\gamma \, \lambda_{\mathrm{Max}}\{M\} \, \|\tilde{\boldsymbol{q}}\| \, \|\dot{\boldsymbol{q}}\|$$

and consequently the Lyapunov function candidate (7.8) satisfies the inequality

$$V(\tilde{\boldsymbol{q}}, \dot{\boldsymbol{q}}) \geq \frac{1}{2} \begin{bmatrix} \|\tilde{\boldsymbol{q}}\| \\ \|\dot{\boldsymbol{q}}\| \end{bmatrix}^T \begin{bmatrix} \lambda_{\min}\{K_p\} & -\gamma \, \lambda_{\mathrm{Max}}\{M\} \\ -\gamma \, \lambda_{\mathrm{Max}}\{M\} & \lambda_{\min}\{M\} \end{bmatrix} \begin{bmatrix} \|\tilde{\boldsymbol{q}}\| \\ \|\dot{\boldsymbol{q}}\| \end{bmatrix}$$

Since by assumption K_p is positive definite – $\lambda_{\min}\{K_p\} > 0$ – and γ is supposed to satisfy (7.9) it follows that $V(\tilde{\boldsymbol{q}}, \dot{\boldsymbol{q}})$ is a positive definite function and moreover it is radially unbounded.

7.2.2 Time Derivative of the Lyapunov Function

The time derivative of the Lyapunov function candidate (7.8) along the trajectories of the closed-loop system (7.2) may be written as

$$\begin{aligned}\dot{V}(\tilde{\boldsymbol{q}}, \dot{\boldsymbol{q}}) &= \dot{\boldsymbol{q}}^T \left[K_p \tilde{\boldsymbol{q}} - K_v \dot{\boldsymbol{q}} - C(\boldsymbol{q}, \dot{\boldsymbol{q}})\dot{\boldsymbol{q}}\right] + \frac{1}{2}\dot{\boldsymbol{q}}^T \dot{M}(\boldsymbol{q})\dot{\boldsymbol{q}} \\ &\quad - \left[K_p \tilde{\boldsymbol{q}}\right]^T \dot{\boldsymbol{q}} + \gamma \dot{\boldsymbol{q}}^T \mathrm{Sech}^2(\tilde{\boldsymbol{q}})^T M(\boldsymbol{q})\dot{\boldsymbol{q}} - \gamma \mathbf{tanh}(\tilde{\boldsymbol{q}})^T \dot{M}(\boldsymbol{q})\dot{\boldsymbol{q}} \\ &\quad - \gamma \mathbf{tanh}(\tilde{\boldsymbol{q}})^T \left[K_p \tilde{\boldsymbol{q}} - K_v \dot{\boldsymbol{q}} - C(\boldsymbol{q}, \dot{\boldsymbol{q}})\dot{\boldsymbol{q}}\right]\end{aligned}$$

where we used Equation (4.14) in

$$\frac{d}{dt}\{\mathbf{tanh}(\tilde{\boldsymbol{q}})\} = -\mathrm{Sech}^2(\tilde{\boldsymbol{q}})\dot{\boldsymbol{q}}$$

that is, $\mathrm{Sech}^2(\tilde{\boldsymbol{q}}) := \mathrm{diag}\{\mathrm{sech}^2(\tilde{q}_i)\}$ where

$$\mathrm{sech}(\tilde{q}_i) := \frac{1}{e^{\tilde{q}_i} + e^{-\tilde{q}_i}}$$

and therefore, $\mathrm{Sech}^2(\tilde{\boldsymbol{q}})$ is a diagonal matrix whose elements, $\mathrm{sech}^2(\tilde{q}_i)$, are positive and smaller than 1.

Using Property 4.2, which establishes that $\dot{\boldsymbol{q}}^T\left[\frac{1}{2}\dot{M} - C\right]\dot{\boldsymbol{q}} = 0$ and $\dot{M}(\boldsymbol{q}) = C(\boldsymbol{q}, \dot{\boldsymbol{q}}) + C(\boldsymbol{q}, \dot{\boldsymbol{q}})^T$, the time derivative of the Lyapunov function candidate yields

$$\begin{aligned}\dot{V}(\tilde{\boldsymbol{q}}, \dot{\boldsymbol{q}}) &= -\dot{\boldsymbol{q}}^T K_v \dot{\boldsymbol{q}} + \gamma \dot{\boldsymbol{q}}^T \mathrm{Sech}^2(\tilde{\boldsymbol{q}})^T M(\boldsymbol{q})\dot{\boldsymbol{q}} - \gamma \mathbf{tanh}(\tilde{\boldsymbol{q}})^T K_p \tilde{\boldsymbol{q}} \\ &\quad + \gamma \mathbf{tanh}(\tilde{\boldsymbol{q}})^T K_v \dot{\boldsymbol{q}} - \gamma \mathbf{tanh}(\tilde{\boldsymbol{q}})^T C(\boldsymbol{q}, \dot{\boldsymbol{q}})^T \dot{\boldsymbol{q}}\,. \end{aligned} \tag{7.11}$$

We now proceed to upper-bound $\dot{V}(\tilde{\boldsymbol{q}}, \dot{\boldsymbol{q}})$ by a negative definite function of the states $\tilde{\boldsymbol{q}}$ and $\dot{\boldsymbol{q}}$. To that end, it is convenient to find upper-bounds for each term of (7.11).

The first term of (7.11) may be trivially bounded by

$$-\dot{\boldsymbol{q}}^T K_v \dot{\boldsymbol{q}} \leq -\lambda_{\min}\{K_v\}\|\dot{\boldsymbol{q}}\|^2 .$$

To upper-bound the second term of (7.11) we use $|\mathrm{sech}^2(x)| \leq 1$, so

$$\left\|\mathrm{Sech}^2(\tilde{\boldsymbol{q}})\dot{\boldsymbol{q}}\right\| \leq \|\dot{\boldsymbol{q}}\| .$$

From this argument we also have

$$\gamma \dot{\boldsymbol{q}}^T \mathrm{Sech}^2(\tilde{\boldsymbol{q}})^T M(\boldsymbol{q})\dot{\boldsymbol{q}} \leq \gamma \lambda_{\mathrm{Max}}\{M\} \|\dot{\boldsymbol{q}}\|^2 .$$

On the other hand, note that in view of (7.7), the following inequality also holds true since K_p is a diagonal positive definite matrix,

$$\gamma \mathbf{tanh}(\tilde{\boldsymbol{q}})^T K_p \tilde{\boldsymbol{q}} \geq \gamma \lambda_{\min}\{K_p\} \|\mathbf{tanh}(\tilde{\boldsymbol{q}})\|^2$$

which in turn, implies the key inequality

$$-\gamma \mathbf{tanh}(\tilde{\boldsymbol{q}})^T K_p \tilde{\boldsymbol{q}} \leq -\gamma \lambda_{\min}\{K_p\} \|\mathbf{tanh}(\tilde{\boldsymbol{q}})\|^2 .$$

A bound on $\gamma \mathbf{tanh}(\tilde{\boldsymbol{q}})^T K_v \dot{\boldsymbol{q}}$ that is obtained directly is

$$\gamma \mathbf{tanh}(\tilde{\boldsymbol{q}})^T K_v \dot{\boldsymbol{q}} \leq \gamma \lambda_{\mathrm{Max}}\{K_v\} \|\dot{\boldsymbol{q}}\| \|\mathbf{tanh}(\tilde{\boldsymbol{q}})\| .$$

The upper-bound on the term $-\gamma \mathbf{tanh}(\tilde{\boldsymbol{q}})^T C(\boldsymbol{q}, \dot{\boldsymbol{q}})^T \dot{\boldsymbol{q}}$ must be carefully selected. Notice that

$$\begin{aligned} -\gamma \mathbf{tanh}(\tilde{\boldsymbol{q}})^T C(\boldsymbol{q}, \dot{\boldsymbol{q}})^T \dot{\boldsymbol{q}} &= -\gamma \dot{\boldsymbol{q}}^T C(\boldsymbol{q}, \dot{\boldsymbol{q}}) \mathbf{tanh}(\tilde{\boldsymbol{q}}) \\ &\leq \gamma \|\dot{\boldsymbol{q}}\| \|C(\boldsymbol{q}, \dot{\boldsymbol{q}}) \mathbf{tanh}(\tilde{\boldsymbol{q}})\| . \end{aligned}$$

Then, considering Property 4.2 but in its variant that establishes the existence of a constant k_{C_1} such that $\|C(\boldsymbol{q}, \boldsymbol{x})\boldsymbol{y}\| \leq k_{C_1} \|\boldsymbol{x}\| \|\boldsymbol{y}\|$ for all $\boldsymbol{q}, \boldsymbol{x}, \boldsymbol{y} \in \mathbb{R}^n$, we obtain

$$-\gamma \mathbf{tanh}(\tilde{\boldsymbol{q}})^T C(\boldsymbol{q}, \dot{\boldsymbol{q}})^T \dot{\boldsymbol{q}} \leq \gamma k_{C_1} \|\dot{\boldsymbol{q}}\|^2 \|\mathbf{tanh}(\tilde{\boldsymbol{q}})\| .$$

Making use of the inequality (7.6) of $\mathbf{tanh}(\tilde{\boldsymbol{q}})$ which says that $\|\mathbf{tanh}(\tilde{\boldsymbol{q}})\| \leq \sqrt{n}$ for all $\tilde{\boldsymbol{q}} \in \mathbb{R}^n$, we obtain

$$-\gamma \mathbf{tanh}(\tilde{\boldsymbol{q}})^T C(\boldsymbol{q}, \dot{\boldsymbol{q}})^T \dot{\boldsymbol{q}} \leq \gamma \sqrt{n}\, k_{C_1} \|\dot{\boldsymbol{q}}\|^2 .$$

The previous bounds yield that the time derivative $\dot{V}(\tilde{\boldsymbol{q}}, \dot{\boldsymbol{q}})$ in (7.11), satisfies

$$\dot{V}(\tilde{\boldsymbol{q}}, \dot{\boldsymbol{q}}) \leq -\gamma \begin{bmatrix} \|\mathbf{tanh}(\tilde{\boldsymbol{q}})\| \\ \|\dot{\boldsymbol{q}}\| \end{bmatrix}^T Q \begin{bmatrix} \|\mathbf{tanh}(\tilde{\boldsymbol{q}})\| \\ \|\dot{\boldsymbol{q}}\| \end{bmatrix} \tag{7.12}$$

where

$$Q = \begin{bmatrix} \lambda_{\min}\{K_p\} & -\frac{1}{2}\lambda_{\mathrm{Max}}\{K_v\} \\ -\frac{1}{2}\lambda_{\mathrm{Max}}\{K_v\} & \frac{1}{\gamma}\lambda_{\min}\{K_v\} - \sqrt{n}\, k_{C1} - \lambda_{\mathrm{Max}}\{M\} \end{bmatrix} .$$

The two following conditions guarantee that the matrix Q is positive definite, hence, these conditions are sufficient to ensure that $\dot{V}(\tilde{\boldsymbol{q}}, \dot{\boldsymbol{q}})$ is a negative definite function,

$$\lambda_{\min}\{K_p\} > 0$$

and

$$\frac{4\lambda_{\min}\{K_p\}\lambda_{\min}\{K_v\}}{\lambda_{\mathrm{Max}}^2\{K_v\} + 4\lambda_{\min}\{K_p\}[\sqrt{n} k_{C1} + \lambda_{\mathrm{Max}}\{M\}]} > \gamma .$$

The first condition is trivially satisfied since K_p is assumed to be diagonal positive definite. The second condition also holds due to the upper-bound (7.10) imposed on γ.

According to the arguments above, there always exists a strictly positive constant γ such that the function $V(\tilde{\boldsymbol{q}}, \dot{\boldsymbol{q}})$, given by (7.8) is positive definite, while $\dot{V}(\tilde{\boldsymbol{q}}, \dot{\boldsymbol{q}})$ expressed as (7.12), is negative definite. For this reason, $V(\tilde{\boldsymbol{q}}, \dot{\boldsymbol{q}})$ is a strict Lyapunov function.

Finally, Theorem 2.4 allows one to establish global asymptotic stability of the origin. It is important to underline that it is not necessary to know the value of γ but only to know that it exists. This has been done to validate the result on global asymptotic stability that was stated.

7.3 Conclusions

Let us restate the most important conclusion from the analyses done in this chapter.

Consider the PD control law with gravity compensation for n-DOF robots and assume that the desired position $\boldsymbol{q}_d$ is constant.

- If the symmetric matrices K_p and K_v of the PD control law with gravity compensation are positive definite, then the origin of the closed-loop equation, expressed in terms of the state vector $\begin{bmatrix}\tilde{\boldsymbol{q}}^T & \dot{\boldsymbol{q}}^T\end{bmatrix}^T$, is a globally asymptotically stable equilibrium. Consequently, for any initial condition $\boldsymbol{q}(0), \dot{\boldsymbol{q}}(0) \in \mathbb{R}^n$, we have $\lim_{t\to\infty} \tilde{\boldsymbol{q}}(t) = \boldsymbol{0} \in \mathbb{R}^n$.

Bibliography

PD control with gravity compensation for robot manipulators was originally analyzed in

- Takegaki M., Arimoto S., 1981, *"A new feedback method for dynamic control of manipulators"*, Transactions ASME, Journal of Dynamic Systems, Measurement and Control, Vol. 103, pp. 119–125.

The following texts present also the proof of global asymptotic stability for the PD control law with gravity compensation of robot manipulators

- Spong M., Vidyasagar M., 1989, *"Robot dynamics and control"*, John Wiley and Sons.
- Yoshikawa T., 1990, *"Foundations of robotics: Analysis and control"*, The MIT Press.

A particularly simple proof of stability for the PD controller with gravity compensation which makes use of La Salle's theorem is presented in

- Paden B., Panja R., 1988, *"Globally asymptotically stable PD+ controller for robot manipulators"*, International Journal of Control, Vol. 47, No. 6, pp. 1697–1712.

The analysis of the PD control with gravity compensation for the case in which the desired joint position $\boldsymbol{q}_d$ is time-varying is presented in

- Kawamura S., Miyazaki F., Arimoto S., 1988, *"Is a local linear PD feedback control law effective for trajectory tracking of robot motion?"*, in Proceedings of the 1988 IEEE International Conference on Robotics and Automation, Philadelphia, PA., pp. 1335–1340, April.

Problems

1. Consider the PD control with gravity compensation for robots. Let $\boldsymbol{q}_d(t)$ be the desired joint position.

   Assume that there exists a constant vector $\boldsymbol{x} \in \mathbb{R}^n$ such that

$$\boldsymbol{x} - K_p^{-1}\left[M(\boldsymbol{q}_d - \boldsymbol{x})\ddot{\boldsymbol{q}}_d + C(\boldsymbol{q}_d - \boldsymbol{x}, \dot{\boldsymbol{q}}_d)\dot{\boldsymbol{q}}_d\right] = \boldsymbol{0} \in \mathbb{R}^n .$$

   a) Show that $\begin{bmatrix}\tilde{\boldsymbol{q}}^T & \dot{\tilde{\boldsymbol{q}}}^T\end{bmatrix}^T = \begin{bmatrix}\boldsymbol{x}^T & \boldsymbol{0}^T\end{bmatrix}^T \in \mathbb{R}^{2n}$ is an equilibrium of the closed-loop equation.

2. Consider the model of an ideal pendulum studied in Example 2.2 (see page 30)

$$J\ddot{q} + mgl\,\sin(q) = \tau .$$

   The PD control law with gravity compensation is in this case

$$\tau = k_p\tilde{q} + k_v\dot{\tilde{q}} + mgl\,\sin(q)$$

   where k_p and k_v are positive constants.

   a) Obtain the closed-loop equation in terms of the state vector $\begin{bmatrix}\tilde{q} & \dot{\tilde{q}}\end{bmatrix}^T$. Is this equation linear in the state ?

   b) Assume that the desired position is $q_d(t) = \alpha t$ where α is any real constant. Show that

$$\lim_{t\to\infty} \tilde{q}(t) = 0 .$$

3. Verify the expression of $\dot{V}(\tilde{\boldsymbol{q}}, \dot{\boldsymbol{q}})$ obtained in (7.4).

4. Consider the 3-DOF Cartesian robot studied in Example 3.4 (see page 69) and shown in Figure 3.5. Its dynamic model is given by

$$\begin{aligned}(m_1+m_2+m_3)\ddot{q}_1+(m_1+m_2+m_3)g &= \tau_1\\ (m_1+m_2)\ddot{q}_2 &= \tau_2\\ m_1\ddot{q}_3 &= \tau_3\,.\end{aligned}$$

Assume that the desired position $\boldsymbol{q}_d$ is constant. Consider using the PD controller with gravity compensation,

$$\boldsymbol{\tau} = K_p\tilde{\boldsymbol{q}} - K_v\dot{\boldsymbol{q}} + \boldsymbol{g}(\boldsymbol{q})$$

where K_p, K_v are positive definite matrices.

a) Obtain $\boldsymbol{g}(\boldsymbol{q})$. Verify that $\boldsymbol{g}(\boldsymbol{q}) = \boldsymbol{g}$ is a constant vector.
b) Define $\tilde{\boldsymbol{q}} = [\tilde{q}_1 \;\; \tilde{q}_2 \;\; \tilde{q}_3]^T$. Obtain the closed-loop equation. Is the closed-loop equation linear in the state ?
c) Is the origin the unique equilibrium of the closed-loop equation?
d) Show that the origin is a globally asymptotically stable equilibrium point.

5. Consider the following variant of PD control with gravity compensation[2]

$$\boldsymbol{\tau} = K_p\tilde{\boldsymbol{q}} - M(\boldsymbol{q})K_v\dot{\boldsymbol{q}} + \boldsymbol{g}(\boldsymbol{q})$$

where $\boldsymbol{q}_d$ is constant, K_p is a symmetric positive definite matrix and $K_v = \mathrm{diag}\{k_v\}$ with $k_v > 0$.

a) Obtain the closed-loop equation in terms of the state vector $\left[\tilde{\boldsymbol{q}}^T \;\; \dot{\boldsymbol{q}}^T\right]^T$.
b) Verify that the origin is a unique equilibrium.
c) Show that the origin is a globally asymptotically stable equilibrium point.

6. Consider the PD control law with gravity compensation where the matrix K_v is a function of time, *i.e.*

$$\boldsymbol{\tau} = K_p\tilde{\boldsymbol{q}} - K_v(t)\dot{\boldsymbol{q}} + \boldsymbol{g}(\boldsymbol{q})$$

and where $\boldsymbol{q}_d$ is constant, K_p is a positive definite matrix and $K_v(t)$ is also positive definite for all $t \geq 0$.

a) Obtain the closed-loop equation in terms of the state vector $\left[\tilde{\boldsymbol{q}}^T \;\; \dot{\boldsymbol{q}}^T\right]^T$. Is the closed-loop equation autonomous?
b) Verify that the origin is the only equilibrium point.
c) Show that the origin is a stable equilibrium.

7. Is the matrix $\mathrm{Sech}^2(\boldsymbol{x})$ positive definite?

---

[2] This problem is taken from Craig J. J., 1989, "*Introduction to robotics: Mechanics and control*", Second edition, Addison–Wesley.

8

PD Control with Desired Gravity Compensation

We have seen that the position control objective for robot manipulators (whose dynamic model includes the gravitational torques vector $\boldsymbol{g}(\boldsymbol{q})$), may be achieved globally by PD control with gravity compensation. The corresponding control law given by Equation (7.1) requires that its design symmetric matrices K_p and K_v be positive definite. On the other hand, this controller uses explicitly in its control law the gravitational torques vector $\boldsymbol{g}(\boldsymbol{q})$ of the dynamic robot model to be controlled.

Nevertheless, it is worth remarking that even in the scenario of position control, where the desired joint position $\boldsymbol{q}_d \in \mathbb{R}^n$ is constant, in the implementation of the PD control law with gravity compensation it is necessary to evaluate, on-line, the vector $\boldsymbol{g}(\boldsymbol{q}(t))$. In general, the elements of the vector $\boldsymbol{g}(\boldsymbol{q})$ involve trigonometric functions of the joint positions $\boldsymbol{q}$, whose evaluations, realized mostly by digital equipment (*e.g.* ordinary personal computers) take a longer time than the evaluation of the 'PD-part' of the control law. In certain applications, the (high) sampling frequency specified may not allow one to evaluate $\boldsymbol{g}(\boldsymbol{q}(t))$ permanently. Naturally, an *ad hoc* solution to this situation is to implement the control law at two sampling frequencies: a high frequency for the evaluation of the PD-part, and a low frequency for the evaluation of $\boldsymbol{g}(\boldsymbol{q}(t))$. An alternative solution consists in using a variant of this controller, the so-called *PD control with desired gravity compensation*. The study of this controller is precisely the subject of the present chapter.

The PD control law with desired gravity compensation is given by

$$\boldsymbol{\tau} = K_p\tilde{\boldsymbol{q}} + K_v\dot{\tilde{\boldsymbol{q}}} + \boldsymbol{g}(\boldsymbol{q}_d) \tag{8.1}$$

where $K_p, K_v \in \mathbb{R}^{n\times n}$ are symmetric positive definite matrices chosen by the designer. As is customary, the position error is denoted by $\tilde{\boldsymbol{q}} = \boldsymbol{q}_d - \boldsymbol{q} \in \mathbb{R}^n$, where $\boldsymbol{q}_d$ stands for the desired joint position. Figure 8.1 presents the block-diagram of the PD control law with desired gravity compensation for robot manipulators. Notice that the only difference with respect to the PD controller with gravity compensation (7.1) is that the term $\boldsymbol{g}(\boldsymbol{q}_d)$ replaces $\boldsymbol{g}(\boldsymbol{q})$. The practical convenience of this controller is evident when the desired position $\boldsymbol{q}_d(t)$ is periodic or constant. Indeed, the vector $\boldsymbol{g}(\boldsymbol{q}_d)$, which depends on $\boldsymbol{q}_d$ and not on $\boldsymbol{q}$, may be evaluated off-line once $\boldsymbol{q}_d$ has been defined and therefore, it is not necessary to evaluate $\boldsymbol{g}(\boldsymbol{q})$ in real time.

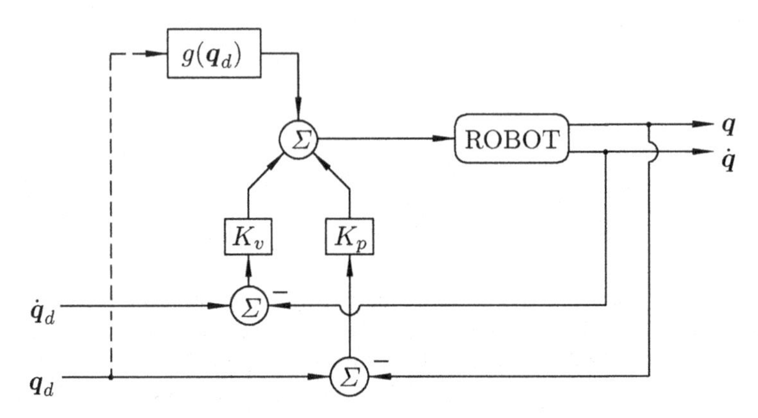

Figure 8.1. Block-diagram: PD control with desired gravity compensation

The closed-loop equation we get by combining the equation of the robot model (II.1) and the equation of the controller (8.1) is

$$M(\boldsymbol{q})\ddot{\boldsymbol{q}} + C(\boldsymbol{q}, \dot{\boldsymbol{q}})\dot{\boldsymbol{q}} + \boldsymbol{g}(\boldsymbol{q}) = K_p\tilde{\boldsymbol{q}} + K_v\dot{\tilde{\boldsymbol{q}}} + \boldsymbol{g}(\boldsymbol{q}_d)$$

or equivalently, in terms of the state vector $\begin{bmatrix}\tilde{\boldsymbol{q}}^T & \dot{\tilde{\boldsymbol{q}}}^T\end{bmatrix}^T$,

$$\frac{d}{dt}\begin{bmatrix}\tilde{\boldsymbol{q}} \\ \dot{\tilde{\boldsymbol{q}}}\end{bmatrix} = \begin{bmatrix}\dot{\tilde{\boldsymbol{q}}} \\ \ddot{\boldsymbol{q}}_d - M(\boldsymbol{q})^{-1}\left[K_p\tilde{\boldsymbol{q}} + K_v\dot{\tilde{\boldsymbol{q}}} - C(\boldsymbol{q}, \dot{\boldsymbol{q}})\dot{\boldsymbol{q}} + \boldsymbol{g}(\boldsymbol{q}_d) - \boldsymbol{g}(\boldsymbol{q})\right]\end{bmatrix}$$

which represents a nonautonomous nonlinear differential equation. The necessary and sufficient condition for the origin $\begin{bmatrix}\tilde{\boldsymbol{q}}^T & \dot{\tilde{\boldsymbol{q}}}^T\end{bmatrix}^T = \boldsymbol{0} \in \mathbb{R}^{2n}$ to be an equilibrium of the closed-loop equation, is that the desired joint position $\boldsymbol{q}_d$ satisfies

$$M(\boldsymbol{q}_d)\ddot{\boldsymbol{q}}_d + C(\boldsymbol{q}_d, \dot{\boldsymbol{q}}_d)\dot{\boldsymbol{q}}_d = \boldsymbol{0} \in \mathbb{R}^n$$

or equivalently, that $\boldsymbol{q}_d(t)$ be a solution of

$$\frac{d}{dt}\begin{bmatrix}\boldsymbol{q}_d \\ \dot{\boldsymbol{q}}_d\end{bmatrix} = \begin{bmatrix}\dot{\boldsymbol{q}}_d \\ -M(\boldsymbol{q}_d)^{-1}\left[C(\boldsymbol{q}_d, \dot{\boldsymbol{q}}_d)\dot{\boldsymbol{q}}_d\right]\end{bmatrix}$$

for any initial condition $\left[\boldsymbol{q}_d(0)^T \ \ \dot{\boldsymbol{q}}_d(0)^T\right]^T \in \mathbb{R}^{2n}$.

Obviously, in the scenario where the desired position $\boldsymbol{q}_d(t)$ does not satisfy the established condition, the origin may not be an equilibrium point of the closed-loop equation and therefore, it may not be expected to satisfy the motion control objective, that is, to drive the position error $\tilde{\boldsymbol{q}}(t)$ asymptotically to zero.

A sufficient condition for the origin $\left[\tilde{\boldsymbol{q}}^T \ \ \dot{\tilde{\boldsymbol{q}}}^T\right]^T = \boldsymbol{0} \in \mathbb{R}^{2n}$ to be an equilibrium point of the closed-loop equation is that the desired joint position $\boldsymbol{q}_d$ be a constant vector. In what is left of this chapter we assume that this is the case.

As we show below, this controller may verify the position objective globally, that is,

$$\lim_{t \to \infty} \boldsymbol{q}(t) = \boldsymbol{q}_d$$

where $\boldsymbol{q}_d \in \mathbb{R}^n$ is a any constant vector and the robot may start off from any configuration. We emphasize that the controller "may achieve" the position control objective under the condition that K_p is chosen sufficiently 'large'. Later on in this chapter, we quantify 'large'.

Considering the desired position $\boldsymbol{q}_d$ to be constant, the closed-loop equation may be written in terms of the new state vector $\left[\tilde{\boldsymbol{q}}^T \ \ \dot{\boldsymbol{q}}^T\right]^T$ as

$$\frac{d}{dt}\begin{bmatrix} \tilde{\boldsymbol{q}} \\ \dot{\boldsymbol{q}} \end{bmatrix} = \begin{bmatrix} -\dot{\boldsymbol{q}} \\ M(\boldsymbol{q})^{-1}\left[K_p\tilde{\boldsymbol{q}} - K_v\dot{\boldsymbol{q}} - C(\boldsymbol{q},\dot{\boldsymbol{q}})\dot{\boldsymbol{q}} + \boldsymbol{g}(\boldsymbol{q}_d) - \boldsymbol{g}(\boldsymbol{q})\right] \end{bmatrix} \tag{8.2}$$

that is, in the form of a nonlinear autonomous differential equation whose origin $\left[\tilde{\boldsymbol{q}}^T \ \ \dot{\boldsymbol{q}}^T\right]^T = \boldsymbol{0} \in \mathbb{R}^{2n}$ is an equilibrium point. Nevertheless, besides the origin, there may exist other equilibria. Indeed, there are as many equilibria as solutions in $\tilde{\boldsymbol{q}}$, may have the equation

$$K_p\tilde{\boldsymbol{q}} = \boldsymbol{g}(\boldsymbol{q}_d - \tilde{\boldsymbol{q}}) - \boldsymbol{g}(\boldsymbol{q}_d)\,. \tag{8.3}$$

Naturally, the explicit solutions of (8.3) are hard to obtain. Nevertheless, as we show that later, if K_p is taken sufficiently "large", then $\tilde{\boldsymbol{q}} = \boldsymbol{0} \in \mathbb{R}^n$ is the unique solution.

Example 8.1. Consider the model of the ideal pendulum studied in Example 2.2 (see page 30)

$$J\ddot{q} + mgl\,\sin(q) = \tau$$

where we identify $g(q) = mgl\,\sin(q)$.

In this case, the expression (8.3) takes the form

$$k_p\tilde{q} = mgl\ [\sin(q_d - \tilde{q}) - \sin(q_d)]\ . \tag{8.4}$$

For the sake of illustration, consider the following numerical values,

$$\begin{array}{ll} J = 1 & mgl = 1 \\ k_p = 0.25 & q_d = \pi/2\,. \end{array}$$

Either via a graphical method or numerical algorithms, one may verify that Equation (8.4) possess exactly three solutions in $\tilde{q}$. The approximated values of these solutions are: 0 (rad), -0.51 (rad) and -4.57 (rad). This means that the PD control law with desired gravity compensation in closed loop with the model of the ideal pendulum has as equilibria,

$$\begin{bmatrix} \tilde{q} \\ \dot{q} \end{bmatrix} \in \left\{ \begin{bmatrix} 0 \\ 0 \end{bmatrix}, \begin{bmatrix} -0.51 \\ 0 \end{bmatrix}, \begin{bmatrix} -4.57 \\ 0 \end{bmatrix} \right\}.$$

Consider now a larger value for k_p (sufficiently "large"), *e.g.*

$$k_p = 1.25$$

In this scenario, it may be verified numerically that Equation (8.4) has a unique solution at $\tilde{q} = 0$ (rad). This means that the PD control law with desired gravity compensation in closed loop with the model of the ideal pendulum, has the origin as its unique equilibrium, *i.e.*

$$\begin{bmatrix} \tilde{q} \\ \dot{q} \end{bmatrix} = \begin{bmatrix} 0 \\ 0 \end{bmatrix} \in \mathbb{R}^2\,.$$

The rest of the chapter focuses on:

- boundedness of solutions;
- unicity of the equilibrium;
- global asymptotic stability.

The studies presented here are limited to the case of robots whose joints are all revolute.

8.1 Boundedness of Position and Velocity Errors, $\tilde{q}$ and $\dot{q}$

Assuming that the design matrices K_p and K_v are positive definite (without assuming that K_p is sufficiently "large"), and of course, for a desired constant position $\boldsymbol{q}_d$ to this point, we only know that the closed-loop Equation (8.2) has an equilibrium at the origin, but there might also be other equilibria. In spite of this, we show by using Lemma 2.2 that both, the position error $\tilde{\boldsymbol{q}}(t)$ and the velocity error $\dot{\boldsymbol{q}}(t)$ remain bounded for all initial conditions $\left[\tilde{\boldsymbol{q}}(0)^T \;\; \dot{\boldsymbol{q}}(0)^T\right]^T \in \mathbb{R}^{2n}$.

Define the function (later on, we show that it is non-negative definite)

$$\begin{aligned} V(\tilde{\boldsymbol{q}}, \dot{\boldsymbol{q}}) = \mathcal{K}(\boldsymbol{q}, \dot{\boldsymbol{q}}) + \mathcal{U}(\boldsymbol{q}) - k_{\mathcal{U}} + \frac{1}{2}\tilde{\boldsymbol{q}}^T K_p \tilde{\boldsymbol{q}} \\ + \tilde{\boldsymbol{q}}^T \boldsymbol{g}(\boldsymbol{q}_d) + \frac{1}{2}\boldsymbol{g}(\boldsymbol{q}_d)^T K_p^{-1} \boldsymbol{g}(\boldsymbol{q}_d) \end{aligned}$$

where $\mathcal{K}(\boldsymbol{q}, \dot{\boldsymbol{q}})$ and $\mathcal{U}(\boldsymbol{q})$ denote the kinetic and potential energy functions of the robot, and the constant $k_{\mathcal{U}}$ is defined as (see Property 4.3)

$$k_{\mathcal{U}} = \min_{q}\{\mathcal{U}(\boldsymbol{q})\}.$$

The function $V(\tilde{\boldsymbol{q}}, \dot{\boldsymbol{q}})$ may be written as

$$V(\tilde{\boldsymbol{q}}, \dot{\boldsymbol{q}}) = \dot{\boldsymbol{q}}^T P(\tilde{\boldsymbol{q}}) \dot{\boldsymbol{q}} + h(\tilde{\boldsymbol{q}}) \tag{8.5}$$

where

$$\begin{aligned} P(\tilde{\boldsymbol{q}}) &:= \frac{1}{2} M(\boldsymbol{q}_d - \tilde{\boldsymbol{q}}) \\ h(\tilde{\boldsymbol{q}}) &:= \mathcal{U}(\boldsymbol{q}_d - \tilde{\boldsymbol{q}}) - k_{\mathcal{U}} + \frac{1}{2}\tilde{\boldsymbol{q}}^T K_p \tilde{\boldsymbol{q}} + \tilde{\boldsymbol{q}}^T \boldsymbol{g}(\boldsymbol{q}_d) + \frac{1}{2}\boldsymbol{g}(\boldsymbol{q}_d)^T K_p^{-1} \boldsymbol{g}(\boldsymbol{q}_d). \end{aligned}$$

Since we assumed that the robot has only revolute joints, $\mathcal{U}(\boldsymbol{q}) - k_{\mathcal{U}} \geq 0$ for all $\boldsymbol{q} \in \mathbb{R}^n$. On the other hand, we have

$$\frac{1}{2}\tilde{\boldsymbol{q}}^T K_p \tilde{\boldsymbol{q}} + \tilde{\boldsymbol{q}}^T \boldsymbol{g}(\boldsymbol{q}_d) + \frac{1}{2}\boldsymbol{g}(\boldsymbol{q}_d)^T K_p^{-1} \boldsymbol{g}(\boldsymbol{q}_d),$$

may be written as

$$\frac{1}{2}\begin{bmatrix} \tilde{\boldsymbol{q}} \\ \boldsymbol{g}(\boldsymbol{q}_d) \end{bmatrix}^T \begin{bmatrix} K_p & I \\ I & K_p^{-1} \end{bmatrix} \begin{bmatrix} \tilde{\boldsymbol{q}} \\ \boldsymbol{g}(\boldsymbol{q}_d) \end{bmatrix}$$

which is non-negative for all $\tilde{\boldsymbol{q}}, \boldsymbol{q}_d \in \mathbb{R}^n$. Therefore, the function $h(\tilde{\boldsymbol{q}})$ is also non-negative. Naturally, since the kinetic energy $\frac{1}{2}\dot{\boldsymbol{q}}^T M(\boldsymbol{q})\dot{\boldsymbol{q}}$ is a positive definite function of $\dot{\boldsymbol{q}}$, then the function $V(\tilde{\boldsymbol{q}}, \dot{\boldsymbol{q}})$ is non-negative for all $\tilde{\boldsymbol{q}}, \dot{\boldsymbol{q}} \in \mathbb{R}^n$.

The time derivative of $V(\tilde{\boldsymbol{q}}, \dot{\boldsymbol{q}})$ is

$$\dot{V}(\tilde{\boldsymbol{q}}, \dot{\boldsymbol{q}}) = \dot{\boldsymbol{q}}^T M(\boldsymbol{q})\ddot{\boldsymbol{q}} + \frac{1}{2}\dot{\boldsymbol{q}}^T \dot{M}(\boldsymbol{q})\dot{\boldsymbol{q}} + \dot{\boldsymbol{q}}^T \boldsymbol{g}(\boldsymbol{q}) + \tilde{\boldsymbol{q}}^T K_p \dot{\tilde{\boldsymbol{q}}} + \dot{\tilde{\boldsymbol{q}}}^T \boldsymbol{g}(\boldsymbol{q}_d) \tag{8.6}$$

where we used (3.20), *i.e.* $\boldsymbol{g}(\boldsymbol{q}) = \frac{\partial}{\partial \boldsymbol{q}}\mathcal{U}(\boldsymbol{q})$. Solving for $M(\boldsymbol{q})\ddot{\boldsymbol{q}}$ in the closed-loop Equation (8.2) and substituting in (8.6) we get

$$\dot{V}(\tilde{\boldsymbol{q}}, \dot{\boldsymbol{q}}) = \dot{\boldsymbol{q}}^T K_p \tilde{\boldsymbol{q}} - \dot{\boldsymbol{q}}^T K_v \dot{\boldsymbol{q}} + \dot{\boldsymbol{q}}^T \boldsymbol{g}(\boldsymbol{q}_d) + \tilde{\boldsymbol{q}}^T K_p \dot{\tilde{\boldsymbol{q}}} + \dot{\tilde{\boldsymbol{q}}}^T \boldsymbol{g}(\boldsymbol{q}_d) \tag{8.7}$$

where the term $\dot{\boldsymbol{q}}^T\left[\frac{1}{2}\dot{M} - C\right]\dot{\boldsymbol{q}}$ was eliminated by virtue of Property 4.2. Recalling that the vector $\boldsymbol{q}_d$ is constant and that $\tilde{\boldsymbol{q}} = \boldsymbol{q}_d - \boldsymbol{q}$, then $\dot{\tilde{\boldsymbol{q}}} = -\dot{\boldsymbol{q}}$. Incorporating this in Equation (8.7) we obtain

$$\dot{V}(\tilde{\boldsymbol{q}}, \dot{\boldsymbol{q}}) = -\dot{\boldsymbol{q}}^T K_v \dot{\boldsymbol{q}}\,. \tag{8.8}$$

Using $V(\tilde{\boldsymbol{q}}, \dot{\boldsymbol{q}})$ and $\dot{V}(\tilde{\boldsymbol{q}}, \dot{\boldsymbol{q}})$ given by (8.5) and (8.8) respectively, and invoking Lemma 2.2 (*cf.* page 52), we conclude that both, $\dot{\boldsymbol{q}}(t)$ and $\tilde{\boldsymbol{q}}(t)$ are also bounded and that the velocities vector $\dot{\boldsymbol{q}}(t)$, is square integrable, *i.e.*

$$\int_0^\infty \|\dot{\boldsymbol{q}}(t)\|^2 dt < \infty\,. \tag{8.9}$$

As a matter of fact, it may be shown that the velocity $\dot{\boldsymbol{q}}$ is not only bounded, but that it also tends asymptotically to zero. For this, notice from (8.2) that

$$\ddot{\boldsymbol{q}} = M(\boldsymbol{q})^{-1}\left[K_p\tilde{\boldsymbol{q}} - K_v\dot{\boldsymbol{q}} + \boldsymbol{g}(\boldsymbol{q}_d) - \boldsymbol{g}(\boldsymbol{q}) - C(\boldsymbol{q}, \dot{\boldsymbol{q}})\dot{\boldsymbol{q}}\right]\,. \tag{8.10}$$

Since $\dot{\boldsymbol{q}}(t)$ and $\tilde{\boldsymbol{q}}(t)$ were shown to be bounded then it follows from Properties 4.2 and 4.3 that $C(\boldsymbol{q}(t), \dot{\boldsymbol{q}}(t))\dot{\boldsymbol{q}}(t)$ and $\boldsymbol{g}(\boldsymbol{q}(t))$ are also bounded. On the other hand, $M(\boldsymbol{q})^{-1}$ is a bounded matrix (from Property 4.1), and finally, from (8.10) we conclude that the accelerations vector $\ddot{\boldsymbol{q}}(t)$ is also bounded and therefore, from (8.9) and Lemma 2.2, we conclude that

$$\lim_{t\to\infty} \dot{\tilde{\boldsymbol{q}}}(t) = \lim_{t\to\infty} \dot{\boldsymbol{q}}(t) = \boldsymbol{0}\,.$$

For the sake of completeness we show next how to compute explicit upper-bounds on the position and velocity errors. Taking into account that $V(\tilde{\boldsymbol{q}}, \dot{\boldsymbol{q}})$ is a non-negative function that decreases along trajectories (*i.e.* $\frac{d}{dt}V(\tilde{\boldsymbol{q}}, \dot{\boldsymbol{q}}) \leq 0$), we have

$$0 \leq V(\tilde{\boldsymbol{q}}(t), \dot{\boldsymbol{q}}(t)) \leq V(\tilde{\boldsymbol{q}}(0), \dot{\boldsymbol{q}}(0))$$

for all $t \geq 0$. Consequently, considering the definition of $V(\tilde{\boldsymbol{q}}, \dot{\boldsymbol{q}})$ we deduce immediately that

$$\frac{1}{2}\tilde{\boldsymbol{q}}(t)^T K_p \tilde{\boldsymbol{q}}(t) + \tilde{\boldsymbol{q}}(t)^T \boldsymbol{g}(\boldsymbol{q}_d) + \frac{1}{2}\boldsymbol{g}(\boldsymbol{q}_d)^T K_p^{-1} \boldsymbol{g}(\boldsymbol{q}_d) \leq V(\tilde{\boldsymbol{q}}(0), \dot{\boldsymbol{q}}(0)) \tag{8.11}$$

$$\frac{1}{2}\dot{\boldsymbol{q}}(t)^T M(\boldsymbol{q}(t))\dot{\boldsymbol{q}}(t) \leq V(\tilde{\boldsymbol{q}}(0), \dot{\boldsymbol{q}}(0)) \tag{8.12}$$

for all $t \geq 0$, and where

$$V(\tilde{\boldsymbol{q}}(0), \dot{\boldsymbol{q}}(0)) = \frac{1}{2}\dot{\boldsymbol{q}}(0)^T M(\boldsymbol{q}(0))\dot{\boldsymbol{q}}(0) + \mathcal{U}(\boldsymbol{q}(0)) - k_{\mathcal{U}} + \frac{1}{2}\tilde{\boldsymbol{q}}(0)^T K_p \tilde{\boldsymbol{q}}(0) + \boldsymbol{g}(\boldsymbol{q}_d)^T \tilde{\boldsymbol{q}}(0) + \frac{1}{2}\boldsymbol{g}(\boldsymbol{q}_d)^T K_p^{-1} \boldsymbol{g}(\boldsymbol{q}_d) .$$

The value of $V(\tilde{\boldsymbol{q}}(0), \dot{\boldsymbol{q}}(0))$ may be obtained if we know the inertia matrix $M(\boldsymbol{q})$ and the vector of gravitational torques $\boldsymbol{g}(\boldsymbol{q})$. Naturally, we assume here that the position $\boldsymbol{q}(t)$, the velocity $\dot{\boldsymbol{q}}(t)$ and, in particular at the instant $t = 0$, are measured by appropriate instruments physically collocated for this purpose on the robot.

We obtain next, explicit bounds on $\|\tilde{\boldsymbol{q}}\|$ and $\|\dot{\boldsymbol{q}}\|$ as a function of the initial conditions. We first notice that

$$\frac{\lambda_{\min}\{K_p\}}{2}\|\tilde{\boldsymbol{q}}\|^2 - \|\boldsymbol{g}(\boldsymbol{q}_d)\|\,\|\tilde{\boldsymbol{q}}\| \leq \frac{1}{2}\tilde{\boldsymbol{q}}^T K_p \tilde{\boldsymbol{q}} + \boldsymbol{g}(\boldsymbol{q}_d)^T \tilde{\boldsymbol{q}} + \underbrace{\frac{1}{2}\boldsymbol{g}(\boldsymbol{q}_d)^T K_p^{-1} \boldsymbol{g}(\boldsymbol{q}_d)}_{c}$$

where we used the fact that $c \geq 0$ and that for all vectors $\boldsymbol{x}$ and $\boldsymbol{y} \in \mathbb{R}^n$ we have $-\boldsymbol{x}^T\boldsymbol{y} \leq \left|\boldsymbol{x}^T\boldsymbol{y}\right| \leq \|\boldsymbol{x}\|\,\|\boldsymbol{y}\|$, so $-\|\boldsymbol{x}\|\,\|\boldsymbol{y}\| \leq \boldsymbol{x}^T\boldsymbol{y}$. Taking (8.11) into account, we have

$$\frac{\lambda_{\min}\{K_p\}}{2}\|\tilde{\boldsymbol{q}}\|^2 - \|\boldsymbol{g}(\boldsymbol{q}_d)\|\,\|\tilde{\boldsymbol{q}}\| - V(\tilde{\boldsymbol{q}}(0), \dot{\boldsymbol{q}}(0)) \leq 0$$

from which we finally obtain

$$\|\tilde{\boldsymbol{q}}(t)\| \leq \frac{\|\boldsymbol{g}(\boldsymbol{q}_d)\| + \sqrt{\|\boldsymbol{g}(\boldsymbol{q}_d)\|^2 + 2\lambda_{\min}\{K_p\}V(\tilde{\boldsymbol{q}}(0), \dot{\boldsymbol{q}}(0))}}{\lambda_{\min}\{K_p\}} \tag{8.13}$$

for all $t \geq 0$.

On the other hand, it is clear from (8.12) that

$$\|\dot{\boldsymbol{q}}(t)\|^2 \leq \frac{2V(\tilde{\boldsymbol{q}}(0), \dot{\boldsymbol{q}}(0))}{\lambda_{\min}\{M(\boldsymbol{q})\}} \tag{8.14}$$

for all $t \geq 0$. The expressions (8.13) and (8.14) establish the bounds we were looking for.

Example 8.2. Consider again the model of the ideal pendulum from Example 8.1

$$J\ddot{q} + mgl\,\sin(q) = \tau,$$

where we clearly identify $M(q) = J$ and $g(q) = mgl\,\sin(q)$. As has been shown before in Example 2.11 (see page 45), the potential energy function is given by

$$\mathcal{U}(q) = mgl[1 - \cos(q)] .$$

Since $\min_q\{\mathcal{U}(q)\} = 0$, the constant $k_{\mathcal{U}}$ takes the value of zero. Consider the numerical values used in Example 8.1

$$\begin{array}{ll} J = 1 & mgl = 1 \\ k_p = 0.25 & k_v = 0.50 \\ q_d = \pi/2 . & \end{array}$$

Assume that we use PD control with desired gravity compensation to control the ideal pendulum from the initial conditions $q(0) = 0$ and $\dot{q}(0) = 0$.

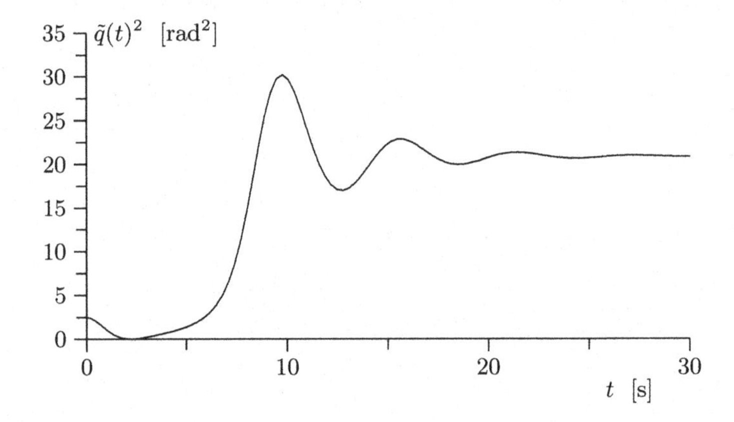

Figure 8.2. PD control with desired gravity compensation: graph of the position error $\tilde{q}(t)^2$

With the previous values it is easy to verify that

$$\begin{aligned} g(q_d) &= mgl \sin(\pi/2) = 1 \\ V(\tilde{q}(0), \dot{q}(0)) &= \frac{1}{2}k_p\tilde{q}^2(0) + mgl\tilde{q}(0) + \frac{1}{2k_p}(mgl)^2 = 3.87 . \end{aligned}$$

According to the bounds (8.13) and (8.14) and taking into account the previous information, we get

$$\begin{aligned} \tilde{q}^2(t) &\leq \left[\frac{mgl + \sqrt{[mgl + k_p\tilde{q}(0)]^2 + (mgl)^2}}{k_p}\right]^2 \\ &\leq 117.79 \ [\mathrm{rad}^2] \end{aligned} \tag{8.15}$$

$$\dot{q}^2(t) \leq \frac{2}{J}\left[\frac{k_p}{2}\tilde{q}^2(0) + mgl\tilde{q}(0) + \frac{1}{2k_p}(mgl)^2\right]$$

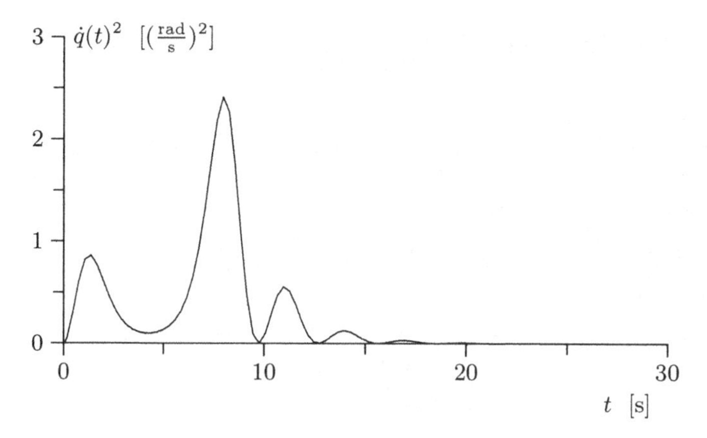

Figure 8.3. PD control with desired gravity compensation: graph of velocity, $\dot{q}(t)^2$

$$\leq 7.75 \left[\frac{\text{rad}}{\text{s}}\right]^2, \tag{8.16}$$

for all $t \geq 0$. Figures 8.2 and 8.3 show the plots of $\tilde{q}(t)^2$ and $\dot{q}(t)^2$ respectively, obtained by simulation. We clearly appreciate from the plots that both variables satisfy the inequalities (8.15) and (8.16). Finally, it is interesting to observe from Figure 8.2 that $\lim_{t\to\infty} \tilde{q}(t)^2 = 20.88$ (evidence from simulation shows that $\lim_{t\to\infty} \tilde{q}(t) = -4.57$) and $\lim_{t\to\infty} \dot{q}^2(t) = 0$ and therefore

$$\lim_{t\to\infty} \begin{bmatrix} \tilde{q}(t) \\ \dot{q}(t) \end{bmatrix} = \begin{bmatrix} -4.57 \\ 0 \end{bmatrix}.$$

This means that the solutions tend precisely to one among the three equilibria computed in Example 8.1, but which do not correspond to the origin. The moral of this example is that PD control with desired gravity compensation may fail to meet the position control objective.

To summarize the developments above we make the following remarks. Consider the PD control law with desired gravity compensation for robots with revolute joints. Assume that the design matrices K_p and K_v are positive definite. If the desired joint position $\boldsymbol{q}_d(t)$ is a constant vector, then:

- the position error $\tilde{\boldsymbol{q}}(t)$ and the velocity $\dot{\boldsymbol{q}}(t)$ are bounded. Maximal bounds on their norms are given by the expressions (8.13) and (8.14) respectively.
- $\lim_{t\to\infty} \dot{\boldsymbol{q}}(t) = \boldsymbol{0} \in \mathbb{R}^n$.

8.2 Unicity of Equilibrium

For robots having only revolute joints, we show that with the choice of K_p sufficiently "large", we can guarantee unicity of the equilibrium for the closed-loop Equation (8.2). To that end, we use here the *contraction mapping theorem* (*cf.* Theorem 2.1 on page 26).

The equilibria of the closed-loop Equation (8.2) satisfy $\begin{bmatrix}\tilde{\boldsymbol{q}}^T & \dot{\boldsymbol{q}}^T\end{bmatrix}^T = \begin{bmatrix}\tilde{\boldsymbol{q}}^T & \boldsymbol{0}^T\end{bmatrix}^T \in \mathbb{R}^{2n}$ where $\tilde{\boldsymbol{q}} \in \mathbb{R}^n$ solves (8.3),

$$\begin{aligned}\tilde{\boldsymbol{q}} &= K_p^{-1}\left[\boldsymbol{g}(\boldsymbol{q}_d - \tilde{\boldsymbol{q}}) - \boldsymbol{g}(\boldsymbol{q}_d)\right] \\ &= \boldsymbol{f}(\tilde{\boldsymbol{q}}, \boldsymbol{q}_d)\,.\end{aligned}$$

Naturally, $\tilde{\boldsymbol{q}} = \boldsymbol{0} \in \mathbb{R}^n$ is a trivial solution of $\tilde{\boldsymbol{q}} = \boldsymbol{f}(\tilde{\boldsymbol{q}}, \boldsymbol{q}_d)$, but as has been illustrated above in Example 8.1, there may exist other solutions.

If the function $\boldsymbol{f}(\tilde{\boldsymbol{q}}, \boldsymbol{q}_d)$ satisfies the condition of the contraction mapping theorem, that is, if $\boldsymbol{f}(\tilde{\boldsymbol{q}}, \boldsymbol{q}_d)$ is Lipschitz (*cf.* page 101) with Lipschitz constant strictly smaller than 1, then the equation $\tilde{\boldsymbol{q}} = \boldsymbol{f}(\tilde{\boldsymbol{q}}, \boldsymbol{q}_d)$ has a unique solution $\tilde{\boldsymbol{q}}^*$ and consequently, the unique equilibrium of the closed-loop Equation (8.2) is $\begin{bmatrix}\tilde{\boldsymbol{q}}^T & \dot{\boldsymbol{q}}^T\end{bmatrix}^T = \begin{bmatrix}\tilde{\boldsymbol{q}}^{*T} & \boldsymbol{0}^T\end{bmatrix}^T \in \mathbb{R}^{2n}$.

Now, notice that for all vectors $\boldsymbol{x}, \boldsymbol{y} \in \mathbb{R}^n$

$$\begin{aligned}\left\|\boldsymbol{f}(\boldsymbol{x}, \boldsymbol{q}_d) - \boldsymbol{f}(\boldsymbol{y}, \boldsymbol{q}_d)\right\| &= \left\|K_p^{-1}\boldsymbol{g}(\boldsymbol{q}_d - \boldsymbol{x}) - K_p^{-1}\boldsymbol{g}(\boldsymbol{q}_d - \boldsymbol{y})\right\| \\ &= \left\|K_p^{-1}\left\{\boldsymbol{g}(\boldsymbol{q}_d - \boldsymbol{x}) - \boldsymbol{g}(\boldsymbol{q}_d - \boldsymbol{y})\right\}\right\| \\ &\leq \lambda_{\text{Max}}\{K_p^{-1}\}\left\|\boldsymbol{g}(\boldsymbol{q}_d - \boldsymbol{x}) - \boldsymbol{g}(\boldsymbol{q}_d - \boldsymbol{y})\right\|\,.\end{aligned}$$

On the other hand, using the fact that $\lambda_{\text{Max}}\{A^{-1}\} = 1/\lambda_{\text{min}}\{A\}$ for any symmetric positive definite matrix A, and Property 4.3 that guarantees the existence of a positive constant k_g such that $\|\boldsymbol{g}(\boldsymbol{x}) - \boldsymbol{g}(\boldsymbol{y})\| \leq k_g \|\boldsymbol{x} - \boldsymbol{y}\|$, we have

$$\left\|\boldsymbol{f}(\boldsymbol{x}, \boldsymbol{q}_d) - \boldsymbol{f}(\boldsymbol{y}, \boldsymbol{q}_d)\right\| \leq \frac{k_g}{\lambda_{\text{min}}\{K_p\}}\left\|\boldsymbol{x} - \boldsymbol{y}\right\|,$$

which, according to the contraction mapping theorem, implies that a sufficient condition for unicity of the solution of $\boldsymbol{f}(\tilde{\boldsymbol{q}}, \boldsymbol{q}_d) - \tilde{\boldsymbol{q}} = \boldsymbol{0}$ or equivalently of

$$K_p^{-1}\left[\boldsymbol{g}(\boldsymbol{q}_d - \tilde{\boldsymbol{q}}) - \boldsymbol{g}(\boldsymbol{q}_d)\right] - \tilde{\boldsymbol{q}} = \boldsymbol{0}$$

and consequently, for the unicity of the equilibrium of the closed-loop equation, is that K_p be chosen so as to satisfy

$$\lambda_{\text{min}}\{K_p\} > k_g\,. \tag{8.17}$$

Therefore, assuming that K_p is chosen so that $\lambda_{\min}\{K_p\} > k_g$, then the unique equilibrium of the closed-loop Equation (8.2) is the origin, $\begin{bmatrix}\tilde{\boldsymbol{q}}^T & \dot{\boldsymbol{q}}^T\end{bmatrix}^T = \begin{bmatrix}\boldsymbol{0}^T & \boldsymbol{0}^T\end{bmatrix}^T \in \mathbb{R}^{2n}$.

8.3 Global Asymptotic Stability

The objective of the present section is to show that the assumption that the matrix K_p satisfies the condition (8.17) is actually also sufficient to guarantee that the origin is globally asymptotically stable for the closed-loop Equation (8.2). To that end we use as usual, Lyapunov's direct method but complemented with La Salle's theorem. This proof is taken from the works cited at the end of the chapter.

First, we present a lemma on positive definite functions of particular relevance to ultimately propose a Lyapunov function candidate.[1]

Lemma 8.1. *Consider the function $f : \mathbb{R}^n \to \mathbb{R}$ given by*

$$f(\tilde{\boldsymbol{q}}) = \mathcal{U}(\boldsymbol{q}_d - \tilde{\boldsymbol{q}}) - \mathcal{U}(\boldsymbol{q}_d) + \boldsymbol{g}(\boldsymbol{q}_d)^T\tilde{\boldsymbol{q}} + \frac{1}{\varepsilon}\tilde{\boldsymbol{q}}^T K_p \tilde{\boldsymbol{q}} \tag{8.18}$$

where $K_p = K_p{}^T > 0$, $\boldsymbol{q}_d \in \mathbb{R}^n$ is a constant vector, ε is a real positive constant number and $\mathcal{U}(\boldsymbol{q})$ is the potential energy function of the robot. If

$$\frac{2}{\varepsilon}K_p + \frac{\partial \boldsymbol{g}(\boldsymbol{q}_d - \tilde{\boldsymbol{q}})}{\partial(\boldsymbol{q}_d - \tilde{\boldsymbol{q}})} > 0$$

for all $\boldsymbol{q}_d, \tilde{\boldsymbol{q}} \in \mathbb{R}^n$, then $f(\tilde{\boldsymbol{q}})$ is a globally positive definite function. The previous condition is satisfied if

$$\lambda_{\min}\{K_p\} > \frac{\varepsilon}{2}k_g$$

where k_g has been defined in Property 4.3, and in turn is such that

$$k_g \geq \left\| \frac{\partial \boldsymbol{g}(\boldsymbol{q})}{\partial \boldsymbol{q}} \right\| .$$

Due to the importance of the above-stated lemma, we present next a detailed proof.

Proof. It consists in establishing that $f(\tilde{\boldsymbol{q}})$ has a global minimum at $\tilde{\boldsymbol{q}} = \boldsymbol{0} \in \mathbb{R}^n$. For this, we use the following result which is well known in optimization techniques. Let $f : \mathbb{R}^n \to \mathbb{R}$ be a function with continuous partial derivatives up to at least the second order. The function $f(\boldsymbol{x})$ has a global minimum at $\boldsymbol{x} = \boldsymbol{0} \in \mathbb{R}^n$ if

[1] See also Example B.2 in Appendix B.

1. The gradient vector of the function $f(\boldsymbol{x})$, evaluated at $\boldsymbol{x} = \boldsymbol{0} \in \mathbb{R}^n$ is zero, *i.e.*
$$\frac{\partial}{\partial \boldsymbol{x}} f(\boldsymbol{0}) = \boldsymbol{0} \in \mathbb{R}^n .$$
2. The Hessian matrix of the function $f(\boldsymbol{x})$, evaluated at each $\boldsymbol{x} \in \mathbb{R}^n$, is positive definite, *i.e.*
$$H(\boldsymbol{x}) = \frac{\partial^2}{\partial x_i \partial x_j} f(\boldsymbol{x}) > 0 .$$

The gradient of $f(\tilde{\boldsymbol{q}})$ with respect to $\tilde{\boldsymbol{q}}$ is

$$\frac{\partial}{\partial \tilde{\boldsymbol{q}}} f(\tilde{\boldsymbol{q}}) = \frac{\partial \mathcal{U}(\boldsymbol{q}_d - \tilde{\boldsymbol{q}})}{\partial \tilde{\boldsymbol{q}}} + \boldsymbol{g}(\boldsymbol{q}_d) + \frac{2}{\varepsilon} K_p \tilde{\boldsymbol{q}} .$$

Recalling from (3.20) that $\boldsymbol{g}(\boldsymbol{q}) = \partial U(\boldsymbol{q})/\partial \boldsymbol{q}$ and that[2]

$$\frac{\partial}{\partial \tilde{\boldsymbol{q}}} \mathcal{U}(\boldsymbol{q}_d - \tilde{\boldsymbol{q}}) = \frac{\partial (\boldsymbol{q}_d - \tilde{\boldsymbol{q}})}{\partial \tilde{\boldsymbol{q}}}^T \frac{\partial \mathcal{U}(\boldsymbol{q}_d - \tilde{\boldsymbol{q}})}{\partial (\boldsymbol{q}_d - \tilde{\boldsymbol{q}})}$$

we finally obtain

$$\frac{\partial}{\partial \tilde{\boldsymbol{q}}} f(\tilde{\boldsymbol{q}}) = -\boldsymbol{g}(\boldsymbol{q}_d - \tilde{\boldsymbol{q}}) + \boldsymbol{g}(\boldsymbol{q}_d) + \frac{2}{\varepsilon} K_p \tilde{\boldsymbol{q}} .$$

Clearly the gradient of $f(\tilde{\boldsymbol{q}})$ is zero for $\tilde{\boldsymbol{q}} = \boldsymbol{0} \in \mathbb{R}^n$. Indeed, one can show that if $\lambda_{\min}\{K_p\} > \frac{\varepsilon}{2} k_g$ the gradient of $f(\tilde{\boldsymbol{q}})$ is zero only at $\tilde{\boldsymbol{q}} = \boldsymbol{0} \in \mathbb{R}^n$. The proof of this claim is similar to the proof of unicity of the equilibrium in Section 8.2.

The Hessian matrix $H(\tilde{\boldsymbol{q}})$ (which by the way, is symmetric) of $f(\tilde{\boldsymbol{q}})$, defined as

$$H(\tilde{\boldsymbol{q}}) = \frac{\partial}{\partial \tilde{\boldsymbol{q}}} \left[\frac{\partial f(\tilde{\boldsymbol{q}})}{\partial \tilde{\boldsymbol{q}}} \right] = \begin{bmatrix} \frac{\partial^2 f(\tilde{\boldsymbol{q}})}{\partial \tilde{q}_1 \partial \tilde{q}_1} & \frac{\partial^2 f(\tilde{\boldsymbol{q}})}{\partial \tilde{q}_1 \partial \tilde{q}_2} & \cdots & \frac{\partial^2 f(\tilde{\boldsymbol{q}})}{\partial \tilde{q}_1 \partial \tilde{q}_n} \\ \frac{\partial^2 f(\tilde{\boldsymbol{q}})}{\partial \tilde{q}_2 \partial \tilde{q}_1} & \frac{\partial^2 f(\tilde{\boldsymbol{q}})}{\partial \tilde{q}_2 \partial \tilde{q}_2} & \cdots & \frac{\partial^2 f(\tilde{\boldsymbol{q}})}{\partial \tilde{q}_2 \partial \tilde{q}_n} \\ \vdots & \vdots & \ddots & \vdots \\ \frac{\partial^2 f(\tilde{\boldsymbol{q}})}{\partial \tilde{q}_n \partial \tilde{q}_1} & \frac{\partial^2 f(\tilde{\boldsymbol{q}})}{\partial \tilde{q}_n \partial \tilde{q}_2} & \cdots & \frac{\partial^2 f(\tilde{\boldsymbol{q}})}{\partial \tilde{q}_n \partial \tilde{q}_n} \end{bmatrix}$$

[2] Let $f : \mathbb{R}^n \to \mathbb{R}$, $\boldsymbol{g} : \mathbb{R}^n \to \mathbb{R}^n$, $\boldsymbol{x}, \boldsymbol{y} \in \mathbb{R}^n$ and $\boldsymbol{x} = \boldsymbol{g}(\boldsymbol{y})$. Then,

$$\frac{\partial f(\boldsymbol{x})}{\partial \boldsymbol{y}} = \left[\frac{\partial \boldsymbol{g}(\boldsymbol{y})}{\partial \boldsymbol{y}} \right]^T \frac{\partial f(\boldsymbol{x})}{\partial \boldsymbol{x}} .$$

corresponds to[3]

$$H(\tilde{\boldsymbol{q}}) = \frac{\partial \boldsymbol{g}(\boldsymbol{q}_d - \tilde{\boldsymbol{q}})}{\partial (\boldsymbol{q}_d - \tilde{\boldsymbol{q}})} + \frac{2}{\varepsilon} K_p \,.$$

Hence, $f(\tilde{\boldsymbol{q}})$ has a (global) minimum at $\tilde{\boldsymbol{q}} = \boldsymbol{0} \in \mathbb{R}^n$ if $H(\tilde{\boldsymbol{q}}) > 0$ for all $\tilde{\boldsymbol{q}} \in \mathbb{R}^n$, in other words, if the symmetric matrix

$$\frac{\partial \boldsymbol{g}(\boldsymbol{q})}{\partial \boldsymbol{q}} + \frac{2}{\varepsilon} K_p \tag{8.19}$$

is positive definite for all $\boldsymbol{q} \in \mathbb{R}^n$.

Here, we use the following result whose proof is given in Example B.2 of Appendix B. Let $A, B \in \mathbb{R}^{n \times n}$ be symmetric matrices. Assume also that the matrix A is positive definite but possibly not B. If $\lambda_{\min}\{A\} > \|B\|$, then the matrix $A + B$ is positive definite. Defining $A = \frac{2}{\varepsilon} K_p$, $B = \frac{\partial \boldsymbol{g}(\boldsymbol{q})}{\partial \boldsymbol{q}}$, and using the result previously mentioned, we conclude that the matrix (8.19) is positive definite if

$$\lambda_{\min}\{K_p\} > \frac{\varepsilon}{2} \left\| \frac{\partial \boldsymbol{g}(\boldsymbol{q})}{\partial \boldsymbol{q}} \right\| \,. \tag{8.20}$$

Since the constant k_g satisfies $k_g \geq \left\| \frac{\partial \boldsymbol{g}(\boldsymbol{q})}{\partial \boldsymbol{q}} \right\|$, then the condition (8.20) is implied by

$$\lambda_{\min}\{K_p\} > \frac{\varepsilon}{2} k_g \,.$$

Therefore, if $\lambda_{\min}\{K_p\} > \frac{\varepsilon}{2} k_g$, then $f(\tilde{\boldsymbol{q}})$ has only one global minimum[4] at $\tilde{\boldsymbol{q}} = \boldsymbol{0} \in \mathbb{R}^n$. Moreover, $f(\boldsymbol{0}) = 0 \in \mathbb{R}$, then $f(\tilde{\boldsymbol{q}})$ is a globally positive definite function. ◇◇◇

We present next, the stability analysis of the closed-loop Equation (8.2) for which we assume that K_p is sufficiently "large" in the sense that its smallest eigenvalue satisfies

$$\lambda_{\min}\{K_p\} > k_g \,.$$

As has been shown in Section 8.2, with this choice of K_p, the closed-loop equation has a unique equilibrium at the origin $\begin{bmatrix} \tilde{\boldsymbol{q}}^T & \dot{\boldsymbol{q}}^T \end{bmatrix}^T = \boldsymbol{0} \in \mathbb{R}^{2n}$.

To study the stability of the latter, we consider the Lyapunov function candidate

$$V(\tilde{\boldsymbol{q}}, \dot{\boldsymbol{q}}) = \frac{1}{2} \dot{\boldsymbol{q}}^T M(\boldsymbol{q}_d - \tilde{\boldsymbol{q}}) \dot{\boldsymbol{q}} + f(\tilde{\boldsymbol{q}}) \tag{8.21}$$

[3] Let $\boldsymbol{f}, \boldsymbol{g} : \mathbb{R}^n \to \mathbb{R}^n$, $\boldsymbol{x}, \boldsymbol{y} \in \mathbb{R}^n$ and $\boldsymbol{x} = \boldsymbol{g}(\boldsymbol{y})$. Then

$$\frac{\partial \boldsymbol{f}(\boldsymbol{x})}{\partial \boldsymbol{y}} = \frac{\partial \boldsymbol{f}(\boldsymbol{x})}{\partial \boldsymbol{x}} \frac{\partial \boldsymbol{g}(\boldsymbol{y})}{\partial \boldsymbol{y}} .$$

[4] It is worth emphasizing that it is not redundant to speak of a *unique* global minimum.

where $f(\tilde{q})$ is given in (8.18) with $\varepsilon = 2$. In other words, this Lyapunov function candidate may be written as

$$\begin{aligned} V(\tilde{q}, \dot{q}) = & \frac{1}{2}\dot{q}^T M(q_d - \tilde{q})\dot{q} + \mathcal{U}(q_d - \tilde{q}) - \mathcal{U}(q_d) \\ & + g(q_d)^T\tilde{q} + \frac{1}{2}\tilde{q}^T K_p \tilde{q} \,. \end{aligned}$$

The previous function is globally positive definite since it is the sum of a globally positive definite term $\dot{q}$: $\dot{q}^T M(q)\dot{q}$, and another globally positive definite term of $\tilde{q}$: $f(\tilde{q})$.

The time derivative of $V(\tilde{q}, \dot{q})$ is given by

$$\begin{aligned} \dot{V}(\tilde{q}, \dot{q}) = & \dot{q}^T M(q)\ddot{q} + \frac{1}{2}\dot{q}^T \dot{M}(q)\dot{q} \\ & + \dot{q}^T g(q_d - \tilde{q}) - g(q_d)^T\dot{q} - \tilde{q}^T K_p \dot{q} \,, \end{aligned}$$

where we used $g(q_d - \tilde{q}) = \partial\mathcal{U}(q_d - \tilde{q})/\partial(q_d - \tilde{q})$ and also

$$\begin{aligned} \frac{d}{dt}\mathcal{U}(q_d - \tilde{q}) & = \dot{\tilde{q}}^T \frac{\partial\mathcal{U}(q_d - \tilde{q})}{\partial\tilde{q}} \\ & = \dot{\tilde{q}}^T \frac{\partial(q_d - \tilde{q})}{\partial\tilde{q}}^T \frac{\partial\mathcal{U}(q_d - \tilde{q})}{\partial(q_d - \tilde{q})} \\ & = \dot{\tilde{q}}^T(-I)g(q_d - \tilde{q}) \\ & = \dot{q}^T g(q_d - \tilde{q}) \,. \end{aligned}$$

Solving for $M(q)\ddot{q}$ from the closed-loop Equation (8.2) and substituting its value, we get

$$\dot{V}(\tilde{q}, \dot{q}) = -\dot{q}^T K_v \dot{q}$$

where we also used Property 4.2 to eliminate $\dot{q}^T\left[\frac{1}{2}\dot{M} - C\right]\dot{q}$. Since $-\dot{V}(\tilde{q}, \dot{q})$ is a positive semidefinite function, the origin is stable (*cf.* Theorem 2.2).

Since the closed-loop Equation (8.2) is autonomous, we may explore the application of La Salle's Theorem (*cf.* Theorem 2.7) to analyze the global asymptotic stability of the origin.

To that end, notice that the set Ω is here given by

$$\begin{aligned} \Omega & = \left\{ x \in \mathbb{R}^{2n} : \dot{V}(x) = 0 \right\} \\ & = \left\{ x = \begin{bmatrix} \tilde{q} \\ \dot{q} \end{bmatrix} \in \mathbb{R}^{2n} : \dot{V}(\tilde{q}, \dot{q}) = 0 \right\} \\ & = \{ \tilde{q} \in \mathbb{R}^n, \dot{q} = \mathbf{0} \in \mathbb{R}^n \} \,. \end{aligned}$$

Observe that $\dot{V}(\tilde{\boldsymbol{q}},\dot{\boldsymbol{q}}) = 0$ if and only if $\dot{\boldsymbol{q}} = \boldsymbol{0}$. For a solution $\boldsymbol{x}(t)$ to belong to Ω for all $t \geq 0$, it is necessary and sufficient that $\dot{\boldsymbol{q}}(t) = \boldsymbol{0}$ for all $t \geq 0$. Therefore, it must also hold that $\ddot{\boldsymbol{q}}(t) = \boldsymbol{0}$ for all $t \geq 0$. Taking this into account, we conclude from the closed-loop Equation (8.2) that if $\boldsymbol{x}(t) \in \Omega$ for all $t \geq 0$, then

$$\boldsymbol{0} = M(\boldsymbol{q}_d - \tilde{\boldsymbol{q}}(t))^{-1}\left[K_p\tilde{\boldsymbol{q}}(t) + \boldsymbol{g}(\boldsymbol{q}_d) - \boldsymbol{g}(\boldsymbol{q}_d - \tilde{\boldsymbol{q}}(t)\right].$$

Moreover, since K_p has been chosen so that $\lambda_{\min}\{K_p\} > k_g$ hence, $\tilde{\boldsymbol{q}}(t) = \boldsymbol{0}$ for all $t \geq 0$ is its unique solution. Therefore, $\begin{bmatrix}\tilde{\boldsymbol{q}}(0)^T & \dot{\boldsymbol{q}}(0)^T\end{bmatrix}^T = \boldsymbol{0} \in \mathbb{R}^{2n}$ is the unique initial condition in Ω for which $\boldsymbol{x}(t) \in \Omega$ for all $t \geq 0$. Thus, from La Salle's theorem (*cf.* Theorem 2.7), it follows that the latter is enough to guarantee global asymptotic stability of the origin $\begin{bmatrix}\tilde{\boldsymbol{q}}^T & \dot{\boldsymbol{q}}^T\end{bmatrix}^T = \boldsymbol{0} \in \mathbb{R}^{2n}$.

In particular, we have

$$\lim_{t\to\infty} \tilde{\boldsymbol{q}}(t) = \boldsymbol{0},$$
$$\lim_{t\to\infty} \dot{\boldsymbol{q}}(t) = \boldsymbol{0},$$

that is, the position control objective is achieved.

We present next an example with the purpose of showing the performance achieved under PD control with desired gravity compensation on a 2-DOF robot.

Example 8.3. Consider the 2-DOF *prototype robot* studied in Chapter 5 and illustrated in Figure 5.2.

The components of the gravitational torques vector $\boldsymbol{g}(\boldsymbol{q})$ are given by

$$g_1(\boldsymbol{q}) = [m_1 l_{c1} + m_2 l_1] g \sin(q_1) + m_2 l_{c2} g \sin(q_1 + q_2)$$
$$g_2(\boldsymbol{q}) = m_2 l_{c2} g \sin(q_1 + q_2).$$

According to Property 4.3, the constant k_g may be obtained as (see also Example 9.2)

$$\begin{aligned} k_g &= n\left[\max_{i,j,q}\left|\frac{\partial g_i(\boldsymbol{q})}{\partial q_j}\right|\right] \\ &= n\left[[m_1 l_{c1} + m_2 l_1]g + m_2 l_{c2} g\right] \\ &= 23.94\ \left[\text{kg m}^2/\text{s}^2\right]. \end{aligned}$$

Consider the PD control law with desired gravity compensation of the robot shown in Figure 5.2 for position control, and where the design matrices are taken positive definite and such that

$$\lambda_{\min}\{K_p\} > k_g \,.$$

In particular, we pick

$$\begin{aligned} K_p &= \mathrm{diag}\{k_p\} = \mathrm{diag}\{30\} \quad [\mathrm{Nm/rad}]\,, \\ K_v &= \mathrm{diag}\{k_v\} = \mathrm{diag}\{7,\ 3\} \quad [\mathrm{Nm\ s/rad}]\,. \end{aligned}$$

The components of the control input $\boldsymbol{\tau}$ are given by

$$\begin{aligned} \tau_1 &= k_p\tilde{q}_1 - k_v\dot{q}_1 + g_1(\boldsymbol{q}_d)\,, \\ \tau_2 &= k_p\tilde{q}_2 - k_v\dot{q}_2 + g_2(\boldsymbol{q}_d)\,. \end{aligned}$$

The initial conditions corresponding to the positions and velocities, are set to

$$\begin{aligned} q_1(0) &= 0, & q_2(0) &= 0\,, \\ \dot{q}_1(0) &= 0, & \dot{q}_2(0) &= 0\,. \end{aligned}$$

The desired joint positions are chosen as

$$q_{d1} = \pi/10 \ [\mathrm{rad}] \qquad q_{d2} = \pi/30 \ [\mathrm{rad}]\,.$$

In terms of the state vector of the closed-loop equation, the initial state is set to

$$\begin{bmatrix} \tilde{\boldsymbol{q}}(0) \\ \\ \dot{\boldsymbol{q}}(0) \end{bmatrix} = \begin{bmatrix} \pi/10 \\ \pi/30 \\ 0 \\ 0 \end{bmatrix} = \begin{bmatrix} 0.3141 \\ 0.1047 \\ 0 \\ 0 \end{bmatrix} \ [\mathrm{rad}]\,.$$

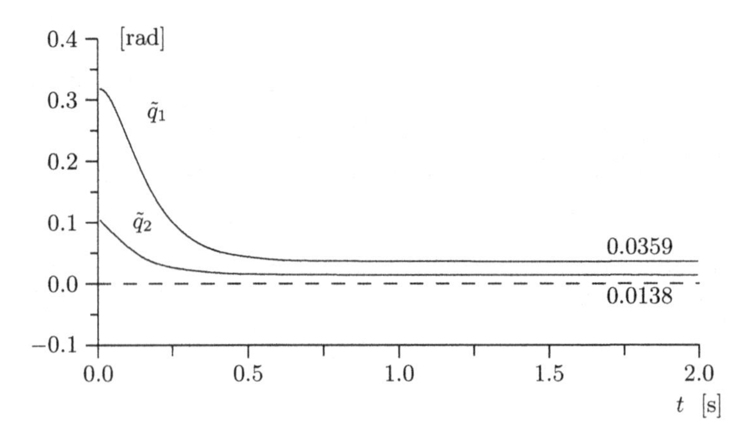

Figure 8.4. Graph of the position errors $\tilde{q}_1$ and $\tilde{q}_2$

Figure 8.4 shows the experimental results. In particular, it shows that the components of the position error vector $\tilde{\boldsymbol{q}}(t)$ tend asymptotically to a small value. They do not vanish due to non–modeled friction effects at the arm joints.

It is interesting to note the little difference between the results shown in Figure 8.4 and those obtained with PD control plus gravity compensation presented in Figure 7.3. ♢

The previous example clearly shows the good performance achieved under PD control with desired gravity compensation for a 2-DOF robot. Certainly, the suggested tuning procedure has been followed carefully, that is, the matrix K_p satisfies $\lambda_{\min}\{K_p\} > k_g$. Naturally, at this point one may ask the question: What if the tuning procedure ($\lambda_{\min}\{K_p\} \leq k_g$) is violated? As was previously shown, if the matrix K_p is positive definite (of course, also with K_v positive definite) then boundedness of the position and velocity errors $\tilde{\boldsymbol{q}}$ and $\dot{\boldsymbol{q}}$ may be guaranteed. Nevertheless, this situation where $k_g \geq \lambda_{\min}\{K_p\}$ yields an interesting dynamic behavior of the closed-loop equation. Phenomena such as bifurcations of equilibria and *catastrophic jumps* may occur. These types of phenomena appear even in the case of one single link with a revolute joint. We present next an example which illustrates these observations.

Example 8.4. Consider the pendulum model studied in Example 2.2 (see page 30),

$$J\ddot{q} + mgl\,\sin(q) = \tau$$

where we identify $g(q) = mgl\,\sin(q)$.

The PD control law with desired gravity compensation applied in the position control problem (q_d constant) is in this case given by

$$\tau = k_p\tilde{q} - k_v\dot{q} + mgl\,\sin(q_d)$$

where $k_v > 0$ and we consider here that k_p is a real number not necessarily positive and not larger than $k_g = mgl$.

The equation that governs the behavior of the control system in closed loop may be described by

$$\frac{d}{dt}\begin{bmatrix}\tilde{q}\\ \dot{q}\end{bmatrix} = \begin{bmatrix} -\dot{q} \\ \frac{1}{J}\left[\,k_p\tilde{q} - k_v\dot{q} + mgl[\sin(q_d) - \sin(q_d - \tilde{q})]\,\right]\end{bmatrix}$$

which is an autonomous differential equation and whose origin $[\tilde{q}\;\;\dot{q}]^T = \boldsymbol{0} \in \mathbb{R}^2$ is an equilibrium regardless of the values of k_p, k_v and q_d. Moreover, given q_d (constant) and defining the set Ω_{q_d} as

$$\Omega_{q_d} = \{\tilde{q} \in \mathbb{R} : k_p\tilde{q} + mgl\;[\sin(q_d) - \sin(q_d - \tilde{q})] = 0 \quad \forall\; k_p\}\,,$$

any vector $[\tilde{q}^*\;\;0]^T \in \mathbb{R}^2$ is also an equilibrium as long as $\tilde{q}^* \in \Omega_{q_d}$.

In the rest of this example we consider the innocuous case when $q_d = 0$, that is when the control objective is to drive the pendulum to

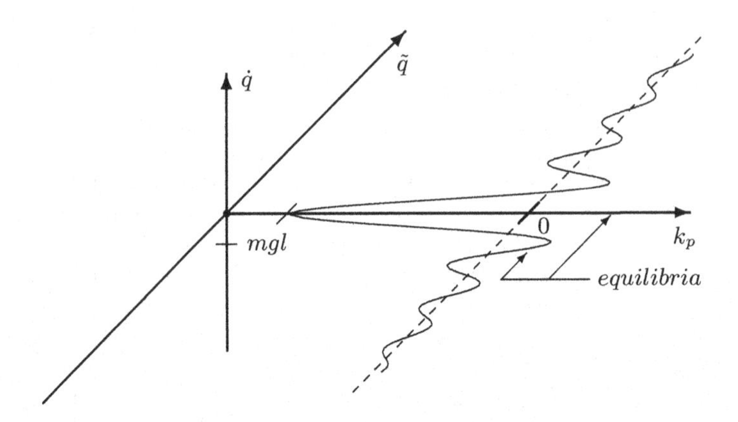

Figure 8.5. Bifurcation diagram

the vertical downward position. In this scenario the set $\Omega_{q_d} = \Omega_0$ is given by

$$\Omega_0 = \left\{ \tilde{q} \in \mathbb{R} : \tilde{q} = \mathrm{sinc}^{-1}\left(-\frac{k_p}{mgl} \right) \right\} ,$$

where the function $\mathrm{sinc}(x) = \frac{\sin(x)}{x}$. Figure 8.5 shows the diagram of equilibria in terms of k_p. Notice that with $k_p = 0$ there are an infinite number of equilibria. In particular, for $k_p = -mgl$ the origin $[\tilde{q} \;\; \dot{q}]^T = \mathbf{0} \in \mathbb{R}^2$ is the unique equilibrium. As a matter of fact, we say that the closed-loop equation has a bifurcation of equilibria for $k_p = -mgl$ since for slightly smaller values than $-mgl$ there exists a unique equilibrium while for values of k_p slightly larger than $-mgl$ there exist three equilibria.

Even though we do not show it here, for values of k_p slightly smaller than $-mgl$, the origin (which is the unique equilibrium) is unstable, while for values slightly larger than $-mgl$ the origin is actually asymptotically stable and the two other equilibria are unstable. This type of phenomenon is called *pitchfork bifurcation.* Figure 8.6 presents several trajectories of the closed-loop equation for $k_p = -11, -4, 3$, where we considered $J = 1$, $mgl = 9.8$ and $k_v = 0.1$

Besides the pitchfork bifurcation at $k_p = -mgl$, there also exists another type of bifurcation for this control system in closed loop: saddle-node bifurcation. In this case, for some values of k_p there exists an isolated equilibrium, and for slightly smaller (resp. larger) values there exist two equilibria, one of which is asymptotically stable and the other unstable, while for values of k_p slightly larger (resp. smaller) there does not exist any equilibrium in the vicinity of the one which exists for the original value of k_p. As a matter of fact, for the closed-loop control system considered here (with $q_d = 0$), the diagram of

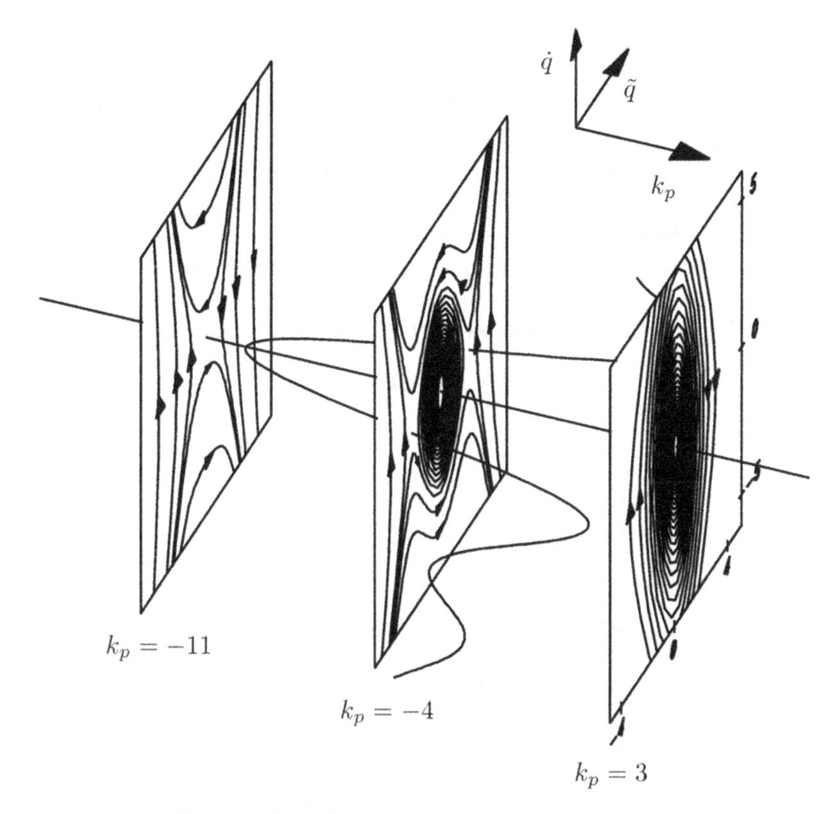

Figure 8.6. Simulation with $k_p = -11, -4, 3$

equilibria shown in Figure 8.5 suggests the possible existence of an infinite number of saddle-node bifurcations.

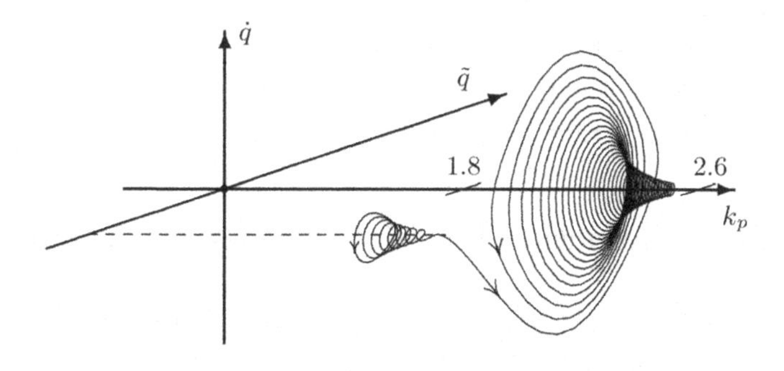

Figure 8.7. Catastrophic jump 1

The closed-loop equation also exhibits another interesting type of phenomenon: catastrophic jumps. This situation may show up when the parameter k_p varies "slowly" passing through values that correspond to saddle-node bifurcations. Briefly, a catastrophic jump occurs when for a small variation (and which moreover is slow with respect to the dynamics determined by the differential equation in question) of k_p, the solution of the closed-loop equation whose tendency is to converge towards a region of the state space, changes abruptly its behavior to go instead towards another region "far away" in the state space. Figures 8.7 and 8.8 show such phenomenon; here we took

$$k_p(t) = 0.01t + 1.8$$

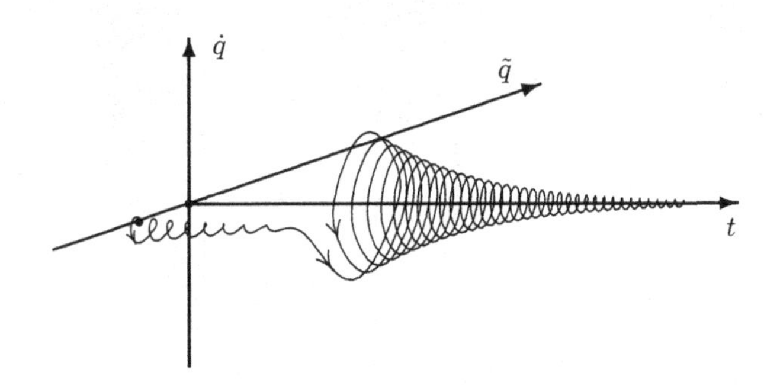

Figure 8.8. Catastrophic jump 2

and we considered again the numerical values: $J = 1$, $mgl = 9.8$ and $k_v = 0.1$, with the initial conditions $q(0) = 4$ [rad] and $\dot{q}(0) = 0$ [rad/s]. When the value of k_p is increased passing through 2.1288, the asymptotically stable equilibrium at $[\tilde{q}\ \dot{q}]^T = [1.43030\pi\ \ 0]^T$, disappears and the system solution "jumps" to the unique (globally) asymptotically stable equilibrium: the origin $[\tilde{q}\ \dot{q}]^T = \mathbf{0} \in \mathbb{R}^2$. ♢

8.4 Lyapunov Function for Global Asymptotic Stability

A Lyapunov function which allows one to show directly global asymptotic stability without using La Salle's theorem is studied in this subsection. The reference corresponding to this topic is given at the end of the chapter. We present next this analysis for which we consider the case of robots having only revolute joints. The reader may, if wished, omit this section and continue his/her reading with the following section.

Consider again the closed-loop Equation (8.2),

$$\frac{d}{dt}\begin{bmatrix}\tilde{q}\\ \dot{q}\end{bmatrix} = \begin{bmatrix}-\dot{q}\\ M(q)^{-1}\left[K_p\tilde{q} - K_v\dot{q} - C(q,\dot{q})\dot{q} + g(q_d) - g(q)\right]\end{bmatrix}. \tag{8.22}$$

As a design hypothesis, we assume here, as in Section 8.3, that the gain position matrix K_p has been chosen to satisfy

$$\lambda_{\min}\{K_p\} > k_g .$$

This selection of K_p satisfies the sufficiency condition obtained in Section 8.2 to guarantee that the origin is the unique equilibrium of the closed-loop Equation (8.22).

To study the stability properties of the origin, consider now the following Lyapunov function candidate, which as a matter of fact, may be regarded as a generalization of the function (8.21),

$$\begin{aligned}V(\tilde{q},\dot{q}) &= \frac{1}{2}\begin{bmatrix}\tilde{q}\\ \dot{q}\end{bmatrix}^T \overbrace{\begin{bmatrix}\frac{2}{\varepsilon_2}K_p & -\frac{\varepsilon_0}{1+\|\tilde{q}\|}M(q)\\ -\frac{\varepsilon_0}{1+\|\tilde{q}\|}M(q) & M(q)\end{bmatrix}}^{P}\begin{bmatrix}\tilde{q}\\ \dot{q}\end{bmatrix}\\ &\quad \underbrace{+\,\mathcal{U}(q) - \mathcal{U}(q_d) + g(q_d)^T\tilde{q} + \frac{1}{\varepsilon_1}\tilde{q}^TK_p\tilde{q}}_{f(\tilde{q})},\\ &= \frac{1}{2}\dot{q}^TM(q)\dot{q} + \mathcal{U}(q) - \mathcal{U}(q_d) + g(q_d)^T\tilde{q}\\ &\quad + \left[\frac{1}{\varepsilon_1} + \frac{1}{\varepsilon_2}\right]\tilde{q}^TK_p\tilde{q} - \frac{\varepsilon_0}{1+\|\tilde{q}\|}\,\tilde{q}^TM(q)\dot{q}\end{aligned} \tag{8.23}$$

where $f(\tilde{q})$ was defined in (8.18) and the constants $\varepsilon_0 > 0$, $\varepsilon_1 > 2$ and $\varepsilon_2 > 2$ are chosen so that

$$\frac{2\lambda_{\min}\{K_p\}}{k_g} > \varepsilon_1 > 2 \tag{8.24}$$

$$\varepsilon_2 = \frac{2\varepsilon_1}{\varepsilon_1 - 2} > 2 \tag{8.25}$$

$$\sqrt{\frac{2\lambda_{\min}\{K_p\}}{\varepsilon_2\beta}} > \varepsilon_0 > 0 \tag{8.26}$$

where β ($\geq \lambda_{\mathrm{Max}}\{M(q)\}$) was defined in Property 4.1. The condition (8.24) guarantees that $f(\tilde{q})$ is a positive definite function (see Lemma 8.1), while (8.26) ensures that P is a positive definite matrix. Finally (8.25) implies that $\frac{1}{\varepsilon_1} + \frac{1}{\varepsilon_2} = \frac{1}{2}$.

Alternatively, to show that the Lyapunov function candidate $V(\tilde{\boldsymbol{q}}, \dot{\boldsymbol{q}})$ is positive definite, first define ε as

$$\varepsilon = \varepsilon(\|\tilde{\boldsymbol{q}}\|) = \frac{\varepsilon_0}{1 + \|\tilde{\boldsymbol{q}}\|}. \tag{8.27}$$

Consequently, Inequality (8.26) implies that the matrix

$$\frac{2}{\varepsilon_2}K_p - \left[\frac{\varepsilon_0}{1 + \|\tilde{\boldsymbol{q}}\|}\right]^2 M(\boldsymbol{q}) = \frac{2}{\varepsilon_2}K_p - \varepsilon^2 M(\boldsymbol{q})$$

is positive definite.

On the other hand, the Lyapunov function candidate (8.23) may be rewritten as

$$\begin{aligned} V(\tilde{\boldsymbol{q}}, \dot{\boldsymbol{q}}) = & \frac{1}{2}\left[-\dot{\boldsymbol{q}} + \varepsilon\tilde{\boldsymbol{q}}\right]^T M(\boldsymbol{q})\left[-\dot{\boldsymbol{q}} + \varepsilon\tilde{\boldsymbol{q}}\right] + \frac{1}{2}\tilde{\boldsymbol{q}}^T\left[\frac{2}{\varepsilon_2}K_p - \varepsilon^2 M(\boldsymbol{q})\right]\tilde{\boldsymbol{q}} \\ & \underbrace{+\,\mathcal{U}(\boldsymbol{q}) - \mathcal{U}(\boldsymbol{q}_d) + \boldsymbol{g}(\boldsymbol{q}_d)^T\tilde{\boldsymbol{q}} + \frac{1}{\varepsilon_1}\tilde{\boldsymbol{q}}^T K_p\tilde{\boldsymbol{q}}}_{f(\tilde{\boldsymbol{q}})}, \end{aligned}$$

which is clearly positive definite since the matrices $M(\boldsymbol{q})$ and $\frac{2}{\varepsilon_2}K_p - \varepsilon^2 M(\boldsymbol{q})$ are positive definite and $f(\tilde{\boldsymbol{q}})$ is also a positive definite function (due to $\lambda_{\min}\{K_p\} > k_g$ and Lemma 8.1).

The time derivative of the Lyapunov function candidate (8.23) along the trajectories of the closed-loop Equation (8.22) takes the form

$$\begin{aligned} \dot{V}(\tilde{\boldsymbol{q}}, \dot{\boldsymbol{q}}) = & \dot{\boldsymbol{q}}^T\left[K_p\tilde{\boldsymbol{q}} - K_v\dot{\boldsymbol{q}} - C(\boldsymbol{q}, \dot{\boldsymbol{q}})\dot{\boldsymbol{q}} + \boldsymbol{g}(\boldsymbol{q}_d) - \boldsymbol{g}(\boldsymbol{q})\right] + \frac{1}{2}\dot{\boldsymbol{q}}^T\dot{M}(\boldsymbol{q})\dot{\boldsymbol{q}} \\ & + \boldsymbol{g}(\boldsymbol{q})^T\dot{\boldsymbol{q}} - \boldsymbol{g}(\boldsymbol{q}_d)^T\dot{\boldsymbol{q}} - \tilde{\boldsymbol{q}}^T K_p\dot{\boldsymbol{q}} + \varepsilon\dot{\boldsymbol{q}}^T M(\boldsymbol{q})\dot{\boldsymbol{q}} - \varepsilon\tilde{\boldsymbol{q}}^T\dot{M}(\boldsymbol{q})\dot{\boldsymbol{q}} \\ & - \varepsilon\tilde{\boldsymbol{q}}^T\left[K_p\tilde{\boldsymbol{q}} - K_v\dot{\boldsymbol{q}} - C(\boldsymbol{q}, \dot{\boldsymbol{q}})\dot{\boldsymbol{q}} + \boldsymbol{g}(\boldsymbol{q}_d) - \boldsymbol{g}(\boldsymbol{q})\right] \\ & - \dot{\varepsilon}\tilde{\boldsymbol{q}}^T M(\boldsymbol{q})\dot{\boldsymbol{q}}, \end{aligned}$$

where we used $\boldsymbol{g}(\boldsymbol{q}) = \frac{\partial\mathcal{U}(\boldsymbol{q})}{\partial\boldsymbol{q}}$. After some simplifications, the time derivative $\dot{V}(\tilde{\boldsymbol{q}}, \dot{\boldsymbol{q}})$ may be written as

$$\begin{aligned} \dot{V}(\tilde{\boldsymbol{q}}, \dot{\boldsymbol{q}}) = & -\dot{\boldsymbol{q}}^T K_v\dot{\boldsymbol{q}} + \dot{\boldsymbol{q}}^T\left[\frac{1}{2}\dot{M}(\boldsymbol{q}) - C(\boldsymbol{q}, \dot{\boldsymbol{q}})\right]\dot{\boldsymbol{q}} + \varepsilon\dot{\boldsymbol{q}}^T M(\boldsymbol{q})\dot{\boldsymbol{q}} \\ & - \varepsilon\tilde{\boldsymbol{q}}^T\left[\dot{M}(\boldsymbol{q}) - C(\boldsymbol{q}, \dot{\boldsymbol{q}})\right]\dot{\boldsymbol{q}} - \varepsilon\tilde{\boldsymbol{q}}^T\left[K_p\tilde{\boldsymbol{q}} - K_v\dot{\boldsymbol{q}}\right] \\ & - \varepsilon\tilde{\boldsymbol{q}}^T\left[\boldsymbol{g}(\boldsymbol{q}_d) - \boldsymbol{g}(\boldsymbol{q})\right] - \dot{\varepsilon}\tilde{\boldsymbol{q}}^T M(\boldsymbol{q})\dot{\boldsymbol{q}}. \end{aligned}$$

Finally, considering Property 4.2, *i.e.* that the matrix $\frac{1}{2}\dot{M}(\boldsymbol{q}) - C(\boldsymbol{q}, \dot{\boldsymbol{q}})$ is skew-symmetric and that $\dot{M}(\boldsymbol{q}) = C(\boldsymbol{q}, \dot{\boldsymbol{q}}) + C(\boldsymbol{q}, \dot{\boldsymbol{q}})^T$, we get

$$\dot{V}(\tilde{\boldsymbol{q}}, \dot{\boldsymbol{q}}) = -\dot{\boldsymbol{q}}^T K_v \dot{\boldsymbol{q}} + \varepsilon \dot{\boldsymbol{q}}^T M(\boldsymbol{q}) \dot{\boldsymbol{q}} - \varepsilon \tilde{\boldsymbol{q}}^T K_p \tilde{\boldsymbol{q}} + \varepsilon \tilde{\boldsymbol{q}}^T K_v \dot{\boldsymbol{q}} \\ - \varepsilon \dot{\boldsymbol{q}}^T C(\boldsymbol{q}, \dot{\boldsymbol{q}}) \tilde{\boldsymbol{q}} - \varepsilon \tilde{\boldsymbol{q}}^T [\boldsymbol{g}(\boldsymbol{q}_d) - \boldsymbol{g}(\boldsymbol{q})] - \dot{\varepsilon} \tilde{\boldsymbol{q}}^T M(\boldsymbol{q}) \dot{\boldsymbol{q}} . \quad (8.28)$$

As is well known, to conclude global asymptotic stability by Lyapunov's direct method, it is sufficient to prove that $\dot{V}(\mathbf{0}, \mathbf{0}) = 0$ and $\dot{V}(\tilde{\boldsymbol{q}}, \dot{\boldsymbol{q}}) < 0$ for all vectors $\left[\tilde{\boldsymbol{q}}^T \ \dot{\boldsymbol{q}}^T\right]^T \neq \mathbf{0} \in \mathbb{R}^{2n}$. These conditions are verified if $\dot{V}(\tilde{\boldsymbol{q}}, \dot{\boldsymbol{q}})$ is a negative definite function. Observe that it is very difficult to ensure from (8.28), that $\dot{V}(\tilde{\boldsymbol{q}}, \dot{\boldsymbol{q}})$ is negative definite. With the aim of finding additional conditions on ε_0 such that $\dot{V}(\tilde{\boldsymbol{q}}, \dot{\boldsymbol{q}})$ is negative definite, we present now the upper-bounds on the following three terms:

- $-\varepsilon \dot{\boldsymbol{q}}^T C(\boldsymbol{q}, \dot{\boldsymbol{q}}) \tilde{\boldsymbol{q}}$
- $-\varepsilon \tilde{\boldsymbol{q}}^T [\boldsymbol{g}(\boldsymbol{q}_d) - \boldsymbol{g}(\boldsymbol{q})]$
- $-\dot{\varepsilon} \tilde{\boldsymbol{q}}^T M(\boldsymbol{q}) \dot{\boldsymbol{q}}$.

First, concerning $-\varepsilon \dot{\boldsymbol{q}}^T C(\boldsymbol{q}, \dot{\boldsymbol{q}}) \tilde{\boldsymbol{q}}$, we have

$$\begin{aligned} -\varepsilon \dot{\boldsymbol{q}}^T C(\boldsymbol{q}, \dot{\boldsymbol{q}}) \tilde{\boldsymbol{q}} &\leq \left| -\varepsilon \dot{\boldsymbol{q}}^T C(\boldsymbol{q}, \dot{\boldsymbol{q}}) \tilde{\boldsymbol{q}} \right| \\ &\leq \varepsilon \|\dot{\boldsymbol{q}}\| \, \|C(\boldsymbol{q}, \dot{\boldsymbol{q}}) \tilde{\boldsymbol{q}}\| \\ &\leq \varepsilon k_{C_1} \|\dot{\boldsymbol{q}}\| \, \|\dot{\boldsymbol{q}}\| \, \|\tilde{\boldsymbol{q}}\| \\ &\leq \varepsilon_0 k_{C_1} \|\dot{\boldsymbol{q}}\|^2 \end{aligned} \quad (8.29)$$

where we took into account Property 4.2, *i.e.* $\|C(\boldsymbol{q}, \boldsymbol{x}) \boldsymbol{y}\| \leq k_{C_1} \|\boldsymbol{x}\| \, \|\boldsymbol{y}\|$, and the definition of ε in (8.27).

Next, concerning the term $-\varepsilon \tilde{\boldsymbol{q}}^T [\boldsymbol{g}(\boldsymbol{q}_d) - \boldsymbol{g}(\boldsymbol{q})]$ we have

$$\begin{aligned} -\varepsilon \tilde{\boldsymbol{q}}^T [\boldsymbol{g}(\boldsymbol{q}_d) - \boldsymbol{g}(\boldsymbol{q})] &\leq \left| -\varepsilon \tilde{\boldsymbol{q}}^T [\boldsymbol{g}(\boldsymbol{q}_d) - \boldsymbol{g}(\boldsymbol{q})] \right| \\ &\leq \varepsilon \|\tilde{\boldsymbol{q}}\| \, \|\boldsymbol{g}(\boldsymbol{q}_d) - \boldsymbol{g}(\boldsymbol{q})\| \\ &\leq \varepsilon k_g \|\tilde{\boldsymbol{q}}\|^2 \end{aligned} \quad (8.30)$$

where we used Property 4.3, *i.e.* $\|\boldsymbol{g}(\boldsymbol{x}) - \boldsymbol{g}(\boldsymbol{y})\| \leq k_g \|\boldsymbol{x} - \boldsymbol{y}\|$.

Finally, for the term $-\dot{\varepsilon} \tilde{\boldsymbol{q}}^T M(\boldsymbol{q}) \dot{\boldsymbol{q}}$, we have

$$\begin{aligned} -\dot{\varepsilon} \tilde{\boldsymbol{q}}^T M(\boldsymbol{q}) \dot{\boldsymbol{q}} &\leq \left| -\dot{\varepsilon} \tilde{\boldsymbol{q}}^T M(\boldsymbol{q}) \dot{\boldsymbol{q}} \right| \\ &= \left| \frac{\varepsilon_0}{\|\tilde{\boldsymbol{q}}\| \left[1 + \|\tilde{\boldsymbol{q}}\|\right]^2} \tilde{\boldsymbol{q}}^T \dot{\boldsymbol{q}} \tilde{\boldsymbol{q}}^T M(\boldsymbol{q}) \dot{\boldsymbol{q}} \right| \\ &\leq \frac{\varepsilon_0}{\|\tilde{\boldsymbol{q}}\| \left[1 + \|\tilde{\boldsymbol{q}}\|\right]^2} \|\tilde{\boldsymbol{q}}\| \, \|\dot{\boldsymbol{q}}\| \, \|\tilde{\boldsymbol{q}}\| \, \|M(\boldsymbol{q}) \dot{\boldsymbol{q}}\| \\ &\leq \frac{\varepsilon_0}{1 + \|\tilde{\boldsymbol{q}}\|} \|\dot{\boldsymbol{q}}\|^2 \lambda_{\mathrm{Max}}\{M(\boldsymbol{q})\} \\ &\leq \varepsilon_0 \beta \|\dot{\boldsymbol{q}}\|^2 \end{aligned} \quad (8.31)$$

where we used again the definition of ε in (8.27) and Property 4.1, *i.e.* $\beta \|\dot{\boldsymbol{q}}\| \geq \lambda_{\mathrm{Max}}\{M(\boldsymbol{q})\} \|\dot{\boldsymbol{q}}\| \geq \|M(\boldsymbol{q})\dot{\boldsymbol{q}}\|$.

From the inequalities (8.29), (8.30) and (8.31), the time derivative $\dot{V}(\tilde{\boldsymbol{q}}, \dot{\boldsymbol{q}})$ in (8.28) reduces to

$$
\begin{aligned}
\dot{V}(\tilde{\boldsymbol{q}}, \dot{\boldsymbol{q}}) \leq & -\dot{\boldsymbol{q}}^T K_v \dot{\boldsymbol{q}} + \varepsilon \dot{\boldsymbol{q}}^T M(\boldsymbol{q}) \dot{\boldsymbol{q}} - \varepsilon \tilde{\boldsymbol{q}}^T K_p \tilde{\boldsymbol{q}} + \varepsilon \tilde{\boldsymbol{q}}^T K_v \dot{\boldsymbol{q}} \\
& + \varepsilon_0 k_{C_1} \|\dot{\boldsymbol{q}}\|^2 + \varepsilon k_g \|\tilde{\boldsymbol{q}}\|^2 + \varepsilon_0 \beta \|\dot{\boldsymbol{q}}\|^2 ,
\end{aligned}
$$

which in turn may be written as

$$
\begin{aligned}
\dot{V}(\tilde{\boldsymbol{q}}, \dot{\boldsymbol{q}}) \leq & - \begin{bmatrix} \tilde{\boldsymbol{q}} \\ \dot{\boldsymbol{q}} \end{bmatrix}^T \begin{bmatrix} \varepsilon K_p & -\frac{\varepsilon}{2} K_v \\ -\frac{\varepsilon}{2} K_v & \frac{1}{2} K_v \end{bmatrix} \begin{bmatrix} \tilde{\boldsymbol{q}} \\ \dot{\boldsymbol{q}} \end{bmatrix} + \varepsilon k_g \|\tilde{\boldsymbol{q}}\|^2 \\
& - \frac{1}{2} \left[\lambda_{\min}\{K_v\} - 2\varepsilon_0 (k_{C_1} + 2\beta) \right] \|\dot{\boldsymbol{q}}\|^2 ,
\end{aligned} \tag{8.32}
$$

where we used

$$
-\dot{\boldsymbol{q}}^T K_v \dot{\boldsymbol{q}} \leq -\frac{1}{2} \dot{\boldsymbol{q}}^T K_v \dot{\boldsymbol{q}} - \frac{\lambda_{\min}\{K_v\}}{2} \|\dot{\boldsymbol{q}}\|^2
$$

and $\varepsilon \dot{\boldsymbol{q}}^T M(\boldsymbol{q}) \dot{\boldsymbol{q}} \leq \varepsilon_0 \beta \|\dot{\boldsymbol{q}}\|^2$.

Finally, from (8.32) we get

$$
\begin{aligned}
\dot{V}(\tilde{\boldsymbol{q}}, \dot{\boldsymbol{q}}) \leq & - \varepsilon \begin{bmatrix} \|\tilde{\boldsymbol{q}}\| \\ \|\dot{\boldsymbol{q}}\| \end{bmatrix}^T \overbrace{\begin{bmatrix} \lambda_{\min}\{K_p\} - k_g & -\frac{1}{2}\lambda_{\mathrm{Max}}\{K_v\} \\ -\frac{1}{2}\lambda_{\mathrm{Max}}\{K_v\} & \frac{1}{2\varepsilon_0}\lambda_{\min}\{K_v\} \end{bmatrix}}^{Q} \begin{bmatrix} \|\tilde{\boldsymbol{q}}\| \\ \|\dot{\boldsymbol{q}}\| \end{bmatrix} \\
& - \frac{1}{2} \underbrace{\left[\lambda_{\min}\{K_v\} - 2\varepsilon_0 (k_{C_1} + 2\beta) \right]}_{\delta} \|\dot{\boldsymbol{q}}\|^2 .
\end{aligned} \tag{8.33}
$$

Next, from the latter inequality we find immediately the conditions on ε_0 for $\dot{V}(\tilde{\boldsymbol{q}}, \dot{\boldsymbol{q}})$ to be negative definite. To that end, we first require to guarantee that the matrix Q is positive definite and that $\delta > 0$. The matrix Q is positive definite if it holds that

$$
\lambda_{\min}\{K_p\} > k_g , \tag{8.34}
$$

$$
\frac{2\lambda_{\min}\{K_v\}(\lambda_{\min}\{K_p\} - k_g)}{\lambda_{\mathrm{Max}}^2\{K_v\}} > \varepsilon_0 , \tag{8.35}
$$

and we have $\delta > 0$ if

$$
\frac{\lambda_{\min}\{K_v\}}{2[k_{C_1} + 2\beta]} > \varepsilon_0 . \tag{8.36}
$$

Observe that (8.34) is verified since K_p was assumed to be picked so as to satisfy $\lambda_{\min}\{K_p\} > \frac{\varepsilon_1}{2} k_g$ with $\varepsilon_1 > 2$. It is important to stress that the constant ε_0 is only needed for the purposes of stability analysis and it is not required to know its actual numerical value. Choosing ε_0 so as to satisfy simultaneously (8.35) and (8.36), we have $\lambda_{\min}\{Q\} > 0$. Under this scenario, we get from (8.33) that

$$\begin{aligned}\dot{V}(\tilde{\boldsymbol{q}}, \dot{\boldsymbol{q}}) &\leq -\frac{\varepsilon_0}{1 + \|\tilde{\boldsymbol{q}}\|} \lambda_{\min}\{Q\} \left[\|\tilde{\boldsymbol{q}}\|^2 + \|\dot{\boldsymbol{q}}\|^2 \right] - \frac{\delta}{2} \|\dot{\boldsymbol{q}}\|^2, \\ &\leq -\varepsilon_0 \lambda_{\min}\{Q\} \frac{\|\tilde{\boldsymbol{q}}\|^2}{1 + \|\tilde{\boldsymbol{q}}\|} - \frac{\delta}{2} \|\dot{\boldsymbol{q}}\|^2,\end{aligned}$$

which is a negative definite function. Finally, using Lyapunov's direct method (*cf.* Theorem 2.4), we conclude that the origin $\left[\tilde{\boldsymbol{q}}^T \;\; \dot{\boldsymbol{q}}^T\right]^T = \boldsymbol{0} \in \mathbb{R}^{2n}$ is a globally asymptotically stable equilibrium of the closed-loop equation.

8.5 Conclusions

The conclusions drawn from the analysis presented in this chapter can be summarized as follows.

Consider PD control with desired gravity compensation for n-DOF robots. Assume that the desired position $\boldsymbol{q}_d$ is constant.

- If the symmetric matrices K_p and K_v of the PD control law with desired gravity compensation are positive definite and moreover $\lambda_{\min}\{K_p\} > k_g$, then the origin of the closed-loop equation, expressed in terms of the state vector $\left[\tilde{\boldsymbol{q}}^T \;\; \dot{\boldsymbol{q}}^T\right]^T$, is globally asymptotically stable. Consequently, for any initial condition $\boldsymbol{q}(0), \dot{\boldsymbol{q}}(0) \in \mathbb{R}^n$ we have $\lim_{t\to\infty} \tilde{\boldsymbol{q}}(t) = \boldsymbol{0} \in \mathbb{R}^n$.

Bibliography

PD control with desired gravity compensation is the subject of study in

- Takegaki M., Arimoto S., 1981, *"A new feedback method for dynamic control of manipulators"*, Journal of Dynamic Systems, Measurement, and Control, Vol. 103, pp. 119–125.
- Arimoto S., Miyazaki F., 1986, *"Stability and robustness of PD feedback control with gravity compensation for robot manipulators"*, in F. Paul and D. Youcef–Toumi (ed.), Robotics: Theory and Applications, DSC Vol. 3.
- Tomei P., 1991, *"Adaptive PD controller for robot manipulators"*, IEEE Transactions on Robotics and Automation, Vol. 7, No. 4, August, pp. 565–570.

- Kelly R., 1997, *"PD control with desired gravity compensation of robotic manipulators: A review"*, The International Journal of Robotics Research, Vol. 16, No. 5, pp. 660–672.

Topics on bifurcation of equilibria may be consulted in

- Parker T. S., Chua L. O., 1989, *"Practical numerical algorithms for chaotic systems"*, Springer-Verlag.
- Guckenheimer J., Holmes P., 1990, *"Nonlinear oscillations, dynamical systems, and bifurcation of vector fields"*, Springer-Verlag.
- Wiggins S., 1990, *"Introduction to applied nonlinear dynamical systems and chaos"*, Springer-Verlag.
- Hale J. K., Koçak H., 1991, *"Dynamics and bifurcations"*, Springer-Verlag.
- Jackson E. A., 1991, *"Perspectives of nonlinear dynamics"*, Vol. 1, Cambridge University Press.

Study of the Lyapunov function for global asymptotic stability presented in Section 8.4, is taken from

- Kelly R., 1993, *"Comments on: Adaptive PD controller for robot manipulators"*, IEEE Transactions on Robotics and Automation, Vol. 9, No. 1, February, pp. 117–119.

Problems

1. Consider the model of the ideal pendulum studied in Example 8.2

$$J\ddot{q} + mgl\,\sin(q) = \tau$$

with the following numerical values

$$J = 1, \qquad mgl = 1, \qquad q_d = \pi/2\,.$$

Consider the use of PD control with desired gravity compensation and suppose the following initial conditions: $q(0) = 0$ and $\dot{q}(0) = 0$. From this, we have $\tilde{q}(0) = \pi/2$. Assume that $k_p = 4/\pi$.

   a) Obtain an upper-bound on $\dot{q}(t)^2$.
   Hint: Use (8.16).

2. Consider the PD control law with desired gravity compensation for the pendulum described by the equation

$$J\ddot{q} + mgl\,\sin(q) = \tau\,.$$

The equilibria of the closed-loop equation are $[\tilde{q} \;\; \dot{q}]^T = [s \;\; 0]^T$ where s is the solution of

$$k_p s + mgl\,[\sin(q_d) - \sin(q_d - s)] = 0\,.$$

a) Show that s satisfies

$$|s| \leq \frac{2mgl}{k_p}\,.$$

b) Simulate the system in closed loop with the following numerical values: $J = 1$, $m = l = 1$, $g = 10$, $k_p = 1/4$, $k_v = 1$ and with initial conditions: $q(0) = \pi/8$ and $\dot{q}(0) = 0$. For the desired angular position $q_d = \pi/2$, verify by simulation that $\lim_{t\to\infty} q(t) \neq q_d$.

c) Obtain by simulation the approximate value of $\lim_{t\to\infty} q(t)$. Verify that $\lim_{t\to\infty} |q_d - q(t)| \leq \frac{2mgl}{k_p}$.

3. Consider the model of the ideal pendulum studied in Example 8.1

$$J\ddot{q} + mgl\,\sin(q) = \tau$$

with the following numerical values:

$$J = 1, \qquad mgl = 1, \qquad q_d = \pi/2\,.$$

Consider the use of PD control with desired gravity compensation.

a) Obtain the closed-loop equation in terms of the state vector $[\tilde{q} \;\; \dot{q}]^T$.

b) Compute the value of the constant k_g (see Property 4.3).

c) In Example 8.1 we showed that the closed-loop equation had three equilibria for $k_p = 0.25$. Compute a value for k_p that guarantees that the origin $[\tilde{q} \;\; \dot{q}]^T = \mathbf{0} \in \mathbb{R}^2$ is the unique equilibrium point of the closed-loop equation.

d) Consider the previously obtained value for k_p. Show that the origin is a stable equilibrium point.

Hint: Show that

$$\begin{aligned} f(\tilde{q}) &= -mgl\,\cos(q_d - \tilde{q}) + mgl\,\cos(q_d) \\ &\quad + mgl\,\sin(q_d)\tilde{q} + \frac{1}{2}k_p\tilde{q}^2 \end{aligned}$$

is a positive definite function. Use the Lyapunov function candidate

$$V(\tilde{q}, \dot{q}) = \frac{1}{2}J\dot{q}^2 + f(\tilde{q})\,.$$

e) Use La Salle's theorem (*cf.* Theorem 2.7) to show that the origin is globally asymptotically stable.

4. Verify the expression for $\dot{V}(\tilde{\boldsymbol{q}}, \dot{\boldsymbol{q}})$ given in Equation (8.7).
5. Consider the model of robots with elastic joints, (3.27) and (3.28), but with an additional term for friction at the links $\boldsymbol{f}(\dot{\boldsymbol{q}})$ that satisfies $\dot{\boldsymbol{q}}^T \boldsymbol{f}(\dot{\boldsymbol{q}}) > 0 \;\; \forall \dot{\boldsymbol{q}} \neq \boldsymbol{0}$ and $\boldsymbol{f}(\boldsymbol{0}) = \boldsymbol{0}$, *i.e.*

$$\begin{aligned} M(\boldsymbol{q})\ddot{\boldsymbol{q}} + C(\boldsymbol{q}, \dot{\boldsymbol{q}})\dot{\boldsymbol{q}} + \boldsymbol{g}(\boldsymbol{q}) + \boldsymbol{f}(\dot{\boldsymbol{q}}) + K(\boldsymbol{q} - \boldsymbol{\theta}) &= \boldsymbol{0}, \\ J\ddot{\boldsymbol{\theta}} - K(\boldsymbol{q} - \boldsymbol{\theta}) &= \boldsymbol{\tau}. \end{aligned}$$

It is assumed that only the vector of positions $\boldsymbol{\theta}$ of the motors axes, and the corresponding velocities $\dot{\boldsymbol{\theta}}$, are available from measurement. The control objective is that $\boldsymbol{q}(t) \rightarrow \boldsymbol{q}_d$ as $t \rightarrow \infty$ where $\boldsymbol{q}_d$ is constant.

A variant of the PD control with desired gravity compensation is[5]

$$\boldsymbol{\tau} = K_p\tilde{\boldsymbol{\theta}} - K_v\dot{\boldsymbol{\theta}} + \boldsymbol{g}(\boldsymbol{q}_d)$$

where

$$\begin{aligned} \tilde{\boldsymbol{\theta}} &= \boldsymbol{q}_d - \boldsymbol{\theta} + K^{-1}\boldsymbol{g}(\boldsymbol{q}_d), \\ \tilde{\boldsymbol{q}} &= \boldsymbol{q}_d - \boldsymbol{q}, \end{aligned}$$

and $K_p, K_v \in \mathbb{R}^{n\times n}$ are symmetric positive definite matrices. The matrix of elasticity K, is diagonal and positive definite.

a) Verify that the closed-loop equation in terms of the state vector $\left[\tilde{\boldsymbol{q}}^T \;\; \tilde{\boldsymbol{\theta}}^T \;\; \dot{\boldsymbol{q}}^T \;\; \dot{\boldsymbol{\theta}}^T\right]^T$, may be written as

$$\frac{d}{dt}\begin{bmatrix} \tilde{\boldsymbol{q}} \\ \tilde{\boldsymbol{\theta}} \\ \dot{\boldsymbol{q}} \\ \dot{\boldsymbol{\theta}} \end{bmatrix} = \begin{bmatrix} -\dot{\boldsymbol{q}} \\ -\dot{\boldsymbol{\theta}} \\ M(\boldsymbol{q})^{-1}\left[-K(\tilde{\boldsymbol{\theta}} - \tilde{\boldsymbol{q}}) + \boldsymbol{g}(\boldsymbol{q}_d) - C(\boldsymbol{q}, \dot{\boldsymbol{q}})\dot{\boldsymbol{q}} - \boldsymbol{g}(\boldsymbol{q}) - \boldsymbol{f}(\dot{\boldsymbol{q}})\right] \\ J^{-1}\left[K_p\tilde{\boldsymbol{\theta}} - K_v\dot{\boldsymbol{\theta}} + K(\tilde{\boldsymbol{\theta}} - \tilde{\boldsymbol{q}})\right] \end{bmatrix}.$$

b) Verify that the origin is an equilibrium of the closed-loop equation.

c) Show that if $\lambda_{\min}\{K_p\} > k_g$ and $\lambda_{\min}\{K\} > k_g$, then the origin is a stable equilibrium point.

Hint: Use the following Lyapunov function:

$$V(\tilde{\boldsymbol{q}}, \tilde{\boldsymbol{\theta}}, \dot{\boldsymbol{q}}, \dot{\boldsymbol{\theta}}) = V_1(\tilde{\boldsymbol{q}}, \tilde{\boldsymbol{\theta}}, \dot{\boldsymbol{q}}, \dot{\boldsymbol{\theta}}) + \frac{1}{2}\tilde{q}^T K\tilde{q} + V_2(\tilde{q})$$

[5] This controller was proposed and analyzed in Tomei P., 1991, *"A simple PD controller for robots with elastic joints"*, IEEE Transactions on Automatic Control, Vol. 36, No. 10, October.

where

$$
\begin{aligned}
V_1(\tilde{\boldsymbol{q}}, \tilde{\boldsymbol{\theta}}, \dot{\boldsymbol{q}}, \dot{\boldsymbol{\theta}}) &= \frac{1}{2}\dot{\boldsymbol{q}}^T M(\boldsymbol{q})\dot{\boldsymbol{q}} + \frac{1}{2}\dot{\boldsymbol{\theta}}^T J\dot{\boldsymbol{\theta}} + \frac{1}{2}\tilde{\boldsymbol{\theta}}^T K_p\tilde{\boldsymbol{\theta}} \\
&\quad + \frac{1}{2}\tilde{\boldsymbol{\theta}}^T K\tilde{\boldsymbol{\theta}} - \tilde{\boldsymbol{\theta}}^T K\tilde{\boldsymbol{q}}\,, \\
V_2(\tilde{\boldsymbol{q}}) &= \mathcal{U}(\boldsymbol{q}_d - \tilde{\boldsymbol{q}}) - \mathcal{U}(\boldsymbol{q}_d) + \tilde{\boldsymbol{q}}^T \boldsymbol{g}(\boldsymbol{q}_d)\,,
\end{aligned}
$$

and verify that

$$
\dot{V}(\tilde{\boldsymbol{q}}, \tilde{\boldsymbol{\theta}}, \dot{\boldsymbol{q}}, \dot{\boldsymbol{\theta}}) = -\dot{\boldsymbol{\theta}}^T K_v\dot{\boldsymbol{\theta}} - \dot{\boldsymbol{q}}^T \boldsymbol{f}(\dot{\boldsymbol{q}})\,.
$$

9

PID Control

With PD control we are able to achieve the position control objective for robots whose models do not contain the gravitational term (*i.e.* $\boldsymbol{g}(\boldsymbol{q}) = \mathbf{0}$). In this case, the tuning procedure[1] for PD control is trivial since it is enough to select the design matrices K_p and K_v to be symmetric and positive definite (see Chapter 6).

In the case where the robot model contains the vector of gravitational torques (*i.e.* $\boldsymbol{g}(\boldsymbol{q}) \neq \mathbf{0}$) and if in particular, $\boldsymbol{g}(\boldsymbol{q}_d) \neq \mathbf{0}$, where $\boldsymbol{q}_d$ is the joint desired position, then the position control objective cannot be achieved by means of a simple PD control law. As a matter of fact it may happen that the position error $\tilde{\boldsymbol{q}}$ tends to a constant vector but which is always different from the vector $\mathbf{0} \in \mathbb{R}^n$. Then, from an automatic control viewpoint and with the aim of satisfying the position control objective, in this case it seems natural to introduce an *Integral* component to the PD control to drive the position error to zero. This reasoning justifies the application of Proportional Integral Derivative (PID) control to robot manipulators.

The PID control law is given by

$$\boldsymbol{\tau} = K_p\tilde{\boldsymbol{q}} + K_v\dot{\tilde{\boldsymbol{q}}} + K_i\int_0^t \tilde{\boldsymbol{q}}(\sigma)\, d\sigma \tag{9.1}$$

where the design matrices $K_p, K_v, K_i \in \mathbb{R}^{n\times n}$, which are respectively called "position, velocity and integral gains", are symmetric positive definite and suitably selected. Figure 9.1 shows the block-diagram of the PID control for robot manipulators.

Nowadays, most industrial robot manipulators are controlled by PID controllers. The wide use of robot manipulators in everyday applications, is testament to the performance that can be achieved in a large variety of applications

[1] By 'tuning procedure' the reader should interpret the process of determining the numerical values of the design parameters of the control law, which guarantee the achievement of the control objective.

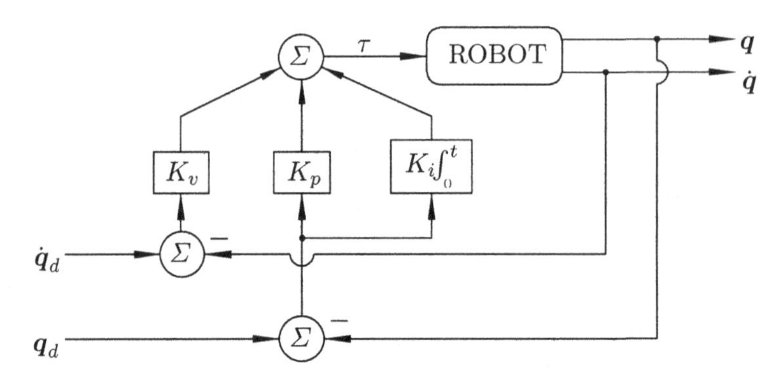

Figure 9.1. Block-diagram: PID control

when using PID control. However, in contrast to PD control, the tuning procedure for PID controllers, that is, the procedure to choose appropriate positive definite matrices K_p, K_v and K_i, is far from trivial.

In practice, the tuning of PID controllers is easier for robots whose transmission system includes reduction mechanisms such as gears or bands. The use of these reductions effectively increases the torque or force produced by the actuators, and therefore, these are able to drive links of considerably large masses. In principle, this has the consequence that large accelerations may be reached for 'light' links. Nevertheless, the presence of reduction mechanisms, such as gears and bands, may introduce undesired physical phenomena that hamper the performance of the robot in its required task. Among these phenomena we cite vibrations due to backlash among the teeth of the gears, positioning errors and energy waste caused by friction in the gears, positioning errors caused by vibrations and elasticity of the bands and by gear torsions. In spite of all these the use of reduction mechanisms is common in most robot manipulators. This has a positive impact on the tuning task of controllers, and more particularly of PID controllers. Indeed, as has been shown in Chapter 3 the complete dynamic of the robot with high-reduction-ratio transmissions is basically characterized by the model of the actuators themselves, which are often modeled by linear differential equations. Thus, in this scenario the differential equation that governs the behavior of the closed-loop system becomes linear and therefore, the tuning of the controller becomes relatively simple. This last topic is not treated here since it is well documented in the literature; the interested reader is invited to see the texts cited at the end of the chapter. Here, we consider the more general nonlinear case.

In the introduction to Part II we assumed that the considered robot actuators were ideal sources of torques and forces. Under this condition, the dynamic model of a robot of n DOF is given by (3.18), *i.e.*

$$M(\boldsymbol{q})\ddot{\boldsymbol{q}} + C(\boldsymbol{q}, \dot{\boldsymbol{q}})\dot{\boldsymbol{q}} + \boldsymbol{g}(\boldsymbol{q}) = \boldsymbol{\tau}\,, \tag{9.2}$$

where the vector of gravitational torques $\boldsymbol{g}(\boldsymbol{q})$ is clearly present. In this chapter, we assume that the joints of the robot are all revolute.

We study the control system formed by the PID controller given by Equation (9.1) and the robot model (9.2). This study is slightly more complex than that for the PID control of robots with high-reduction actuators. Specifically, we obtain a tuning procedure for PID control that guarantees achievement of the position control objective, locally. In other words, for the desired constant position $\boldsymbol{q}_d$, tuning ensures that $\lim_{t\to\infty} \tilde{\boldsymbol{q}}(t) = \mathbf{0}$, as long as the initial position error $\tilde{\boldsymbol{q}}(0)$ and the initial velocity error $\dot{\boldsymbol{q}}(0)$ are sufficiently "small". From an analytic viewpoint this is done by proving local asymptotic stability of the origin of the equation that describes the behavior of the closed-loop system. For this analysis, we use the following information drawn from Properties 4.1, 4.2, and 4.3:

- The matrix $\frac{1}{2}\dot{M}(\boldsymbol{q}) - C(\boldsymbol{q}, \dot{\boldsymbol{q}})$ is skew-symmetric.
- There exists a non-negative constant k_{C_1} such that for all $\boldsymbol{x}, \boldsymbol{y}, \boldsymbol{z} \in \mathbb{R}^n$, we have
$$\|C(\boldsymbol{x}, \boldsymbol{y})\boldsymbol{z}\| \leq k_{C_1} \|\boldsymbol{y}\| \|\boldsymbol{z}\| .$$
- There exists a non-negative constant k_g such that for all $\boldsymbol{x}, \boldsymbol{y} \in \mathbb{R}^n$, we have
$$\|\boldsymbol{g}(\boldsymbol{x}) - \boldsymbol{g}(\boldsymbol{y})\| \leq k_g \|\boldsymbol{x} - \boldsymbol{y}\| ,$$
where $k_g \geq \left\| \frac{\partial \boldsymbol{g}(\boldsymbol{q})}{\partial \boldsymbol{q}} \right\|$ for all $\boldsymbol{q} \in \mathbb{R}^n$.

The integral action of the PID control law (9.1) introduces an additional state variable that is denoted here by $\boldsymbol{\xi}$, and whose time derivative is $\dot{\boldsymbol{\xi}} = \tilde{\boldsymbol{q}}$.

The PID control law may be expressed via the two following equations:

$$\boldsymbol{\tau} = K_p\tilde{\boldsymbol{q}} + K_v\dot{\tilde{\boldsymbol{q}}} + K_i\boldsymbol{\xi} \tag{9.3}$$
$$\dot{\boldsymbol{\xi}} = \tilde{\boldsymbol{q}} . \tag{9.4}$$

The closed-loop equation is obtained by substituting the control action $\boldsymbol{\tau}$ from (9.3) in the robot model (9.2), *i.e.*

$$M(\boldsymbol{q})\ddot{\boldsymbol{q}} + C(\boldsymbol{q}, \dot{\boldsymbol{q}})\dot{\boldsymbol{q}} + \boldsymbol{g}(\boldsymbol{q}) = K_p\tilde{\boldsymbol{q}} + K_v\dot{\tilde{\boldsymbol{q}}} + K_i\boldsymbol{\xi}$$
$$\dot{\boldsymbol{\xi}} = \tilde{\boldsymbol{q}},$$

which may be written in terms of the state vector $\begin{bmatrix}\boldsymbol{\xi}^T & \tilde{\boldsymbol{q}}^T & \dot{\tilde{\boldsymbol{q}}}^T\end{bmatrix}^T$, as

$$\frac{d}{dt}\begin{bmatrix}\boldsymbol{\xi} \\ \tilde{\boldsymbol{q}} \\ \dot{\tilde{\boldsymbol{q}}}\end{bmatrix} = \begin{bmatrix}\tilde{\boldsymbol{q}} \\ \dot{\tilde{\boldsymbol{q}}} \\ \ddot{\boldsymbol{q}}_d - M(\boldsymbol{q})^{-1}\left[K_p\tilde{\boldsymbol{q}} + K_v\dot{\tilde{\boldsymbol{q}}} + K_i\boldsymbol{\xi} - C(\boldsymbol{q}, \dot{\boldsymbol{q}})\dot{\boldsymbol{q}} - \boldsymbol{g}(\boldsymbol{q})\right]\end{bmatrix} . \tag{9.5}$$

The equilibria of the equation above, if any, have the form $\begin{bmatrix}\boldsymbol{\xi}^T & \tilde{\boldsymbol{q}}^T & \dot{\tilde{\boldsymbol{q}}}^T\end{bmatrix}^T = \begin{bmatrix}\boldsymbol{\xi}^{*T} & \boldsymbol{0}^T & \boldsymbol{0}^T\end{bmatrix}^T$ where

$$\boldsymbol{\xi}^* = K_i^{-1}\left[M(\boldsymbol{q}_d)\ddot{\boldsymbol{q}}_d + C(\boldsymbol{q}_d, \dot{\boldsymbol{q}}_d)\dot{\boldsymbol{q}}_d + \boldsymbol{g}(\boldsymbol{q}_d)\right]$$

must be a constant vector. Certainly, for $\boldsymbol{\xi}^*$ to be a constant vector, if the desired joint position $\boldsymbol{q}_d$ is time-varying, it may not be arbitrary but should have a very particular form. One way to obtain a $\boldsymbol{q}_d$ for which $\boldsymbol{\xi}^*$ is constant, is by solving the differential equations

$$\frac{d}{dt}\begin{bmatrix}\boldsymbol{q}_d \\ \dot{\boldsymbol{q}}_d\end{bmatrix} = \begin{bmatrix}\dot{\boldsymbol{q}}_d \\ M(\boldsymbol{q}_d)^{-1}\left[\boldsymbol{\tau}_0 - C(\boldsymbol{q}_d, \dot{\boldsymbol{q}}_d)\dot{\boldsymbol{q}}_d - \boldsymbol{g}(\boldsymbol{q}_d)\right]\end{bmatrix}, \qquad \begin{bmatrix}\boldsymbol{q}_d(0) \\ \dot{\boldsymbol{q}}_d(0)\end{bmatrix} \in \mathbb{R}^{2n} \tag{9.6}$$

where $\boldsymbol{\tau}_0 \in \mathbb{R}^n$ is a constant vector. This way $\boldsymbol{\xi}^* = K_i^{-1}\boldsymbol{\tau}_0$. In particular, if $\boldsymbol{\tau}_0 = \boldsymbol{0} \in \mathbb{R}^n$ then the origin of the closed-loop Equation (9.5), is an equilibrium. Notice that the solution of (9.6) is simply the position $\boldsymbol{q}$ and velocity $\dot{\boldsymbol{q}}$ when one applies a constant torque $\boldsymbol{\tau} = \boldsymbol{\tau}_0$ to the robot in question. In general, it is not possible to obtain an expression in closed form for $\boldsymbol{q}_d$, so Equation (9.6) must be solved numerically. Nevertheless, the resulting desired position $\boldsymbol{q}_d$, may have a capricious form and therefore be of little utility. This is illustrated in the following example.

Example 9.1. Consider the Pelican prototype robot studied in Chapter 5, and shown in Figure 5.2.

Considering $\boldsymbol{\tau}_0 = \boldsymbol{0} \in \mathbb{R}^2$ and the initial condition $[q_{d1}\ q_{d2}\ \dot{q}_{d1}\ \dot{q}_{d2}]^T = [-\pi/20\ \ \pi/20\ \ 0\ \ 0]^T$; the numerical solution of (9.6) for $\boldsymbol{q}_d(t)$, is shown in Figure 9.2.

With $\boldsymbol{q}_d(t)$, whose two components are shown in Figure 9.2, the origin $\begin{bmatrix}\boldsymbol{\xi}^T & \tilde{\boldsymbol{q}}^T & \dot{\tilde{\boldsymbol{q}}}^T\end{bmatrix}^T = \boldsymbol{0} \in \mathbb{R}^6$ is an equilibrium of the closed-loop equation formed by the PID control and the robot in question. ♢

In the case where the desired joint position $\boldsymbol{q}_d$ is an arbitrary function of time, and does not tend to a constant vector value, then the closed-loop equation has no equilibrium. In such cases, we cannot study the stability in the sense of Lyapunov and in particular, one may not expect that the position error $\tilde{\boldsymbol{q}}$ tend to zero. In the best case scenario, and under the hypothesis that the initial position and velocity errors $\tilde{\boldsymbol{q}}(0)$ and $\dot{\boldsymbol{q}}(0)$ are small, the position error $\tilde{\boldsymbol{q}}$ remains bounded. The formal proof of these claims is established in the works cited at the end of the chapter.

Let us come back to our discussion on the determination of an equilibrium for the closed-loop system. We said that a sufficient condition for the existence

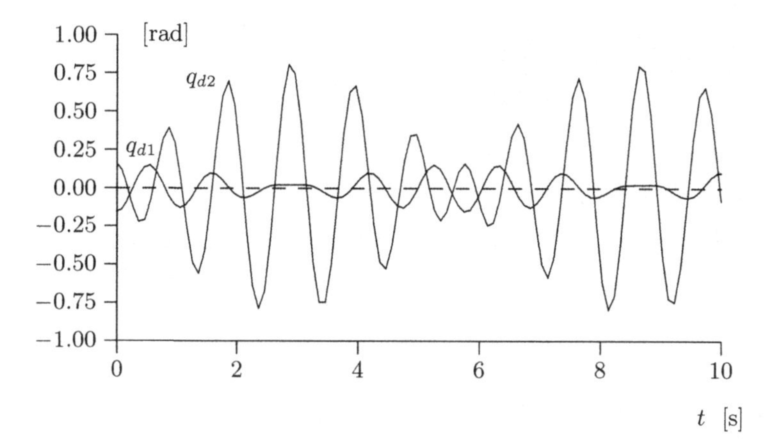

Figure 9.2. Desired joint positions

and unicity of the equilibrium for the closed-loop Equation (9.5) is that the desired position $\boldsymbol{q}_d(t)$ be constant. Denoting by $\boldsymbol{q}_d$ such a constant vector the equilibrium is

$$\begin{bmatrix} \boldsymbol{\xi} \\ \tilde{\boldsymbol{q}} \\ \dot{\tilde{\boldsymbol{q}}} \end{bmatrix} = \begin{bmatrix} {K_i}^{-1}\, \boldsymbol{g}(\boldsymbol{q}_d) \\ \mathbf{0} \\ \mathbf{0} \end{bmatrix} \in \mathbb{R}^{3n} .$$

This equilibrium may of course, be translated to the origin via a suitable change of variable, *e.g.* defining

$$\boldsymbol{z} = \boldsymbol{\xi} - {K_i}^{-1} \boldsymbol{g}(\boldsymbol{q}_d) .$$

Then, the corresponding closed-loop equation may be expressed in terms of the state vector $\begin{bmatrix} \boldsymbol{z}^T & \tilde{\boldsymbol{q}}^T & \dot{\boldsymbol{q}}^T \end{bmatrix}^T$ as

$$\frac{d}{dt}\begin{bmatrix} \boldsymbol{z} \\ \tilde{\boldsymbol{q}} \\ \dot{\boldsymbol{q}} \end{bmatrix} = \begin{bmatrix} \tilde{\boldsymbol{q}} \\ -\dot{\boldsymbol{q}} \\ M(\boldsymbol{q})^{-1}\left[K_p\tilde{\boldsymbol{q}} - K_v\dot{\boldsymbol{q}} + K_i\boldsymbol{z} + \boldsymbol{g}(\boldsymbol{q}_d) - C(\boldsymbol{q},\dot{\boldsymbol{q}})\dot{\boldsymbol{q}} - \boldsymbol{g}(\boldsymbol{q})\right] \end{bmatrix} . \tag{9.7}$$

Notice that the previous equation is autonomous and its unique equilibrium is the origin $\begin{bmatrix} \boldsymbol{z}^T & \tilde{\boldsymbol{q}}^T & \dot{\boldsymbol{q}}^T \end{bmatrix}^T = \mathbf{0} \in \mathbb{R}^{3n}$.

For the sequel, we find it convenient to adopt the following global change of variables,

$$\begin{bmatrix} \boldsymbol{w} \\ \tilde{\boldsymbol{q}} \\ \dot{\boldsymbol{q}} \end{bmatrix} = \begin{bmatrix} \alpha I & I & 0 \\ 0 & I & 0 \\ 0 & 0 & I \end{bmatrix} \begin{bmatrix} \boldsymbol{z} \\ \tilde{\boldsymbol{q}} \\ \dot{\boldsymbol{q}} \end{bmatrix} \tag{9.8}$$

with $\alpha > 0$.

The closed-loop Equation (9.7) may be expressed as a function of the new variables as

$$\frac{d}{dt}\begin{bmatrix} \boldsymbol{w} \\ \tilde{\boldsymbol{q}} \\ \dot{\boldsymbol{q}} \end{bmatrix} =$$

$$\begin{bmatrix} \alpha\, \tilde{\boldsymbol{q}} - \dot{\boldsymbol{q}} \\ -\dot{\boldsymbol{q}} \\ M(\boldsymbol{q})^{-1}\left[\left[K_p - \frac{1}{\alpha}K_i\right]\tilde{\boldsymbol{q}} - K_v\dot{\boldsymbol{q}} + \frac{1}{\alpha}K_i\boldsymbol{w} + \boldsymbol{g}(\boldsymbol{q}_d) - C(\boldsymbol{q},\dot{\boldsymbol{q}})\dot{\boldsymbol{q}} - \boldsymbol{g}(\boldsymbol{q})\right] \end{bmatrix}. \tag{9.9}$$

These equations are autonomous and the origin of the state space, $\left[\boldsymbol{w}^T \;\; \tilde{\boldsymbol{q}}^T \;\; \dot{\boldsymbol{q}}^T\right]^T = \boldsymbol{0} \in \mathbb{R}^{3n}$ is the unique equilibrium of the closed-loop system. Moreover, due to the globality of the change of variable (9.8), the stability features of this equilibrium correspond to those for the equilibrium $\left[\boldsymbol{z}^T \;\; \tilde{\boldsymbol{q}}^T \;\; \dot{\boldsymbol{q}}^T\right]^T = \boldsymbol{0} \in \mathbb{R}^{3n}$ of Equation (9.7).

If we can find conditions on the matrices K_p, K_v and K_i of the PID controller such that the origin of the closed-loop Equation (9.9) is asymptotically stable, *i.e.* such that the origin is stable and at least for sufficiently small values of the initial states $\boldsymbol{z}(0)$, $\tilde{\boldsymbol{q}}(0)$ and $\dot{\boldsymbol{q}}(0)$, the state – particularly $\tilde{\boldsymbol{q}}(t)$ – tend asymptotically to zero, then we are able to conclude that the position control objective is achieved (at least locally). Therefore, based on this argument we use Lyapunov's direct method via La Salle's theorem to establish conditions under which the choice of design matrices for the PID controller guarantee asymptotic stability of the origin of the closed-loop Equation (9.9).

Thus, let us study the stability of the origin of the equation closed-loop (9.9). This equation governs the behavior of n-DOF robot manipulators under PID control in the case of a constant desired position $\boldsymbol{q}_d$. Specifically, we shall see that if the matrices K_p and K_v are sufficiently "large" and the integral gain matrix K_i sufficiently "small" in the following sense

$$\frac{\lambda_{\min}\{M\}\lambda_{\min}\{K_v\}}{\lambda_{\mathrm{Max}}{}^2\{M\}} > \frac{\lambda_{\mathrm{Max}}\{K_i\}}{\lambda_{\min}\{K_p\} - k_g}, \tag{9.10}$$

and moreover

$$\lambda_{\min}\{K_p\} > k_g \tag{9.11}$$

then, we can guarantee that the position control objective is achieved locally. As a matter of fact, the larger the margin with which the inequality (9.10) holds, the larger the domain of attraction[2] of the equilibrium will be. The property of asymptotic stability with domain of attraction arbitrarily large is commonly referred to as *semiglobal* asymptotic stability.

9.1 Lyapunov Function Candidate

To study the stability of the origin of the state space, we use Lyapunov's direct method by proposing the following Lyapunov function candidate

$$\begin{aligned} V(\tilde{\boldsymbol{q}}, \dot{\boldsymbol{q}}, \boldsymbol{w}) &= \frac{1}{2}\begin{bmatrix} \boldsymbol{w} \\ \tilde{\boldsymbol{q}} \\ \dot{\boldsymbol{q}} \end{bmatrix}^T \begin{bmatrix} \frac{1}{\alpha}K_i & 0 & 0 \\ 0 & \alpha K_v & -\alpha M(\boldsymbol{q}) \\ 0 & -\alpha M(\boldsymbol{q}) & M(\boldsymbol{q}) \end{bmatrix} \begin{bmatrix} \boldsymbol{w} \\ \tilde{\boldsymbol{q}} \\ \dot{\boldsymbol{q}} \end{bmatrix} \\ &\quad + \frac{1}{2}\tilde{\boldsymbol{q}}^T \left[K_p - \frac{1}{\alpha}K_i\right] \tilde{\boldsymbol{q}} + \mathcal{U}(\boldsymbol{q}_d - \tilde{\boldsymbol{q}}) - \mathcal{U}(\boldsymbol{q}_d) + \tilde{\boldsymbol{q}}^T \boldsymbol{g}(\boldsymbol{q}_d) \end{aligned} \tag{9.12}$$

where $\mathcal{U}(\boldsymbol{q})$ denotes as usual, the robot's potential energy and α is the positive constant used in the definition of the variable transformation (9.8). For the sequel we assume that this number satisfies

$$\frac{\lambda_{\min}\{M\}\lambda_{\min}\{K_v\}}{{\lambda_{\mathrm{Max}}}^2\{M\}} > \alpha > \frac{\lambda_{\mathrm{Max}}\{K_i\}}{\lambda_{\min}\{K_p\} - k_g}. \tag{9.13}$$

Following the tuning guide given by (9.10), there always exists α that satisfies the above.

For the Lyapunov function candidate (9.12) to be a Lyapunov function we must verify first that it is positive definite. For this, consider the following terms from this function,

$$\frac{1}{2}\tilde{\boldsymbol{q}}^T \left[K_p - \frac{1}{\alpha}K_i\right] \tilde{\boldsymbol{q}} + \mathcal{U}(\boldsymbol{q}_d - \tilde{\boldsymbol{q}}) - \mathcal{U}(\boldsymbol{q}_d) + \tilde{\boldsymbol{q}}^T \boldsymbol{g}(\boldsymbol{q}_d).$$

According to Example B.2 of Appendix B and taking $\varepsilon = 2$, we conclude that the function defined as the sum of these terms is positive definite, globally for all $\tilde{\boldsymbol{q}}$ if

[2] Roughly, for the purposes of this book, a domain of attraction may be considered as the set of all possible initial conditions which generate trajectories that tend to the equilibrium asymptotically.

$$\lambda_{\min}\{K_p - \frac{1}{\alpha}K_i\} > k_g\,.$$

In its turn, this inequality is implied by[3]

$$\lambda_{\min}\{K_p\} - \frac{1}{\alpha}\lambda_{\mathrm{Max}}\{K_i\} > k_g\,,$$

which holds due to the lower-bound condition imposed on α, in accordance with (9.13). Therefore, we may claim that the Lyapunov function candidate (9.12) satisfies

$$V(\tilde{\boldsymbol{q}}, \dot{\boldsymbol{q}}, \boldsymbol{w}) \geq \frac{1}{2}\begin{bmatrix}\boldsymbol{w}\\ \tilde{\boldsymbol{q}}\\ \dot{\boldsymbol{q}}\end{bmatrix}^T \begin{bmatrix}\frac{1}{\alpha}K_i & 0 & 0\\ 0 & \alpha K_v & -\alpha M(\boldsymbol{q})\\ 0 & -\alpha M(\boldsymbol{q}) & M(\boldsymbol{q})\end{bmatrix}\begin{bmatrix}\boldsymbol{w}\\ \tilde{\boldsymbol{q}}\\ \dot{\boldsymbol{q}}\end{bmatrix}.$$

On the other hand using the inequalities

$$\frac{1}{\alpha}\boldsymbol{w}^T K_i \boldsymbol{w} \geq \frac{1}{\alpha}\lambda_{\min}\{K_i\}\left\|\boldsymbol{w}\right\|^2,$$

$$\alpha\,\tilde{\boldsymbol{q}}^T K_v \tilde{\boldsymbol{q}} \geq \alpha\,\lambda_{\min}\{K_v\}\left\|\tilde{\boldsymbol{q}}\right\|^2,$$

$$\dot{\boldsymbol{q}}^T M(\boldsymbol{q})\dot{\boldsymbol{q}} \geq \lambda_{\min}\{M\}\left\|\dot{\boldsymbol{q}}\right\|^2,$$

$$-\alpha\tilde{\boldsymbol{q}}^T M(\boldsymbol{q})\dot{\boldsymbol{q}} \geq -\alpha\lambda_{\mathrm{Max}}\{M\}\left\|\tilde{\boldsymbol{q}}\right\|\left\|\dot{\boldsymbol{q}}\right\|,$$

we come to the following lower-bound for the Lyapunov function candidate,

$$V(\tilde{\boldsymbol{q}}, \dot{\boldsymbol{q}}, \boldsymbol{w}) \geq \frac{\alpha}{2}\begin{bmatrix}\|\boldsymbol{w}\|\\ \|\tilde{\boldsymbol{q}}\|\\ \|\dot{\boldsymbol{q}}\|\end{bmatrix}^T \begin{bmatrix}\frac{1}{\alpha^2}\lambda_{\min}\{K_i\} & 0 & 0\\ 0 & \lambda_{\min}\{K_v\} & -\lambda_{\mathrm{Max}}\{M\}\\ 0 & -\lambda_{\mathrm{Max}}\{M\} & \frac{1}{\alpha}\lambda_{\min}\{M\}\end{bmatrix}\begin{bmatrix}\|\boldsymbol{w}\|\\ \|\tilde{\boldsymbol{q}}\|\\ \|\dot{\boldsymbol{q}}\|\end{bmatrix}.$$

This shows that $V(\tilde{\boldsymbol{q}}, \dot{\boldsymbol{q}}, \boldsymbol{w})$ is globally positive definite, since α is supposed to satisfy (9.13), in particular because

[3] Theorem of Weyl (R. A. Horn and C. R. Johnson, 1985, "Matrix Analysis", Cambridge University Press, pp. 181.). For matrices A and B symmetric it holds that

$$\lambda_{\min}\{A+B\} \geq \lambda_{\min}\{A\} + \lambda_{\min}\{B\}\,.$$

On the other hand, since $\lambda_{\min}\{A\} = -\lambda_{\mathrm{Max}}\{-A\}$, we have

$$\lambda_{\min}\{A-B\} \geq \lambda_{\min}\{A\} - \lambda_{\mathrm{Max}}\{B\}$$

$$\frac{\lambda_{\min}\{M\}\lambda_{\min}\{K_v\}}{\lambda_{\mathrm{Max}}{}^2\{M\}} > \alpha\,,$$

which implies that the determinant of the 'southeastern' 2×2 sub-block in the matrix above is positive for all $\boldsymbol{q}$.

9.2 Time Derivative of the Lyapunov Function Candidate

Once we have established the conditions under which the Lyapunov function candidate is (globally) positive definite, we may proceed to compute its time derivative. After some straightforward algebraic computations which involve the use of the skew-symmetry of the matrix $\frac{1}{2}\dot{M}(\boldsymbol{q}) - C(\boldsymbol{q}, \dot{\boldsymbol{q}})$ and the equality $\dot{M}(\boldsymbol{q}) = C(\boldsymbol{q}, \dot{\boldsymbol{q}}) + C(\boldsymbol{q}, \dot{\boldsymbol{q}})^T$ established in Property 4.2, the time derivative of the Lyapunov function candidate (9.12) along the solutions of the closed-loop Equation (9.9), may be written as

$$\begin{aligned} \dot{V}(\tilde{\boldsymbol{q}}, \dot{\boldsymbol{q}}, \boldsymbol{w}) = & -\dot{\boldsymbol{q}}^T \left[K_v - \alpha M(\boldsymbol{q})\right] \dot{\boldsymbol{q}} - \tilde{\boldsymbol{q}}^T \left[\alpha K_p - K_i\right] \tilde{\boldsymbol{q}} \\ & - \alpha \tilde{\boldsymbol{q}}^T C(\boldsymbol{q}, \dot{\boldsymbol{q}})^T \dot{\boldsymbol{q}} - \alpha \tilde{\boldsymbol{q}}^T \left[\boldsymbol{g}(\boldsymbol{q}_d) - \boldsymbol{g}(\boldsymbol{q})\right] \end{aligned} \tag{9.14}$$

where we also used (3.20), *i.e.* $\boldsymbol{g}(\boldsymbol{q}) = \partial \mathcal{U}(\boldsymbol{q})/\partial \boldsymbol{q}$.

From standard properties of symmetric positive definite matrices we conclude also that the first two terms of the derivative of the Lyapunov function candidate above, satisfy the following inequalities:

$$-\dot{\boldsymbol{q}}^T \left[K_v - \alpha M(\boldsymbol{q})\right] \dot{\boldsymbol{q}} \leq - \left[\lambda_{\min}\{K_v\} - \alpha\ \lambda_{\mathrm{Max}}\{M\}\right] \|\dot{\boldsymbol{q}}\|^2$$

and

$$-\tilde{\boldsymbol{q}}^T \left[\alpha K_p - K_i\right] \tilde{\boldsymbol{q}} \leq - \left[\alpha\ \lambda_{\min}\{K_p\} - \lambda_{\mathrm{Max}}\{K_i\}\right] \|\tilde{\boldsymbol{q}}\|^2$$

respectively.

On the other hand, from Properties 4.2 and 4.3 and using:

- $\|C(\boldsymbol{x}, \boldsymbol{y})\boldsymbol{z}\| \leq k_{C_1} \|\boldsymbol{y}\| \|\boldsymbol{z}\|$,
- $\|\boldsymbol{g}(\boldsymbol{x}) - \boldsymbol{g}(\boldsymbol{y})\| \leq k_g \|\boldsymbol{x} - \boldsymbol{y}\|$,

we may show that the last two terms in the expression of the derivative of the Lyapunov function candidate satisfy

$$-\alpha \tilde{\boldsymbol{q}}^T C(\boldsymbol{q}, \dot{\boldsymbol{q}})^T \dot{\boldsymbol{q}} \leq \alpha\ k_{C_1} \|\tilde{\boldsymbol{q}}\| \|\dot{\boldsymbol{q}}\|^2$$

and

$$-\alpha \tilde{\boldsymbol{q}}^T [\boldsymbol{g}(\boldsymbol{q}_d) - \boldsymbol{g}(\boldsymbol{q})] \leq \alpha\ k_g \|\tilde{\boldsymbol{q}}\|^2 .$$

Consequently, it also holds that

$$\dot{V}(\tilde{\boldsymbol{q}}, \dot{\boldsymbol{q}}, \boldsymbol{w}) \leq - \begin{bmatrix} \|\tilde{\boldsymbol{q}}\| \\ \|\dot{\boldsymbol{q}}\| \end{bmatrix}^T \begin{bmatrix} Q_{11} & 0 \\ 0 & Q_{22}(\tilde{\boldsymbol{q}}) \end{bmatrix} \begin{bmatrix} \|\tilde{\boldsymbol{q}}\| \\ \|\dot{\boldsymbol{q}}\| \end{bmatrix} \tag{9.15}$$

where

$$Q_{11} = \alpha \; [\lambda_{\min}\{K_p\} - k_g] - \lambda_{\mathrm{Max}}\{K_i\},$$
$$Q_{22}(\tilde{\boldsymbol{q}}) = \lambda_{\min}\{K_v\} - \alpha \; [\lambda_{\mathrm{Max}}\{M\} + k_{C_1} \|\tilde{\boldsymbol{q}}\|] \,.$$

We show next that there exists a ball $\mathcal{D}$ of radius $\eta > 0$ centered at the origin of the state space, that is,

$$\mathcal{D} := \left\{ \tilde{\boldsymbol{q}}, \dot{\boldsymbol{q}}, \boldsymbol{w} \in \mathbb{R}^n : \left\| \begin{matrix} \boldsymbol{w} \\ \tilde{\boldsymbol{q}} \\ \dot{\boldsymbol{q}} \end{matrix} \right\| < \eta \right\}$$

on which $\dot{V}(\tilde{\boldsymbol{q}}, \dot{\boldsymbol{q}}, \boldsymbol{w})$ is negative semidefinite.

With this goal in mind, notice that we have from the condition (9.13) imposed on α that

$$\frac{\lambda_{\min}\{M\}\lambda_{\min}\{K_v\} \, [\lambda_{\min}\{K_p\} - k_g]}{\lambda_{\mathrm{Max}}\{K_i\}} > \lambda^2_{\mathrm{Max}}\{M\} \,.$$

Since obviously $\lambda_{\mathrm{Max}}\{M\} \geq \lambda_{\min}\{M\}$, then

$$\frac{\lambda_{\min}\{K_v\} \, [\lambda_{\min}\{K_p\} - k_g]}{\lambda_{\mathrm{Max}}\{K_i\}} > \lambda_{\mathrm{Max}}\{M\}$$

and knowing that $k_{C_1} \geq 0$, it readily follows that

$$\frac{1}{k_{C_1}} \left[\frac{\lambda_{\min}\{K_v\} \, [\lambda_{\min}\{K_p\} - k_g]}{\lambda_{\mathrm{Max}}\{K_i\}} - \lambda_{\mathrm{Max}}\{M\} \right] \geq 0 \,. \tag{9.16}$$

It is now convenient to define the radius η of the ball $\mathcal{D}$ centered at the origin of the state space to be exactly the term on the left-hand side of the inequality (9.16), that is

$$\eta := \frac{1}{k_{C_1}} \left[\frac{\lambda_{\min}\{K_v\} \, [\lambda_{\min}\{K_p\} - k_g]}{\lambda_{\mathrm{Max}}\{K_i\}} - \lambda_{\mathrm{Max}}\{M\} \right] \,. \tag{9.17}$$

It is important to observe that the radius η is not only positive but it increases as $\lambda_{\min}\{K_p\}$ and $\lambda_{\min}\{K_v\}$ increase and as $\lambda_{\mathrm{Max}}\{K_i\}$ decreases, while always satisfying the condition (9.13) on α.

We now show that $\dot{V}(\tilde{\boldsymbol{q}}, \dot{\boldsymbol{q}}, \boldsymbol{w})$ is negative semidefinite on the ball $\mathcal{D}$. On this region of the state space we have

$$\|\tilde{\boldsymbol{q}}\| < \eta .$$

Incorporating the definition (9.17) of η we get

$$\|\tilde{\boldsymbol{q}}\| < \frac{1}{k_{C_1}} \left[\frac{\lambda_{\min}\{K_v\} \left[\lambda_{\min}\{K_p\} - k_g\right]}{\lambda_{\text{Max}}\{K_i\}} - \lambda_{\text{Max}}\{M\} \right]$$

which, after some simple manipulations, yields the inequality

$$\frac{\lambda_{\min}\{K_v\}}{\lambda_{\text{Max}}\{M\} + k_{C_1} \|\tilde{\boldsymbol{q}}\|} > \frac{\lambda_{\text{Max}}\{K_i\}}{\lambda_{\min}\{K_p\} - k_g}$$

which is valid on the region $\mathcal{D}$.

Since α is any real number that satisfies the condition (9.13) and in particular it satisfies

$$\alpha > \frac{\lambda_{\text{Max}}\{K_i\}}{\lambda_{\min}\{K_p\} - k_g} ,$$

it can always be picked to satisfy also,

$$\frac{\lambda_{\min}\{K_v\}}{\lambda_{\text{Max}}\{M\} + k_{C_1} \|\tilde{\boldsymbol{q}}\|} > \alpha > \frac{\lambda_{\text{Max}}\{K_i\}}{\lambda_{\min}\{K_p\} - k_g}$$

on the region $\mathcal{D}$. It is precisely these inequalities that captures the conditions on α such that the scalars Q_{11} and $Q_{22}(\tilde{\boldsymbol{q}})$ in the derivative (9.15) of the Lyapunov function candidate are strictly positive. This implies that $\dot{V}(\tilde{\boldsymbol{q}}, \dot{\boldsymbol{q}}, \boldsymbol{w})$ is negative semidefinite on the ball $\mathcal{D}$ with radius η defined by (9.17).

9.3 Asymptotic Stability

Summarizing, so far we have shown that if the matrices K_p, K_v and K_i of the PID control law satisfy the conditions (9.10) and (9.11), then $V(\tilde{\boldsymbol{q}}, \dot{\boldsymbol{q}}, \boldsymbol{w})$ is a globally positive definite function while $\dot{V}(\tilde{\boldsymbol{q}}, \dot{\boldsymbol{q}}, \boldsymbol{w})$ is negative semidefinite – locally, though the domain of validity may be arbitrarily enlarged by an appropriate choice of the control gains. Therefore, according to Theorem 2.2, the origin of the closed-loop Equation (9.9), is a stable equilibrium. Since $\dot{V}(\tilde{\boldsymbol{q}}, \dot{\boldsymbol{q}}, \boldsymbol{w})$ is not negative definite but only negative semidefinite, Theorem 2.4 cannot be invoked to prove asymptotic stability. Yet, since the closed-loop Equation (9.9) is autonomous, we may use a local version of La Salle's theorem (*i.e.* a local version of Theorem 2.7).

Since the function $\dot{V}(\tilde{\boldsymbol{q}}, \dot{\boldsymbol{q}}, \boldsymbol{w})$ obtained in (9.15) is negative semidefinite but only locally, Theorem 2.7 may not be invoked directly. Even though we do not show this here[4], it is also true that if in the statement of Theorem 2.7 we

[4] See any of the cited texts on nonlinear systems.

replace "$\dot{V}(\boldsymbol{x}) \leq 0$ for all $\boldsymbol{x} \in \mathbb{R}^n$", by "$\dot{V}(\boldsymbol{x}) \leq 0$ for all $\boldsymbol{x} \in \mathbb{R}^n$ sufficiently small in norm" then, one can guarantee local asymptotic stability. We may proceed to claim that in this case the set Ω is given by

$$\begin{aligned}\Omega &= \left\{\boldsymbol{x} \in \mathbb{R}^{3n} : \dot{V}(\boldsymbol{x}) = 0\right\} \\ &= \left\{\boldsymbol{x} = \begin{bmatrix}\boldsymbol{w} \\ \tilde{\boldsymbol{q}} \\ \dot{\boldsymbol{q}}\end{bmatrix} \in \mathbb{R}^{3n} \ : \ \dot{V}(\tilde{\boldsymbol{q}}, \dot{\boldsymbol{q}}, \boldsymbol{w}) = 0\right\} \\ &= \{\boldsymbol{w} \in \mathbb{R}^n, \tilde{\boldsymbol{q}} = \boldsymbol{0} \in \mathbb{R}^n, \dot{\boldsymbol{q}} = \boldsymbol{0} \in \mathbb{R}^n\} \ .\end{aligned}$$

Observe that from (9.14) and with the restrictions imposed on the choice of α, it follows that $\dot{V}(\tilde{\boldsymbol{q}}, \dot{\boldsymbol{q}}, \boldsymbol{w}) = 0$ if and only if $\tilde{\boldsymbol{q}} = \boldsymbol{0}$ and $\dot{\boldsymbol{q}} = \boldsymbol{0}$. For a solution $\boldsymbol{x}(t)$ to belong to Ω for all $t \geq 0$, it is necessary and sufficient that $\tilde{\boldsymbol{q}} = \boldsymbol{0}$ and $\dot{\boldsymbol{q}}(t) = \boldsymbol{0}$ for all $t \geq 0$. Therefore it must also hold that $\ddot{\boldsymbol{q}}(t) = \boldsymbol{0}$ for all $t \geq 0$. Taking this into consideration, we conclude from the closed-loop Equation (9.9) that if $\boldsymbol{x}(t) \in \Omega$ for all $t \geq 0$, then

$$\begin{aligned}\dot{\boldsymbol{w}}(t) &= \boldsymbol{0} \\ \boldsymbol{0} &= \frac{1}{\alpha} M(\boldsymbol{q}_d)^{-1} K_i \boldsymbol{w}(t) \, ,\end{aligned}$$

which, since $M(\boldsymbol{q}_d)$ is positive definite, implies that $\boldsymbol{w}(t) = \boldsymbol{0}$ for all $t \geq 0$. Therefore, $\left[\boldsymbol{w}(0)^T \quad \tilde{\boldsymbol{q}}(0)^T \quad \dot{\boldsymbol{q}}(0)^T\right]^T = \boldsymbol{0} \in \mathbb{R}^{3n}$ is the only initial condition in Ω for which $\boldsymbol{x}(t) \in \Omega$ for all $t \geq 0$. Finally, we conclude from all this that the origin of the closed-loop Equation (9.9) is locally asymptotically stable. It is worth emphasizing that in the particular situation of robots for which $k_{C_1} = 0$, *e.g.* a pendulum, some Cartesian robots or with transmission by levers, the function $\dot{V}(\tilde{\boldsymbol{q}}, \dot{\boldsymbol{q}}, \boldsymbol{w})$ obtained in (9.15), is globally negative semidefinite. In this situation, the asymptotic stability of the origin is also global.

The analysis developed so far applies to the classic PID control law given by (9.3) and (9.4). We have also assumed that the joint desired position $\boldsymbol{q}_d$ is constant. In the practical implementation of the PID controller on ordinary personal computers, the desired position $\boldsymbol{q}_d$ may be a piecewise constant function, being constant in between two sampling times. Therefore, before and after the sampling times, the desired position $\boldsymbol{q}_d$ takes constant values which are usually different, and at the sampling instant there occurs an abrupt change of magnitude. For this reason, the position error $\tilde{\boldsymbol{q}} = \boldsymbol{q}_d - \boldsymbol{q}$, and the control action $\boldsymbol{\tau} = K_p\tilde{\boldsymbol{q}} - K_v\dot{\boldsymbol{q}} + K_i\boldsymbol{\xi}$ also vary abruptly at the sampling times. Consequently, the control action $\boldsymbol{\tau}$ demands instantaneous variations to the actuators; in this situation the latter may be damaged or, simply, we may obtain unacceptable performance in some tasks.

With the aim of conserving the advantages of PID control but avoiding this undesirable phenomenon, several modifications to the classical PID control have been proposed in the literature. An interesting modification is

$$\boldsymbol{\tau} = -K_p \boldsymbol{q} - K_v \dot{\boldsymbol{q}} + K_i \boldsymbol{\xi}$$

where

$$\boldsymbol{\xi}(t) = \int_0^t \tilde{\boldsymbol{q}}(\sigma) \, d\sigma + \boldsymbol{\xi}(0) \, .$$

This modified PID control preserves exactly the same closed-loop Equation (9.7) that the classical PID control, but with $\boldsymbol{z}$ defined by

$$\boldsymbol{z} = \boldsymbol{\xi} - K_i^{-1} \left[\boldsymbol{g}(\boldsymbol{q}_d) + K_p \boldsymbol{q}_d \right] \, .$$

Therefore, the analysis and conclusions established above also apply to this modified PID control.

9.4 Tuning Procedure

From the stability analysis presented in the previous subsections we can draw a tuning procedure which is fairly simple for PID control. This method yields symmetric matrices K_p, K_v and K_i that guarantee achievement of the position control objective, locally.

The procedure stems from (9.11) and from the condition (9.13) imposed on α and may be summarized in terms of the eigenvalues of the gain matrices as follows:

- $\lambda_{\text{Max}}\{K_i\} \geq \lambda_{\text{min}}\{K_i\} > 0$;
- $\lambda_{\text{Max}}\{K_p\} \geq \lambda_{\text{min}}\{K_p\} > k_g$;
- $\lambda_{\text{Max}}\{K_v\} \geq \lambda_{\text{min}}\{K_v\} > \dfrac{\lambda_{\text{Max}}\{K_i\}}{\lambda_{\text{min}}\{K_p\} - k_g} \cdot \dfrac{{\lambda_{\text{Max}}}^2\{M\}}{\lambda_{\text{min}}\{M\}}$.

It is important to underline that this tuning procedure requires knowledge of the structure of the inertia matrix $M(\boldsymbol{q})$ and of the vector of gravitational torques $\boldsymbol{g}(\boldsymbol{q})$ of the robot in question. This, is necessary to compute $\lambda_{\text{min}}\{M(\boldsymbol{q})\}$, $\lambda_{\text{Max}}\{M(\boldsymbol{q})\}$ and k_g respectively. Nonetheless, since it is sufficient to have upper-bounds on $\lambda_{\text{Max}}\{M(\boldsymbol{q})\}$ and k_g, and a lower-bound for $\lambda_{\text{min}}\{M(\boldsymbol{q})\}$ to use the tuning procedure, it is not necessary to know the exact values of the dynamic parameters of the robot, *e.g.* masses and inertias, but only the bounds.

Next, we present an example in order to show an application of the ideas discussed above.

Example 9.2. Consider the 2-DOF *Pelican robot* shown in Figure 9.3.

The elements of the inertia matrix $M(\boldsymbol{q})$ are

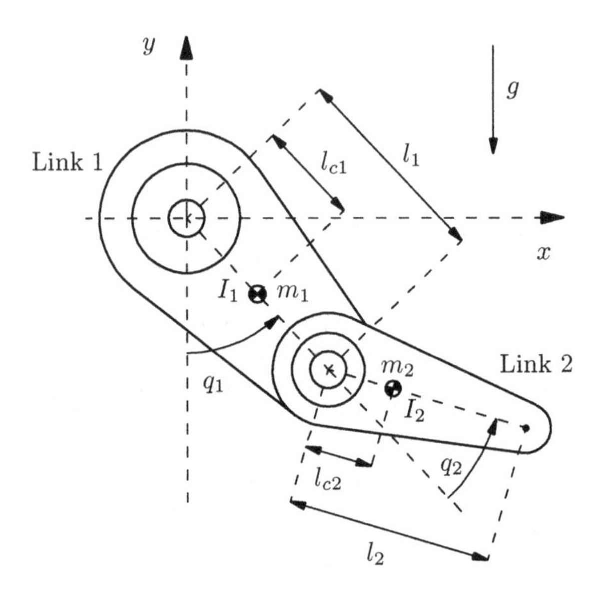

Figure 9.3. Diagram of the Pelican prototype robot

$$
\begin{aligned}
M_{11}(\boldsymbol{q}) &= m_1 l_{c1}^2 + m_2 \left[l_1^2 + l_{c2}^2 + 2 l_1 l_{c2} \cos(q_2) \right] + I_1 + I_2 \\
M_{12}(\boldsymbol{q}) &= m_2 \left[l_{c2}^2 + l_1 l_{c2} \cos(q_2) \right] + I_2 \\
M_{21}(\boldsymbol{q}) &= m_2 \left[l_{c2}^2 + l_1 l_{c2} \cos(q_2) \right] + I_2 \\
M_{22}(\boldsymbol{q}) &= m_2 l_{c2}^2 + I_2 .
\end{aligned}
$$

The components of the gravitational torques vector $\boldsymbol{g}(\boldsymbol{q})$, are given by

$$
\begin{aligned}
g_1(\boldsymbol{q}) &= (m_1 l_{c1} + m_2 l_1) g \sin(q_1) + m_2 l_{c2} g \sin(q_1 + q_2) \\
g_2(\boldsymbol{q}) &= m_2 l_{c2} g \sin(q_1 + q_2).
\end{aligned}
$$

The purpose of this example is to use the tuning procedure presented above in order to determine suitable gain matrices K_p, K_v and K_i. First, we compute the value of k_g and then that of the eigenvalues $\lambda_{\min}\{M(\boldsymbol{q})\}$ and $\lambda_{\mathrm{Max}}\{M(\boldsymbol{q})\}$.

Using Property 4.3, as well as the elements of the vector of gravitational torques $\boldsymbol{g}(\boldsymbol{q})$, we obtain

$$
\begin{aligned}
k_g &= n \left(\mathrm{Max}_{\ i,j,q} \left| \frac{\partial g_i(\boldsymbol{q})}{\partial q_j} \right| \right) \\
&= n(m_1 l_{c1} + m_2 l_1 + m_2 l_{c2}) g \\
&= 23.94 \quad \left[\mathrm{kg\ m^2/s^2} \right]
\end{aligned}
$$

where we also used the numerical values of the robot parameters listed in Table 5.1 on page 115.

We proceed next to compute $\lambda_{\min}\{M(\boldsymbol{q})\}$ and $\lambda_{\text{Max}}\{M(\boldsymbol{q})\}$, which for this robot depend only on q_2; we do this by evaluating the matrix $M(\boldsymbol{q})$ for a set of values of q_2 between 0 and 2π, and extracting the corresponding eigenvalues. The values obtained are

$$\lambda_{\min}\{M(\boldsymbol{q})\} = 0.011 \quad \left[\text{kg m}^2\right],$$
$$\lambda_{\text{Max}}\{M(\boldsymbol{q})\} = 0.361, \quad \left[\text{kg m}^2\right]$$

which correspond to $q_2 = 0$.

With the numerical values of $\lambda_{\min}\{M(\boldsymbol{q})\}$, $\lambda_{\text{Max}}\{M(\boldsymbol{q})\}$ and k_g, and following the tuning procedure, we finally determine the following matrices:

$$K_i = \text{diag}\{1.5\} \quad [\text{Nm} \,/\, (\text{rad s})],$$
$$K_p = \text{diag}\{30\} \quad [\text{Nm} \,/\, \text{rad}],$$
$$K_v = \text{diag}\{7, 3\} \quad [\text{Nm s} \,/\, \text{rad}].$$

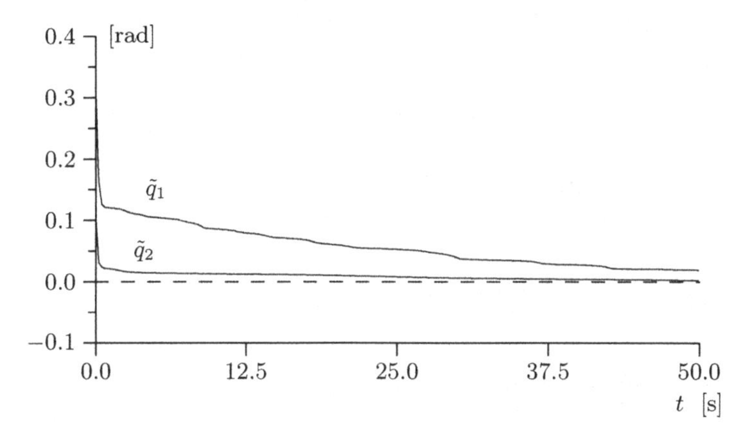

Figure 9.4. Graphs of position errors $\tilde{q}_1$ and $\tilde{q}_2$

Figure 9.4 depicts the position errors against time; notice that the scale on the ordinates axis spans up to 50 s which is longer than the intervals used elsewhere. We may conclude from this, that the transient response is slower than those obtained with PD control with gravity compensation (see Figure 7.3) and PD control with desired gravity compensation (see Figure 8.4). This is due to the fact that the control gains of the PID control law were chosen according with the tuning procedure exposed in this chapter, which limits the maximal eigenvalue of K_i by a relatively small upper-bound.

However, as may be appreciated from Figure 9.5, if the tuning procedure is violated the performance of PID control improves to parallel

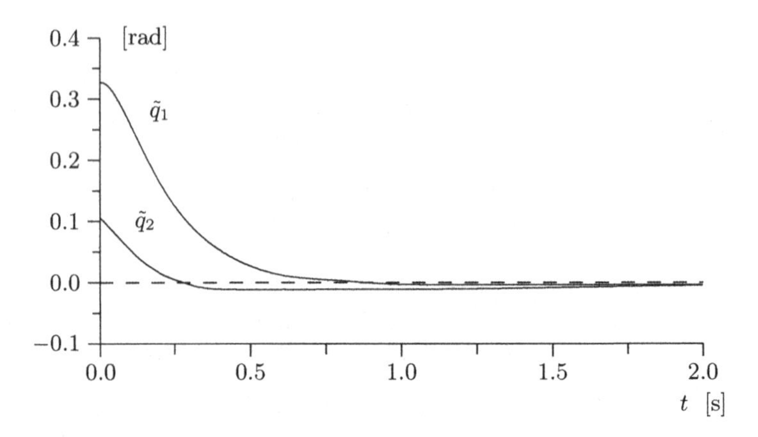

Figure 9.5. Graphs of position errors $\tilde{q}_1$ and $\tilde{q}_2$

the respective performances of PD control with gravity compensation and PD control with desired gravity compensation. The latter experimental results have been obtained with the following gains:

$$K_p = \begin{bmatrix} 30 & 0 \\ 0 & 30 \end{bmatrix} \quad [\text{Nm} \,/\, \text{rad}]\,,$$

$$K_v = \begin{bmatrix} 7 & 0 \\ 0 & 3 \end{bmatrix} \quad [\text{Nm s} \,/\, \text{rad}]\,,$$

$$K_i = \begin{bmatrix} 70 & 0 \\ 0 & 100 \end{bmatrix} \quad [\text{Nm} \,/\, (\text{rad s})]\,,$$

that is, K_p and K_v have the same values than in the cases of the PD-based control laws. ♢

9.5 Conclusions

We may summarize the ideas of this chapter as follows. Consider PID control for n-DOF robots and assume that the desired position $\boldsymbol{q}_d$ is constant.

- If the symmetric matrices K_p, K_v and K_i of the PID control law satisfy
  (i) $\lambda_{\text{Max}}\{K_i\} \geq \lambda_{\text{min}}\{K_i\} > 0$
  (ii) $\lambda_{\text{Max}}\{K_p\} \geq \lambda_{\text{min}}\{K_p\} > k_g$
  (iii) $\lambda_{\text{Max}}\{K_v\} \geq \lambda_{\text{min}}\{K_v\} > \dfrac{\lambda_{\text{Max}}\{K_i\}}{\lambda_{\text{min}}\{K_p\} - k_g} \cdot \dfrac{{\lambda_{\text{Max}}}^2\{M\}}{\lambda_{\text{min}}\{M\}}$

then the origin of the closed-loop equation, expressed in terms of the state vector $\left[\boldsymbol{w}^T \ \tilde{\boldsymbol{q}}^T \ \dot{\boldsymbol{q}}^T\right]^T$ is locally asymptotically stable. Consequently, if the initial conditions are "sufficiently small", we have $\lim_{t\to\infty} \tilde{\boldsymbol{q}}(t) = \mathbf{0}$. Nevertheless, this conclusion holds also for "large" initial conditions as long as $\lambda_{\min}\{K_p\}$ and $\lambda_{\min}\{K_v\}$ also are "large" and $\lambda_{\text{Max}}\{K_i\}$ are "small".

If the terms of centrifugal and Coriolis forces, $C(\boldsymbol{q}, \dot{\boldsymbol{q}})$ do not appear in the robot dynamic model then, the asymptotic stability is global.

Bibliography

The following works had a strong influence on the final form that this chapter took.

- Arimoto S., Miyazaki F., 1984, *"Stability and robustness of PID feedback control for robot manipulators of sensory capability"*, in M. Brady and R. P. Paul (ed.), Robotics Research: First International Symposium, MIT Press, pp. 783–799.
- Ortega R., Loría A., Nicklasson P. J., Sira-Ramírez H., 1998, *"Passivity-based control of Euler-Lagrange systems mechanical, electrical and electromechanical applications"*, Springer-Verlag: London, Communications and Control Engg. Series.
- Kelly R., 1995, *"A tuning procedure for stable PID control of robot manipulators"*, Robotica, Vol. 13, Part 2, March–April, pp. 141–148.
- Meza J. L., Santibáñez V., 2001, *"Análisis simple de estabilidad asintótica semiglobal de un regulador PID lineal para robots manipuladores"*, 3er. Congreso Mexicano de Robótica, Querétaro, Qro., Mexico.

The study of PID control for robots with dynamic models which are "dominated" by the dynamics of the actuators (DC motors), is presented in the following texts.

- Paul R., 1981, *"Robot manipulators: Mathematics, programming, and control"*, The MIT Press.
- Spong M., Vidyasagar M., 1989, *"Robot dynamics and control"*, John Wiley and Sons.

The analysis of PID control for robots where the desired joint position $\boldsymbol{q}_d$ is a function of time, is treated in the following works:

- Samson C., 1987, *"Robust control of a class of non–linear systems and application to robotics"*, International Journal of Adaptive Control and Signal Processing, Vol. 1, pp. 49–68.

- Kawamura S., Miyazaki F., Arimoto S., 1988, *"Is a local linear PD feedback control law effective for trajectory tracking of robot motion ?"*, in Proceedings of the 1988 IEEE International Conference on Robotics and Automation, Philadelphia, PA., pp. 1335–1340, April.
- Wen J. T., 1989, *"PID control of robot manipulators"*, *Tech. Report*, Rensselaer Polytechnic Institute, June.
- Wen J. T., Murphy S., 1990, *"PID control for robot manipulators"*, CIRSSE Document # 54, Rensselaer Polytechnic Institute, May.
- Qu Z., Dorsey J., 1991, *"Robust PID control of robots"*, International Journal of Robotics and Automation, Vol. 6, No. 4, pp. 228–235.
- Rocco P., 1996, *"Stability of PID control for industrial robot arms"*, IEEE Transactions on Robotics and Automation, Vol. 12, No. 4, pp. 606–614.
- Cervantes I., Alvarez–Ramirez J., 2001, *"On the PID tracking control of robot manipulators"*, Systems and Control Letters, Vol. 42, pp. 37–46.
- Choi Y., Chung W. K., 2004, "PID trajectory tracking control for mechanical systems", *Lecture Notes in Control and Information Sciences*, No. 298, Springer-Verlag: London.

A proof of semi-global asymptotic stability, that is, when the domain of attraction may be enlarged at wish, by increasing the control gains, is presented in

- Ortega R., Loría A., Kelly R., 1995, *"A semiglobally stable output feedback PI^2D regulator for robot manipulators"*, IEEE Transactions on Automatic Control, Vol. 40, No. 8, pp. 1432–1436.
- Alvarez–Ramirez J., Cervantes I., Kelly R., 2000, *"PID regulation of robot manipulators: stability and performance"*, Systems and Control Letters, Vol. 41, pp. 73–83.

Studies on controllers with small modifications to the structure of the PID control have been reported in Ortega, Loría, Kelly (1995) above and in

- Kelly R., 1998, *"Global positioning of robot manipulators via PD control plus a class of nonlinear integral actions"*, IEEE Transactions on Automatic Control, Vol. 43, No. 7, pp. 934–938.

Problems

1. Consider the model of an ideal pendulum studied in Example 2.2 (see page 30)
$$J\ddot{q} + mgl\,\sin(q) = \tau\,.$$
Assume that the PID control

$$\tau = k_p\tilde{q} + k_v\dot{\tilde{q}} + k_i \int_0^t \tilde{q}(\sigma)\, d\sigma$$

is used to drive the position q to a desired constant position q_d.

a) Obtain the closed-loop equation in terms of the state vector $[w \;\; \tilde{q} \;\; \dot{q}]^T$, where

$$w = \alpha \int_0^t \tilde{q}(\sigma)\, d\sigma - \alpha \frac{mgl}{k_i} \sin(q_d) + \tilde{q}$$

with $\alpha > 0$.

b) Show, by means of a candidate Lyapunov function that if

- $k_i > 0$,
- $k_p > mgl$,
- $k_v > \dfrac{k_i \; J}{k_p - mgl}$,

then the origin of the closed-loop equation is globally asymptotically stable.

c) Does this imply that $\lim_{t\to\infty} \tilde{q}(t) = 0$ for all $\tilde{q}(0) \in \mathbb{R}$?

2. Verify the expression of $\dot{V}(\tilde{\boldsymbol{q}}, \dot{\boldsymbol{q}}, \boldsymbol{z})$ given in Equation (9.14).

3. Consider the PID control law *with gravity compensation* for the position control problem (*i.e.* with constant desired position, $\boldsymbol{q}_d$) for n-DOF robots and whose control law is

$$\begin{aligned}
\boldsymbol{\tau} &= K_p\tilde{\boldsymbol{q}} + K_v\dot{\tilde{\boldsymbol{q}}} + K_i\boldsymbol{\xi} + \boldsymbol{g}(\boldsymbol{q}) \\
\dot{\boldsymbol{\xi}} &= \tilde{\boldsymbol{q}} \,.
\end{aligned}$$

a) Obtain the closed-loop equation in terms of the state vector $\left[\boldsymbol{\xi}^T \;\; \tilde{\boldsymbol{q}}^T \;\; \dot{\boldsymbol{q}}^T\right]^T$. Verify that the origin is an equilibrium of the closed-loop equation.

b) Explain why the following tuning policy guarantees local asymptotic stability of the origin.

- $\lambda_{\rm Max}\{K_i\} \geq \lambda_{\rm min}\{K_i\} > 0$,
- $\lambda_{\rm Max}\{K_p\} \geq \lambda_{\rm min}\{K_p\} > 0$,
- $\lambda_{\rm Max}\{K_v\} \geq \lambda_{\rm min}\{K_v\} > \dfrac{\lambda_{\rm Max}\{K_i\}}{\lambda_{\rm min}\{K_p\}} \cdot \dfrac{{\lambda_{\rm Max}}^2\{M\}}{\lambda_{\rm min}\{M\}}$.

Hint: The only difference with the tuning procedure established above is the absence of k_g in the expressions for K_p and K_v.

4. Consider the PID control law given by (9.1). Show that this is equivalent to

$$\boldsymbol{\tau} = K_p'\tilde{\boldsymbol{q}} - K_v\dot{\boldsymbol{q}} + K_i' \int_0^t \left[\alpha\tilde{\boldsymbol{q}}(\sigma) + \dot{\tilde{\boldsymbol{q}}}(\sigma)\right] d\sigma$$

where

$$K_p' = K_p - \frac{1}{\alpha} K_i$$
$$K_i' = \frac{1}{\alpha} K_i$$

for all $\alpha \neq 0$.

5. Consider the linear multivariable system described by the equation:

$$M\ddot{\boldsymbol{x}} + C\dot{\boldsymbol{x}} + \boldsymbol{g} = \boldsymbol{u}$$

where $\boldsymbol{x} \in \mathbb{R}^n$ is part of the state and at the same time, the output of the system, $\boldsymbol{u} \in \mathbb{R}^n$ is the input, $M, C \in \mathbb{R}^{n\times n}$ are symmetric positive definite constant matrices and $\boldsymbol{g} \in \mathbb{R}^n$ is constant as well.

Consider the application of PID control to drive the state trajectories $\boldsymbol{x}(t)$ to a constant vector $\boldsymbol{x}_d \in \mathbb{R}^n$, *i.e.*

$$\boldsymbol{u} = K_p\tilde{\boldsymbol{x}} - K_v\dot{\boldsymbol{x}} + K_i\boldsymbol{\xi}$$
$$\dot{\boldsymbol{\xi}} = \tilde{\boldsymbol{x}}$$

where $\tilde{\boldsymbol{x}} = \boldsymbol{x}_d - \boldsymbol{x}$.

a) Obtain the closed-loop equation expressed in terms of the state $\begin{bmatrix}\boldsymbol{\xi}^T & \tilde{\boldsymbol{x}}^T & \dot{\boldsymbol{x}}^T\end{bmatrix}^T$.

b) Verify that the point

$$\begin{bmatrix}\boldsymbol{\xi} \\ \tilde{\boldsymbol{x}} \\ \dot{\boldsymbol{x}}\end{bmatrix} = \begin{bmatrix}K_i^{-1}g \\ 0 \\ 0\end{bmatrix} \in \mathbb{R}^{3n}$$

of the closed-loop equation is the unique equilibrium.

c) Use the tuning procedure presented in this chapter to suggest a policy of tuning that guarantees asymptotic stability of the equilibrium for the closed-loop equation. Can it be shown that the asymptotic stability is actually global ?

Part III

Motion Control

Introduction to Part III

Consider the dynamic model of a robot manipulator with n degrees of freedom, rigid links, no friction at the joints and with ideal actuators, (3.18), which we repeat here for ease of reference:

$$M(\boldsymbol{q})\ddot{\boldsymbol{q}} + C(\boldsymbol{q}, \dot{\boldsymbol{q}})\dot{\boldsymbol{q}} + \boldsymbol{g}(\boldsymbol{q}) = \boldsymbol{\tau} \,. \tag{III.1}$$

In terms of the state vector $\begin{bmatrix}\boldsymbol{q}^T & \dot{\boldsymbol{q}}^T\end{bmatrix}^T$ these equations are rewritten as

$$\frac{d}{dt}\begin{bmatrix}\boldsymbol{q} \\ \dot{\boldsymbol{q}}\end{bmatrix} = \begin{bmatrix}\dot{\boldsymbol{q}} \\ M(\boldsymbol{q})^{-1}\left[\boldsymbol{\tau}(t) - C(\boldsymbol{q}, \dot{\boldsymbol{q}})\dot{\boldsymbol{q}} - \boldsymbol{g}(\boldsymbol{q})\right]\end{bmatrix}$$

where $M(\boldsymbol{q}) \in \mathbb{R}^{n\times n}$ is the inertia matrix, $C(\boldsymbol{q}, \dot{\boldsymbol{q}})\dot{\boldsymbol{q}} \in \mathbb{R}^n$ is the vector of centrifugal and Coriolis forces, $\boldsymbol{g}(\boldsymbol{q}) \in \mathbb{R}^n$ is the vector of gravitational torques and $\boldsymbol{\tau} \in \mathbb{R}^n$ is a vector of external forces and torques applied at the joints. The vectors $\boldsymbol{q}, \dot{\boldsymbol{q}}, \ddot{\boldsymbol{q}} \in \mathbb{R}^n$ denote the position, velocity and joint acceleration respectively.

The problem of motion control, tracking control, for robot manipulators may be formulated in the following terms. Consider the dynamic model of an n-DOF robot (III.1). Given a set of vectorial bounded functions $\boldsymbol{q}_d$, $\dot{\boldsymbol{q}}_d$ and $\ddot{\boldsymbol{q}}_d$ referred to as *desired* joint positions, velocities and accelerations we wish to find a vectorial function $\boldsymbol{\tau}$ such that the positions $\boldsymbol{q}$, associated to the robot's joint coordinates follow $\boldsymbol{q}_d$ accurately.

In more formal terms, the *objective of motion control* consists in finding $\boldsymbol{\tau}$ such that

$$\lim_{t\to\infty} \tilde{\boldsymbol{q}}(t) = \boldsymbol{0}$$

where $\tilde{\boldsymbol{q}} \in \mathbb{R}^n$ stands for the joint position errors vector or is simply called position error, and is defined by

$$\tilde{\boldsymbol{q}}(t) := \boldsymbol{q}_d(t) - \boldsymbol{q}(t) \,.$$

Considering the previous definition, the vector $\dot{\tilde{q}}(t) = \dot{q}_d(t) - \dot{q}(t)$ stands for the velocity error. The control objective is achieved if the manipulator's joint variables follow asymptotically the trajectory of the desired motion.

The computation of the vector $\boldsymbol{\tau}$ involves in general, a vectorial nonlinear function of $\boldsymbol{q}$, $\dot{\boldsymbol{q}}$ and $\ddot{\boldsymbol{q}}$. This function is called "control law" or simply, "controller". It is important to recall that robot manipulators are equipped with sensors to measure position and velocity at each joint henceforth, the vectors $\boldsymbol{q}$ and $\dot{\boldsymbol{q}}$ are measurable and may be used by the controllers. In some robots, only measurement of joint position is available and joint velocities may be estimated. In general, a motion control law may be expressed as

$$\boldsymbol{\tau} = \boldsymbol{\tau}\left(\boldsymbol{q}, \dot{\boldsymbol{q}}, \ddot{\boldsymbol{q}}, \boldsymbol{q}_d, \dot{\boldsymbol{q}}_d, \ddot{\boldsymbol{q}}_d, M(\boldsymbol{q}), C(\boldsymbol{q}, \dot{\boldsymbol{q}}), \boldsymbol{g}(\boldsymbol{q})\right).$$

However, for practical purposes it is desirable that the controller does not depend on the joint acceleration $\ddot{\boldsymbol{q}}$ since accelerometers are usually highly sensitive to noise.

Figure III.1 presents the block-diagram of a robot in closed loop with a motion controller.

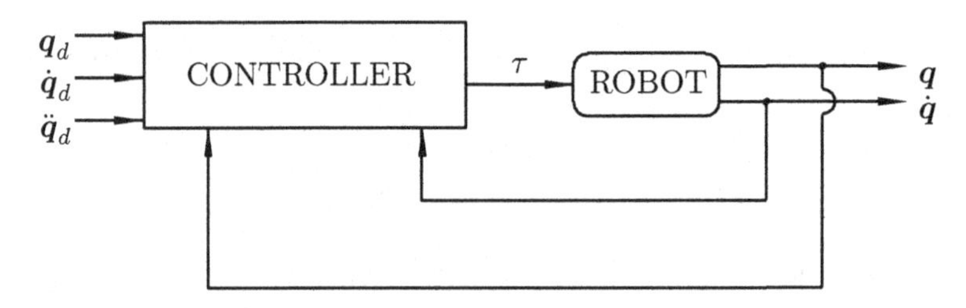

Figure III.1. Motion control: closed-loop system

In this third part of the textbook we carry out the stability analysis of a group of motion controllers for robot manipulators. As for the position control problem, the methodology to analyze the stability may be summarized in the following steps.

1. Derivation of the closed-loop dynamic equation. Such an equation is obtained by replacing the control action control $\boldsymbol{\tau}$ in the dynamic model of the manipulator. In general, the closed-loop equation is a nonautonomous nonlinear ordinary differential equation since $\boldsymbol{q}_d = \boldsymbol{q}_d(t)$.
2. Representation of the closed-loop equation in the state-space form,

$$\frac{d}{dt}\begin{bmatrix} \boldsymbol{q}_d - \boldsymbol{q} \\ \dot{\boldsymbol{q}}_d - \dot{\boldsymbol{q}} \end{bmatrix} = \boldsymbol{f}(\boldsymbol{q}, \dot{\boldsymbol{q}}, \boldsymbol{q}_d, \dot{\boldsymbol{q}}_d, \ddot{\boldsymbol{q}}_d, M(\boldsymbol{q}), C(\boldsymbol{q}, \dot{\boldsymbol{q}}), \boldsymbol{g}(\boldsymbol{q}))\,.$$

This closed-loop equation may be regarded as a dynamic system whose inputs are $\boldsymbol{q}_d, \dot{\boldsymbol{q}}_d$ and $\ddot{\boldsymbol{q}}_d$, and whose outputs are the state vectors $\tilde{\boldsymbol{q}} = \boldsymbol{q}_d - \boldsymbol{q}$ and $\dot{\tilde{\boldsymbol{q}}} = \dot{\boldsymbol{q}}_d - \dot{\boldsymbol{q}}$. Figure III.2 shows the corresponding block-diagram.

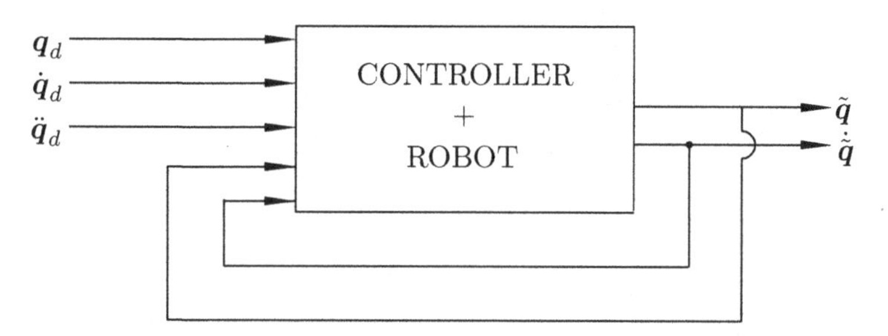

Figure III.2. Motion control closed-loop system in its input–output representation

3. Study of the existence and possible unicity of the equilibrium for the closed-loop equation

$$\frac{d}{dt}\begin{bmatrix}\tilde{\boldsymbol{q}}\\ \dot{\tilde{\boldsymbol{q}}}\end{bmatrix} = \tilde{\boldsymbol{f}}(t, \tilde{\boldsymbol{q}}, \dot{\tilde{\boldsymbol{q}}}) \qquad \text{(III.2)}$$

   where $\tilde{\boldsymbol{f}}$ is obtained by replacing $\boldsymbol{q}$ with $\boldsymbol{q}_d(t) - \tilde{\boldsymbol{q}}$ and $\dot{\boldsymbol{q}}$ with $\dot{\boldsymbol{q}}_d(t) - \dot{\tilde{\boldsymbol{q}}}$. Whence the dependence of $\tilde{\boldsymbol{f}}$ on t. That is, the closed-loop system equation is nonautonomous.

   Thus, for Equation (III.2) we want to verify that the origin, $[\tilde{\boldsymbol{q}}^T, \dot{\tilde{\boldsymbol{q}}}^T]^T = \boldsymbol{0} \in \mathbb{R}^{2n}$ is an equilibrium and whether it is unique.
4. Proposal of a Lyapunov function candidate to study the stability of any equilibrium of interest for the closed-loop equation, by using the Theorems 2.2, 2.3 and 2.4. In particular, verification of the required properties, *i.e.* positivity and, negativity of the time derivative. Notice that in this case, we cannot use La Salle's theorem (*cf.* Theorem 2.7) since the closed-loop system is described, in general, by a nonautonomous differential equation.
5. Alternatively to step 4, in the case that the proposed Lyapunov function candidate appears to be inappropriate (that is, if it does not satisfy all of the required conditions) to establish the stability properties of the equilibrium under study, we may use Lemma 2.2 by proposing a positive definite function whose characteristics allow one to determine the qualitative behavior of the solutions of the closed-loop equation. In particular, the convergence of part of the state.

The rest of this third part is divided in three chapters. The controllers that we consider are, in order,

- Computed torque control and computed torque+ control.
- PD control with compensation and PD+ control.

- Feedforward control and PD plus feedforward control.

For references regarding the problem of motion control of robot manipulators see the Introduction of Part II on page 139.

10

Computed-torque Control and Computed-torque+ Control

In this chapter we study the motion controllers:

- Computed-torque control and
- Computed-torque+ control.

Computed-torque control allows one to obtain a linear closed-loop equation in terms of the state variables. This fact has no precedent in the study of the controllers studied in this text so far. On the other hand, computed-torque+ control is characterized for being a *dynamic* controller, that is, its complete control law includes additional state variables. Finally, it is worth anticipating that both of these controllers satisfy the motion control objective with a trivial choice of their design parameters.

The contents of this chapter have been taken from the references cited at the end. The reader interested in going deeper into the material presented here is invited to consult these and the references therein.

10.1 Computed-torque Control

The dynamic model (III.1) that characterizes the behavior of robot manipulators is in general, composed of nonlinear functions of the state variables (joint positions and velocities). This feature of the dynamic model might lead us to believe that given any controller, the differential equation that models the control system in closed loop should also be composed of nonlinear functions of the corresponding state variables. This intuition is confirmed for the case of all the control laws studied in previous chapters. Nevertheless, there exists a controller which is also nonlinear in the state variables but which leads to a closed-loop control system which is described by a linear differential equation. This controller is capable of fulfilling the motion control objective, globally and moreover with a trivial selection of its design parameters. It receives the name computed-torque control.

The computed-torque control law is given by

$$\tau = M(q)\left[\ddot{q}_d + K_v\dot{\tilde{q}} + K_p\tilde{q}\right] + C(q,\dot{q})\dot{q} + g(q)\,, \tag{10.1}$$

where K_v and K_p are symmetric positive definite design matrices and $\tilde{q} = q_d - q$ denotes as usual, the position error.

Notice that the control law (10.1) contains the terms $K_p\tilde{q} + K_v\dot{\tilde{q}}$ which are of the PD type. However, these terms are actually premultiplied by the inertia matrix $M(q_d - \tilde{q})$. Therefore this is not a linear controller as the PD, since the position and velocity gains are not constant but they depend explicitly on the position error $\tilde{q}$. This may be clearly seen when expressing the computed-torque control law given by (10.1) as

$$\tau = M(q_d - \tilde{q})K_p\tilde{q} + M(q_d - \tilde{q})K_v\dot{\tilde{q}} + M(q)\ddot{q}_d + C(q,\dot{q})\dot{q} + g(q)\,.$$

Computed-torque control was one of the first *model-based* motion control approaches created for manipulators, that is, in which one makes explicit use of the knowledge of the matrices $M(q)$, $C(q,\dot{q})$ and of the vector $g(q)$. Furthermore, observe that the desired trajectory of motion $q_d(t)$, and its derivatives $\dot{q}_d(t)$ and $\ddot{q}_d(t)$, as well as the position and velocity measurements $q(t)$ and $\dot{q}(t)$, are used to compute the control action (10.1).

The block-diagram that corresponds to computed-torque control of robot manipulators is presented in Figure 10.1.

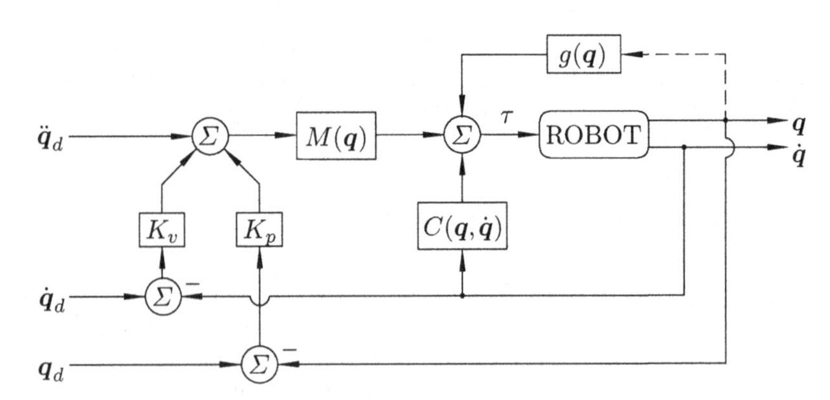

Figure 10.1. Block-diagram: computed-torque control

The closed-loop equation is obtained by substituting the control action τ from (10.1) in the equation of the robot model (III.1) to obtain

$$M(q)\ddot{q} = M(q)\left[\ddot{q}_d + K_v\dot{\tilde{q}} + K_p\tilde{q}\right]\,. \tag{10.2}$$

Since $M(\boldsymbol{q})$ is a positive definite matrix (Property 4.1) and therefore it is also invertible, Equation (10.2) reduces to

$$\ddot{\tilde{\boldsymbol{q}}} + K_v \dot{\tilde{\boldsymbol{q}}} + K_p \tilde{\boldsymbol{q}} = \boldsymbol{0}$$

which in turn, may be expressed in terms of the state vector $\begin{bmatrix} \tilde{\boldsymbol{q}}^T & \dot{\tilde{\boldsymbol{q}}}^T \end{bmatrix}^T$ as

$$\begin{aligned} \frac{d}{dt}\begin{bmatrix} \tilde{\boldsymbol{q}} \\ \dot{\tilde{\boldsymbol{q}}} \end{bmatrix} &= \begin{bmatrix} \dot{\tilde{\boldsymbol{q}}} \\ -K_p\tilde{\boldsymbol{q}} - K_v\dot{\tilde{\boldsymbol{q}}} \end{bmatrix} \\ &= \begin{bmatrix} 0 & I \\ -K_p & -K_v \end{bmatrix} \begin{bmatrix} \tilde{\boldsymbol{q}} \\ \dot{\tilde{\boldsymbol{q}}} \end{bmatrix}, \end{aligned} \tag{10.3}$$

where I is the identity matrix of dimension n.

It is important to remark that the closed-loop Equation (10.3) is represented by a linear autonomous differential equation, whose unique equilibrium point is given by $\begin{bmatrix} \tilde{\boldsymbol{q}}^T & \dot{\tilde{\boldsymbol{q}}}^T \end{bmatrix}^T = \boldsymbol{0} \in \mathbb{R}^{2n}$. The unicity of the equilibrium follows from the fact that the matrix K_p is designed to be positive definite and therefore nonsingular.

Since the closed-loop Equation (10.3) is linear and autonomous, its solutions may be obtained in closed form and be used to conclude about the stability of the origin. Nevertheless, for pedagogical purposes we proceed to analyze the stability of the origin as an equilibrium point of the closed-loop equation. We do this using Lyapunov's direct method.

To that end, we start by introducing the constant ε satisfying

$$\lambda_{\min}\{K_v\} > \varepsilon > 0\,.$$

Multiplying by $\boldsymbol{x}^T\boldsymbol{x}$ where $\boldsymbol{x} \in \mathbb{R}^n$ is any nonzero vector, we obtain $\lambda_{\min}\{K_v\}\boldsymbol{x}^T\boldsymbol{x} > \varepsilon\boldsymbol{x}^T\boldsymbol{x}$. Since K_v is by design, a symmetric matrix then $\boldsymbol{x}^T K_v \boldsymbol{x} \geq \lambda_{\min}\{K_v\}\boldsymbol{x}^T\boldsymbol{x}$ and therefore,

$$\boldsymbol{x}^T\left[K_v - \varepsilon I\right]\boldsymbol{x} > 0 \qquad \forall\, \boldsymbol{x} \neq \boldsymbol{0} \in \mathbb{R}^n\,.$$

This means that the matrix $K_v - \varepsilon I$ is positive definite, *i.e.*

$$K_v - \varepsilon I > 0\,. \tag{10.4}$$

Considering all this, the positivity of the matrix K_p and that of the constant ε we conclude that

$$K_p + \varepsilon K_v - \varepsilon^2 I > 0\,. \tag{10.5}$$

Consider next the Lyapunov function candidate

$$V(\tilde{q}, \dot{\tilde{q}}) = \frac{1}{2} \begin{bmatrix} \tilde{q} \\ \dot{\tilde{q}} \end{bmatrix}^T \begin{bmatrix} K_p + \varepsilon K_v & \varepsilon I \\ \varepsilon I & I \end{bmatrix} \begin{bmatrix} \tilde{q} \\ \dot{\tilde{q}} \end{bmatrix}$$
$$= \frac{1}{2} \left[\dot{\tilde{q}} + \varepsilon \tilde{q}\right]^T \left[\dot{\tilde{q}} + \varepsilon \tilde{q}\right] + \frac{1}{2} \tilde{q}^T [K_p + \varepsilon K_v - \varepsilon^2 I] \, \tilde{q} \tag{10.6}$$

where the constant ε satisfies (10.4) and of course, also (10.5). From this, it follows that the function (10.6) is globally positive definite. This may be more clear if we rewrite the Lyapunov function candidate $V(\tilde{q}, \dot{\tilde{q}})$ in (10.6) as

$$V(\tilde{q}, \dot{\tilde{q}}) = \frac{1}{2} \dot{\tilde{q}}^T \dot{\tilde{q}} + \frac{1}{2} \tilde{q}^T [K_p + \varepsilon K_v] \, \tilde{q} + \varepsilon \tilde{q}^T \dot{\tilde{q}} \, .$$

Evaluating the total time derivative of $V(\tilde{q}, \dot{\tilde{q}})$ we get

$$\dot{V}(\tilde{q}, \dot{\tilde{q}}) = \ddot{\tilde{q}}^T \dot{\tilde{q}} + \tilde{q}^T [K_p + \varepsilon K_v] \, \dot{\tilde{q}} + \varepsilon \dot{\tilde{q}}^T \dot{\tilde{q}} + \varepsilon \tilde{q}^T \ddot{\tilde{q}} \, .$$

Substituting $\ddot{\tilde{q}}$ from the closed-loop Equation (10.3) in the previous expression and making some simplifications we obtain

$$\dot{V}(\tilde{q}, \dot{\tilde{q}}) = -\dot{\tilde{q}}^T [K_v - \varepsilon I] \, \dot{\tilde{q}} - \varepsilon \tilde{q}^T K_p \tilde{q}$$
$$= - \begin{bmatrix} \tilde{q} \\ \dot{\tilde{q}} \end{bmatrix}^T \begin{bmatrix} \varepsilon K_p & 0 \\ 0 & K_v - \varepsilon I \end{bmatrix} \begin{bmatrix} \tilde{q} \\ \dot{\tilde{q}} \end{bmatrix} . \tag{10.7}$$

Now, since ε is chosen so that $K_v - \varepsilon I > 0$, and since K_p is by design positive definite, the function $\dot{V}(\tilde{q}, \dot{\tilde{q}})$ in (10.7) is globally negative definite. In view of Theorem 2.4, we conclude that the origin $\begin{bmatrix} \tilde{q}^T & \dot{\tilde{q}}^T \end{bmatrix}^T = \mathbf{0} \in \mathbb{R}^{2n}$ of the closed-loop equation is globally uniformly asymptotically stable and therefore

$$\lim_{t \to \infty} \dot{\tilde{q}}(t) = \mathbf{0}$$
$$\lim_{t \to \infty} \tilde{q}(t) = \mathbf{0}$$

from which it follows that the motion control objective is achieved. As a matter of fact, since Equation (10.3) is linear and autonomous this is equivalent to global exponential stability of the origin.

For practical purposes, the design matrices K_p and K_v may be chosen diagonal. This means that the closed-loop Equation (10.3) represents a decoupled multivariable linear system that is, the dynamic behavior of the errors of each joint position is governed by second-order linear differential equations which are independent of each other. In this scenario the selection of the matrices K_p and K_v may be made specifically as

$$K_p = \text{diag}\left\{\omega_1^2, \cdots, \omega_n^2\right\}$$
$$K_v = \text{diag}\left\{2\omega_1, \cdots, 2\omega_n\right\}.$$

With this choice, each joint responds as a critically damped linear system with bandwidth ω_i. The bandwidth ω_i defines the velocity of the joint in question and consequently, the decay exponential rate of the errors $\tilde{\boldsymbol{q}}(t)$ and $\dot{\tilde{\boldsymbol{q}}}(t)$. Therefore, in view of these expressions we may not only guarantee the control objective but we may also govern the performance of the closed-loop control system.

Example 10.1. Consider the equation of a pendulum of length l and mass m concentrated at its tip, subject to the action of gravity g and to which is applied a torque τ at the axis of rotation that is,

$$ml^2\ddot{q} + mgl\,\sin(q) = \tau,$$

where q is the angular position with respect to the vertical. For this example we have $M(q) = ml^2$, $C(q,\dot{q}) = 0$ and $g(q) = mgl\,\sin(q)$. The computed-torque control law (10.1), is given by

$$\tau = ml^2\left[\ddot{q}_d + k_v\dot{\tilde{q}} + k_p\tilde{q}\right] + mgl\,\sin(q),$$

with $k_v > 0$, $k_p > 0$. With this control strategy it is guaranteed that the motion control objective is achieved globally.

Next, we present the experimental results obtained for the Pelican prototype presented in Chapter 5 under computed-torque control.

Example 10.2. Consider the Pelican prototype robot studied in Chapter 5, and shown in Figure 5.2. Consider the computed-torque control law (10.1) on this robot for motion control.

The desired reference trajectory, $\boldsymbol{q}_d(t)$, is given by Equation (5.7). The desired velocities and accelerations $\dot{\boldsymbol{q}}_d(t)$ and $\ddot{\boldsymbol{q}}_d(t)$, were analytically found, and they correspond to Equations (5.8) and (5.9), respectively.

The symmetric positive definite matrices K_p and K_v are chosen as

$$K_p = \text{diag}\{\omega_1^2,\ \omega_2^2\} = \text{diag}\{1500,\ 14000\} \quad [1/\text{s}]$$
$$K_v = \text{diag}\{2\omega_1,\ 2\omega_2\} = \text{diag}\{77.46,\ 236.64\} \quad \left[1/\text{s}^2\right],$$

where we used $\omega_1 = 38.7$ [rad/s] and $\omega_2 = 118.3$ [rad/s].

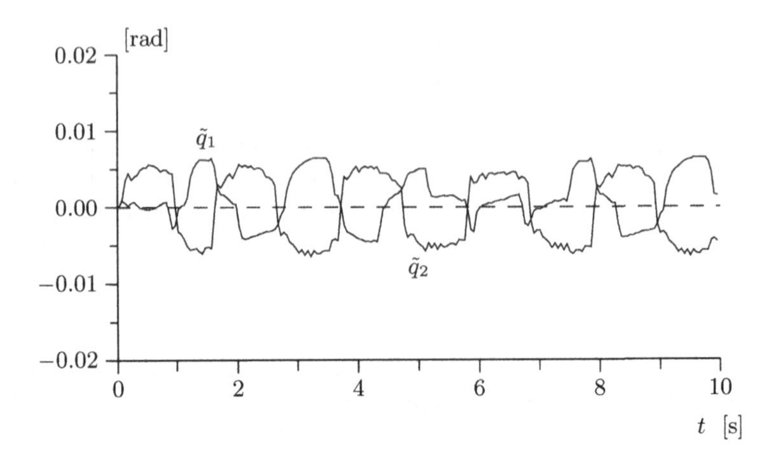

Figure 10.2. Graph of position errors against time

The initial conditions which correspond to the positions and velocities, are chosen as

$$\begin{aligned} q_1(0) &= 0, & q_2(0) &= 0 \\ \dot{q}_1(0) &= 0, & \dot{q}_2(0) &= 0\,. \end{aligned}$$

Figure 10.2 shows the experimental position errors. The steady-state position errors are not zero due to the friction effects of the actual robot which nevertheless, are neglected in the analysis. ♢

10.2 Computed-torque+ Control

Most of the controllers analyzed so far in this textbook, both for position as well as for motion control, have the common structural feature that they use static state feedback (of joint positions and velocities). The exception to this rule are the PID control and the controllers that do not require measurement of velocities, studied in Chapter 13.

In this section[1] we study a motion controller which uses dynamic state feedback. As we show next, this controller basically consists in one part that

[1] The material of this section may appear advanced to some readers; in particular, for a senior course on robot control since it makes use of results involving concepts such as '*functional spaces*', material exposed in Appendix A and reserved for the advanced student. Therefore, the material may be skipped if convenient without affecting the continuity of the exposition of motion controllers. The material is adapted from the corresponding references cited as usual, at the end of the chapter.

is exactly equal to the computed-torque control law given by the expression (10.1), and a second part that includes dynamic terms. Due to this characteristic, this controller was originally called *computed-torque control with compensation*, however, in the sequel we refer to it simply as computed-torque+.

The reason to include the computed-torque+ control as subject of study in this text is twofold. First, the motion controllers analyzed previously use static state feedback; hence, it is interesting to study a motion controller whose structure uses dynamic state feedback. Secondly, computed-torque+ control may be easily generalized to consider an *adaptive* version of it, which allows one to deal with uncertainties in the model (*cf.* Part IV).

The equation corresponding to the computed-torque+ controller is given by

$$\boldsymbol{\tau} = M(\boldsymbol{q})\left[\ddot{\boldsymbol{q}}_d + K_v\dot{\tilde{\boldsymbol{q}}} + K_p\tilde{\boldsymbol{q}}\right] + C(\boldsymbol{q},\dot{\boldsymbol{q}})\dot{\boldsymbol{q}} + \boldsymbol{g}(\boldsymbol{q}) - C(\boldsymbol{q},\dot{\boldsymbol{q}})\boldsymbol{\nu} \tag{10.8}$$

where K_v and K_p are symmetric positive definite design matrices, the vector $\tilde{\boldsymbol{q}} = \boldsymbol{q}_d - \boldsymbol{q}$ denotes as usual, the position error and the vector $\boldsymbol{\nu} \in \mathbb{R}^n$ is obtained by filtering the errors of position $\tilde{\boldsymbol{q}}$ and velocity $\dot{\tilde{\boldsymbol{q}}}$, that is,

$$\boldsymbol{\nu} = -\frac{bp}{p+\lambda}\dot{\tilde{\boldsymbol{q}}} - \frac{b}{p+\lambda}\left[K_v\dot{\tilde{\boldsymbol{q}}} + K_p\tilde{\boldsymbol{q}}\right], \tag{10.9}$$

where p is the differential operator (*i.e.* $p := \frac{d}{dt}$) and λ, b are positive design constants. For simplicity, and with no loss of generality, we take $b = 1$.

Notice that the difference between the computed-torque and computed-torque+ control laws given by (10.1) and (10.8) respectively, resides exclusively in that the latter contains the additional term $C(\boldsymbol{q},\dot{\boldsymbol{q}})\boldsymbol{\nu}$.

The implementation of computed-torque+ control expressed by (10.8) and (10.9) requires knowledge of the matrices $M(\boldsymbol{q})$, $C(\boldsymbol{q},\dot{\boldsymbol{q}})$ and of the vector $\boldsymbol{g}(\boldsymbol{q})$ as well as of the desired motion trajectory $\boldsymbol{q}_d(t)$, $\dot{\boldsymbol{q}}_d(t)$ and $\ddot{\boldsymbol{q}}_d(t)$ and measurement of the positions $\boldsymbol{q}(t)$ and of the velocities $\dot{\boldsymbol{q}}(t)$. It is assumed that $C(\boldsymbol{q},\dot{\boldsymbol{q}})$ in the control law (10.8) was obtained by using the Christoffel symbols (*cf.* Equation 3.21). The block-diagram corresponding to computed-torque+ control is presented in Figure 10.3.

Due to the presence of the vector $\boldsymbol{\nu}$ in (10.8) the computed-torque+ control law is *dynamic*, that is, the control action $\boldsymbol{\tau}$ depends not only on the actual values of the state vector formed by $\boldsymbol{q}$ and $\dot{\boldsymbol{q}}$, but also on its *past* values. This fact has as a consequence that we need additional state variables to completely characterize the control law. Indeed, the expression (10.9) in the state space form is a linear autonomous system given by

$$\frac{d}{dt}\begin{bmatrix}\boldsymbol{\xi}_1 \\ \boldsymbol{\xi}_2\end{bmatrix} = \begin{bmatrix}-\lambda I & 0 \\ 0 & -\lambda I\end{bmatrix}\begin{bmatrix}\boldsymbol{\xi}_1 \\ \boldsymbol{\xi}_2\end{bmatrix} + \begin{bmatrix}K_p & K_v \\ 0 & -\lambda I\end{bmatrix}\begin{bmatrix}\tilde{\boldsymbol{q}} \\ \dot{\tilde{\boldsymbol{q}}}\end{bmatrix} \tag{10.10}$$

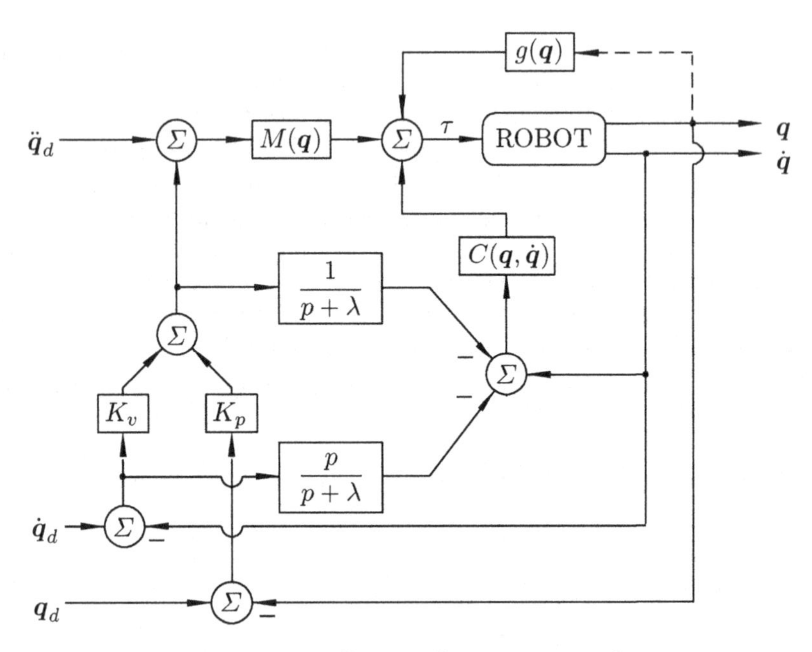

Figure 10.3. Computed-torque+ control

$$\boldsymbol{\nu} = [-I \quad -I] \begin{bmatrix} \boldsymbol{\xi}_1 \\ \boldsymbol{\xi}_2 \end{bmatrix} - [0 \quad I] \begin{bmatrix} \tilde{\boldsymbol{q}} \\ \dot{\tilde{\boldsymbol{q}}} \end{bmatrix} \tag{10.11}$$

where $\boldsymbol{\xi}_1, \boldsymbol{\xi}_2 \in \mathbb{R}^n$ are the new state variables.

To derive the closed-loop equation we combine first the dynamic equation of the robot (III.1) with that of the controller (10.8) to obtain the expression

$$M(\boldsymbol{q}) \left[\ddot{\tilde{\boldsymbol{q}}} + K_v \dot{\tilde{\boldsymbol{q}}} + K_p \tilde{\boldsymbol{q}}\right] - C(\boldsymbol{q}, \dot{\boldsymbol{q}})\boldsymbol{\nu} = \mathbf{0}\,. \tag{10.12}$$

In terms of the state vector $\left[\tilde{\boldsymbol{q}}^T \;\; \dot{\tilde{\boldsymbol{q}}}^T \;\; \boldsymbol{\xi}_1^T \;\; \boldsymbol{\xi}_2^T\right]^T$, the equations (10.12), (10.10) and (10.11) allow one to obtain the closed-loop equation

$$\frac{d}{dt} \begin{bmatrix} \tilde{\boldsymbol{q}} \\ \dot{\tilde{\boldsymbol{q}}} \\ \boldsymbol{\xi}_1 \\ \boldsymbol{\xi}_2 \end{bmatrix} = \begin{bmatrix} \dot{\tilde{\boldsymbol{q}}} \\ -M(\boldsymbol{q})^{-1} C(\boldsymbol{q}, \dot{\boldsymbol{q}}) \left[\boldsymbol{\xi}_1 + \boldsymbol{\xi}_2 + \dot{\tilde{\boldsymbol{q}}}\right] - K_v \dot{\tilde{\boldsymbol{q}}} - K_p \tilde{\boldsymbol{q}} \\ -\lambda \boldsymbol{\xi}_1 + K_p \tilde{\boldsymbol{q}} + K_v \dot{\tilde{\boldsymbol{q}}} \\ -\lambda \boldsymbol{\xi}_2 - \lambda \dot{\tilde{\boldsymbol{q}}} \end{bmatrix}, \tag{10.13}$$

of which the origin $\left[\tilde{\boldsymbol{q}}^T \;\; \dot{\tilde{\boldsymbol{q}}}^T \;\; \boldsymbol{\xi}_1^T \;\; \boldsymbol{\xi}_2^T\right]^T = \mathbf{0} \in \mathbb{R}^{4n}$ is an equilibrium point.

The study of global asymptotic stability of the origin of the closed-loop Equation (10.13) is actually an open problem in the robot control academic community. Nevertheless we can show that the functions $\tilde{\boldsymbol{q}}(t)$, $\dot{\tilde{\boldsymbol{q}}}(t)$ and $\boldsymbol{\nu}(t)$ are bounded and, using Lemma 2.2, that the motion control objective is verified.

To analyze the control system we first proceed to write it in a different but equivalent form. For this, notice that the expression for $\boldsymbol{\nu}$ given in (10.9) allows one to derive

$$\dot{\boldsymbol{\nu}} + \lambda\boldsymbol{\nu} = -\left[\ddot{\tilde{\boldsymbol{q}}} + K_v\dot{\tilde{\boldsymbol{q}}} + K_p\tilde{\boldsymbol{q}}\right]. \tag{10.14}$$

Incorporating (10.14) in (10.12) we get

$$M(\boldsymbol{q})\left[\dot{\boldsymbol{\nu}} + \lambda\boldsymbol{\nu}\right] + C(\boldsymbol{q}, \dot{\boldsymbol{q}})\boldsymbol{\nu} = \boldsymbol{0}. \tag{10.15}$$

The previous equation is the starting point in the analysis that we present next. Consider now the following non-negative function

$$V(t, \boldsymbol{\nu}, \tilde{\boldsymbol{q}}) = \frac{1}{2}\boldsymbol{\nu}^T M(\boldsymbol{q}_d - \tilde{\boldsymbol{q}})\boldsymbol{\nu} \geq 0,$$

which, even though it does not satisfy the conditions to be a Lyapunov function candidate for the closed-loop Equation (10.13), it is useful in the proofs that we present below. Specifically, $V(\boldsymbol{\nu}, \tilde{\boldsymbol{q}})$ may not be a Lyapunov function candidate for the closed-loop Equation (10.13) since it is not a positive definite function of the whole state, that is, considering *all* the state variables $\tilde{\boldsymbol{q}}$, $\dot{\tilde{\boldsymbol{q}}}$, $\boldsymbol{\xi}_1$ and $\boldsymbol{\xi}_2$. Notice that it does not even depend on all the state variables.

The derivative with respect to time of $V(\boldsymbol{\nu}, \tilde{\boldsymbol{q}})$ is given by

$$\dot{V}(\boldsymbol{\nu}, \tilde{\boldsymbol{q}}) = \boldsymbol{\nu}^T M(\boldsymbol{q})\dot{\boldsymbol{\nu}} + \frac{1}{2}\boldsymbol{\nu}^T \dot{M}(\boldsymbol{q})\boldsymbol{\nu}.$$

Solving for $M(\boldsymbol{q})\dot{\boldsymbol{\nu}}$ in Equation (10.15) and substituting in the previous equation we obtain

$$\dot{V}(\boldsymbol{\nu}, \tilde{\boldsymbol{q}}) = -\boldsymbol{\nu}^T\lambda M(\boldsymbol{q})\boldsymbol{\nu} \leq 0 \tag{10.16}$$

where the term $\boldsymbol{\nu}^T\left[\frac{1}{2}\dot{M} - C\right]\boldsymbol{\nu}$ was canceled by virtue of Property 4.2. Now, considering $V(\boldsymbol{\nu}, \tilde{\boldsymbol{q}})$ and (10.16) we see that

$$\dot{V}(\boldsymbol{\nu}, \tilde{\boldsymbol{q}}) = -2\lambda V(\boldsymbol{\nu}, \tilde{\boldsymbol{q}}),$$

which in turn implies that

$$V(\boldsymbol{\nu}(t), \tilde{\boldsymbol{q}}(t)) = V(\boldsymbol{\nu}(0), \tilde{\boldsymbol{q}}(0))e^{-2\lambda t}.$$

Invoking Property 4.1 that there exists a constant $\alpha > 0$ such that $M(\boldsymbol{q}) \geq \alpha I$, we obtain

$$\alpha\, \boldsymbol{\nu}(t)^T\boldsymbol{\nu}(t) \le \boldsymbol{\nu}(t)^T M(\boldsymbol{q}(t))\boldsymbol{\nu}(t) = 2V(\boldsymbol{\nu}(t), \tilde{\boldsymbol{q}}(t))$$

from which we finally get

$$\underbrace{\boldsymbol{\nu}(t)^T\boldsymbol{\nu}(t)}_{\|\boldsymbol{\nu}(t)\|^2} \le \frac{2V(\boldsymbol{\nu}(0), \tilde{\boldsymbol{q}}(0))}{\alpha} e^{-2\lambda t}. \tag{10.17}$$

This means that that $\boldsymbol{\nu}(t) \to \mathbf{0}$ exponentially.

On the other hand, the Equation (10.14) may also be written as

$$(p+\lambda)\boldsymbol{\nu} = -\left[p^2 I + pK_v + K_p\right]\tilde{\boldsymbol{q}}$$

or in equivalent form as

$$\tilde{\boldsymbol{q}} = -(p+\lambda)\left[p^2 I + pK_v + K_p\right]^{-1}\boldsymbol{\nu}. \tag{10.18}$$

Since $\lambda > 0$, while K_v and K_p are positive definite symmetric matrices, Equation (10.14) written in the form above defines a linear dynamic system which is exponentially stable and strictly proper (*i.e.* where the degree of the denominator is strictly larger than that of the numerator). The input to this system is $\boldsymbol{\nu}$ which tends to zero exponentially fast, and its output $\tilde{\boldsymbol{q}}$. So we invoke the fact that a stable strictly proper filter with an exponentially decaying input produces an exponentially decaying output[2], that is,

$$\lim_{t\to\infty} \tilde{q}(t) = \mathbf{0},$$

which means that the motion control objective is verified.

It is interesting to remark that the equation of the computed-torque+ controller (10.8), reduces to the computed-torque controller given by (10.1) in the particular case of manipulators that do not have the centrifugal and forces matrix $C(\boldsymbol{q}, \dot{\boldsymbol{q}})$. Such is the case for example, of Cartesian manipulators.

Next, we present the experimentation results obtained for the computed-torque+ control on the Pelican robot.

Example 10.3. Consider the 2-DOF *prototype robot* studied in Chapter 5, and shown in Figure 5.2.

Consider the computed-torque+ control law given by (10.8), (10.10) and (10.11) applied to this robot.

The desired trajectories are those used in the previous examples, that is, the robot must track the position, velocity and acceleration trajectories $\boldsymbol{q}_d(t)$, $\dot{\boldsymbol{q}}_d(t)$ and $\ddot{\boldsymbol{q}}_d(t)$ given by Equations (5.7)–(5.9).

---

[2] The technical details of why the latter is true rely on the use of Corollary A.2 which is reserved to the advanced reader.

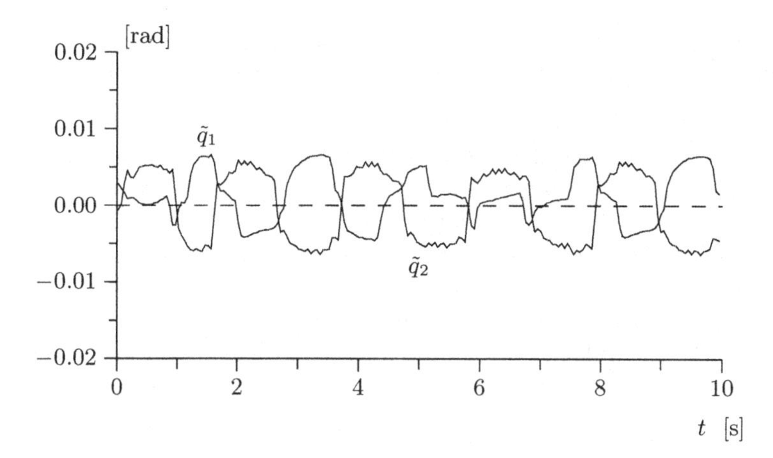

Figure 10.4. Graph of position errors against time

The symmetric positive definite matrices K_p and K_v, and the constant λ are taken as

$$\begin{aligned} K_p &= \mathrm{diag}\{\omega_1^2,\ \omega_2^2\} = \mathrm{diag}\{1500,\ 14000\} \quad [1/\mathrm{s}] \\ K_v &= \mathrm{diag}\{2\omega_1,\ 2\omega_2\} = \mathrm{diag}\{77.46,\ 236.64\} \quad [1/\mathrm{s}^2] \\ \lambda &= 60\,. \end{aligned}$$

The initial conditions of the controller state variables are fixed at

$$\boldsymbol{\xi}_1(0) = 0, \quad \boldsymbol{\xi}_2(0) = 0\,.$$

The initial conditions corresponding to the actual positions and velocities are set to

$$\begin{aligned} q_1(0) &= 0, \quad q_2(0) = 0 \\ \dot{q}_1(0) &= 0, \quad \dot{q}_2(0) = 0\,. \end{aligned}$$

Figure 10.4 shows the experimental tracking position errors. It is interesting to remark that the plots presented in Figure 10.2 obtained with the computed-torque control law, present a considerable similarity to those of Figure 10.4. ♢

10.3 Conclusions

The conclusions drawn from the analysis presented in this chapter may be summarized as follows.

- For any choice of the symmetric positive definite matrices K_p and K_v, the origin of the closed-loop equation by computed-torque control expressed in terms of the state vector $\left[\tilde{\boldsymbol{q}}^T \ \dot{\tilde{\boldsymbol{q}}}^T\right]^T$, is globally uniformly asymptotically stable. Therefore, computed-torque control satisfies the motion control objective, globally. Consequently, for any initial position error $\tilde{\boldsymbol{q}}(0) \in \mathbb{R}^n$ velocity error $\dot{\tilde{\boldsymbol{q}}}(0) \in \mathbb{R}^n$, we have $\lim_{t\to\infty} \tilde{\boldsymbol{q}}(t) = \mathbf{0}$.
- For any selection of the symmetric positive definite matrices K_p and K_v, and any positive constant λ, computed-torque+ control satisfies the motion control objective, globally. Consequently, for any initial position error $\tilde{\boldsymbol{q}}(0) \in \mathbb{R}^n$ and velocity error $\dot{\tilde{\boldsymbol{q}}}(0) \in \mathbb{R}^n$, and for any initial condition of the controller $\boldsymbol{\xi}_1(0) \in \mathbb{R}^n$, $\boldsymbol{\xi}_2(0) \in \mathbb{R}^n$, we have $\lim_{t\to\infty} \tilde{\boldsymbol{q}}(t) = \mathbf{0}$.

Bibliography

Computed-torque control is analyzed in the following texts.

- Fu K., Gonzalez R., Lee C., 1987, *"Robotics: Control, sensing, vision and intelligence"*, McGraw–Hill.
- Craig J., 1989, *"Introduction to robotics: Mechanics and control"*, Addison–Wesley.
- Spong M., Vidyasagar M., 1989, *"Robot dynamics and control"*, John Wiley and Sons.
- Yoshikawa T., 1990, *"Foundations of robotics: Analysis and control"*, The MIT Press.

The stability analysis for the computed-torque controller as presented in Section 10.1 follows the guidelines of

- Wen J. T., Bayard D., 1988, *"New class of control law for robotic manipulators. Part 1: Non-adaptive case"*, International Journal of Control, Vol. 47, No. 5, pp. 1361–1385.

The computed-torque+ control law as presented here is an adaptation from its original adaptive form, proposed in

- Kelly R., Carelli R., 1988. *"Unified approach to adaptive control of robotic manipulators"*, Proceedings of the 27th IEEE Conference on Decision and Control, Austin, TX., December, Vol. 1, pp. 1598–1603.
- Kelly R., Carelli R., Ortega R., 1989. *"Adaptive motion control design of robot manipulators: An input-output approach"*, International Journal of Control, Vol. 50, No. 6, September, pp. 2563–2581.

- Kelly R., 1990, *"Adaptive computed torque plus compensation control for robot manipulators "*, Mechanism and Machine Theory, Vol. 25, No. 2, pp. 161–165.

Problems

1. Consider the Cartesian robot 2-DOF shown in Figure 10.5.

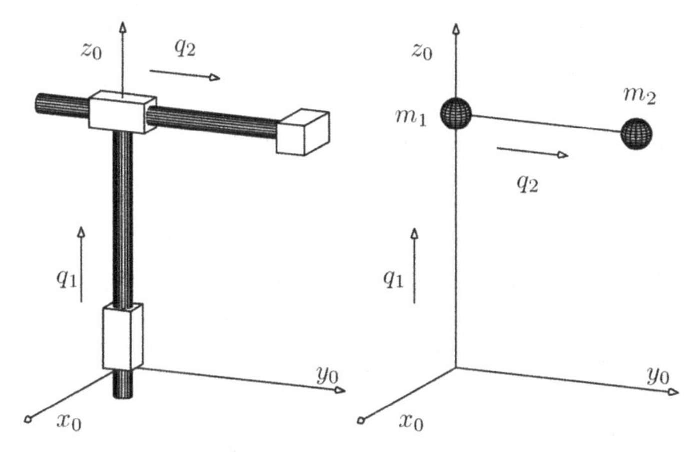

Figure 10.5. Problem 1. Cartesian 2-DOF robot.

   a) Obtain the dynamic model and specifically determine explicitly $M(\boldsymbol{q})$, $C(\boldsymbol{q}, \dot{\boldsymbol{q}})$ and $\boldsymbol{g}(\boldsymbol{q})$.
   b) Write the computed-torque control law and give explicitly τ_1 and τ_2.

2. Consider the model of an ideal pendulum with mass m concentrated at the tip, at length l from its axis of rotation, under the control action of a torque τ and a constant external additional torque τ_e,

$$ml^2\ddot{q} + mgl\,\sin(q) = \tau - \tau_e\,.$$

   To control the motion of this device we use a computed-torque controller that is,

$$\tau = ml^2\left[\ddot{q}_d + k_v\dot{\tilde{q}} + k_p\tilde{q}\right] + mgl\,\sin(q),$$

   where $k_p > 0$ and $k_v > 0$. Show that

$$\lim_{t\to\infty}\tilde{q}(t) = \frac{\tau_e}{k_p ml^2}\,.$$

   Hint: Obtain the closed-loop equation in terms of the state vector

$$\begin{bmatrix} \tilde{q} - \dfrac{\tau_e}{k_p m l^2} \\ \dot{\tilde{q}} \end{bmatrix}^T$$

and show that the origin is globally asymptotically stable.

3. Consider the model of an ideal pendulum as described in the previous problem to which is applied a control torque τ, and an external torque τ_e from torsional spring of constant $k_e > 0$ ($\tau_e = k_e q$),

$$ml^2\ddot{q} + mgl\,\sin(q) = \tau - k_e q\,.$$

To control the motion of such a device we use the computed-torque controller

$$\tau = ml^2\left[\ddot{q}_d + k_v\dot{\tilde{q}} + k_p\tilde{q}\right] + mgl\,\sin(q)$$

where $k_p > 0$ and $k_v > 0$. Assume that q_d is constant. Show that

$$\lim_{t\to\infty} \tilde{q}(t) = \frac{k_e}{k_p m l^2 + k_e} q_d\,.$$

Hint: Obtain the closed-loop equation in terms of the state vector

$$\begin{bmatrix} \tilde{q} - \dfrac{k_e}{k_p m l^2 + k_e} q_d \\ \dot{q} \end{bmatrix}$$

and show that the origin is a globally asymptotically stable equilibrium.

4. Consider the model of an ideal pendulum described in Problem 2 under the control action of a torque τ, *i.e.*

$$ml^2\ddot{q} + mgl\,\sin(q) = \tau\,.$$

Assume that the values of the parameters l and g are exactly known, but for the mass m only an approximate value m_0 is known. To control the motion of this device we use computed-torque control where m has been substituted by m_0 since the value of m is assumed unknown, that is,

$$\tau = m_0 l^2\left[\ddot{q}_d + k_v\dot{\tilde{q}} + k_p\tilde{q}\right] + m_0 gl\,\sin(q),$$

where $k_p > 0$ and $k_v > 0$.

a) Obtain the closed-loop equation in terms of the state vector $\begin{bmatrix}\tilde{q} & \dot{\tilde{q}}\end{bmatrix}^T$.

b) Verify that independently of the value of m and m_0 (but with $m \neq 0$), the origin $\begin{bmatrix}\tilde{q} & \dot{\tilde{q}}\end{bmatrix}^T = \mathbf{0} \in \mathbb{R}^2$ is an equilibrium of the closed-loop equation if the desired position $q_d(t)$ satisfies

$$\ddot{q}_d(t) + \frac{g}{l}\sin(q_d(t)) = 0 \quad \forall\, t \geq 0\,.$$

5. Consider the closed-loop equation obtained with the computed-torque+ controller given by Equation (10.13) and whose origin $\begin{bmatrix}\tilde{\boldsymbol{q}}^T & \dot{\tilde{\boldsymbol{q}}}^T & \boldsymbol{\xi}_1^T & \boldsymbol{\xi}_2^T\end{bmatrix}^T = \mathbf{0}$ is an equilibrium. Regarding the variable $\boldsymbol{\nu}$ we shown in (10.17) that

$$\underbrace{\boldsymbol{\nu}(t)^T\boldsymbol{\nu}(t)}_{\|\boldsymbol{\nu}(t)\|^2} \leq \frac{2V(\boldsymbol{\nu}(0), \tilde{\boldsymbol{q}}(0))}{\alpha} e^{-2\lambda t}.$$

On the other hand we have from (10.10) and (10.18)

$$\begin{aligned}
\boldsymbol{\xi}_1 &= \frac{1}{p+\lambda}\left[K_p\tilde{\boldsymbol{q}} + K_v\dot{\tilde{\boldsymbol{q}}}\right] \\
\boldsymbol{\xi}_2 &= -\frac{\lambda}{p+\lambda}\dot{\tilde{\boldsymbol{q}}} \\
\tilde{\boldsymbol{q}} &= -(p+\lambda)\left[p^2 I + pK_v + K_p\right]^{-1}\boldsymbol{\nu}
\end{aligned}$$

where K_p and K_v are symmetric positive definite matrices. Assume that the robot has only revolute joints.

a) May we also conclude that $\tilde{\boldsymbol{q}}(t)$, $\dot{\tilde{\boldsymbol{q}}}(t)$, $\boldsymbol{\xi}_1(t)$ and $\boldsymbol{\xi}_2(t)$ tend exponentially to zero ?

b) Would the latter imply that the origin is globally exponentially stable?

6. In this chapter it was shown that the origin of the robot system in closed loop with the computed-torque controller is globally uniformly asymptotically stable. Since the closed-loop system is linear autonomous, it was observed that this is equivalent to global exponential stability. Verify this claim using Theorem 2.5.

11

PD+ Control and PD Control with Compensation

As we have seen in Chapter 10 the motion control objective for robot manipulators may be achieved globally by means of computed-torque control. Computed-torque control belongs to the so-called class of *feedback linearizing* controllers. Roughly, the technique of feedback linearization in its simplest form consists in applying a control law such that the closed-loop equations are linear. Historically, the motivation to develop feedback-linearization based controllers is that the stability theory of linear systems is far more developed than that of nonlinear systems. In particular, the tuning of the gains of such controllers is trivial since the resulting system is described by linear differential equation.

While computed-torque control was one of the first model-based controllers for robot manipulators, and rapidly gained popularity it has the disadvantages of other feedback-linearizing controllers: first, it requires a considerable computing load since the torque has to be computed on-line so that the closed-loop system equations become linear and autonomous, and second, it relies on a very accurate knowledge of the system. This second feature may be of significant importance since the computed-torque control law contains the vector, of centrifugal and Coriolis forces vector, $C(\boldsymbol{q},\dot{\boldsymbol{q}})\dot{\boldsymbol{q}}$, which contains quadratic terms of the components of the joint velocities. The consequence of this is that high order nonlinearities appear in the control law and therefore, in the case of model uncertainty, the control law introduces undesirable high order nonlinearities in the equations of the closed-loop system. Moreover, even in the case that the model is accurately known, the control law increases proportionally to the square of certain components of the vector of joint velocities hence, these demanded large control actions may drive the actuators into saturation.

In this chapter we present two controllers whose control laws are based on the dynamic equations of the system but which also involve certain nonlinearities that are evaluated along the desired trajectories, *i.e.* the desired motion. These control systems are presented in increasing order of complexity with respect to their stability analyses. They are:

- PD control with compensation and
- PD+ control.

These controllers have been thoroughly studied in the literature and, as for the other chapters, the corresponding references are given at the end. For clarity of exposition each of these controllers is treated in separate sections.

11.1 PD Control with Compensation

In 1987 an *adaptive* controller to solve the motion control problem of robot manipulators was reported in the literature. This controller, which over the years has become increasingly popular within the academic environment, is often referred to by the names of its creators: 'Slotine and Li controller'. The related references are presented at the end of the chapter. While this controller is the subject of inquiry in Chapter 16 its 'non-adaptive' version is studied in this first section of the present chapter in its non-adaptive version. From a purely structural viewpoint, the control law of this controller is formed by a 'PD' term plus a 'compensation' term hence, one could also call it "PD plus compensation control".

The material of this section has been taken from the references cited at the end of the text. The PD control law with compensation may be written as

$$\boldsymbol{\tau} = K_p\tilde{\boldsymbol{q}} + K_v\dot{\tilde{\boldsymbol{q}}} + M(\boldsymbol{q})\left[\ddot{\boldsymbol{q}}_d + \Lambda\dot{\tilde{\boldsymbol{q}}}\right] + C(\boldsymbol{q},\dot{\boldsymbol{q}})\left[\dot{\boldsymbol{q}}_d + \Lambda\tilde{\boldsymbol{q}}\right] + \boldsymbol{g}(\boldsymbol{q}), \qquad (11.1)$$

where $K_p, K_v \in \mathbb{R}^{n\times n}$ are symmetric positive definite design matrices, $\tilde{\boldsymbol{q}} = \boldsymbol{q}_d - \boldsymbol{q}$ denotes the position error and Λ is defined as

$$\Lambda = {K_v}^{-1}K_p\,.$$

Notice that Λ is the product of two symmetric positive definite matrices. Even though in general this matrix may or may not be symmetric or positive definite, it is always nonsingular. This characteristic of Λ will be of utility later on. It is assumed that the centrifugal and Coriolis matrix $C(\boldsymbol{q},\dot{\boldsymbol{q}})$ is built using the Christoffel symbols (*cf.* Equation 3.21).

Observe that the first two terms on the right-hand side of the control law (11.1) correspond to the PD control law. PD control with compensation is model-based, that is, the control law explicitly uses the terms from the model of the robot (III.1), $M(\boldsymbol{q})$, $C(\boldsymbol{q},\dot{\boldsymbol{q}})$ and $\boldsymbol{g}(\boldsymbol{q})$. Figure 11.1 presents the block-diagram that corresponds to the PD control law with compensation.

The closed-loop equation is obtained by substituting the control action $\boldsymbol{\tau}$ from the control law (11.1) in the equation of the robot model (III.1), to obtain

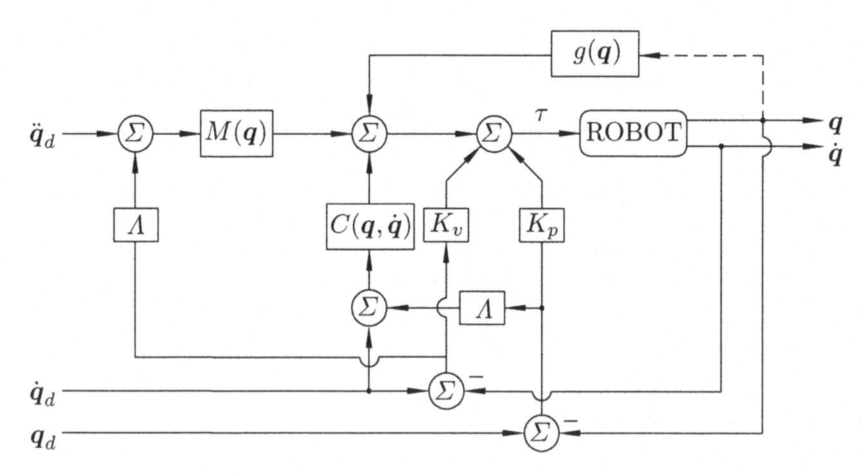

Figure 11.1. Block-diagram: PD control with compensation

$$M(\boldsymbol{q})\left[\ddot{\tilde{\boldsymbol{q}}} + \Lambda\dot{\tilde{\boldsymbol{q}}}\right] + C(\boldsymbol{q}, \dot{\boldsymbol{q}})\left[\dot{\tilde{\boldsymbol{q}}} + \Lambda\tilde{\boldsymbol{q}}\right] = -K_p\tilde{\boldsymbol{q}} - K_v\dot{\tilde{\boldsymbol{q}}},$$

which may be expressed in terms of the state vector $\left[\tilde{\boldsymbol{q}}^T \quad \dot{\tilde{\boldsymbol{q}}}^T\right]^T$ as

$$\frac{d}{dt}\begin{bmatrix}\tilde{\boldsymbol{q}} \\ \dot{\tilde{\boldsymbol{q}}}\end{bmatrix} = \begin{bmatrix}\dot{\tilde{\boldsymbol{q}}} \\ M(\boldsymbol{q})^{-1}\left[-K_p\tilde{\boldsymbol{q}} - K_v\dot{\tilde{\boldsymbol{q}}} - C(\boldsymbol{q}, \dot{\boldsymbol{q}})\left[\dot{\tilde{\boldsymbol{q}}} + \Lambda\tilde{\boldsymbol{q}}\right]\right] - \Lambda\dot{\tilde{\boldsymbol{q}}}\end{bmatrix}, \tag{11.2}$$

that is, this differential equation is nonautonomous and has the origin $\left[\tilde{\boldsymbol{q}}^T \quad \dot{\tilde{\boldsymbol{q}}}^T\right]^T = \boldsymbol{0} \in \mathbb{R}^{2n}$ as an equilibrium point.

The stability analysis of the origin of the closed-loop equation may be carried out by considering the Lyapunov function candidate

$$V(t, \tilde{\boldsymbol{q}}, \dot{\tilde{\boldsymbol{q}}}) = \frac{1}{2}\begin{bmatrix}\tilde{\boldsymbol{q}} \\ \dot{\tilde{\boldsymbol{q}}}\end{bmatrix}^T \begin{bmatrix}2K_p + \Lambda^T M(\boldsymbol{q}_d - \tilde{\boldsymbol{q}})\Lambda & \Lambda^T M(\boldsymbol{q}_d - \tilde{\boldsymbol{q}}) \\ M(\boldsymbol{q}_d - \tilde{\boldsymbol{q}})\Lambda & M(\boldsymbol{q}_d - \tilde{\boldsymbol{q}})\end{bmatrix}\begin{bmatrix}\tilde{\boldsymbol{q}} \\ \dot{\tilde{\boldsymbol{q}}}\end{bmatrix}.$$

At first sight it may not appear evident that this Lyapunov function candidate is positive definite but it may help that we rewrite it in the following form:

$$V(t, \tilde{\boldsymbol{q}}, \dot{\tilde{\boldsymbol{q}}}) = \frac{1}{2}\left[\dot{\tilde{\boldsymbol{q}}} + \Lambda\tilde{\boldsymbol{q}}\right]^T M(\boldsymbol{q})\left[\dot{\tilde{\boldsymbol{q}}} + \Lambda\tilde{\boldsymbol{q}}\right] + \tilde{\boldsymbol{q}}^T K_p \tilde{\boldsymbol{q}} \tag{11.3}$$

which is equivalent to

$$V(t, \tilde{\boldsymbol{q}}, \dot{\tilde{\boldsymbol{q}}}) = \frac{1}{2}\begin{bmatrix}\dot{\tilde{\boldsymbol{q}}} \\ \tilde{\boldsymbol{q}}\end{bmatrix}^T \underbrace{\begin{bmatrix}I & \Lambda^T \\ 0 & I\end{bmatrix}\begin{bmatrix}2K_p & 0 \\ 0 & M(\boldsymbol{q})\end{bmatrix}\begin{bmatrix}I & 0 \\ \Lambda & I\end{bmatrix}}_{B^T A B}\begin{bmatrix}\dot{\tilde{\boldsymbol{q}}} \\ \tilde{\boldsymbol{q}}\end{bmatrix}. \tag{11.4}$$

And we see from Lemma 2.1 (*cf.* page 24) that the matrix product $B^T AB$, above, is positive definite since K_p and $M(\boldsymbol{q})$ are positive definite[1].

Also, it is apparent from (11.4) that $V(t, \tilde{\boldsymbol{q}}, \dot{\tilde{\boldsymbol{q}}})$ satisfies

$$V(t, \tilde{\boldsymbol{q}}, \dot{\tilde{\boldsymbol{q}}}) \geq \frac{1}{2}\lambda_{\min}\{B^T AB\} \left[\|\dot{\tilde{\boldsymbol{q}}}\|^2 + \|\tilde{\boldsymbol{q}}\|^2\right]$$

so it is also radially unbounded. Correspondingly, since the inertia matrix is bounded uniformly in $\boldsymbol{q}$, from (11.3), we have

$$V(t, \tilde{\boldsymbol{q}}, \dot{\tilde{\boldsymbol{q}}}) \leq \frac{1}{2}\lambda_{\mathrm{Max}}\{M\} \left\|\dot{\tilde{\boldsymbol{q}}} + \Lambda\tilde{\boldsymbol{q}}\right\|^2 + \lambda_{\mathrm{Max}}\{K_p\} \|\tilde{\boldsymbol{q}}\|^2$$

hence, $V(t, \tilde{\boldsymbol{q}}, \dot{\tilde{\boldsymbol{q}}})$ is also decrescent.

It is interesting to mention that the function (11.3) may be regarded as an extension of the Lyapunov function (11.9) used in the study of the PD+ control. Indeed, both functions are the same if, as well as in the control laws, Λ is set to zero.

The time derivative of the Lyapunov function candidate (11.3) is

$$\begin{aligned}\dot{V}(\tilde{\boldsymbol{q}}, \dot{\tilde{\boldsymbol{q}}}) = &\left[\dot{\tilde{\boldsymbol{q}}} + \Lambda\tilde{\boldsymbol{q}}\right]^T M(\boldsymbol{q}) \left[\ddot{\tilde{\boldsymbol{q}}} + \Lambda\dot{\tilde{\boldsymbol{q}}}\right] + \frac{1}{2}\left[\dot{\tilde{\boldsymbol{q}}} + \Lambda\tilde{\boldsymbol{q}}\right]^T \dot{M}(\boldsymbol{q}) \left[\dot{\tilde{\boldsymbol{q}}} + \Lambda\tilde{\boldsymbol{q}}\right] \\ &+ 2\tilde{\boldsymbol{q}}^T K_p \dot{\tilde{\boldsymbol{q}}}\,. \end{aligned} \tag{11.5}$$

Solving for $M(\boldsymbol{q})\ddot{\tilde{\boldsymbol{q}}}$ in the closed-loop Equation (11.2) and substituting in the previous equation, we obtain

$$\dot{V}(t, \tilde{\boldsymbol{q}}, \dot{\tilde{\boldsymbol{q}}}) = -\left[\dot{\tilde{\boldsymbol{q}}} + \Lambda\tilde{\boldsymbol{q}}\right]^T K_v \left[\dot{\tilde{\boldsymbol{q}}} + \Lambda\tilde{\boldsymbol{q}}\right] + 2\tilde{\boldsymbol{q}}^T K_p \dot{\tilde{\boldsymbol{q}}},$$

where we canceled the term

$$\left[\dot{\tilde{\boldsymbol{q}}} + \Lambda\tilde{\boldsymbol{q}}\right]^T \left[\frac{1}{2}\dot{M}(\boldsymbol{q}) - C(\boldsymbol{q}, \dot{\boldsymbol{q}})\right] \left[\dot{\tilde{\boldsymbol{q}}} + \Lambda\tilde{\boldsymbol{q}}\right]$$

by virtue of Property 4.2. Now, using $K_p = K_v\Lambda$, the equation of $\dot{V}(t, \tilde{\boldsymbol{q}}, \dot{\tilde{\boldsymbol{q}}})$ reduces finally to

$$\begin{aligned}\dot{V}(t, \tilde{\boldsymbol{q}}, \dot{\tilde{\boldsymbol{q}}}) &= -\dot{\tilde{\boldsymbol{q}}}^T K_v \dot{\tilde{\boldsymbol{q}}} - \tilde{\boldsymbol{q}}^T \Lambda^T K_v \Lambda \tilde{\boldsymbol{q}} \\ &= -\begin{bmatrix} \tilde{\boldsymbol{q}} \\ \dot{\tilde{\boldsymbol{q}}} \end{bmatrix}^T \begin{bmatrix} \Lambda^T K_v \Lambda & 0 \\ 0 & K_v \end{bmatrix} \begin{bmatrix} \tilde{\boldsymbol{q}} \\ \dot{\tilde{\boldsymbol{q}}} \end{bmatrix}. \end{aligned} \tag{11.6}$$

Recalling again Lemma 2.1, since K_v is symmetric positive definite and Λ is nonsingular, we conclude that $\Lambda^T K_v \Lambda$ is positive definite. Consequently,

[1] See Remark 2.1 on page 25.

$\dot{V}(t, \tilde{\boldsymbol{q}}, \dot{\tilde{\boldsymbol{q}}})$ given by (11.6) is indeed a globally negative definite function. Since moreover the Lyapunov function candidate (11.3) is globally positive definite, from Theorem 2.4 we conclude immediately global uniform asymptotic stability of the equilibrium $\begin{bmatrix}\tilde{\boldsymbol{q}}^T & \dot{\tilde{\boldsymbol{q}}}^T\end{bmatrix}^T = \mathbf{0} \in \mathbb{R}^{2n}$. Consequently, for any initial position and velocity error we have

$$\lim_{t\to\infty} \dot{\tilde{\boldsymbol{q}}}(t) = \lim_{t\to\infty} (\dot{\boldsymbol{q}}_d(t) - \dot{\boldsymbol{q}}(t)) = \mathbf{0}$$

$$\lim_{t\to\infty} \tilde{\boldsymbol{q}}(t) = \lim_{t\to\infty} (\boldsymbol{q}_d(t) - \boldsymbol{q}(t)) = \mathbf{0},$$

thus, the motion control objective is verified.

Next, we present some experimental results for the PD control with compensation on the Pelican robot.

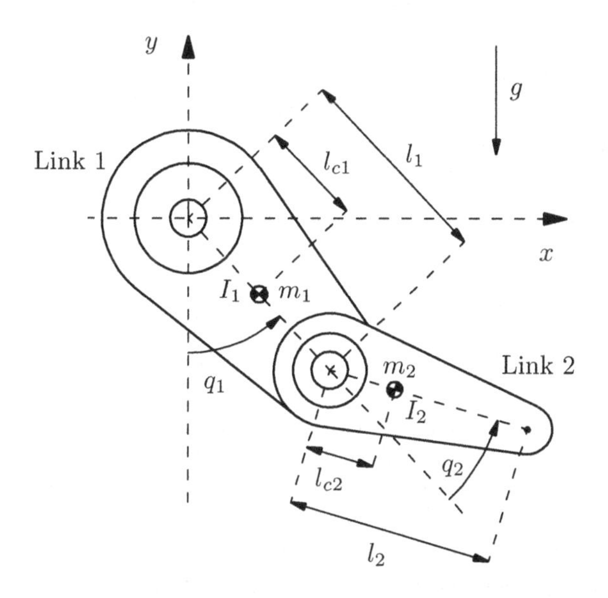

Figure 11.2. Diagram of the Pelican robot

Example 11.1. Consider the Pelican robot presented in Chapter 5, and shown in Figure 11.2. The numerical values of its parameters are listed in Table 5.1.

Consider this robot under PD control with compensation (11.1). It is desired that the robot tracks the trajectories $\boldsymbol{q}_d(t)$, $\dot{\boldsymbol{q}}_d(t)$ and $\ddot{\boldsymbol{q}}_d(t)$ represented by Equations (5.7)–(5.9).

The symmetric positive definite matrices K_p and K_v are chosen so that

$$K_p = \text{diag}\{200, 150\} \quad [\text{N m/rad}],$$
$$K_v = \text{diag}\{3\} \quad [\text{N m s/rad}],$$

and therefore $\Lambda = K_v^{-1}K_p = \text{diag}\{66.6, 50\}$ [1/s].

The initial conditions corresponding to the positions and velocities are

$$q_1(0) = 0, \qquad q_2(0) = 0$$
$$\dot{q}_1(0) = 0, \qquad \dot{q}_2(0) = 0.$$

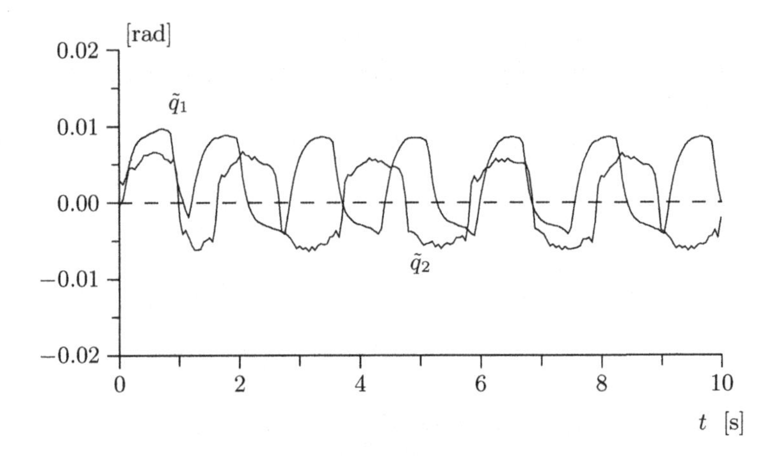

Figure 11.3. Graph of position errors against time

Figure 11.3 shows that the experimental tracking position errors $\tilde{q}(t)$ remain acceptably small. Although in view of the stability analysis of the control system we could expect that the tracking errors vanish, a number of practical aspects – neglected in the theoretical analysis – are responsible for the resulting behavior; for instance, the fact of digitally implementing the robot control system, the sampling period, the fact of estimating (and not measuring) velocities and, most importantly, in the case of the Pelican, friction at the joints. ♢

11.2 PD+ Control

PD+ control is without doubt one of the simplest control laws that may be used in the control of robot manipulators with a formal guarantee of the achievement of the motion control objective, *globally*. The PD+ control law is given by

$$\boldsymbol{\tau} = K_p\tilde{\boldsymbol{q}} + K_v\dot{\tilde{\boldsymbol{q}}} + M(\boldsymbol{q})\ddot{\boldsymbol{q}}_d + C(\boldsymbol{q},\dot{\boldsymbol{q}})\dot{\boldsymbol{q}}_d + \boldsymbol{g}(\boldsymbol{q}) \tag{11.7}$$

where $K_p, K_v \in \mathbb{R}^{n\times n}$ are symmetric positive definite matrices chosen by the designer and as is customary, $\tilde{\boldsymbol{q}} = \boldsymbol{q}_d - \boldsymbol{q}$ denotes the position error. The centrifugal and Coriolis matrix $C(\boldsymbol{q},\dot{\boldsymbol{q}})$ is assumed to be chosen by using the Christoffel symbols (*cf.* Equation 3.21). This ensures that the matrix $\frac{1}{2}\dot{M}(\boldsymbol{q}) - C(\boldsymbol{q},\dot{\boldsymbol{q}})$ is skew-symmetric; a feature that will be useful in the stability analysis.

The practical implementation of PD+ control requires the exact knowledge of the model of the manipulator, that is, of $M(\boldsymbol{q})$, $C(\boldsymbol{q},\dot{\boldsymbol{q}})$ and $\boldsymbol{g}(\boldsymbol{q})$. In addition, it is necessary to know the desired trajectories $\boldsymbol{q}_d(t)$, $\dot{\boldsymbol{q}}_d(t)$ and $\ddot{\boldsymbol{q}}_d(t)$ as well as to have the measurements $\boldsymbol{q}(t)$ and $\dot{\boldsymbol{q}}(t)$. Figure 11.4 depicts the corresponding block-diagram of the PD+ control for robot manipulators.

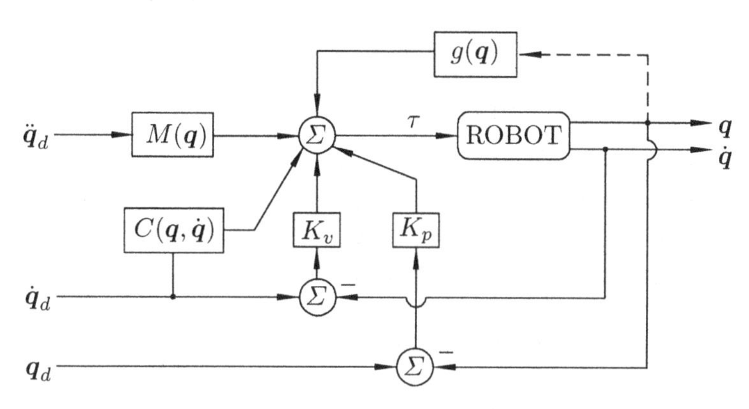

Figure 11.4. Block-diagram: PD+ control

Notice that in the particular case of position control, that is, when $\dot{\boldsymbol{q}}_d = \ddot{\boldsymbol{q}}_d = \boldsymbol{0} \in \mathbb{R}^n$, PD+ control described by (11.7) is equivalent to PD control with gravity compensation, (7.1).

The equation which governs the behavior in closed loop is obtained by substituting the control action $\boldsymbol{\tau}$ of the control law (11.7) in the equation of the robot model (III.1) to get

$$M(\boldsymbol{q})\ddot{\tilde{\boldsymbol{q}}} + C(\boldsymbol{q},\dot{\boldsymbol{q}})\dot{\tilde{\boldsymbol{q}}} = -K_p\tilde{\boldsymbol{q}} - K_v\dot{\tilde{\boldsymbol{q}}}\,.$$

Notice that the closed-loop equation may be written in terms of the state $\begin{bmatrix}\tilde{\boldsymbol{q}}^T & \dot{\tilde{\boldsymbol{q}}}^T\end{bmatrix}^T$ as

$$\frac{d}{dt}\begin{bmatrix}\tilde{\boldsymbol{q}}\\ \dot{\tilde{\boldsymbol{q}}}\end{bmatrix} = \begin{bmatrix}\dot{\tilde{\boldsymbol{q}}}\\ M(\boldsymbol{q}_d - \tilde{\boldsymbol{q}})^{-1}\left[-K_p\tilde{\boldsymbol{q}} - K_v\dot{\tilde{\boldsymbol{q}}} - C(\boldsymbol{q}_d - \tilde{\boldsymbol{q}}, \dot{\boldsymbol{q}}_d - \dot{\tilde{\boldsymbol{q}}})\dot{\tilde{\boldsymbol{q}}}\right]\end{bmatrix}, \tag{11.8}$$

which is a nonlinear differential equation – in general, nonautonomous. The latter is due to the fact that this equation depends explicitly on the functions of time: $\boldsymbol{q}_d(t)$ and $\dot{\boldsymbol{q}}_d(t)$.

Moreover, it is immediate to see the only equilibrium point of the closed-loop equation is the origin $\begin{bmatrix}\tilde{\boldsymbol{q}}^T & \dot{\tilde{\boldsymbol{q}}}^T\end{bmatrix}^T = \boldsymbol{0} \in \mathbb{R}^{2n}$. Therefore, if $\boldsymbol{q}(0) = \boldsymbol{q}_d(0)$ and $\dot{\boldsymbol{q}}(0) = \dot{\boldsymbol{q}}_d(0)$, then $\boldsymbol{q}(t) = \boldsymbol{q}_d(t)$ and $\dot{\boldsymbol{q}}(t) = \dot{\boldsymbol{q}}_d(t)$ for all $t \geq 0$. Notice that the latter follows simply from the concept of equilibrium without needing to invoke any other argument. However, to draw conclusions for the case when $\boldsymbol{q}(0) \neq \boldsymbol{q}_d(0)$ or $\dot{\boldsymbol{q}}(0) \neq \dot{\boldsymbol{q}}_d(0)$ it is necessary to proceed to the stability analysis of the equilibrium.

To analyze the stability of the origin consider now the following Lyapunov function candidate

$$\begin{aligned} V(t, \tilde{\boldsymbol{q}}, \dot{\tilde{\boldsymbol{q}}}) &= \frac{1}{2}\begin{bmatrix}\tilde{\boldsymbol{q}} \\ \dot{\tilde{\boldsymbol{q}}}\end{bmatrix}^T \begin{bmatrix} K_p & 0 \\ 0 & M(\boldsymbol{q}_d - \tilde{\boldsymbol{q}}) \end{bmatrix} \begin{bmatrix}\tilde{\boldsymbol{q}} \\ \dot{\tilde{\boldsymbol{q}}}\end{bmatrix} \\ &= \frac{1}{2}\dot{\tilde{\boldsymbol{q}}}^T M(\boldsymbol{q})\dot{\tilde{\boldsymbol{q}}} + \frac{1}{2}\tilde{\boldsymbol{q}}^T K_p \tilde{\boldsymbol{q}}, \end{aligned} \tag{11.9}$$

which is positive definite since both the inertia matrix $M(\boldsymbol{q})$ and the matrix of position (or proportional) gains, K_p, are positive definite.

Taking the time derivative of (11.9) we obtain

$$\dot{V}(t, \tilde{\boldsymbol{q}}, \dot{\tilde{\boldsymbol{q}}}) = \dot{\tilde{\boldsymbol{q}}}^T M(\boldsymbol{q})\ddot{\tilde{\boldsymbol{q}}} + \frac{1}{2}\dot{\tilde{\boldsymbol{q}}}^T \dot{M}(\boldsymbol{q})\dot{\tilde{\boldsymbol{q}}} + \tilde{\boldsymbol{q}}^T K_p \dot{\tilde{\boldsymbol{q}}}\,.$$

Solving for $M(\boldsymbol{q})\ddot{\tilde{\boldsymbol{q}}}$ of the closed-loop Equation (11.8) and substituting in the previous equation,

$$\begin{aligned} \dot{V}(t, \tilde{\boldsymbol{q}}, \dot{\tilde{\boldsymbol{q}}}) &= -\dot{\tilde{\boldsymbol{q}}}^T K_v \dot{\tilde{\boldsymbol{q}}} \\ &= -\begin{bmatrix}\tilde{\boldsymbol{q}} \\ \dot{\tilde{\boldsymbol{q}}}\end{bmatrix}^T \begin{bmatrix} 0 & 0 \\ 0 & K_v \end{bmatrix} \begin{bmatrix}\tilde{\boldsymbol{q}} \\ \dot{\tilde{\boldsymbol{q}}}\end{bmatrix} \leq 0 \end{aligned} \tag{11.10}$$

where the term $\dot{\tilde{\boldsymbol{q}}}^T\left[\frac{1}{2}\dot{M} - C\right]\dot{\tilde{\boldsymbol{q}}}$ has been canceled by virtue of Property 4.2. From Theorem 2.3 we immediately conclude stability of the origin $\begin{bmatrix}\tilde{\boldsymbol{q}}^T & \dot{\tilde{\boldsymbol{q}}}^T\end{bmatrix}^T = \boldsymbol{0} \in \mathbb{R}^{2n}$ and, by Theorem 2.3, the state remains bounded.

Notice that the expression (11.10) is similar to that obtained for the Lyapunov function used to analyze the stability of the robot in closed loop with PD control with gravity compensation (*cf.* Inequality 7.4). For that controller we used La Salle's theorem to conclude global asymptotic stability. With this under consideration one might also be tempted to conclude global asymptotic stability for the origin of the closed-loop system with the PD+ controller that is, for the origin of (11.8). Nevertheless, this procedure would be incorrect since we remind the reader that the closed-loop Equation (11.8) is nonautonomous due to the presence of $\boldsymbol{q}_d = \boldsymbol{q}_d(t)$. Hence, Theorem (2.7) cannot be used.

Alternatively, we may use Lemma 2.2 to conclude that the position and velocity errors are bounded and the velocity error is square-integrable, *i.e.* it satisfies:

$$\int_0^\infty \left\| \dot{\tilde{\boldsymbol{q}}}(t) \right\|^2 dt < \infty . \tag{11.11}$$

Taking into account these observations, we show next that the velocity error $\dot{\tilde{\boldsymbol{q}}}$ tends asymptotically to zero. For this, notice from the closed-loop Equation (11.8) that

$$\ddot{\tilde{\boldsymbol{q}}} = M(\boldsymbol{q})^{-1} \left[-K_p \tilde{\boldsymbol{q}} - K_v \dot{\tilde{\boldsymbol{q}}} - C(\boldsymbol{q}, \dot{\boldsymbol{q}}) \dot{\tilde{\boldsymbol{q}}} \right] \tag{11.12}$$

where the terms on the right-hand side are bounded due to the following. We know that $\tilde{\boldsymbol{q}}(t)$ and $\dot{\tilde{\boldsymbol{q}}}(t)$ are bounded and that $M(\boldsymbol{q})$ and $C(\boldsymbol{q}, \dot{\boldsymbol{q}})$ are bounded matrices provided that their arguments are also bounded. Now, due to the boundedness of $\tilde{\boldsymbol{q}}$ and $\dot{\tilde{\boldsymbol{q}}}$, we have $\boldsymbol{q}(t) = -\tilde{\boldsymbol{q}}(t) + \boldsymbol{q}_d(t)$, and $\dot{\boldsymbol{q}}(t) = -\dot{\tilde{\boldsymbol{q}}}(t) + \dot{\boldsymbol{q}}_d(t)$ are also bounded since the desired position and velocity $\boldsymbol{q}_d$ and $\dot{\boldsymbol{q}}_d$ are bounded vector functions. Under these conditions, the acceleration error $\ddot{\tilde{\boldsymbol{q}}}(t)$ in (11.12) is a bounded vector function of time. The latter, together with (11.11) and Lemma 2.2, imply that

$$\lim_{t \to \infty} \dot{\tilde{\boldsymbol{q}}}(t) = \lim_{t \to \infty} (\dot{\boldsymbol{q}}_d(t) - \dot{\boldsymbol{q}}(t)) = \boldsymbol{0} \in \mathbb{R}^n .$$

Unfortunately, from the study sketched above, it is not possible to draw any immediate conclusion about the asymptotic behavior of the position error $\tilde{\boldsymbol{q}}$. For this, we need to show not only stability of the origin, as has already been done, but we also need to prove asymptotic stability. As mentioned above, La Salle's theorem (*cf.* Theorem 2.7) cannot be used to study global asymptotic stability since the closed-loop Equation (11.8) is nonautonomous. However, we stress that one may show that the origin of (11.8) is global uniform asymptotic stability by other means. For instance, invoking the so-called *Matrosov's theorem* which applies nonautonomous differential equations specifically in the case that the derivative of the Lyapunov function is only negative semidefinite. The study of this theorem is beyond the scope of this text, hence reader is invited to see the references cited at the end of the chapter for more details on this subject.

Yet for the sake of completeness, we present in the next subsection an alternative analysis of global uniform asymptotic stability by means of a Lyapunov function which has a negative definite derivative.

Example 11.2. Consider the model of an ideal pendulum of length l with mass m concentrated at its tip, subject to the action of gravity g

and to which a torque τ is applied at the axis of rotation (see Example 2.2) that is,

$$ml^2\ddot{q} + mgl\ \sin(q) = \tau$$

where we identify $M(q) = ml^2$, $C(q,\dot{q}) = 0$ and $g(q) = mgl\ \sin(q)$. For this example, the PD+ control law given by (11.7) becomes

$$\tau = k_p\tilde{q} + k_v\dot{\tilde{q}} + ml^2\ddot{q}_d + mgl\ \sin(q)$$

where k_p and k_v are real positive numbers. The closed-loop equation is

$$ml^2\ddot{\tilde{q}} + k_v\dot{\tilde{q}} + k_p\tilde{q} = 0$$

which constitutes a linear autonomous differential equation whose unique equilibrium $\begin{bmatrix}\tilde{q} & \dot{\tilde{q}}\end{bmatrix}^T = \mathbf{0} \in \mathbb{R}^2$, as can easily be shown, is globally exponentially stable. ♢

Next, we present experimental results obtained for the Pelican 2-DOF robot under PD+ control.

Example 11.3. Consider the 2-DOF *prototype robot* studied in Chapter 5, and shown in Figure 11.2.

Consider the application of PD control+ (11.7) to this robot. The joint desired trajectories of position, velocity and acceleration, $\boldsymbol{q}_d(t)$, $\dot{\boldsymbol{q}}_d(t)$ and $\ddot{\boldsymbol{q}}_d(t)$, are given by Equations (5.7)–(5.9) on page 128.

The symmetric positive definite matrices K_p and K_v are chosen as

$$K_p = \text{diag}\{200,\ 150\} \quad [\text{N m/rad}]$$
$$K_v = \text{diag}\{3\} \quad [\text{N m s/rad}]\ .$$

The initial conditions corresponding to the positions and velocities, are fixed at

$$q_1(0) = 0, \qquad q_2(0) = 0$$
$$\dot{q}_1(0) = 0, \qquad \dot{q}_2(0) = 0\,.$$

Figure 11.5 presents the experimental steady state tracking position errors $\tilde{\boldsymbol{q}}(t)$, which, in view of the practical aspects mentioned in Example 11.1 (mainly friction phenomena), do not vanish. ♢

Comparing the experimental results in Figures 11.5 and 11.3, we see that PD control with compensation behaves better than PD+ control, in the sense that the tracking error $|\tilde{q}_1|$ satisfies a smaller bound.

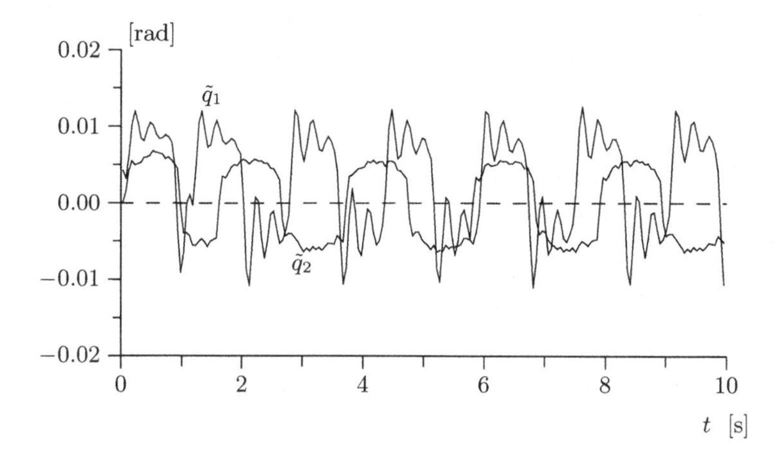

Figure 11.5. Graph of the position errors against time

11.2.1 Lyapunov Function for Asymptotic Stability

We present next an alternative stability analysis for the origin of the closed-loop Equation (11.8). This study has been taken from the literature and its reference is cited at the end of the chapter. The advantage of the study we present in this section is that we use a Lyapunov function that allows one to conclude directly global uniform asymptotic stability.

In the particular case that all the joints of the robot manipulator are revolute, it may be shown that the origin $\begin{bmatrix}\tilde{\boldsymbol{q}}^T & \dot{\tilde{\boldsymbol{q}}}^T\end{bmatrix}^T = \boldsymbol{0} \in \mathbb{R}^{2n}$ is globally uniformly asymptotically stable. This can be done by using a Lyapunov function somewhat more complex than the one proposed in (11.9) and exploiting the properties presented in Chapter 4.

As stated in Property 4.1, the fact that a robot manipulator has only revolute joints, implies that the largest eigenvalue $\lambda_{\text{Max}}\{M(\boldsymbol{q})\}$ of the inertia matrix is bounded. On the other hand, in this study we assume that the desired joint velocity $\dot{\boldsymbol{q}}_d(t)$ is a bounded vector, but the bound does not need to be known.

We present now the formal stability analysis. To study the stability properties of the origin $\begin{bmatrix}\tilde{\boldsymbol{q}}^T & \dot{\tilde{\boldsymbol{q}}}^T\end{bmatrix}^T = \boldsymbol{0} \in \mathbb{R}^{2n}$ of the closed-loop Equation (11.8), consider the following Lyapunov function candidate:

$$V(t,\tilde{\boldsymbol{q}},\dot{\tilde{\boldsymbol{q}}}) = \frac{1}{2}\begin{bmatrix}\tilde{\boldsymbol{q}}\\ \dot{\tilde{\boldsymbol{q}}}\end{bmatrix}^T \begin{bmatrix} K_p & \dfrac{\varepsilon_0}{1+\|\tilde{\boldsymbol{q}}\|}M(\boldsymbol{q}) \\ \dfrac{\varepsilon_0}{1+\|\tilde{\boldsymbol{q}}\|}M(\boldsymbol{q}) & M(\boldsymbol{q})\end{bmatrix}\begin{bmatrix}\tilde{\boldsymbol{q}}\\ \dot{\tilde{\boldsymbol{q}}}\end{bmatrix} \tag{11.13}$$

$$= W(t, \tilde{q}, \dot{\tilde{q}}) + \underbrace{\frac{\varepsilon_0}{1 + \|\tilde{q}\|}}_{\varepsilon(\tilde{q})} \tilde{q}^T M(q) \dot{\tilde{q}}$$

where $W(t, \tilde{q}, \dot{\tilde{q}})$ is the Lyapunov function (11.9). The positive constant ε_0 is chosen so as to satisfy simultaneously the three inequalities:

- $\sqrt{\dfrac{\lambda_{\min}\{K_p\}}{\lambda_{\text{Max}}\{M(q)\}}} > \varepsilon_0 > 0\,;$
- $\dfrac{\lambda_{\min}\{K_v\}}{2\left(k_{C_1} + 2\lambda_{\text{Max}}\{M(q)\}\right)} > \varepsilon_0 > 0\,;$
- $\dfrac{2\lambda_{\min}\{K_p\}\lambda_{\min}\{K_v\}}{\left(\lambda_{\text{Max}}\{K_v\} + k_{C_1}\,\|\dot{q}_d\|_{\text{Max}}\right)^2} > \varepsilon_0 > 0\,;$

where $k_{C_1} > 0$ is a constant such that $\|C(q, \dot{q})\tilde{q}\| \leq k_{C_1}\,\|\dot{q}\|\,\|\tilde{q}\|$ for all $q \in \mathbb{R}^n$ (by Property 4.2) and $\|\dot{q}_d\|_{\text{Max}}$ is the largest value of $\|\dot{q}_d(t)\|$. It is important to underline that in the study presented here, it is only required to guarantee the existence of $\varepsilon_0 > 0$ and it is not necessary to know its actual numerical value. In particular, it is not necessary to know $\|\dot{q}_d\|_{\text{Max}}$. Fortunately, the existence of $\varepsilon_0 > 0$ is guaranteed since in the inequalities above, upper-bounds on ε_0 exist (that is, they are finite) and are strictly positive.

Next, we show that the Lyapunov function candidate (11.13) is positive definite provided that

$$\sqrt{\frac{\lambda_{\min}\{K_p\}}{\lambda_{\text{Max}}\{M(q)\}}} > \varepsilon_0 > 0 \qquad \forall\ q \in \mathbb{R}^n\,. \tag{11.14}$$

To show that under the previous condition the function (11.13) is positive definite, we proceed in two steps. First, notice that since $\varepsilon_0^2 \geq \varepsilon^2$ where $\varepsilon = \varepsilon_0/(1 + \|\tilde{q}\|)$, Inequality (11.14) implies that

$$\frac{\lambda_{\min}\{K_p\}}{\lambda_{\text{Max}}\{M(q)\}} > \varepsilon^2 > 0 \qquad \forall\ q \in \mathbb{R}^n\,.$$

This in turn, implies that the matrix $K_p - \varepsilon^2 M(q)$ is positive definite, *i.e.*

$$K_p - \varepsilon^2 M(q) > 0\,.$$

On the other hand, the function (11.13) may be rewritten as

$$V(t, \tilde{q}, \dot{\tilde{q}}) = \frac{1}{2}\left[\dot{\tilde{q}} + \varepsilon\tilde{q}\right]^T M(q)\left[\dot{\tilde{q}} + \varepsilon\tilde{q}\right] + \frac{1}{2}\tilde{q}^T\left[K_p - \varepsilon^2 M(q)\right]\tilde{q}$$

which is positive definite since so are $M(q)$ and $K_p - \varepsilon^2 M(q)$.

To show that the function defined in (11.13) is also decrescent, notice that $V(t, \tilde{q}, \dot{\tilde{q}})$ satisfies

$$V(t, \tilde{q}, \dot{\tilde{q}}) \leq \frac{1}{2} \begin{bmatrix} \|\tilde{q}\| \\ \|\dot{\tilde{q}}\| \end{bmatrix}^T \begin{bmatrix} \lambda_{\mathrm{Max}}\{K_p\} & \varepsilon_0 \, \lambda_{\mathrm{Max}}\{M\} \\ \varepsilon_0 \, \lambda_{\mathrm{Max}}\{M\} & \lambda_{\mathrm{Max}}\{M\} \end{bmatrix} \begin{bmatrix} \|\tilde{q}\| \\ \|\dot{\tilde{q}}\| \end{bmatrix}, \tag{11.15}$$

for which we have used the fact that

$$\left| \frac{\varepsilon_0}{1 + \|\tilde{q}\|} \right| \|\tilde{q}\| \leq \varepsilon_0 \|\tilde{q}\| .$$

The matrix on the right-hand side of inequality (11.15) is positive definite in view of the condition on ε_0,

$$\sqrt{\frac{\lambda_{\mathrm{Max}}\{K_p\}}{\lambda_{\mathrm{Max}}\{M(q)\}}} > \varepsilon_0 > 0 \qquad \forall \ q \in \mathbb{R}^n ,$$

which implies that the determinant of this matrix is positive, and which trivially holds under hypothesis (11.14). Thus, the function (11.13) is positive definite, radially unbounded and decrescent.

It is interesting to remark that the Lyapunov function candidate (11.13) is very similar to that used for the study of the PD control law with gravity compensation (7.8). On the other hand, the function (11.13) may be considered as a more general version of the previous Lyapunov function (11.9) which corresponds to the case $\varepsilon_0 = 0$.

Following the study of stability, the time derivative of the Lyapunov function candidate (11.13) is given by

$$\begin{aligned} \dot{V}(t, \tilde{q}, \dot{\tilde{q}}) &= \dot{W}(t, \tilde{q}, \dot{\tilde{q}}) + \varepsilon(\tilde{q})\dot{\tilde{q}}^T M(q)\dot{\tilde{q}} + \varepsilon(\tilde{q})\tilde{q}^T \dot{M}(q)\dot{\tilde{q}} \\ &\quad + \varepsilon(\tilde{q})\tilde{q}^T M(q)\ddot{\tilde{q}} + \dot{\varepsilon}(\tilde{q})\tilde{q}^T M(q)\dot{\tilde{q}} \end{aligned} \tag{11.16}$$

where $\dot{W}(t, \tilde{q}, \dot{\tilde{q}})$ corresponds to the right-hand side of (11.10), that is

$$\dot{W}(t, \tilde{q}, \dot{\tilde{q}}) = -\dot{\tilde{q}}^T K_v \dot{\tilde{q}} .$$

Taking into account the previous expression, substituting $M(q)\ddot{\tilde{q}}$ from the closed-loop Equation (11.8) and rearranging terms, Equation (11.16) becomes

$$\begin{aligned} \dot{V}(t, \tilde{q}, \dot{\tilde{q}}) &= -\dot{\tilde{q}}^T K_v \dot{\tilde{q}} + \varepsilon(\tilde{q})\dot{\tilde{q}}^T M(q)\dot{\tilde{q}} + \varepsilon(\tilde{q})\tilde{q}^T \Big[\dot{M}(q) - C(q, \dot{q})\Big] \dot{\tilde{q}} \\ &\quad - \varepsilon(\tilde{q})\tilde{q}^T \Big[K_p \tilde{q} + K_v \dot{\tilde{q}}\Big] + \dot{\varepsilon}(\tilde{q})\tilde{q}^T M(q)\dot{\tilde{q}} . \end{aligned} \tag{11.17}$$

Now, considering Property 4.2 which establishes $\dot{M}(q) = C(q, \dot{q}) + C(q, \dot{q})^T$, Equation (11.17) becomes

$$\dot{V}(\tilde{\boldsymbol{q}},\dot{\tilde{\boldsymbol{q}}}) = -\dot{\tilde{\boldsymbol{q}}}^T K_v \dot{\tilde{\boldsymbol{q}}} + \varepsilon(\tilde{\boldsymbol{q}})\dot{\tilde{\boldsymbol{q}}}^T M(\boldsymbol{q})\dot{\tilde{\boldsymbol{q}}} + \underbrace{\varepsilon(\tilde{\boldsymbol{q}})\dot{\tilde{\boldsymbol{q}}}^T C(\boldsymbol{q},\dot{\boldsymbol{q}})\tilde{\boldsymbol{q}}}_{a(\tilde{\boldsymbol{q}},\dot{\tilde{\boldsymbol{q}}})}$$
$$- \varepsilon(\tilde{\boldsymbol{q}})\tilde{\boldsymbol{q}}^T\Big[K_p\tilde{\boldsymbol{q}} + K_v\dot{\tilde{\boldsymbol{q}}}\Big] + \underbrace{\dot{\varepsilon}(\tilde{\boldsymbol{q}})\tilde{\boldsymbol{q}}^T M(\boldsymbol{q})\dot{\tilde{\boldsymbol{q}}}}_{b(\tilde{\boldsymbol{q}},\dot{\tilde{\boldsymbol{q}}})}\,. \tag{11.18}$$

Next, we proceed to obtain upper-bounds on the terms $a(\tilde{\boldsymbol{q}},\dot{\tilde{\boldsymbol{q}}})$ and $b(\tilde{\boldsymbol{q}},\dot{\tilde{\boldsymbol{q}}})$. Regarding the term $a(\tilde{\boldsymbol{q}},\dot{\tilde{\boldsymbol{q}}})$ we have

$$\begin{aligned}
a(\tilde{\boldsymbol{q}},\dot{\tilde{\boldsymbol{q}}}) &\le \left|a(\tilde{\boldsymbol{q}},\dot{\tilde{\boldsymbol{q}}})\right| = \left|\varepsilon\dot{\tilde{\boldsymbol{q}}}^T C(\boldsymbol{q},\dot{\boldsymbol{q}})\tilde{\boldsymbol{q}}\right| \\
&\le \varepsilon\left\|\dot{\tilde{\boldsymbol{q}}}\right\|\left\|C(\boldsymbol{q},\dot{\boldsymbol{q}})\tilde{\boldsymbol{q}}\right\| \\
&\le \varepsilon k_{C_1}\left\|\dot{\tilde{\boldsymbol{q}}}\right\|\left\|\dot{\boldsymbol{q}}\right\|\left\|\tilde{\boldsymbol{q}}\right\| \\
&\le \varepsilon k_{C_1}\left\|\dot{\tilde{\boldsymbol{q}}}\right\|\left(\left\|\dot{\tilde{\boldsymbol{q}}}\right\| + \left\|\dot{\boldsymbol{q}}_d\right\|\right)\left\|\tilde{\boldsymbol{q}}\right\| \\
&= \frac{\varepsilon_0}{1+\|\tilde{\boldsymbol{q}}\|}k_{C_1}\left\|\tilde{\boldsymbol{q}}\right\|\left\|\dot{\tilde{\boldsymbol{q}}}\right\|^2 + \varepsilon k_{C_1}\left\|\tilde{\boldsymbol{q}}\right\|\left\|\dot{\tilde{\boldsymbol{q}}}\right\|\left\|\dot{\boldsymbol{q}}_d\right\| \\
&\le \varepsilon_0 k_{C_1}\left\|\dot{\tilde{\boldsymbol{q}}}\right\|^2 + \varepsilon k_{C_1}\left\|\tilde{\boldsymbol{q}}\right\|\left\|\dot{\tilde{\boldsymbol{q}}}\right\|\left\|\dot{\boldsymbol{q}}_d\right\|
\end{aligned}$$

where once again, we used Property 4.2 (*i.e.* that $\|C(\boldsymbol{q},\dot{\boldsymbol{q}})\tilde{\boldsymbol{q}}\| \le k_{C_1}\|\dot{\boldsymbol{q}}\|\,\|\tilde{\boldsymbol{q}}\|$ for all $\boldsymbol{q}\in\mathbb{R}^n$).

Next, regarding the term $b(\tilde{\boldsymbol{q}},\dot{\tilde{\boldsymbol{q}}})$, first notice that

$$\dot{\varepsilon}(\tilde{\boldsymbol{q}}) = -\frac{\varepsilon_0}{\|\tilde{\boldsymbol{q}}\|\left[1+\|\tilde{\boldsymbol{q}}\|\right]^2}\tilde{\boldsymbol{q}}^T\dot{\tilde{\boldsymbol{q}}}\,.$$

Taking this into account we obtain

$$\begin{aligned}
b(\tilde{\boldsymbol{q}},\dot{\tilde{\boldsymbol{q}}}) &\le \left|b(\tilde{\boldsymbol{q}},\dot{\tilde{\boldsymbol{q}}})\right| = \left|\dot{\varepsilon}(\tilde{\boldsymbol{q}})\tilde{\boldsymbol{q}}^T M(\boldsymbol{q})\dot{\tilde{\boldsymbol{q}}}\right| \\
&\le \frac{\varepsilon_0}{\|\tilde{\boldsymbol{q}}\|\left[1+\|\tilde{\boldsymbol{q}}\|\right]^2}\left|\tilde{\boldsymbol{q}}^T\dot{\tilde{\boldsymbol{q}}}\right|\left|\tilde{\boldsymbol{q}}^T M(\boldsymbol{q})\dot{\tilde{\boldsymbol{q}}}\right| \\
&\le \frac{\varepsilon_0}{\|\tilde{\boldsymbol{q}}\|\left[1+\|\tilde{\boldsymbol{q}}\|\right]^2}\left\|\tilde{\boldsymbol{q}}\right\|\left\|\dot{\tilde{\boldsymbol{q}}}\right\|\left|\tilde{\boldsymbol{q}}^T M(\boldsymbol{q})\dot{\tilde{\boldsymbol{q}}}\right| \\
&\le \frac{\varepsilon_0}{\|\tilde{\boldsymbol{q}}\|\left[1+\|\tilde{\boldsymbol{q}}\|\right]^2}\left\|\tilde{\boldsymbol{q}}\right\|\left\|\dot{\tilde{\boldsymbol{q}}}\right\|\left\|\tilde{\boldsymbol{q}}\right\|\left\|M(\boldsymbol{q})\dot{\tilde{\boldsymbol{q}}}\right\| \\
&\le \frac{\varepsilon_0}{\|\tilde{\boldsymbol{q}}\|\left[1+\|\tilde{\boldsymbol{q}}\|\right]^2}\left\|\tilde{\boldsymbol{q}}\right\|\left\|\dot{\tilde{\boldsymbol{q}}}\right\|\left\|\tilde{\boldsymbol{q}}\right\|\sqrt{\lambda_{\mathrm{Max}}\{M(\boldsymbol{q})^T M(\boldsymbol{q})\}}\left\|\dot{\tilde{\boldsymbol{q}}}\right\| \\
&\le \underbrace{\frac{\varepsilon_0}{1+\|\tilde{\boldsymbol{q}}\|}}_{\varepsilon(\tilde{\boldsymbol{q}})}\left\|\dot{\tilde{\boldsymbol{q}}}\right\|^2\lambda_{\mathrm{Max}}\{M(\boldsymbol{q})\}
\end{aligned}$$

where $\lambda_{\text{Max}}\{M\} = \sqrt{\lambda_{\text{Max}}\{M^T M\}}$, since $M(\boldsymbol{q})$ is a symmetric positive definite matrix function.

Keeping in mind the upper-bounds on $a(\tilde{\boldsymbol{q}}, \dot{\tilde{\boldsymbol{q}}})$ and $b(\tilde{\boldsymbol{q}}, \dot{\tilde{\boldsymbol{q}}})$, the derivative $\dot{V}(t, \tilde{\boldsymbol{q}}, \dot{\tilde{\boldsymbol{q}}})$ of the Lyapunov function given by Equation (11.18), may be upper-bounded as

$$\begin{aligned} \dot{V}(t, \tilde{\boldsymbol{q}}, \dot{\tilde{\boldsymbol{q}}}) \le & -\dot{\tilde{\boldsymbol{q}}}^T K_v \dot{\tilde{\boldsymbol{q}}} + \varepsilon \dot{\tilde{\boldsymbol{q}}}^T M \dot{\tilde{\boldsymbol{q}}} + \varepsilon_0 k_{C_1} \left\| \dot{\tilde{\boldsymbol{q}}} \right\|^2 + \varepsilon k_{C_1} \|\tilde{\boldsymbol{q}}\| \left\| \dot{\tilde{\boldsymbol{q}}} \right\| \|\dot{\boldsymbol{q}}_d\| \\ & - \varepsilon \tilde{\boldsymbol{q}}^T \Big[K_p \tilde{\boldsymbol{q}} + K_v \dot{\tilde{\boldsymbol{q}}} \Big] + \varepsilon \lambda_{\text{Max}}\{M\} \left\| \dot{\tilde{\boldsymbol{q}}} \right\|^2 \end{aligned} \tag{11.19}$$

where for simplicity in the notation we omitted the arguments of $\varepsilon(\tilde{\boldsymbol{q}})$ and $M(\boldsymbol{q})$. Using the inequalities

- $-\dot{\tilde{\boldsymbol{q}}}^T K_v \dot{\tilde{\boldsymbol{q}}} \le -\frac{1}{2} \dot{\tilde{\boldsymbol{q}}}^T K_v \dot{\tilde{\boldsymbol{q}}} - \dfrac{\lambda_{\min}\{K_v\}}{2} \left\| \dot{\tilde{\boldsymbol{q}}} \right\|^2$
- $\varepsilon \lambda_{\text{Max}}\{M\} \left\| \dot{\tilde{\boldsymbol{q}}} \right\|^2 \le \varepsilon_0 \lambda_{\text{Max}}\{M\} \left\| \dot{\tilde{\boldsymbol{q}}} \right\|^2$
- $\varepsilon \dot{\tilde{\boldsymbol{q}}}^T M(\boldsymbol{q}) \dot{\tilde{\boldsymbol{q}}} \le \varepsilon \lambda_{\text{Max}}\{M\} \left\| \dot{\tilde{\boldsymbol{q}}} \right\|^2 \le \varepsilon_0 \lambda_{\text{Max}}\{M\} \left\| \dot{\tilde{\boldsymbol{q}}} \right\|^2$,

we see that Inequality (11.19) may be rewritten in the form

$$\begin{aligned} \dot{V}(t, \tilde{\boldsymbol{q}}, \dot{\tilde{\boldsymbol{q}}}) \le & - \begin{bmatrix} \tilde{\boldsymbol{q}} \\ \dot{\tilde{\boldsymbol{q}}} \end{bmatrix}^T \begin{bmatrix} \varepsilon K_p & \frac{\varepsilon}{2} K_v \\ \frac{\varepsilon}{2} K_v & \frac{1}{2} K_v \end{bmatrix} \begin{bmatrix} \tilde{\boldsymbol{q}} \\ \dot{\tilde{\boldsymbol{q}}} \end{bmatrix} + \varepsilon k_{C_1} \|\tilde{\boldsymbol{q}}\| \left\| \dot{\tilde{\boldsymbol{q}}} \right\| \|\dot{\boldsymbol{q}}_d\| \\ & - \frac{1}{2} \left[\lambda_{\min}\{K_v\} - 2\varepsilon_0 (k_{C_1} + 2\lambda_{\text{Max}}\{M\}) \right] \left\| \dot{\tilde{\boldsymbol{q}}} \right\|^2 . \end{aligned}$$

Notice that in addition $\dot{V}(t, \tilde{\boldsymbol{q}}, \dot{\tilde{\boldsymbol{q}}})$ may be upper-bounded in the following manner:

$$\begin{aligned} \dot{V}(t, \tilde{\boldsymbol{q}}, \dot{\tilde{\boldsymbol{q}}}) \le & -\varepsilon \underbrace{\begin{bmatrix} \|\tilde{\boldsymbol{q}}\| \\ \left\|\dot{\tilde{\boldsymbol{q}}}\right\| \end{bmatrix}^T Q \begin{bmatrix} \|\tilde{\boldsymbol{q}}\| \\ \left\|\dot{\tilde{\boldsymbol{q}}}\right\| \end{bmatrix}}_{h(\|\tilde{\boldsymbol{q}}\|, \|\dot{\tilde{\boldsymbol{q}}}\|)} \\ & - \frac{1}{2} \underbrace{\left[\lambda_{\min}\{K_v\} - 2\varepsilon_0 \left(k_{C_1} + 2\lambda_{\text{Max}}\{M\} \right) \right]}_{\delta} \left\| \dot{\tilde{\boldsymbol{q}}} \right\|^2 , \end{aligned} \tag{11.20}$$

where the symmetric matrix Q is given by

$$Q = \begin{bmatrix} \lambda_{\min}\{K_p\} & -\frac{1}{2} \left(\lambda_{\text{Max}}\{K_v\} + k_{C_1} \|\dot{\boldsymbol{q}}_d\| \right) \\ -\frac{1}{2} \left(\lambda_{\text{Max}}\{K_v\} + k_{C_1} \|\dot{\boldsymbol{q}}_d\| \right) & \frac{1}{2\varepsilon_0} \lambda_{\min}\{K_v\} \end{bmatrix}$$

and where we used

- $-\frac{\varepsilon}{2}\tilde{q}^T K_v \dot{\tilde{q}} \leq \frac{\varepsilon}{2}\left|\tilde{q}^T K_v \dot{\tilde{q}}\right| \leq \frac{\varepsilon}{2}\|\tilde{q}\|\left\|K_v \dot{\tilde{q}}\right\| \leq \frac{\varepsilon}{2}\|\tilde{q}\|\sqrt{\lambda_{\mathrm{Max}}\{K_v^T K_v\}}\left\|\dot{\tilde{q}}\right\|$
- $-\frac{1}{2\varepsilon}\lambda_{\min}\{K_v\}\left\|\dot{\tilde{q}}\right\|^2 = -\frac{1+\|\tilde{q}\|}{2\varepsilon_0}\lambda_{\min}\{K_v\}\left\|\dot{\tilde{q}}\right\|^2 \leq -\frac{1}{2\varepsilon_0}\lambda_{\min}\{K_v\}\left\|\dot{\tilde{q}}\right\|^2.$

To guarantee that $\dot{V}(t, \tilde{q}, \dot{\tilde{q}})$ is a negative definite function, it is necessary to pick ε_0 appropriately. On one hand, it is required that $\delta > 0$, that is

$$\frac{\lambda_{\min}\{K_v\}}{2\left(k_{C_1} + 2\lambda_{\mathrm{Max}}\{M\}\right)} > \varepsilon_0 .$$

On the other hand, it is also required that the matrix Q be positive definite. The latter is guaranteed if

$$\frac{2\lambda_{\min}\{K_p\}\lambda_{\min}\{K_v\}}{\left(\lambda_{\mathrm{Max}}\{K_v\} + k_{C_1}\|\dot{q}_d\|_{\mathrm{Max}}\right)^2} > \varepsilon_0,$$

where $\|\dot{q}_d\|_{\mathrm{Max}} \geq \|\dot{q}_d(t)\|$ for all $t \geq 0$, since the previous inequality implies that

$$\frac{1}{2\varepsilon_0}\lambda_{\min}\{K_p\}\lambda_{\min}\{K_v\} - \frac{1}{4}\left(\lambda_{\mathrm{Max}}\{K_v\} + k_{C_1}\|\dot{q}_d\|\right)^2 > 0 ,$$

which actually corresponds to the determinant of Q.

To summarize, the inequality in (11.20) involving $\dot{V}(t, \tilde{q}, \dot{\tilde{q}})$, may be written as

$$\dot{V}(t, \tilde{q}, \dot{\tilde{q}}) \leq -\varepsilon(\tilde{q})\, h(\|\tilde{q}\|, \|\dot{\tilde{q}}\|) - c\|\dot{\tilde{q}}\|^2,$$

where, with the choice we made for ε_0, it follows that $h(\|\tilde{q}\|, \left\|\dot{\tilde{q}}\right\|)$ is a positive definite function and $c > 0$. The function $\dot{V}(t, \tilde{q}, \dot{\tilde{q}})$ is negative definite since $\varepsilon(\tilde{q})\, h(\|\tilde{q}\|, \|\dot{\tilde{q}}\|) - c\|\dot{\tilde{q}}\|^2$ is a positive definite function of q and $\dot{q}$. In particular,

- $\dot{V}(t, \mathbf{0}, \mathbf{0}) = 0$
- $\dot{V}(t, \tilde{q}, \dot{\tilde{q}}) < 0 \qquad \forall \begin{bmatrix} \|\tilde{q}\| \\ \|\dot{\tilde{q}}\| \end{bmatrix} \neq \mathbf{0}$
- $\dot{V}(t, \tilde{q}, \dot{\tilde{q}}) \to -\infty$ when $\|\tilde{q}\|^2 + \|\dot{\tilde{q}}\|^2 \to \infty .$

Thus, using Theorem 2.4 we conclude that the origin $\left[\tilde{q}^T \ \dot{\tilde{q}}^T\right]^T = \mathbf{0} \in \mathbb{R}^{2n}$ is globally uniformly asymptotically stable.

11.3 Conclusions

We may summarize the ideas exposed in this chapter as follows:

- For any selection of the symmetric positive definite matrices K_p and K_v, the origin of the closed-loop equation of robots with the PD control law with compensation, expressed in terms of the state vector $\left[\tilde{\boldsymbol{q}}^T \ \dot{\tilde{\boldsymbol{q}}}^T\right]^T$, is globally uniformly asymptotically stable. Therefore, the PD control law with compensation satisfies the motion control objective, globally. This implies in particular, that for any initial position error $\tilde{\boldsymbol{q}}(0) \in \mathbb{R}^n$ and any velocity error $\dot{\tilde{\boldsymbol{q}}}(0) \in \mathbb{R}^n$, we have $\lim_{t\to\infty} \tilde{\boldsymbol{q}}(t) = \boldsymbol{0}$.
- For any choice of the symmetric positive definite matrices K_p and K_v, the origin of the closed-loop equation of a robot with the PD+ control law, expressed in terms of the state vector $\left[\tilde{\boldsymbol{q}}^T \ \dot{\tilde{\boldsymbol{q}}}^T\right]^T$, is globally uniformly asymptotically stable. Therefore, PD+ control satisfies the motion control objective, globally. In particular, for any initial position error $\tilde{\boldsymbol{q}}(0) \in \mathbb{R}^n$ and velocity error $\dot{\tilde{\boldsymbol{q}}}(0) \in \mathbb{R}^n$, we have $\lim_{t\to\infty} \tilde{\boldsymbol{q}}(t) = \boldsymbol{0}$.

Bibliography

The structure of the PD control law with compensation has been proposed and studied in

- Slotine J. J., Li W., 1987. *"On the adaptive control of robot manipulators"*, The International Journal of Robotics Research, Vol. 6, No. 3, pp. 49–59.
- Slotine J. J., Li W., 1988. *"Adaptive manipulator control: A case study"*, IEEE Transactions on Automatic Control, Vol. AC-33, No. 11, November, pp. 995–1003.
- Slotine J. J., Li W., 1991, *"Applied nonlinear control"*, Prentice-Hall.

The Lyapunov function (11.3) for the analysis of global uniform asymptotic stability for the PD control law with compensation was proposed in

- Spong M., Ortega R., Kelly R., 1990, *"Comments on "Adaptive manipulator control: A case study"*, IEEE Transactions on Automatic Control, Vol. 35, No. 6, June, pp.761–762.
- Egeland O., Godhavn J. M., 1994, *"A note on Lyapunov stability for an adaptive robot controller"*, IEEE Transactions on Automatic Control, Vol. 39, No. 8, August, pp. 1671–1673.

The structure of the PD+ control law was proposed in

- Koditschek D. E., 1984, *"Natural motion for robot arms"*, Proceedings of the IEEE 23th Conference on Decision and Control, Las Vegas, NV., December, pp. 733–735.

PD+ control was originally presented in[2]

- Paden B., Panja R., 1988, *"Globally asymptotically stable PD+ controller for robot manipulators"*, International Journal of Control, Vol. 47, No. 6, pp. 1697–1712.

The material in Subsection 11.2.1 on the Lyapunov function to show global uniform asymptotic stability is taken from

- Whitcomb L. L., Rizzi A., Koditschek D. E., 1993, *"Comparative experiments with a new adaptive controller for robot arms"*, IEEE Transactions on Robotics and Automation, Vol. 9, No. 1, February, pp. 59–70.

Problems

1. Consider the model of an ideal pendulum studied in Example 2.2 (see page 30)
$$J\ddot{q} + mgl\,\sin(q) = \tau\,.$$
Assume that we apply the PD controller with compensation
$$\tau = k_p\tilde{q} + k_v\dot{\tilde{q}} + J[\ddot{q}_d + \lambda\dot{\tilde{q}}] + mgl\,\sin(q)$$
where $\lambda = k_p/k_v$, k_p and k_v are positive numbers.
   a) Obtain the closed-loop equation in terms of the state vector $\begin{bmatrix}\tilde{q} & \dot{\tilde{q}}\end{bmatrix}^T$. Verify that the origin is its unique equilibrium point.
   b) Show that the origin $\begin{bmatrix}\tilde{q} & \dot{\tilde{q}}\end{bmatrix}^T = \mathbf{0} \in \mathbb{R}^2$ is globally asymptotically stable.
   Hint: Use the Lyapunov function candidate
$$V(\tilde{q},\dot{\tilde{q}}) = \frac{1}{2}J\left[\dot{\tilde{q}} + \lambda\tilde{q}\right]^2 + k_p\tilde{q}^2\,.$$
2. Consider PD+ control for the ideal pendulum presented in Example 11.2. Propose a Lyapunov function candidate to show that the origin $\begin{bmatrix}\tilde{q} & \dot{\tilde{q}}\end{bmatrix} = [0\ \ 0]^T = \mathbf{0} \in \mathbb{R}^2$ of the closed-loop equation
$$ml^2\ddot{\tilde{q}} + k_v\dot{\tilde{q}} + k_p\tilde{q} = 0$$
is a globally asymptotically stable equilibrium point.

[2] This, together with PD control with compensation were the first controls with rigorous proofs of global uniform asymptotic stability proposed for the motion control problem.

3. Consider the model of the pendulum from Example 3.8 and illustrated in Figure 3.13,

$$\left(J_m + \frac{J_L}{r^2}\right)\ddot{q} + \left(f_m + \frac{f_L}{r^2} + \frac{K_a K_b}{R_a}\right)\dot{q} + \frac{k_L}{r^2}\sin(q) = \frac{K_a}{rR_a}v$$

where

- v is the armature voltage (input)
- q is the angular position of the pendulum with respect to the vertical (output),

and the rest of the parameters are constants related to the electrical and mechanical parts of the system and which are positive and known.

It is desired to drive the angular position $q(t)$ to a constant value q_d. For this, we propose to use the following control law of type PD+[3],

$$v = \frac{rR_a}{K_a}\left(k_p\tilde{q} - k_v\dot{q} + \frac{k_L}{r^2}\sin(q)\right)$$

with k_p and k_v positive design constants and $\tilde{q}(t) = q_d - q(t)$.

a) Obtain the closed-loop equation in terms of the state $[\tilde{q} \ \ \dot{q}]^T$.

b) Verify that the origin is an equilibrium and propose a Lyapunov function to demonstrate its stability.

c) Could it be possible to show as well that the origin is actually globally asymptotically stable?

4. Consider the control law

$$\boldsymbol{\tau} = K_p\tilde{\boldsymbol{q}} + K_v\dot{\tilde{\boldsymbol{q}}} + M(\boldsymbol{q})\ddot{\boldsymbol{q}}_d + C(\boldsymbol{q},\dot{\boldsymbol{q}}_d)\dot{\boldsymbol{q}} + \boldsymbol{g}(\boldsymbol{q})\,.$$

a) Point out the difference with respect to the PD+ control law given by Equation (11.7)

b) Show that in reality, the previous controller is equivalent to the PD+ controller.

Hint: Use Property 4.2.

5. Verify Equation (11.6) by use of (11.5).

[3] Notice that since the task here is position control, in this case the controller is simply of type PD with gravity compensation.

12

Feedforward Control and PD Control plus Feedforward

The practical implementation of controllers for robot manipulators is typically carried out using digital technology. The way these control systems operate consists basically of the following stages:

- sampling of the joint position $\boldsymbol{q}$ (and of the velocity $\dot{\boldsymbol{q}}$);
- computation of the control action $\boldsymbol{\tau}$ from the control law;
- the 'order' to apply this control action is sent to the actuators.

In certain applications where it is required that the robot realize repetitive tasks at high velocity, the previous stages must be executed at a high cadence. The *bottleneck* in time-consumption terms, is the computation of the control action $\boldsymbol{\tau}$. Naturally, a reduction in the time for computation of $\boldsymbol{\tau}$ has the advantage of a higher processing frequency and hence a larger potential for the execution of 'fast' tasks. This is the main reason for the interest in controllers that require "little" computing power. In particular, this is the case for controllers that use information based on the desired positions, velocities, and accelerations $\boldsymbol{q}_d(t)$, $\dot{\boldsymbol{q}}_d(t)$, and $\ddot{\boldsymbol{q}}_d(t)$ respectively. Indeed, in repetitive tasks the desired position $\boldsymbol{q}_d(t)$ and its time derivatives happen to be vectorial periodic functions of time and moreover they are known once the task has been specified. Once the processing frequency has been established, the terms in the control law that depend exclusively on the form of these functions, may be computed and stored in memory, in a look-up table. During computation of the control action, these precomputed terms are simply *collected out* of memory, thereby reducing the computational burden.

In this chapter we consider two control strategies which have been suggested in the literature and which make wide use of precomputed terms in their respective control laws:

- feedforward control;
- PD control plus feedforward.

Each of these controllers is the subject of a section in the present chapter.

12.1 Feedforward Control

Among the conceptually simplest control strategies that may be used to control a dynamic system we find the so-called open-loop control, where the controller is simply the inverse dynamics model of the system evaluated along the desired reference trajectories.

For the case of linear dynamic systems, this control technique may be roughly sketched as follows. Consider the linear system described by

$$\dot{\boldsymbol{x}} = A\boldsymbol{x} + \boldsymbol{u}$$

where $\boldsymbol{x} \in \mathbb{R}^n$ is the state vector and at same time the *output* of the system, $A \in \mathbb{R}^{n\times n}$ is a matrix whose eigenvalues $\lambda_i\{A\}$ have negative real part, and $\boldsymbol{u} \in \mathbb{R}^n$ is the *input* to the system. Assume that we specify a vectorial function $\boldsymbol{x}_d$ as well as its time derivative $\dot{\boldsymbol{x}}_d$ to be bounded. The control goal is that $\boldsymbol{x}(t) \to \boldsymbol{x}_d(t)$ when $t \to \infty$. In other words, defining the errors vector $\tilde{\boldsymbol{x}} = \boldsymbol{x}_d - \boldsymbol{x}$, the control problem consists in designing a controller that allows one to determine the input $\boldsymbol{u}$ to the system so that $\lim_{t\to\infty} \tilde{\boldsymbol{x}}(t) = \boldsymbol{0}$. The solution to this control problem using the inverse dynamic model approach consists basically in substituting $\boldsymbol{x}$ and $\dot{\boldsymbol{x}}$ with $\boldsymbol{x}_d$ and $\dot{\boldsymbol{x}}_d$ in the equation of the system to control, and then solving for $\boldsymbol{u}$, *i.e.*

$$\boldsymbol{u} = \dot{\boldsymbol{x}}_d - A\boldsymbol{x}_d\,.$$

In this manner, the system formed by the linear system to control and the previous controller satisfies

$$\dot{\tilde{\boldsymbol{x}}} = A\tilde{\boldsymbol{x}}$$

which in turn is a linear system of the new state vector $\tilde{\boldsymbol{x}}$ and moreover we know from linear systems theory that since the eigenvalues of the matrix A have negative real parts, then $\lim_{t\to\infty} \tilde{\boldsymbol{x}}(t) = \boldsymbol{0}$ for all $\tilde{\boldsymbol{x}}(0) \in \mathbb{R}^n$.

In robot control, this strategy provides the supporting arguments to the following argument. If we apply a torque $\boldsymbol{\tau}$ at the input of the robot, the behavior of its outputs $\boldsymbol{q}$ and $\dot{\boldsymbol{q}}$ is governed by (III.1), *i.e.*

$$\frac{d}{dt}\begin{bmatrix}\boldsymbol{q}\\ \dot{\boldsymbol{q}}\end{bmatrix} = \begin{bmatrix}\dot{\boldsymbol{q}}\\ M(\boldsymbol{q})^{-1}\left[\boldsymbol{\tau} - C(\boldsymbol{q},\dot{\boldsymbol{q}})\dot{\boldsymbol{q}} - \boldsymbol{g}(\boldsymbol{q})\right]\end{bmatrix}. \tag{12.1}$$

If we wish that the behavior of the outputs $\boldsymbol{q}$ and $\dot{\boldsymbol{q}}$ be equal to that specified by $\boldsymbol{q}_d$ and $\dot{\boldsymbol{q}}_d$ respectively, it seems reasonable to replace $\boldsymbol{q}$, $\dot{\boldsymbol{q}}$ and $\ddot{\boldsymbol{q}}$ by $\boldsymbol{q}_d$, $\dot{\boldsymbol{q}}_d$, and $\ddot{\boldsymbol{q}}_d$ in the Equation (12.1) and to solve for $\boldsymbol{\tau}$. This reasoning leads to the equation of the feedforward controller, given by

$$\boldsymbol{\tau} = M(\boldsymbol{q}_d)\ddot{\boldsymbol{q}}_d + C(\boldsymbol{q}_d,\dot{\boldsymbol{q}}_d)\dot{\boldsymbol{q}}_d + \boldsymbol{g}(\boldsymbol{q}_d)\,. \tag{12.2}$$

Notice that the control action $\boldsymbol{\tau}$ does not depend on $\boldsymbol{q}$ nor on $\dot{\boldsymbol{q}}$, that is, it is an open loop control. Moreover, such a controller does not possess any design parameter. As with any other open-loop control strategy, this approach needs the precise knowledge of the dynamic system to be controlled, that is, of the dynamic model of the manipulator and specifically, of the structure of the matrices $M(\boldsymbol{q})$, $C(\boldsymbol{q}, \dot{\boldsymbol{q}})$ and of the vector $\boldsymbol{g}(\boldsymbol{q})$ as well as knowledge of their parameters (masses, inertias etc.). For this reason it is said that the feedforward control is (robot-) 'model-based'. The interest in a controller of this type resides in the advantages that it offers in implementation. Indeed, having determined $\boldsymbol{q}_d$, $\dot{\boldsymbol{q}}_d$ and $\ddot{\boldsymbol{q}}_d$ (in particular for repetitive tasks), one may determine the terms $M(\boldsymbol{q}_d)$, $C(\boldsymbol{q}_d, \dot{\boldsymbol{q}}_d)$ and $\boldsymbol{g}(\boldsymbol{q}_d)$ *off-line* and easily compute the control action $\boldsymbol{\tau}$ according to Equation (12.2). This motivates the qualifier "feedforward" in the name of this controller.

Nonetheless, one should not forget that a controller of this type has the intrinsic disadvantages of open-loop control systems, *e.g.* lack of robustness with respect to parametric and structural uncertainties, performance degradation in the presence of external perturbations, etc. In Figure 12.1 we present the block-diagram corresponding to a robot under feedforward control.

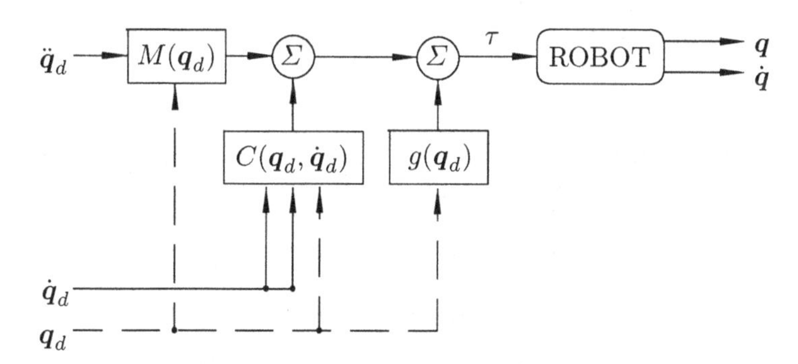

Figure 12.1. Block-diagram: feedforward control

The behavior of the control system is described by an equation obtained by substituting the equation of the controller (12.2) in the model of the robot (III.1), that is

$$M(\boldsymbol{q})\ddot{\boldsymbol{q}} + C(\boldsymbol{q}, \dot{\boldsymbol{q}})\dot{\boldsymbol{q}} + \boldsymbol{g}(\boldsymbol{q}) = M(\boldsymbol{q}_d)\ddot{\boldsymbol{q}}_d + C(\boldsymbol{q}_d, \dot{\boldsymbol{q}}_d)\dot{\boldsymbol{q}}_d + \boldsymbol{g}(\boldsymbol{q}_d)\,. \tag{12.3}$$

To avoid cumbersome notation in this chapter we use from now on and whenever it appears, the following notation

$$\begin{aligned} M &= M(\boldsymbol{q}) \\ M_d &= M(\boldsymbol{q}_d) \\ C &= C(\boldsymbol{q}, \dot{\boldsymbol{q}}) \end{aligned}$$

$$
\begin{aligned}
C_d &= C(\boldsymbol{q}_d, \dot{\boldsymbol{q}}_d) \\
\boldsymbol{g} &= \boldsymbol{g}(\boldsymbol{q}) \\
\boldsymbol{g}_d &= \boldsymbol{g}(\boldsymbol{q}_d) \,.
\end{aligned}
$$

Equation (12.3) may be written in terms of the state vector $\begin{bmatrix}\tilde{\boldsymbol{q}}^T & \dot{\tilde{\boldsymbol{q}}}^T\end{bmatrix}^T$ as

$$
\frac{d}{dt}\begin{bmatrix}\tilde{\boldsymbol{q}} \\ \dot{\tilde{\boldsymbol{q}}}\end{bmatrix} = \begin{bmatrix}\dot{\tilde{\boldsymbol{q}}} \\ -M^{-1}\left[(M_d - M)\ddot{\boldsymbol{q}}_d + C_d\dot{\boldsymbol{q}}_d - C\dot{\boldsymbol{q}} + \boldsymbol{g}_d - \boldsymbol{g}\right]\end{bmatrix},
$$

which represents an ordinary nonlinear nonautonomous differential equation. The origin $\begin{bmatrix}\tilde{\boldsymbol{q}}^T & \dot{\tilde{\boldsymbol{q}}}^T\end{bmatrix}^T = \boldsymbol{0} \in \mathbb{R}^{2n}$ is an equilibrium point of the previous equation but in general, it is not the only one. This is illustrated in the following examples.

Example 12.1. Consider the model of an ideal pendulum of length l with mass m concentrated at the tip and subject to the action of gravity g. Assume that a torque τ is applied at the rotating axis

$$
ml^2\ddot{q} + mgl\,\sin(q) = \tau
$$

where we identify $M(q) = ml^2$, $C(q,\dot{q}) = 0$ and $g(q) = mgl\,\sin(q)$. The feedforward controller (12.2), reduces to

$$
\tau = ml^2\ddot{q}_d + mgl\,\sin(q_d)\,.
$$

The behavior of the system is characterized by Equation (12.3),

$$
ml^2\ddot{q} + mgl\,\sin(q) = ml^2\ddot{q}_d + mgl\,\sin(q_d)
$$

or, in terms of the state $\begin{bmatrix}\tilde{q} & \dot{\tilde{q}}\end{bmatrix}^T$, by

$$
\frac{d}{dt}\begin{bmatrix}\tilde{q} \\ \dot{\tilde{q}}\end{bmatrix} = \begin{bmatrix}\dot{\tilde{q}} \\ -\frac{g}{l}\left[\sin(q_d) - \sin(q_d - \tilde{q})\right]\end{bmatrix}.
$$

Clearly the origin $\begin{bmatrix}\tilde{q} & \dot{\tilde{q}}\end{bmatrix}^T = \boldsymbol{0} \in \mathbb{R}^2$ is an equilibrium but so are the points $\begin{bmatrix}\tilde{q} & \dot{\tilde{q}}\end{bmatrix}^T = [2n\pi \;\; 0]^T$ for any integer value that n takes. ♢

The following example presents the study of the feedforward control of a 3-DOF Cartesian robot. The dynamic model of this manipulator is an innocuous linear system.

Example 12.2. Consider the 3-DOF Cartesian robot studied in Example 3.4 (see page 69) and shown in Figure 3.5. Its dynamic model is given by

$$
\begin{aligned}
[m_1 + m_2 + m_3]\ddot{q}_1 + [m_1 + m_2 + m_3]g &= \tau_1 \\
[m_1 + m_2]\, \ddot{q}_2 &= \tau_2 \\
m_1 \ddot{q}_3 &= \tau_3 \,,
\end{aligned}
$$

where we identify

$$
\begin{aligned}
M(\boldsymbol{q}) &= \begin{bmatrix} m_1 + m_2 + m_3 & 0 & 0 \\ 0 & m_1 + m_2 & 0 \\ 0 & 0 & m_1 \end{bmatrix} \\
C(\boldsymbol{q}, \dot{\boldsymbol{q}}) &= 0 \\
\boldsymbol{g}(\boldsymbol{q}) &= \begin{bmatrix} [m_1 + m_2 + m_3]g \\ 0 \\ 0 \end{bmatrix} .
\end{aligned}
$$

Notice that the dynamic model is characterized by a linear differential equation. The "closed-loop" equation[1] obtained with feedforward control is given by

$$
\frac{d}{dt} \begin{bmatrix} \tilde{\boldsymbol{q}} \\ \dot{\tilde{\boldsymbol{q}}} \end{bmatrix} = \begin{bmatrix} 0 & I \\ 0 & 0 \end{bmatrix} \begin{bmatrix} \tilde{\boldsymbol{q}} \\ \dot{\tilde{\boldsymbol{q}}} \end{bmatrix} ,
$$

which has an infinite number of non-isolated equilibria given by

$$
\begin{bmatrix} \tilde{\boldsymbol{q}}^T & \dot{\tilde{\boldsymbol{q}}}^T \end{bmatrix}^T = \begin{bmatrix} \tilde{\boldsymbol{q}}^T & \boldsymbol{0}^T \end{bmatrix}^T \in \mathbb{R}^{2n},
$$

where $\tilde{\boldsymbol{q}}$ is any vector in $\mathbb{R}^n$. Naturally, the origin is an equilibrium but it is not isolated. Consequently, this equilibrium (and actually any other) may not be asymptotically stable even locally. Moreover, due to the linear nature of the equation that characterizes the control system, it may be shown that in this case any equilibrium point is unstable (see problem 12.2). ♢

The previous examples makes it clear that multiple equilibria may coexist for the differential equation that characterizes the behavior of the control system. Moreover, due to the lack of design parameters in the controller, it is impossible to modify either the location or the number of equilibria, and even

[1] Here we write "closed-loop" in quotes since as a matter of fact the control system in itself is a system in open loop.

less, their stability properties, which are determined only by the dynamics of the manipulator. Obviously, a controller whose behavior in robot control has these features is of little utility in real applications. As a matter of fact, its use may yield catastrophic results in certain applications as we show in the following example.

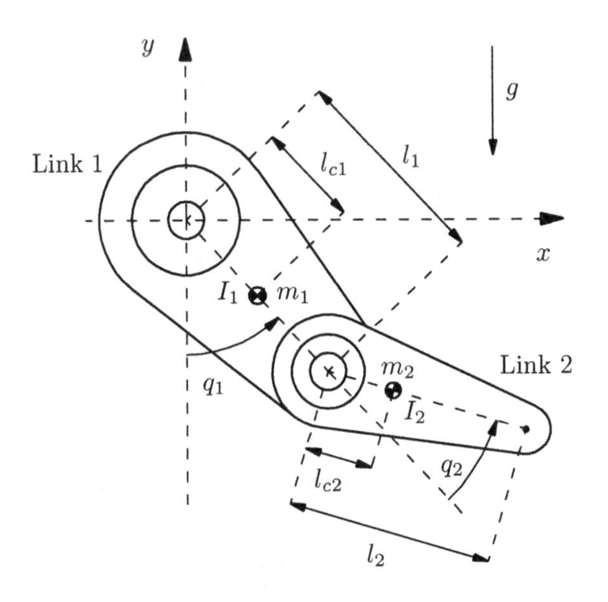

Figure 12.2. Diagram of the Pelican prototype

Example 12.3. Consider the 2-DOF *prototype robot* studied in Chapter 5, and shown in Figure 12.2.

Consider the feedforward control law (12.2) on this robot. The desired trajectory in joint space is given by $\boldsymbol{q}_d(t)$ which is defined in (5.7) and whose graph is depicted in Figure 5.5 (*cf.* page 129).

The initial conditions for positions and velocities are chosen as

$$\begin{aligned} q_1(0) &= 0, & q_2(0) &= 0 \\ \dot{q}_1(0) &= 0, & \dot{q}_2(0) &= 0\,. \end{aligned}$$

Figure 12.3 presents experimental results; it shows the components of position error $\tilde{\boldsymbol{q}}(t)$, which tend to a largely oscillatory behavior. Naturally, this behavior is far from satisfactory.

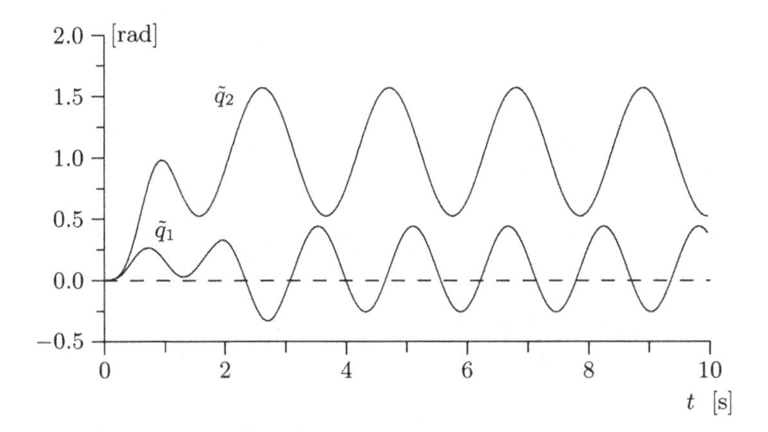

Figure 12.3. Graphs of position errors $\tilde{q}_1$ and $\tilde{q}_2$

So far, we have presented a series of examples that show negative features of the feedforward control given by (12.2). Naturally, these examples might discourage a formal study of stability of the origin as an equilibrium of the differential equation which models the behavior of this control system.

Moreover, a rigorous generic analysis of stability or instability seems to be an impossible task. While we presented in Example 12.2 the case when the origin of the equation which characterizes the control system is unstable, Problem 12.1 addresses the case in which the origin is a stable equilibrium.

The previous observations make it evident that feedforward control, given by (12.2), even with exact knowledge of the model of the robot, may be inadequate to achieve the motion control objective and even that of position control. Therefore, we may conclude that, in spite of the practical motivation to use feedforward control (12.2) should not be applied in robot control.

Feedforward control (12.2) may be modified by the addition, for example, of a *feedback* Proportional–Derivative (PD) term

$$\boldsymbol{\tau} = M(\boldsymbol{q}_d)\ddot{\boldsymbol{q}}_d + C(\boldsymbol{q}_d, \dot{\boldsymbol{q}}_d)\dot{\boldsymbol{q}}_d + \boldsymbol{g}(\boldsymbol{q}_d) + K_p\tilde{\boldsymbol{q}} + K_v\dot{\tilde{\boldsymbol{q}}} \tag{12.4}$$

where K_p and K_v are the gain matrices ($n \times n$) of position and velocity respectively. The controller (12.4) is now a closed-loop controller in view of the explicit feedback of $\boldsymbol{q}$ and $\dot{\boldsymbol{q}}$ used to compute $\tilde{\boldsymbol{q}}$ and $\dot{\tilde{\boldsymbol{q}}}$ respectively. The controller (12.4) is studied in the following section.

12.2 PD Control plus Feedforward

The wide practical interest in incorporating the smallest number of computations in real time to implement a robot controller has been the main motivation for the PD plus feedforward control law, given by

$$\boldsymbol{\tau} = K_p\tilde{\boldsymbol{q}} + K_v\dot{\tilde{\boldsymbol{q}}} + M(\boldsymbol{q}_d)\ddot{\boldsymbol{q}}_d + C(\boldsymbol{q}_d, \dot{\boldsymbol{q}}_d)\dot{\boldsymbol{q}}_d + \boldsymbol{g}(\boldsymbol{q}_d), \tag{12.5}$$

where $K_p, K_v \in \mathbb{R}^{n\times n}$ are symmetric positive definite matrices, called gains of position and velocity respectively. As is customary in this textbook, $\tilde{\boldsymbol{q}} = \boldsymbol{q}_d - \boldsymbol{q}$ stands for the position error. The term 'feedforward' in the name of the controller results from the fact that the control law uses the dynamics of the robot evaluated explicitly at the desired motion trajectory. In the control law (12.5), the centrifugal and Coriolis forces matrix, $C(\boldsymbol{q}, \dot{\boldsymbol{q}})$, is assumed to be computed via the Christoffel symbols (*cf.* Equation 3.21). This allows one to ensure that the matrix $\frac{1}{2}\dot{M}(\boldsymbol{q}) - C(\boldsymbol{q}, \dot{\boldsymbol{q}})$ is skew-symmetric, a property which is fundamental to the stability analysis of the closed-loop control system.

It is assumed that the manipulator has only revolute joints and that the upper-bounds on the norms of desired velocities and accelerations, denoted as $\|\dot{\boldsymbol{q}}_d\|_{\mathrm{M}}$ and $\|\ddot{\boldsymbol{q}}_d\|_{\mathrm{M}}$, are known.

The PD control law plus feedforward given by (12.5) may be regarded as a generalization of the PD control law with gravity precompensation (8.1). Figure 12.4 shows the block-diagram corresponding to the PD control law plus feedforward.

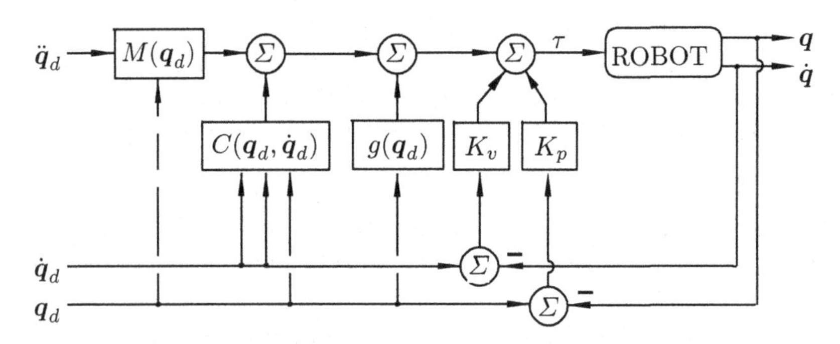

Figure 12.4. Block-diagram: PD control plus feedforward

Reported experiences in the literature of robot motion control using the control law (12.5) detail an excellent performance actually comparable with the performance of the popular *computed-torque* control law, which is presented in Chapter 10. Nevertheless, these comparison results may be misleading since good performance is not only due to the controller structure, but also to appropriate tuning of the controller gains.

The dynamics in closed loop is obtained by substituting the control action $\boldsymbol{\tau}$ from (12.5) in the equation of the robot model (III.1) to get

$$M(\boldsymbol{q})\ddot{\boldsymbol{q}} + C(\boldsymbol{q}, \dot{\boldsymbol{q}})\dot{\boldsymbol{q}} + \boldsymbol{g}(\boldsymbol{q}) = K_p\tilde{\boldsymbol{q}} + K_v\dot{\tilde{\boldsymbol{q}}} + M(\boldsymbol{q}_d)\ddot{\boldsymbol{q}}_d + C(\boldsymbol{q}_d, \dot{\boldsymbol{q}}_d)\dot{\boldsymbol{q}}_d + \boldsymbol{g}(\boldsymbol{q}_d). \tag{12.6}$$

The closed-loop Equation (12.6) may be written in terms of the state vector $\begin{bmatrix}\tilde{\boldsymbol{q}}^T & \dot{\tilde{\boldsymbol{q}}}^T\end{bmatrix}^T$ as

$$\frac{d}{dt}\begin{bmatrix}\tilde{\boldsymbol{q}} \\ \dot{\tilde{\boldsymbol{q}}}\end{bmatrix} = \begin{bmatrix}\dot{\tilde{\boldsymbol{q}}} \\ M(\boldsymbol{q})^{-1}\left[-K_p\tilde{\boldsymbol{q}} - K_v\dot{\tilde{\boldsymbol{q}}} - C(\boldsymbol{q},\dot{\boldsymbol{q}})\dot{\tilde{\boldsymbol{q}}} - \boldsymbol{h}(t,\tilde{\boldsymbol{q}},\dot{\tilde{\boldsymbol{q}}})\right]\end{bmatrix} \tag{12.7}$$

where we remind the reader that $\boldsymbol{h}(t,\tilde{\boldsymbol{q}},\dot{\tilde{\boldsymbol{q}}})$ is the so-called residual dynamics, given by

$$\boldsymbol{h}(t,\tilde{\boldsymbol{q}},\dot{\tilde{\boldsymbol{q}}}) = [M(\boldsymbol{q}_d) - M(\boldsymbol{q})]\ddot{\boldsymbol{q}}_d + [C(\boldsymbol{q}_d,\dot{\boldsymbol{q}}_d) - C(\boldsymbol{q},\dot{\boldsymbol{q}})]\dot{\boldsymbol{q}}_d + \boldsymbol{g}(\boldsymbol{q}_d) - \boldsymbol{g}(\boldsymbol{q})\,.$$

See (4.12).

It is simple to prove that the origin $[\tilde{\boldsymbol{q}}^T \ \ \dot{\tilde{\boldsymbol{q}}}^T]^T = \boldsymbol{0} \in \mathbb{R}^{2n}$ of the state space is an equilibrium, independently of the gain matrices K_p and K_v. However, the number of equilibria of the system in closed loop, *i.e.*(12.7), depends on the proportional gain K_p. This is formally studied in the following section.

12.2.1 Unicity of the Equilibrium

We present sufficient conditions on K_p that guarantee the existence of a unique equilibrium (the origin) for the closed-loop Equation (12.7).

For the case of robots having only revolute joints and with a sufficiently "large" choice of K_p, we can show that the origin $\begin{bmatrix}\tilde{\boldsymbol{q}}^T & \dot{\tilde{\boldsymbol{q}}}^T\end{bmatrix}^T = \boldsymbol{0} \in \mathbb{R}^{2n}$ is the unique equilibrium of the closed-loop Equation (12.7). Indeed, the equilibria are the constant vectors $[\tilde{\boldsymbol{q}}^T \ \ \dot{\tilde{\boldsymbol{q}}}^T]^T = [\tilde{\boldsymbol{q}}^{*T} \ \ \boldsymbol{0}^T]^T \in \mathbb{R}^{2n}$, where $\tilde{\boldsymbol{q}}^* \in \mathbb{R}^n$ is a solution of

$$K_p\tilde{\boldsymbol{q}}^* + \boldsymbol{h}(t,\tilde{\boldsymbol{q}}^*,\boldsymbol{0}) = \boldsymbol{0}\,. \tag{12.8}$$

The previous equation is always satisfied by the trivial solution $\tilde{\boldsymbol{q}}^* = \boldsymbol{0} \in \mathbb{R}^n$, but this does not exclude other vectors $\tilde{\boldsymbol{q}}^*$ from being solutions, depending of course, on the value of the proportional gain K_p. Explicit conditions on the proportional gain to ensure unicity of the equilibrium are presented next. To that end define

$$\boldsymbol{k}(\tilde{\boldsymbol{q}}^*) = K_p^{-1}\boldsymbol{h}(t,\tilde{\boldsymbol{q}}^*,\boldsymbol{0})\,.$$

The idea is to note that any fixed point $\tilde{\boldsymbol{q}}^* \in \mathbb{R}^n$ of $\boldsymbol{k}(\tilde{\boldsymbol{q}}^*)$ is a solution of (12.8). Hence, we wish to find conditions on K_p so that $\boldsymbol{k}(\tilde{\boldsymbol{q}}^*)$ has a unique fixed point. Given that $\tilde{\boldsymbol{q}}^* = \boldsymbol{0}$ is always a fixed point, then this shall be unique.

Notice that for all vectors $\boldsymbol{x},\boldsymbol{y} \in \mathbb{R}^n$, we have

$$\begin{aligned}\|\boldsymbol{k}(\boldsymbol{x})-\boldsymbol{k}(\boldsymbol{y})\| &\leq \left\|K_p^{-1}\left[\boldsymbol{h}(t,\boldsymbol{x},\boldsymbol{0})-\boldsymbol{h}(t,\boldsymbol{y},\boldsymbol{0})\right]\right\| \\ &\leq \lambda_{\mathrm{Max}}\left\{K_p^{-1}\right\}\|\boldsymbol{h}(t,\boldsymbol{x},\boldsymbol{0})-\boldsymbol{h}(t,\boldsymbol{y},\boldsymbol{0})\| \,.\end{aligned}$$

On the other hand, using the definition of the residual dynamics (4.12), we have

$$\begin{aligned}\|\boldsymbol{h}(t,\boldsymbol{x},\boldsymbol{0})-\boldsymbol{h}(t,\boldsymbol{y},\boldsymbol{0})\| \leq\ & \|[M(\boldsymbol{q}_d-\boldsymbol{y})-M(\boldsymbol{q}_d-\boldsymbol{x})]\,\ddot{\boldsymbol{q}}_d\| \\ &+\|[C(\boldsymbol{q}_d-\boldsymbol{y},\dot{\boldsymbol{q}}_d)-C(\boldsymbol{q}_d-\boldsymbol{x},\dot{\boldsymbol{q}}_d)]\,\dot{\boldsymbol{q}}_d\| \\ &+\|\boldsymbol{g}(\boldsymbol{q}_d-\boldsymbol{y})-\boldsymbol{g}(\boldsymbol{q}_d-\boldsymbol{x})\| \,.\end{aligned}$$

From Properties 4.1 to 4.3 we guarantee the existence of constants k_M, k_{C_1}, k_{C_2} and k_g, associated to the inertia matrix $M(\boldsymbol{q})$, to the matrix of centrifugal and Coriolis forces $C(\boldsymbol{q},\dot{\boldsymbol{q}})$, and to the vector of gravitational torques $\boldsymbol{g}(\boldsymbol{q})$ respectively, such that

$$\begin{aligned}\|M(\boldsymbol{x})\boldsymbol{z}-M(\boldsymbol{y})\boldsymbol{z}\| &\leq k_M\|\boldsymbol{x}-\boldsymbol{y}\|\ \|\boldsymbol{z}\|, \\ \|C(\boldsymbol{x},\boldsymbol{z})\boldsymbol{w}-C(\boldsymbol{y},\boldsymbol{v})\boldsymbol{w}\| &\leq k_{C1}\|\boldsymbol{z}-\boldsymbol{v}\|\ \|\boldsymbol{w}\| \\ &\quad +k_{C2}\|\boldsymbol{z}\|\ \|\boldsymbol{x}-\boldsymbol{y}\|\ \|\boldsymbol{w}\|, \\ \|\boldsymbol{g}(\boldsymbol{x})-\boldsymbol{g}(\boldsymbol{y})\| &\leq k_g\|\boldsymbol{x}-\boldsymbol{y}\|,\end{aligned}$$

for all $\boldsymbol{v}$, $\boldsymbol{w}$, $\boldsymbol{x}$, $\boldsymbol{y}$, $\boldsymbol{z}\in\mathbb{R}^n$. Taking into account this fact we obtain

$$\|\boldsymbol{h}(t,\boldsymbol{x},\boldsymbol{0})-\boldsymbol{h}(t,\boldsymbol{y},\boldsymbol{0})\| \leq \left[k_g+k_M\|\ddot{\boldsymbol{q}}_d\|_{\mathrm{M}}+k_{C_2}\|\dot{\boldsymbol{q}}_d\|_{\mathrm{M}}^2\right]\|\boldsymbol{x}-\boldsymbol{y}\| \,.$$

From this and using $\lambda_{\mathrm{Max}}\left\{K_p^{-1}\right\}=1/\lambda_{\min}\left\{K_p\right\}$, since K_p is a symmetric positive definite matrix, we get

$$\|\boldsymbol{k}(\boldsymbol{x})-\boldsymbol{k}(\boldsymbol{y})\| \leq \frac{1}{\lambda_{\min}\left\{K_p\right\}}\left[k_g+k_M\|\ddot{\boldsymbol{q}}_d\|_{\mathrm{M}}+k_{C_2}\|\dot{\boldsymbol{q}}_d\|_{\mathrm{M}}^2\right]\|\boldsymbol{x}-\boldsymbol{y}\| \,.$$

Finally, invoking the contraction mapping theorem (*cf.* Theorem 2.1 on page 26), we conclude that

$$\lambda_{\min}\left\{K_p\right\} > k_g+k_M\|\ddot{\boldsymbol{q}}_d\|_{\mathrm{M}}+k_{C_2}\|\dot{\boldsymbol{q}}_d\|_{\mathrm{M}}^2 \qquad (12.9)$$

is a sufficient condition for $\boldsymbol{k}(\tilde{\boldsymbol{q}}^*)$ to have a unique fixed point, and therefore, for the origin of the state space to be the unique equilibrium of the closed-loop system, *i.e.* Equation (12.7).

As has been shown before, the PD control law plus feedforward, (12.5), reduces to control with desired gravity compensation (8.1) in the case when the desired position $\boldsymbol{q}_d$ is constant. For this last controller, we shown in Section 8.2 that the corresponding closed-loop equation had a unique equilibrium if $\lambda_{\min}\{K_p\}>k_g$. It is interesting to remark that when $\boldsymbol{q}_d$ is constant we recover from (12.9), the previous condition for unicity.

Example 12.4. Consider the model of an ideal pendulum of length l with mass m concentrated at its tip and subject to the action of gravity g. Assume that a torque τ is applied at the axis of rotation, that is

$$ml^2\ddot{q} + mgl\,\sin(q) = \tau\,.$$

The PD control law plus feedforward, (12.5), is in this case

$$\tau = k_p\tilde{q} + k_v\dot{\tilde{q}} + ml^2\ddot{q}_d + mgl\,\sin(q_d)$$

where k_p and k_v are positive design constants. The closed-loop equation is

$$\frac{d}{dt}\begin{bmatrix}\tilde{q}\\ \dot{\tilde{q}}\end{bmatrix} = \begin{bmatrix}\dot{\tilde{q}}\\ -\frac{1}{ml^2}\left[k_p\tilde{q} + k_v\dot{\tilde{q}} + mgl\;[\sin(q_d) - \sin(q_d - \tilde{q})]\right]\end{bmatrix}$$

which has an equilibrium at the origin $\begin{bmatrix}\tilde{q} & \dot{\tilde{q}}\end{bmatrix}^T = \mathbf{0} \in \mathrm{I\!R}^2$. If $q_d(t)$ is constant, there may exist additional equilibria $\begin{bmatrix}\tilde{q} & \dot{\tilde{q}}\end{bmatrix}^T = [\tilde{q}^* \;\; 0]^T \in \mathrm{I\!R}^2$ where $\tilde{q}^*$ is a solution of

$$k_p\tilde{q}^* + mgl\;[\sin(q_d) - \sin(q_d - \tilde{q}^*)] = 0\,.$$

Example 8.1 shows the case when the previous equation has three solutions. For the same example, if k_p is sufficiently large, it was shown that $\tilde{q}^* = 0$ is the unique solution.

We stress that according to Theorem 2.6, if there exist more than one equilibrium, then none of them may be globally uniformly asymptotically stable. ♢

12.2.2 Global Uniform Asymptotic Stability

In this section we present the analysis of the closed-loop Equation (12.6) or equivalently, of (12.7). In this analysis we establish conditions on the design matrices K_p and K_v that guarantee global uniform asymptotic stability of the origin of the state space corresponding to the closed-loop equation. We assume that the symmetric positive definite matrix K_p is also diagonal.

Before studying the stability of the origin $\begin{bmatrix}\tilde{\boldsymbol{q}}^T & \dot{\tilde{\boldsymbol{q}}}^T\end{bmatrix}^T = \mathbf{0} \in \mathrm{I\!R}^{2n}$ of the closed-loop Equation (12.6) or (12.7), it is worth recalling Definition 4.1 of the vectorial tangent hyperbolic function which has the form given in (4.13), *i.e.*

$$\mathbf{tanh}(\boldsymbol{x}) = \begin{bmatrix} \tanh(x_1) \\ \vdots \\ \tanh(x_n) \end{bmatrix} \tag{12.10}$$

with $\boldsymbol{x} \in \mathbb{R}^n$. As stated in Definition 4.1, this function satisfies the following properties for all $\boldsymbol{x}, \dot{\boldsymbol{x}} \in \mathbb{R}^n$

- $\|\mathbf{tanh}(\boldsymbol{x})\| \leq \alpha_1 \|\boldsymbol{x}\|$
- $\|\mathbf{tanh}(\boldsymbol{x})\| \leq \alpha_2$
- $\|\mathbf{tanh}(\boldsymbol{x})\|^2 \leq \alpha_3\, \mathbf{tanh}(\boldsymbol{x})^T \boldsymbol{x}$
- $\left\|\mathrm{Sech}^2(\boldsymbol{x})\dot{\boldsymbol{x}}\right\| \leq \alpha_4 \|\dot{\boldsymbol{x}}\|$

with $\alpha_1, \cdots, \alpha_4 > 0$. For $\mathbf{tanh}(\boldsymbol{x})$ defined as in (4.13), the constants involved are taken as $\alpha_1 = 1, \alpha_2 = \sqrt{n}, \alpha_3 = 1, \alpha_4 = 1$.

In the sequel we assume that given a constant $\gamma > 0$, the matrix K_v is chosen sufficiently "large" in the sense that

$$\lambda_{\mathrm{Max}}\{K_v\} \geq \lambda_{\min}\{K_v\} > k_{h1} + \gamma\, b\,, \tag{12.11}$$

and so is K_p but in the sense that

$$\lambda_{\mathrm{Max}}\{K_p\} \geq \lambda_{\min}\{K_p\} > \alpha_3 \left[\frac{\left[2\,\gamma\, a + k_{h2}\right]^2}{4\,\gamma\left[\lambda_{\min}\{K_v\} - k_{h1} - \gamma b\right]} + k_{h2} \right] \tag{12.12}$$

so that

$$\lambda_{\mathrm{Max}}\{K_p\} \geq \lambda_{\min}\{K_p\} > \gamma^2 \frac{\alpha_1^2\, \lambda_{\mathrm{Max}}^2\{M\}}{\lambda_{\min}\{M\}} \tag{12.13}$$

where k_{h1} and k_{h2} are defined in (4.25) and (4.24), while the constants a and b are given by

$$\begin{aligned} a &= \frac{1}{2}\left[\lambda_{\mathrm{Max}}\{K_v\} + k_{C_1} \|\dot{\boldsymbol{q}}_d\|_{\mathrm{M}} + k_{h1}\right], \\ b &= \alpha_4\, \lambda_{\mathrm{Max}}\{M\} + \alpha_2\, k_{C_1}\,. \end{aligned}$$

Lyapunov Function Candidate

Consider the Lyapunov function candidate[2] (7.3),

$$V(t, \tilde{\boldsymbol{q}}, \dot{\tilde{\boldsymbol{q}}}) = \frac{1}{2}\dot{\tilde{\boldsymbol{q}}}^T M(\boldsymbol{q})\dot{\tilde{\boldsymbol{q}}} + \frac{1}{2}\tilde{\boldsymbol{q}}^T K_p \tilde{\boldsymbol{q}} + \gamma\mathbf{tanh}(\tilde{\boldsymbol{q}})^T M(\boldsymbol{q})\dot{\tilde{\boldsymbol{q}}} \tag{12.14}$$

[2] Notice that $V = V(t, \tilde{\boldsymbol{q}}, \dot{\tilde{\boldsymbol{q}}})$. The dependence of t comes from the fact that, to avoid cumbersome notation, we have abbreviated $M(\boldsymbol{q}_d(t) - \tilde{\boldsymbol{q}})$ to $M(\boldsymbol{q})$.

where $\mathbf{tanh}(\tilde{\boldsymbol{q}})$ is the vectorial tangent hyperbolic function (12.10) and $\gamma > 0$ is a given constant.

To show that the Lyapunov function candidate (12.14) is positive definite and radially unbounded, we first observe that the third term in (12.14) satisfies

$$\begin{aligned}\gamma \mathbf{tanh}(\tilde{\boldsymbol{q}})^T M(\boldsymbol{q})\dot{\tilde{\boldsymbol{q}}} &\leq \gamma \left\|\mathbf{tanh}(\tilde{\boldsymbol{q}})\right\| \left\|M(\boldsymbol{q})\dot{\tilde{\boldsymbol{q}}}\right\| \\ &\leq \gamma\ \lambda_{\text{Max}}\{M\} \left\|\mathbf{tanh}(\tilde{\boldsymbol{q}})\right\| \left\|\dot{\tilde{\boldsymbol{q}}}\right\| \\ &\leq \gamma\ \alpha_1 \lambda_{\text{Max}}\{M\} \left\|\tilde{\boldsymbol{q}}\right\| \left\|\dot{\tilde{\boldsymbol{q}}}\right\|\end{aligned}$$

where we used $\|\mathbf{tanh}(\tilde{\boldsymbol{q}})\| \leq \alpha_1 \|\tilde{\boldsymbol{q}}\|$ in the last step. From this we obtain

$$-\gamma \mathbf{tanh}(\tilde{\boldsymbol{q}})^T M(\boldsymbol{q})\dot{\tilde{\boldsymbol{q}}} \geq -\gamma\ \alpha_1 \lambda_{\text{Max}}\{M\} \left\|\tilde{\boldsymbol{q}}\right\| \left\|\dot{\tilde{\boldsymbol{q}}}\right\| .$$

Therefore, the Lyapunov function candidate (12.14) satisfies the following inequality:

$$V(t, \tilde{\boldsymbol{q}}, \dot{\tilde{\boldsymbol{q}}}) \geq \frac{1}{2} \begin{bmatrix} \|\tilde{\boldsymbol{q}}\| \\ \left\|\dot{\tilde{\boldsymbol{q}}}\right\| \end{bmatrix}^T \begin{bmatrix} \lambda_{\min}\{K_p\} & -\gamma\ \alpha_1\ \lambda_{\text{Max}}\{M\} \\ -\gamma\ \alpha_1\ \lambda_{\text{Max}}\{M\} & \lambda_{\min}\{M\} \end{bmatrix} \begin{bmatrix} \|\tilde{\boldsymbol{q}}\| \\ \left\|\dot{\tilde{\boldsymbol{q}}}\right\| \end{bmatrix}$$

and consequently, it happens to be positive definite and radially unbounded since by assumption, K_p is positive definite (*i.e.* $\lambda_{\min}\{K_p\} > 0$) and we also supposed that it is chosen so as to satisfy (12.13).

Following similar steps to those above one may also show that the Lyapunov function candidate $V(t, \tilde{\boldsymbol{q}}, \dot{\tilde{\boldsymbol{q}}})$ defined in (12.14) is bounded from above by

$$V(t, \tilde{\boldsymbol{q}}, \dot{\tilde{\boldsymbol{q}}}) \leq \frac{1}{2} \begin{bmatrix} \|\tilde{\boldsymbol{q}}\| \\ \left\|\dot{\tilde{\boldsymbol{q}}}\right\| \end{bmatrix}^T \begin{bmatrix} \lambda_{\text{Max}}\{K_p\} & \gamma\ \alpha_1\ \lambda_{\text{Max}}\{M\} \\ \gamma\ \alpha_1\ \lambda_{\text{Max}}\{M\} & \lambda_{\text{Max}}\{M\} \end{bmatrix} \begin{bmatrix} \|\tilde{\boldsymbol{q}}\| \\ \left\|\dot{\tilde{\boldsymbol{q}}}\right\| \end{bmatrix}$$

which is positive definite and radially unbounded since the condition

$$\lambda_{\text{Max}}\{K_p\} > \gamma^2 \alpha_1^2\ \lambda_{\text{Max}}\{M\}$$

is trivially satisfied under hypothesis (12.13) on K_p. This means that $V(t, \tilde{\boldsymbol{q}}, \dot{\tilde{\boldsymbol{q}}})$ is decrescent.

Time Derivative

The time derivative of the Lyapunov function candidate (12.14) along the trajectories of the closed-loop system (12.7) is

$$\dot{V}(t,\tilde{\boldsymbol{q}},\dot{\tilde{\boldsymbol{q}}}) = \dot{\tilde{\boldsymbol{q}}}^T \left[-K_p\tilde{\boldsymbol{q}} - K_v\dot{\tilde{\boldsymbol{q}}} - C(\boldsymbol{q},\dot{\boldsymbol{q}})\dot{\tilde{\boldsymbol{q}}} - \boldsymbol{h}(t,\tilde{\boldsymbol{q}},\dot{\tilde{\boldsymbol{q}}})\right] + \frac{1}{2}\dot{\tilde{\boldsymbol{q}}}^T \dot{M}(\boldsymbol{q})\dot{\tilde{\boldsymbol{q}}}$$
$$+ \tilde{\boldsymbol{q}}^T K_p\dot{\tilde{\boldsymbol{q}}} + \gamma\dot{\tilde{\boldsymbol{q}}}^T \mathrm{Sech}^2(\tilde{\boldsymbol{q}})^T M(\boldsymbol{q})\dot{\tilde{\boldsymbol{q}}} + \gamma\mathbf{tanh}(\tilde{\boldsymbol{q}})^T \dot{M}(\boldsymbol{q})\dot{\tilde{\boldsymbol{q}}}$$
$$+ \gamma\mathbf{tanh}(\tilde{\boldsymbol{q}})^T \left[-K_p\tilde{\boldsymbol{q}} - K_v\dot{\tilde{\boldsymbol{q}}} - C(\boldsymbol{q},\dot{\boldsymbol{q}})\dot{\tilde{\boldsymbol{q}}} - \boldsymbol{h}(t,\tilde{\boldsymbol{q}},\dot{\tilde{\boldsymbol{q}}})\right] .$$

Using Property 4.2 which establishes the skew-symmetry of $\frac{1}{2}\dot{M} - C$ and $\dot{M}(\boldsymbol{q}) = C(\boldsymbol{q},\dot{\boldsymbol{q}}) + C(\boldsymbol{q},\dot{\boldsymbol{q}})^T$, the time derivative of the Lyapunov function candidate yields

$$\dot{V}(t,\tilde{\boldsymbol{q}},\dot{\tilde{\boldsymbol{q}}}) = -\dot{\tilde{\boldsymbol{q}}}^T K_v\dot{\tilde{\boldsymbol{q}}} + \gamma\dot{\tilde{\boldsymbol{q}}}^T \mathrm{Sech}^2(\tilde{\boldsymbol{q}})^T M(\boldsymbol{q})\dot{\tilde{\boldsymbol{q}}} - \gamma\mathbf{tanh}(\tilde{\boldsymbol{q}})^T K_p\tilde{\boldsymbol{q}}$$
$$- \gamma\mathbf{tanh}(\tilde{\boldsymbol{q}})^T K_v\dot{\tilde{\boldsymbol{q}}} + \gamma\mathbf{tanh}(\tilde{\boldsymbol{q}})^T C(\boldsymbol{q},\dot{\boldsymbol{q}})^T\dot{\tilde{\boldsymbol{q}}}$$
$$- \dot{\tilde{\boldsymbol{q}}}^T \boldsymbol{h}(t,\tilde{\boldsymbol{q}},\dot{\tilde{\boldsymbol{q}}}) - \gamma\ \mathbf{tanh}(\tilde{\boldsymbol{q}})^T \boldsymbol{h}(t,\tilde{\boldsymbol{q}},\dot{\tilde{\boldsymbol{q}}}) . \tag{12.15}$$

We now proceed to upper-bound $\dot{V}(t,\tilde{\boldsymbol{q}},\dot{\tilde{\boldsymbol{q}}})$ by a negative definite function in terms of the states $\tilde{\boldsymbol{q}}$ and $\dot{\tilde{\boldsymbol{q}}}$. To that end, it is convenient to find upper-bounds for each term of (12.15).

The first term of (12.15) may be trivially bounded by

$$-\dot{\tilde{\boldsymbol{q}}}^T K_v\dot{\tilde{\boldsymbol{q}}} \leq -\lambda_{\min}\{K_v\}\|\dot{\tilde{\boldsymbol{q}}}\|^2 . \tag{12.16}$$

To upper-bound the second term of (12.15) we first recall that the vectorial tangent hyperbolic function $\mathbf{tanh}(\tilde{\boldsymbol{q}})$ defined in (12.10) satisfies $\left\|\mathrm{Sech}^2(\tilde{\boldsymbol{q}})\dot{\tilde{\boldsymbol{q}}}\right\| \leq \alpha_4 \left\|\dot{\tilde{\boldsymbol{q}}}\right\|$ with $\alpha_4 > 0$. From this, it follows that

$$\gamma\dot{\tilde{\boldsymbol{q}}}^T \mathrm{Sech}^2(\tilde{\boldsymbol{q}})^T M(\boldsymbol{q})\dot{\tilde{\boldsymbol{q}}} \leq \gamma\alpha_4\ \lambda_{\mathrm{Max}}\{M\} \left\|\dot{\tilde{\boldsymbol{q}}}\right\|^2 .$$

On the other hand, notice that in view of the fact that K_p is a diagonal positive definite matrix, and $\|\mathbf{tanh}(\tilde{\boldsymbol{q}})\|^2 \leq \alpha_3\ \mathbf{tanh}(\tilde{\boldsymbol{q}})^T\tilde{\boldsymbol{q}}$, we get

$$\gamma\ \alpha_3\ \mathbf{tanh}(\tilde{\boldsymbol{q}})^T K_p\tilde{\boldsymbol{q}} \geq \gamma\lambda_{\min}\{K_p\}\ \|\mathbf{tanh}(\tilde{\boldsymbol{q}})\|^2$$

which finally leads to the important inequality,

$$-\gamma\mathbf{tanh}(\tilde{\boldsymbol{q}})^T K_p\tilde{\boldsymbol{q}} \leq -\gamma\frac{\lambda_{\min}\{K_p\}}{\alpha_3}\ \|\mathbf{tanh}(\tilde{\boldsymbol{q}})\|^2 .$$

A bound on $\gamma\mathbf{tanh}(\tilde{\boldsymbol{q}})^T K_v\dot{\tilde{\boldsymbol{q}}}$ is obtained straightforwardly and is given by

$$\gamma\mathbf{tanh}(\tilde{\boldsymbol{q}})^T K_v\dot{\tilde{\boldsymbol{q}}} \leq \gamma\lambda_{\mathrm{Max}}\{K_v\} \left\|\dot{\tilde{\boldsymbol{q}}}\right\|\ \|\mathbf{tanh}(\tilde{\boldsymbol{q}})\| .$$

The upper-bound on the term $\gamma\mathbf{tanh}(\tilde{\boldsymbol{q}})^T C(\boldsymbol{q},\dot{\boldsymbol{q}})^T\dot{\tilde{\boldsymbol{q}}}$ must be carefully chosen. Notice that

$$\gamma \mathbf{tanh}(\tilde{\boldsymbol{q}})^T C(\boldsymbol{q}, \dot{\boldsymbol{q}})^T \dot{\tilde{\boldsymbol{q}}} = \gamma \dot{\tilde{\boldsymbol{q}}}^T C(\boldsymbol{q}, \dot{\boldsymbol{q}}) \mathbf{tanh}(\tilde{\boldsymbol{q}}) \\ \leq \gamma \left\| \dot{\tilde{\boldsymbol{q}}} \right\| \left\| C(\boldsymbol{q}, \dot{\boldsymbol{q}}) \mathbf{tanh}(\tilde{\boldsymbol{q}}) \right\| .$$

Considering again Property 4.2 but in its variant that establishes the existence of a constant k_{C_1} such that $\|C(\boldsymbol{q}, \boldsymbol{x})\boldsymbol{y}\| \leq k_{C_1} \|\boldsymbol{x}\| \|\boldsymbol{y}\|$ for all $\boldsymbol{q}, \boldsymbol{x}, \boldsymbol{y} \in \mathbb{R}^n$, we have

$$\begin{aligned} \gamma \mathbf{tanh}(\tilde{\boldsymbol{q}})^T C(\boldsymbol{q}, \dot{\boldsymbol{q}})^T \dot{\tilde{\boldsymbol{q}}} &\leq \gamma k_{C_1} \left\| \dot{\tilde{\boldsymbol{q}}} \right\| \|\dot{\boldsymbol{q}}\| \|\mathbf{tanh}(\tilde{\boldsymbol{q}})\| , \\ &\leq \gamma k_{C_1} \left\| \dot{\tilde{\boldsymbol{q}}} \right\| \left\| \dot{\boldsymbol{q}}_d - \dot{\tilde{\boldsymbol{q}}} \right\| \|\mathbf{tanh}(\tilde{\boldsymbol{q}})\| , \\ &\leq \gamma k_{C_1} \left\| \dot{\tilde{\boldsymbol{q}}} \right\| \|\dot{\boldsymbol{q}}_d\| \|\mathbf{tanh}(\tilde{\boldsymbol{q}})\| \\ &\qquad + \gamma k_{C_1} \left\| \dot{\tilde{\boldsymbol{q}}} \right\| \left\| \dot{\tilde{\boldsymbol{q}}} \right\| \|\mathbf{tanh}(\tilde{\boldsymbol{q}})\| . \end{aligned}$$

Making use of the property that $\|\mathbf{tanh}(\tilde{\boldsymbol{q}})\| \leq \alpha_2$ for all $\tilde{\boldsymbol{q}} \in \mathbb{R}^n$, we get

$$\gamma \mathbf{tanh}(\tilde{\boldsymbol{q}})^T C(\boldsymbol{q}, \dot{\boldsymbol{q}})^T \dot{\tilde{\boldsymbol{q}}} \leq \gamma \, k_{C_1} \|\dot{\boldsymbol{q}}_d\|_{\mathrm{M}} \left\| \dot{\tilde{\boldsymbol{q}}} \right\| \|\mathbf{tanh}(\tilde{\boldsymbol{q}})\| + \gamma \alpha_2 \, k_{C_1} \left\| \dot{\tilde{\boldsymbol{q}}} \right\|^2 .$$

At this point it is only left to find upper-bounds on the two terms which contain $\boldsymbol{h}(t, \tilde{\boldsymbol{q}}, \dot{\tilde{\boldsymbol{q}}})$. This study is based on the use of the characteristics established in Property 4.4 on the vector of residual dynamics $\boldsymbol{h}(t, \tilde{\boldsymbol{q}}, \dot{\tilde{\boldsymbol{q}}})$, which indicates the existence of constants $k_{h1}, k_{h2} \geq 0$ – which may be computed by (4.24) and (4.25) – such that the norm of the residual dynamics satisfies (4.15),

$$\left\| \boldsymbol{h}(t, \tilde{\boldsymbol{q}}, \dot{\tilde{\boldsymbol{q}}}) \right\| \leq k_{h1} \left\| \dot{\tilde{\boldsymbol{q}}} \right\| + k_{h2} \|\mathbf{tanh}(\tilde{\boldsymbol{q}})\| .$$

First, we study the term $-\dot{\tilde{\boldsymbol{q}}}^T \boldsymbol{h}(t, \tilde{\boldsymbol{q}}, \dot{\tilde{\boldsymbol{q}}})$:

$$\begin{aligned} -\dot{\tilde{\boldsymbol{q}}}^T \boldsymbol{h}(t, \tilde{\boldsymbol{q}}, \dot{\tilde{\boldsymbol{q}}}) &\leq \left\| \dot{\tilde{\boldsymbol{q}}} \right\| \left\| \boldsymbol{h}(t, \tilde{\boldsymbol{q}}, \dot{\tilde{\boldsymbol{q}}}) \right\| , \\ &\leq k_{h1} \left\| \dot{\tilde{\boldsymbol{q}}} \right\|^2 + k_{h2} \left\| \dot{\tilde{\boldsymbol{q}}} \right\| \|\mathbf{tanh}(\tilde{\boldsymbol{q}})\| . \end{aligned}$$

The remaining term satisfies

$$\begin{aligned} -\gamma \mathbf{tanh}(\tilde{\boldsymbol{q}})^T \boldsymbol{h}(t, \tilde{\boldsymbol{q}}, \dot{\tilde{\boldsymbol{q}}}) &\leq \gamma \|\mathbf{tanh}(\tilde{\boldsymbol{q}})\| \left\| \boldsymbol{h}(t, \tilde{\boldsymbol{q}}, \dot{\tilde{\boldsymbol{q}}}) \right\| , \\ &\leq \gamma \, k_{h1} \left\| \dot{\tilde{\boldsymbol{q}}} \right\| \|\mathbf{tanh}(\tilde{\boldsymbol{q}})\| + \gamma \, k_{h2} \|\mathbf{tanh}(\tilde{\boldsymbol{q}})\|^2 . \end{aligned} \tag{12.17}$$

The bounds (12.16)–(12.17) yield that the time derivative $\dot{V}(t, \tilde{\boldsymbol{q}}, \dot{\tilde{\boldsymbol{q}}})$ in (12.15), satisfies

$$
\dot{V}(t,\tilde{\boldsymbol{q}},\dot{\boldsymbol{q}}) \leq
-\gamma \begin{bmatrix} \|\mathbf{tanh}(\tilde{\boldsymbol{q}})\| \\ \|\dot{\boldsymbol{q}}\| \end{bmatrix}^T \underbrace{\begin{bmatrix} \frac{\lambda_{\min}\{K_p\}}{\alpha_3} - k_{h2} & -a - \frac{1}{\gamma}\frac{k_{h2}}{2} \\ -a - \frac{1}{\gamma}\frac{k_{h2}}{2} & \frac{1}{\gamma}\left[\lambda_{\min}\{K_v\} - k_{h1}\right] - b \end{bmatrix}}_{R(\gamma)} \begin{bmatrix} \|\mathbf{tanh}(\tilde{\boldsymbol{q}})\| \\ \|\dot{\boldsymbol{q}}\| \end{bmatrix}
\tag{12.18}
$$

where

$$
\begin{aligned}
a &= \frac{1}{2}\left[\lambda_{\mathrm{Max}}\{K_v\} + k_{C_1}\,\|\dot{\boldsymbol{q}}_d\|_{\mathrm{M}} + k_{h1}\right], \\
b &= \alpha_4\ \lambda_{\mathrm{Max}}\{M\} + \alpha_2\ k_{C_1}.
\end{aligned}
$$

According to the theorem of Sylvester, in order for the matrix $R(\gamma)$ to be positive definite it is necessary and sufficient that the component R_{11} and the determinant $\det\{R(\gamma)\}$ be strictly positive. With respect to the first condition we stress that the gain K_p must satisfy

$$
\lambda_{\min}\{K_p\} \geq \alpha_3\ k_{h2}\,. \tag{12.19}
$$

On the other hand, the determinant of $R(\gamma)$ is given by

$$
\begin{aligned}
\det\{R(\gamma)\} = \frac{1}{\gamma}&\left[\frac{\lambda_{\min}\{K_p\}}{\alpha_3} - k_{h2}\right]\left[\lambda_{\min}\{K_v\} - k_{h1}\right] \\
&- \left[\frac{\lambda_{\min}\{K_p\}}{\alpha_3} - k_{h2}\right] b - \left[a + \frac{1}{\gamma}\frac{k_{h2}}{2}\right]^2.
\end{aligned}
$$

The latter must be strictly positive for which it is necessary and sufficient that the gain K_p satisfies

$$
\lambda_{\min}\{K_p\} > \alpha_3\left[\frac{\left[2\ \gamma\ a + k_{h2}\right]^2}{4\ \gamma\left[\lambda_{\min}\{K_v\} - k_{h1} - \gamma b\right]} + k_{h2}\right] \tag{12.20}
$$

while it is sufficient that K_v satisfies

$$
\lambda_{\min}\{K_v\} > k_{h1} + \gamma\ b \tag{12.21}
$$

for the right-hand side of the inequality (12.20) to be positive. Observe that in this case the inequality (12.19) is trivially implied by (12.20).

Notice that the inequalities (12.21) and (12.20) correspond precisely to those in (12.11) and (12.12) as the tuning guidelines for the controller. This means that $R(\gamma)$ is positive definite and therefore, $\dot{V}(t,\tilde{\boldsymbol{q}},\dot{\tilde{\boldsymbol{q}}})$ is globally negative definite.

According to the arguments above, given a positive constant γ we may determine gains K_p and K_v according to (12.11)–(12.13) in a way that the function $V(t, \tilde{\boldsymbol{q}}, \dot{\boldsymbol{q}})$ given by (12.14) is globally positive definite while $\dot{V}(t, \tilde{\boldsymbol{q}}, \dot{\boldsymbol{q}})$ expressed as (12.18) is globally negative definite. For this reason, $V(t, \tilde{\boldsymbol{q}}, \dot{\boldsymbol{q}})$ is a strict Lyapunov function. Theorem 2.4 allows one to establish global uniform asymptotic stability of the origin of the closed-loop system.

Tuning Procedure

The stability analysis presented in previous sections allows one to obtain a tuning procedure for the PD control law plus feedforward. This method determines the smallest eigenvalues of the symmetric design matrices K_p and K_v – with K_p diagonal – which guarantee the achievement of the motion control objective.

The tuning procedure may be summarized as follows.

- Derivation of the dynamic robot model to be controlled. Particularly, computation of $M(\boldsymbol{q})$, $C(\boldsymbol{q}, \dot{\boldsymbol{q}})$ and $\boldsymbol{g}(\boldsymbol{q})$ in closed form.
- Computation of the constants $\lambda_{\rm Max}\{M(\boldsymbol{q})\}$, $\lambda_{\min}\{M(\boldsymbol{q})\}$, k_M, k'_M,k_{C_1}, k_{C_2}, k' and k_g. For this, it is suggested that the information given in Table 4.1 (*cf.* page 109) is used.
- Computation of $\|\ddot{\boldsymbol{q}}_d\|_{\rm Max}$, $\|\dot{\boldsymbol{q}}_d\|_{\rm Max}$ from the specification of a given task to the robot.
- Computation of the constants s_1 and s_2 given respectively by (4.21) and (4.22), *i.e.*

$$s_1 = \left[k_g + k_M \|\ddot{\boldsymbol{q}}_d\|_{\rm M} + k_{C_2} \|\dot{\boldsymbol{q}}_d\|_{\rm M}^2\right],$$

and

$$s_2 = 2\left[k' + k'_M \|\ddot{\boldsymbol{q}}_d\|_{\rm M} + k_{C_1} \|\dot{\boldsymbol{q}}_d\|_{\rm M}^2\right].$$

Computation of k_{h1} and k_{h2} given by (4.24) and (4.25), *i.e.*

  - $k_{h1} \geq k_{C_1} \|\dot{\boldsymbol{q}}_d\|_{\rm M}$;
  - $k_{h2} \geq \dfrac{s_2}{\tanh\left(\frac{s_2}{s_1}\right)}$.

- Computation of the constants a and b given by

$$a = \frac{1}{2}\left[\lambda_{\rm Max}\{K_v\} + k_{C_1} \|\dot{\boldsymbol{q}}_d\|_{\rm M} + k_{h1}\right],$$
$$b = \alpha_4\, \lambda_{\rm Max}\{M\} + \alpha_2\, k_{C_1},$$

  where $\alpha_2 = \sqrt{n}, \alpha_4 = 1$.
- Select $\gamma > 0$ and determine the design matrices K_p and K_v so that their smallest eigenvalues satisfy (12.11)–(12.13), *i.e.*
  - $\lambda_{\min}\{K_v\} > k_{h1} + \gamma\, b$,

- $\lambda_{\min}\{K_p\} > \alpha_3 \left[\frac{[2\,\gamma\,a + k_{h2}]^2}{4\,\gamma\,[\lambda_{\min}\{K_v\} - k_{h1} - \gamma b]} + k_{h2} \right]$,
- $\lambda_{\min}\{K_p\} > \gamma^2 \frac{\alpha_1^2\,\lambda_{\mathrm{Max}}^2\{M\}}{\lambda_{\min}\{M\}}$,

with $\alpha_1 = 1, \alpha_3 = 1$.

Next, we present an example in order to illustrate the ideas presented so far.

Example 12.5. Consider the Pelican prototype robot shown in Figure 12.2, studied in Chapter 5 and in Example 12.3.

The elements of the inertia matrix $M(\boldsymbol{q})$ are

$$
\begin{aligned}
M_{11}(\boldsymbol{q}) &= m_1 l_{c1}^2 + m_2 \left[l_1^2 + l_{c2}^2 + 2 l_1 l_{c2} \cos(q_2) \right] + I_1 + I_2 \\
M_{12}(\boldsymbol{q}) &= m_2 \left[l_{c2}^2 + l_1 l_{c2} \cos(q_2) \right] + I_2 \\
M_{21}(\boldsymbol{q}) &= m_2 \left[l_{c2}^2 + l_1 l_{c2} \cos(q_2) \right] + I_2 \\
M_{22}(\boldsymbol{q}) &= m_2 l_{c2}^2 + I_2 \,.
\end{aligned}
$$

The elements of the centrifugal and Coriolis forces matrix $C(\boldsymbol{q}, \dot{\boldsymbol{q}})$ are given by

$$
\begin{aligned}
C_{11}(\boldsymbol{q}, \dot{\boldsymbol{q}}) &= -m_2 l_1 l_{c2} \sin(q_2) \dot{q}_2 \\
C_{12}(\boldsymbol{q}, \dot{\boldsymbol{q}}) &= -m_2 l_1 l_{c2} \sin(q_2) \left[\dot{q}_1 + \dot{q}_2 \right] \\
C_{21}(\boldsymbol{q}, \dot{\boldsymbol{q}}) &= m_2 l_1 l_{c2} \sin(q_2) \dot{q}_1 \\
C_{22}(\boldsymbol{q}, \dot{\boldsymbol{q}}) &= 0 \,.
\end{aligned}
$$

The elements of the vector of gravitational torques $\boldsymbol{g}(\boldsymbol{q})$ are

$$
\begin{aligned}
g_1(\boldsymbol{q}) &= [m_1 l_{c1} + m_2 l_1] g \sin(q_1) + m_2 l_{c2} g \sin(q_1 + q_2) \\
g_2(\boldsymbol{q}) &= m_2 l_{c2} g \sin(q_1 + q_2) \,.
\end{aligned}
$$

Using the numerical values of the constants given in Table 5.1 (*cf.* page 115) as well as the formulas in Table 4.1 (*cf.* page 109), we get

$$
\begin{aligned}
k_M &= 0.0974 \ \left[\mathrm{kg\ m^2}\right], \\
k_{C_1} &= 0.0487 \ \left[\mathrm{kg\ m^2}\right], \\
k_{C_2} &= 0.0974 \ \left[\mathrm{kg\ m^2}\right], \\
k_g &= 23.94 \ \left[\mathrm{kg\ m^2/s^2}\right], \\
k'_M &= \lambda_{\mathrm{Max}}\{M(\boldsymbol{q})\} = 0.3614 \ \left[\mathrm{kg\ m^2}\right], \\
\lambda_{\min}\{M(\boldsymbol{q})\} &= 0.011 \ \left[\mathrm{kg\ m^2}\right].
\end{aligned}
$$

For the computation of $\lambda_{\rm Max}\{M(\boldsymbol{q})\}$ and $\lambda_{\rm min}\{M(\boldsymbol{q})\}$, see the explanation in Example 9.2 on page 213.

To obtain k', we proceed numerically, that is, we evaluate the norm $\boldsymbol{g}(\boldsymbol{q})$ for a set of values of q_1 and q_2 between 0 and 2π, and extract the maximum. This happens for $q_1 = q_2 = 0$ and the maximum is

$$k' = 7.664 \quad [\mathrm{N\,m}]$$

Consider the PD control law plus feedforward, (12.5), for this robot. As in Example 12.3, the specification of the task for the robot is expressed in terms of the desired trajectory $\boldsymbol{q}_d(t)$ shown in Figure 5.5 and whose analytical expression is given by (5.7). Equations (5.8) and (5.9) correspond to the desired velocity $\dot{\boldsymbol{q}}_d(t)$, and desired acceleration $\ddot{\boldsymbol{q}}_d(t)$ respectively. By numerical simulation, can be verified the following upper-bounds on the norms of the desired velocity and acceleration,

$$\begin{aligned}\|\dot{\boldsymbol{q}}_d\|_{\rm Max} &= 2.33 \;\; [\ \mathrm{rad/s}]\\ \|\ddot{\boldsymbol{q}}_d\|_{\rm Max} &= 9.52 \;\; [\ \mathrm{rad/s^2}]\,.\end{aligned}$$

Using this information and the definitions of the constants from the tuning procedure, we get

$$\begin{aligned}s_1 &= 25.385 \;\; [\mathrm{N\,m}]\,,\\ s_2 &= 22.733 \;\; [\mathrm{N\,m}]\,,\\ k_{h1} &= 0.114 \;\; \left[\mathrm{kg\,m^2/s}\right],\\ k_{h2} &= 31.834 \;\; [\mathrm{N\,m}]\,,\\ a &= 1.614 \;\; \left[\mathrm{kg\,m^2/s}\right],\\ b &= 0.43 \;\; \left[\mathrm{kg\,m^2}\right].\end{aligned}$$

Finally, we set $\gamma = 2$ $[\mathrm{s^{-1}}]$, so that it is only left to fix the design matrices K_p and K_v in accordance with the conditions (12.11)–(12.13). An appropriate choice is

$$\begin{aligned}K_p &= \mathrm{diag}\{200,\ 150\} \;\; [\mathrm{N\,m}]\,,\\ K_v &= \mathrm{diag}\{3\} \;\; [\mathrm{N\,m\,s/rad}]\,.\end{aligned}$$

The initial conditions corresponding to the positions and velocities are chosen as

$$\begin{aligned}q_1(0) &= 0, \qquad & q_2(0) &= 0\\ \dot{q}_1(0) &= 0, \qquad & \dot{q}_2(0) &= 0\,.\end{aligned}$$

Figure 12.5 shows the experimental tracking errors $\tilde{\boldsymbol{q}}(t)$. As pointed out in previous examples, the trajectories $\boldsymbol{q}(t)$ do not vanish as expected due to several aspects always present in real implementations

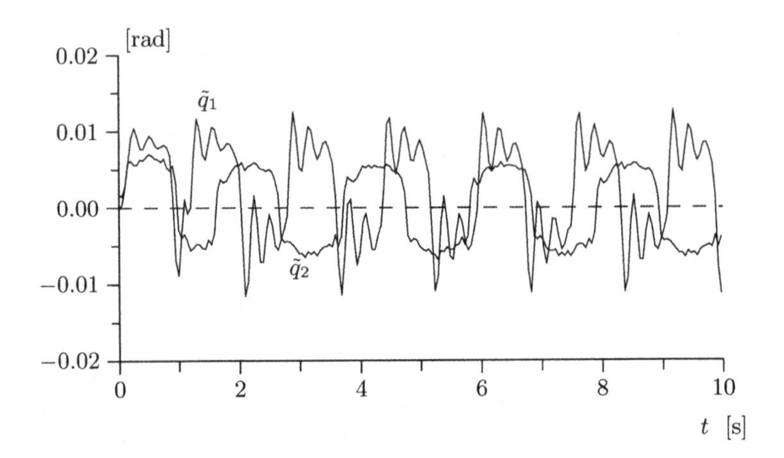

Figure 12.5. Graphs of position errors $\tilde{q}_1$ and $\tilde{q}_2$

– usually neglected in the theoretical analysis – such as digital implementation of the continuous-time closed-loop control system (described by an ordinary differential equation), measurement noise and, most important in our experimental setting, friction at the arm joints. Yet, in contrast to Example 12.3 where the controller did not carry the PD term, the behavior obtained here is satisfactory. ♢

12.3 Conclusions

The conclusions drawn from the analysis presented in this chapter may be summarized in the following terms.

- The feedforward control for n-DOF robots is an open-loop control scheme in open loop. For this reason, it is not advisable and moreover, in general this control is unable to satisfy the motion control objective.
- With PD control plus feedforward we may satisfy the control objective globally for n-DOF robots under the condition that sufficiently "large" design matrices K_p and K_v are used. More significantly, global uniform asymptotic stability of the origin of the closed-loop equations may be achieved under such conditions.

Bibliography

Interesting remarks on feedforward control and PD control plus feedforward may be found in

- Craig J., 1989, *"Introduction to robotics: Mechanics and control"*, Addison–Wesley, Reading, MA.
- Yoshikawa T., 1990, *"Foundations of robotics: Analysis and control"*, The MIT Press.

(Local) asymptotic stability under PD control plus feedforward has been analyzed in

- Paden B., Riedle B. D., 1988, *"A positive–real modification of a class of nonlinear controllers for robot manipulators"*, Proceedings of the American Control Conference, Atlanta, GA, pp. 1782–1785.
- Wen J. T., 1990, *"A unified perspective on robot control: The energy Lyapunov function approach"*, International Journal of Adaptive Control and Signal Processing, Vol. 4, pp. 487–500.
- Kelly R., Salgado R., 1994, *"PD control with computed feedforward of robot manipulators: A design procedure"*, IEEE Transactions on Robotics and Automation, Vol. 10, No. 4, August, pp. 566–571.

The proof of existence of proportional and derivative gains that guarantee global asymptotic stability was reported in

- Santibáñez V., Kelly R., 2001, *"PD control with feedforward compensation for robot manipulators: Analysis and experimentation"*, Robotica, Vol. 19, pp. 11–19.

The following documents present experimental results for the application of PD control plus feedforward on prototype robots.

- Asada H., Kanade T., Takeyama I., 1983, *"Control of a direct–drive arm"*, ASME Journal of Dynamic Systems, Measurement, and control, Vol. 105, pp. 136–142.
- An C., Atkeson C., Hollerbach J., 1988, *"Model-based control of a robot manipulator"*, The MIT Press.
- Khosla P. K., Kanade T., 1988, *"Experimental evaluation of nonlinear feedback and feedforward control schemes for manipulators"*, The International Journal of Robotics Research, Vol. 7, No. 1, pp. 18–28.
- Kokkinis T., Stoughton R., 1991, *"Dynamics and control of a closed-chain robot with application to a new direct-drive robot arm"*, International Journal of Robotics and Automation, Vol. 6, No. 1.
- Tarn T. J., Bejczy A. K., Marth G. T., Ramadarai A. K., 1993, *"Performance comparison of four manipulators servo schemes"*, IEEE Control Systems, Vol. 13, No. 1, February.
- Caccavale F., Chiacchio P., 1994, *"Identification of dynamic parameters and feedforward control for conventional industrial manipulators"*, Control Engineering Practice, Vol. 2, No. 6, pp. 1039–1050.

- Reyes F., Kelly R., 2001, *"Experimental evaluation of model-based controllers on a direct-drive robot arm"*, Mechatronics, Vol. 11, pp. 267–282.

Problems

1. Consider feedforward control of the ideal pendulum studied in Example 12.1. Assume that the desired position $q_d(t)$ is zero for all $t \geq 0$.
   a) Obtain the equation that governs the control system, in terms of $[\tilde{q} \;\; \dot{q}]^T$.
   b) Show that the origin is a stable equilibrium.
   Hint: See Example 2.2, on page 30.
2. Consider feedforward control of the 3-DOF Cartesian robot studied in Example 12.2.
   a) Show that if $\dot{\tilde{\boldsymbol{q}}}(0) \neq \mathbf{0}$, then $\lim_{t\to\infty} \|\tilde{\boldsymbol{q}}(t)\| = \infty$.
3. Consider the model of an ideal pendulum studied in Example 12.1 but now including a term for viscous friction, that is,

$$ml^2\ddot{q} + mgl\,\sin(q) + f\dot{q} = \tau$$

   where $f > 0$ is the friction coefficient. The feedforward controller (12.2) obtained when neglecting the friction term is

$$\tau = ml^2\ddot{q}_d + mgl\,\sin(q_d).$$

   Assume that $q_d(t) = \sin(t)$.
   a) Obtain the equation $\dot{\boldsymbol{x}} = \boldsymbol{f}(t, \boldsymbol{x})$ where $\boldsymbol{x} = [\tilde{q} \;\; \dot{\tilde{q}}]^T$. Does this equation have any equilibria ?
   b) Assume moreover that $q_d(0) = q(0)$ and $\dot{q}_d(0) = \dot{q}(0)$. May it be expected that $\lim_{t\to\infty} \tilde{q}(t) = 0$?
4. Consider a PD control law plus feedforward on the ideal pendulum analyzed in Example 12.4. In this example we derived the closed-loop equation,

$$\frac{d}{dt}\begin{bmatrix} \tilde{q} \\ \dot{\tilde{q}} \end{bmatrix} = \begin{bmatrix} \dot{\tilde{q}} \\ -\frac{1}{ml^2}\left[k_p\tilde{q} + k_v\dot{\tilde{q}} + mgl\;\left(\sin(q_d) - \sin(q_d - \tilde{q})\right)\right] \end{bmatrix}.$$

   Assume now that the desired position is given by

$$q_d(t) = \sin(t)\,.$$

   On the other hand, the design constants k_p and k_v are chosen so that

$$k_v > ml^2$$

$$k_p > \frac{2\left[mgl + ml^2\right]}{\tanh\left(\frac{2\left[mgl+ml^2\right]}{mgl}\right)} + \frac{\left[k_v + \frac{2\left[mgl+ml^2\right]}{\tanh\left(\frac{2\left[mgl+ml^2\right]}{mgl}\right)}\right]^2}{4\left[k_v - ml^2\right]}.$$

a) Show that the origin $\begin{bmatrix}\tilde{q} & \dot{\tilde{q}}\end{bmatrix}^T = \mathbf{0} \in \mathbb{R}^2$ is a globally asymptotically stable equilibrium.

Hint: Use the Lyapunov function candidate

$$V(\tilde{q}, \dot{\tilde{q}}) = \frac{1}{2}ml^2\dot{\tilde{q}}^2 + \frac{1}{2}k_p\tilde{q}^2 + ml^2\tanh(\tilde{q})\dot{\tilde{q}}$$

and verify that its time derivative satisfies

$$\dot{V}(\tilde{q}, \dot{\tilde{q}}) \leq -\begin{bmatrix}|\tanh(\tilde{q})| \\ |\dot{\tilde{q}}|\end{bmatrix}^T R \begin{bmatrix}|\tanh(\tilde{q})| \\ |\dot{\tilde{q}}|\end{bmatrix}$$

where

$$R = \begin{bmatrix} k_p - \frac{2\left[mgl+ml^2\right]}{\tanh\left(\frac{2\left[mgl+ml^2\right]}{mgl}\right)} & -\frac{1}{2}\left[k_v + \frac{2\left[mgl+ml^2\right]}{\tanh\left(\frac{2\left[mgl+ml^2\right]}{mgl}\right)}\right] \\ -\frac{1}{2}\left[k_v + \frac{2\left[mgl+ml^2\right]}{\tanh\left(\frac{2\left[mgl+ml^2\right]}{mgl}\right)}\right] & k_v - ml^2 \end{bmatrix}.$$

5. Consider a PD control law plus feedforward on the 2-DOF robot *prototype* used in Example 12.5. In this example we presented some simulations where $K_p = \text{diag}\{200,\ 150\}$. Verify that with such a selection, the corresponding closed-loop equation has a unique equilibrium.

   Hint: Verify that the condition (12.9) holds.

Part IV

Advanced Topics

Introduction to Part IV

In this last part of the textbook we present some advanced issues on robot control. We deal with topics such as control without velocity measurements and control under model uncertainty. We recommend this part of the text for a second course on robot dynamics and control or for a course on robot control at the first year of graduate level. We assume that the student is familiar with the notion of *functional spaces*, *i.e.* the spaces L_2 and L_∞. If not, we strongly recommend the student to read first Appendix A, which presents additional mathematical baggage necessary to study these last chapters:

- P"D" control with gravity compensation and P"D" control with desired gravity compensation;
- Introduction to adaptive robot control;
- PD control with adaptive gravity compensation;
- PD control with adaptive compensation.

13

P"D" Control with Gravity Compensation and P"D" Control with Desired Gravity Compensation

Robot manipulators are equipped with sensors for the measurement of joint positions and velocities, $\boldsymbol{q}$ and $\dot{\boldsymbol{q}}$ respectively. Physically, position sensors may be from simple variable resistances such as potentiometers to very precise *optical encoders*. On the other hand, the measurement of velocity may be realized through *tachometers*, or in most cases, by numerical approximation of the velocity from the position sensed by the optical encoders. In contrast to the high precision of the position measurements by the optical encoders, the measurement of velocities by the described methods may be quite mediocre in accuracy, specifically for certain intervals of velocity. On certain occasions this may have as a consequence, an unacceptable degradation of the performance of the control system.

The interest in using controllers for robots that do not explicitly require the measurement of velocity, is twofold. First, it is inadequate to feed back a velocity measurement which is possibly of poor quality for certain bands of operation. Second, avoiding the use of velocity measurements removes the need for velocity sensors such as tachometers and therefore, leads to a reduction in production cost while making the robot lighter.

The design of controllers that do not require velocity measurements to control robot manipulators has been a topic of investigation since broached in the decade of the 1990s and to date, many questions remain open. The common idea in the design of such controllers has been to propose state observers to estimate the velocity. Then the so-obtained velocity estimations are incorporated in the controller by replacing the true unavailable velocities. In this way, it has been shown that asymptotic and even exponential stability can be achieved, at least locally. Some important references on this topic are presented at the end of the chapter.

In this chapter we present an alternative to the design of observers to estimate velocity and which is of utility in position control. The idea consists simply in substituting the velocity measurement $\dot{\boldsymbol{q}}$, by the *filtered* position

$\boldsymbol{q}$ through a first-order system of zero relative degree, and whose output is denoted in the sequel, by $\boldsymbol{\vartheta}$.

Specifically, denoting by p the differential operator, *i.e.* $p = \frac{d}{dt}$, the components of $\boldsymbol{\vartheta} \in \mathbb{R}^n$ are given by

$$\begin{bmatrix} \vartheta_1 \\ \vartheta_2 \\ \vdots \\ \vartheta_n \end{bmatrix} = \begin{bmatrix} \frac{b_1 p}{p+a_1} & 0 & \cdots & 0 \\ 0 & \frac{b_2 p}{p+a_2} & \cdots & 0 \\ \vdots & \vdots & \ddots & \vdots \\ 0 & 0 & \cdots & \frac{b_n p}{p+a_n} \end{bmatrix} \begin{bmatrix} q_1 \\ q_2 \\ \vdots \\ q_n \end{bmatrix} \tag{13.1}$$

or in compact form,

$$\boldsymbol{\vartheta} = \mathrm{diag}\left\{ \frac{b_i p}{p+a_i} \right\} \boldsymbol{q}$$

where a_i and b_i are strictly positive real constants but otherwise arbitrary, for $i = 1, 2, \cdots, n$.

A state-space representation of Equation (13.1) is

$$\begin{aligned} \dot{\boldsymbol{x}} &= -A\boldsymbol{x} - AB\boldsymbol{q} \\ \boldsymbol{\vartheta} &= \boldsymbol{x} + B\boldsymbol{q} \end{aligned}$$

where $\boldsymbol{x} \in \mathbb{R}^n$ represents the state vector of the filters, $A = \mathrm{diag}\{a_i\}$ and $B = \mathrm{diag}\{b_i\}$.

In this chapter we study the proposed modification for the following controllers:

- PD control with gravity compensation and
- PD control with desired gravity compensation.

Obviously, the *derivative* part of both control laws is no longer proportional to the *derivative* of the position error $\tilde{\boldsymbol{q}}$; this motivates the quotes around "D" in the names of the controllers. As in other chapters appropriate references are presented at the end of the chapter.

13.1 P"D" Control with Gravity Compensation

The PD control law with gravity compensation (7.1) requires, in its derivative part, measurement of the joint velocity $\dot{\boldsymbol{q}}$ with the purpose of computing the velocity error $\dot{\tilde{\boldsymbol{q}}} = \dot{\boldsymbol{q}}_d - \dot{\boldsymbol{q}}$, and to use the latter in the term $K_v\dot{\tilde{\boldsymbol{q}}}$. Even in the case of position control, that is, when the desired joint position $\boldsymbol{q}_d$ is constant, the measurement of the velocity is needed by the term $K_v\dot{\boldsymbol{q}}$.

A possible modification to the PD control law with gravity compensation consists in replacing the derivative part (D), which is proportional to the derivative of the position error, *i.e.* to the velocity error $\dot{\tilde{q}} = \dot{q}_d - \dot{q}$, by a term proportional to

$$\dot{q}_d - \vartheta$$

where $\vartheta \in \mathbb{R}^n$ is, as said above, the result of filtering the position q by means of a dynamic system of first-order and of zero relative degree.

Specifically, the P"D" control law with gravity compensation is written as

$$\tau = K_p \tilde{q} + K_v [\dot{q}_d - \vartheta] + g(q) \tag{13.2}$$

$$\dot{x} = -Ax - ABq$$

$$\vartheta = x + Bq \tag{13.3}$$

where $K_p, K_v \in \mathbb{R}^{n\times n}$ are diagonal positive definite matrices, $A = \mathrm{diag}\{a_i\}$ and $B = \mathrm{diag}\{b_i\}$ and a_i and b_i are real strictly positive constants but otherwise arbitrary for $i = 1, 2, \cdots, n$.

Figure 13.1 shows the block-diagram corresponding to the robot under P"D" control with gravity compensation. Notice that the measurement of the joint velocity $\dot{q}$ is not required by the controller.

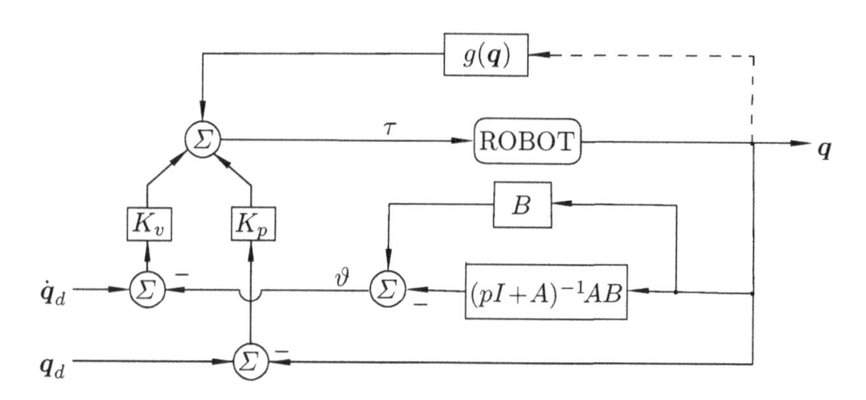

Figure 13.1. Block-diagram: P"D" control with gravity compensation

Define $\xi = x + Bq_d$. The equation that describes the behavior in closed loop may be obtained by combining Equations (III.1) and (13.2)–(13.3), which may be written in terms of the state vector $\begin{bmatrix}\xi^T & \tilde{q}^T & \dot{q}^T\end{bmatrix}^T$ as

$$\frac{d}{dt}\begin{bmatrix}\boldsymbol{\xi}\\ \tilde{\boldsymbol{q}}\\ \dot{\tilde{\boldsymbol{q}}}\end{bmatrix} = \begin{bmatrix}-A\boldsymbol{\xi} + AB\tilde{\boldsymbol{q}} + B\dot{\boldsymbol{q}}_d\\ \dot{\tilde{\boldsymbol{q}}}\\ \ddot{\boldsymbol{q}}_d - M(\boldsymbol{q})^{-1}\left[K_p\tilde{\boldsymbol{q}} + K_v[\dot{\boldsymbol{q}}_d - \boldsymbol{\xi} + B\tilde{\boldsymbol{q}}] - C(\boldsymbol{q},\dot{\boldsymbol{q}})\dot{\boldsymbol{q}}\right]\end{bmatrix}.$$

A sufficient condition for the origin $\begin{bmatrix}\boldsymbol{\xi}^T & \tilde{\boldsymbol{q}}^T & \dot{\tilde{\boldsymbol{q}}}^T\end{bmatrix}^T = \boldsymbol{0} \in \mathbb{R}^{3n}$ to be a unique equilibrium point of the closed-loop equation is that the desired joint position $\boldsymbol{q}_d$ be a constant vector. In what is left of this section we assume that this is the case. Notice that in this scenario, the control law may be expressed as

$$\boldsymbol{\tau} = K_p\tilde{\boldsymbol{q}} - K_v\mathrm{diag}\left\{\frac{b_i p}{p + a_i}\right\}\boldsymbol{q} + \boldsymbol{g}(\boldsymbol{q}),$$

which is close to the PD with gravity compensation control law (7.1), when the desired position $\boldsymbol{q}_d$ is constant. Indeed the only difference is replacement of the velocity $\dot{\boldsymbol{q}}$ by

$$\mathrm{diag}\left\{\frac{b_i p}{p + a_i}\right\}\boldsymbol{q},$$

thereby avoiding the use of the velocity $\dot{\boldsymbol{q}}$ in the control law.

As we show in the following subsections, P"D" control with gravity compensation meets the position control objective, that is,

$$\lim_{t\to\infty}\boldsymbol{q}(t) = \boldsymbol{q}_d$$

where $\boldsymbol{q}_d \in \mathbb{R}^n$ is any constant vector.

Considering the desired position $\boldsymbol{q}_d$ as constant, the closed-loop equation may be rewritten in terms of the new state vector $\begin{bmatrix}\boldsymbol{\xi}^T & \tilde{\boldsymbol{q}}^T & \dot{\boldsymbol{q}}^T\end{bmatrix}^T$ as

$$\frac{d}{dt}\begin{bmatrix}\boldsymbol{\xi}\\ \tilde{\boldsymbol{q}}\\ \dot{\boldsymbol{q}}\end{bmatrix} = \begin{bmatrix}-A\boldsymbol{\xi} + AB\tilde{\boldsymbol{q}}\\ -\dot{\boldsymbol{q}}\\ M(\boldsymbol{q}_d - \tilde{\boldsymbol{q}})^{-1}\left[K_p\tilde{\boldsymbol{q}} - K_v[\boldsymbol{\xi} - B\tilde{\boldsymbol{q}}] - C(\boldsymbol{q}_d - \tilde{\boldsymbol{q}},\dot{\boldsymbol{q}})\dot{\boldsymbol{q}}\right]\end{bmatrix} \tag{13.4}$$

which, in view of the fact that $\boldsymbol{q}_d$ is constant, constitutes an autonomous differential equation. Moreover, the origin $\begin{bmatrix}\boldsymbol{\xi}^T & \tilde{\boldsymbol{q}}^T & \dot{\boldsymbol{q}}^T\end{bmatrix}^T = \boldsymbol{0} \in \mathbb{R}^{3n}$ is the unique equilibrium of this equation.

With the aim of studying the stability of the origin, we consider the Lyapunov function candidate

$$V(\boldsymbol{\xi},\tilde{\boldsymbol{q}},\dot{\boldsymbol{q}}) = \mathcal{K}(\boldsymbol{q},\dot{\boldsymbol{q}}) + \frac{1}{2}\tilde{\boldsymbol{q}}^T K_p\tilde{\boldsymbol{q}} + \frac{1}{2}(\boldsymbol{\xi} - B\tilde{\boldsymbol{q}})^T K_v B^{-1}(\boldsymbol{\xi} - B\tilde{\boldsymbol{q}}) \tag{13.5}$$

where $\mathcal{K}(\boldsymbol{q},\dot{\boldsymbol{q}}) = \frac{1}{2}\dot{\boldsymbol{q}}^T M(\boldsymbol{q})\dot{\boldsymbol{q}}$ is the kinetic energy function corresponding to the robot. Notice that the diagonal matrix $K_v B^{-1}$ is positive definite. Consequently, the function $V(\boldsymbol{\xi},\tilde{\boldsymbol{q}},\dot{\boldsymbol{q}})$ is globally positive definite.

The total time derivative of the Lyapunov function candidate yields

$$\begin{aligned}\dot{V}(\boldsymbol{\xi},\tilde{\boldsymbol{q}},\dot{\boldsymbol{q}}) = {} & \dot{\boldsymbol{q}}^T M(\boldsymbol{q})\ddot{\boldsymbol{q}} + \frac{1}{2}\dot{\boldsymbol{q}}^T \dot{M}(\boldsymbol{q})\dot{\boldsymbol{q}} + \tilde{\boldsymbol{q}}^T K_p \dot{\tilde{\boldsymbol{q}}} \\ & + \left[\boldsymbol{\xi} - B\tilde{\boldsymbol{q}}\right]^T K_v B^{-1}\left[\dot{\boldsymbol{\xi}} - B\dot{\tilde{\boldsymbol{q}}}\right].\end{aligned}$$

Using the closed-loop Equation (13.4) to solve for $\dot{\boldsymbol{\xi}}$, $\dot{\tilde{\boldsymbol{q}}}$ and $M(\boldsymbol{q})\ddot{\boldsymbol{q}}$, and canceling out some terms we obtain

$$\begin{aligned}\dot{V}(\boldsymbol{\xi},\tilde{\boldsymbol{q}},\dot{\boldsymbol{q}}) &= -\left[\boldsymbol{\xi} - B\tilde{\boldsymbol{q}}\right]^T K_v B^{-1} A\left[\boldsymbol{\xi} - B\tilde{\boldsymbol{q}}\right] \\ &= -\begin{bmatrix}\boldsymbol{\xi}\\ \tilde{\boldsymbol{q}}\\ \dot{\boldsymbol{q}}\end{bmatrix}^T \begin{bmatrix} K_v B^{-1} A & -K_v A & 0\\ -K_v A & B K_v A & 0\\ 0 & 0 & 0\end{bmatrix}\begin{bmatrix}\boldsymbol{\xi}\\ \tilde{\boldsymbol{q}}\\ \dot{\boldsymbol{q}}\end{bmatrix}\end{aligned} \tag{13.6}$$

where we used

$$\dot{\boldsymbol{q}}^T\left[\frac{1}{2}\dot{M}(\boldsymbol{q}) - C(\boldsymbol{q},\dot{\boldsymbol{q}})\right]\dot{\boldsymbol{q}} = 0,$$

which follows from Property 4.2.

Clearly, the time derivative $\dot{V}(\boldsymbol{\xi},\tilde{\boldsymbol{q}},\dot{\boldsymbol{q}})$ of the Lyapunov function candidate is globally negative semidefinite. Therefore, invoking Theorem 2.3, we conclude that the origin of the closed-loop Equation (13.4) is stable and that all solutions are bounded.

Since the closed-loop Equation (13.4) is autonomous, La Salle's Theorem 2.7 may be used in a straightforward way to analyze the global asymptotic stability of the origin (*cf.* Problem 3 at the end of the chapter). Nevertheless, we present below, an alternative analysis that also allows one to show global asymptotic stability of the origin of the state-space corresponding to the closed-loop Equation, (13.4). This alternative method of proof, which is longer than via La Salle's theorem, is presented to familiarize the reader with other methods to prove global asymptotic stability; however, we appeal to the material on functional spaces presented in Appendix A.

According to Definition 2.6, since the origin $\begin{bmatrix}\boldsymbol{\xi}^T & \tilde{\boldsymbol{q}}^T & \dot{\boldsymbol{q}}^T\end{bmatrix}^T = \boldsymbol{0} \in \mathbb{R}^{3n}$ is a stable equilibrium, then if $\begin{bmatrix}\boldsymbol{\xi}(t)^T & \tilde{\boldsymbol{q}}(t)^T & \dot{\boldsymbol{q}}(t)^T\end{bmatrix}^T \to \boldsymbol{0} \in \mathbb{R}^{3n}$ as $t \to \infty$ (for all initial conditions), the origin is a globally asymptotically stable equilibrium. It is precisely this property that we show next.

In the development that follows we use additional properties of the dynamic model of robot manipulators. Specifically, assume that $\boldsymbol{q},\dot{\boldsymbol{q}} \in L_\infty^n$. Then,

- $M(\boldsymbol{q})^{-1}, \frac{d}{dt}M(\boldsymbol{q}) \in L_\infty^{n\times n}$
- $C(\boldsymbol{q},\dot{\boldsymbol{q}})\dot{\boldsymbol{q}} \in L_\infty^n$.

If moreover $\ddot{\boldsymbol{q}} \in L_\infty^n$ then,

- $\frac{d}{dt}\left[C(\boldsymbol{q},\dot{\boldsymbol{q}})\dot{\boldsymbol{q}}\right] \in L_\infty^n$.

The Lyapunov function $V(\boldsymbol{\xi},\tilde{\boldsymbol{q}},\dot{\boldsymbol{q}})$ given in (13.5) is positive definite since it is composed of the following three non-negative terms:

- $\frac{1}{2}\dot{\boldsymbol{q}}^T M(\boldsymbol{q})\dot{\boldsymbol{q}}$
- $\frac{1}{2}\tilde{\boldsymbol{q}}^T K_p \tilde{\boldsymbol{q}}$
- $\frac{1}{2}[\boldsymbol{\xi} - B\tilde{\boldsymbol{q}}]^T K_v B^{-1}[\boldsymbol{\xi} - B\tilde{\boldsymbol{q}}]$.

Since the time derivative $\dot{V}(\boldsymbol{\xi},\tilde{\boldsymbol{q}},\dot{\boldsymbol{q}})$ expressed in (13.6) is negative semidefinite, the Lyapunov function $V(\boldsymbol{\xi},\tilde{\boldsymbol{q}},\dot{\boldsymbol{q}})$ is bounded along the trajectories. Therefore, the three non-negative terms above are also bounded along trajectories. From this conclusion we have

$$\dot{\boldsymbol{q}},\ \tilde{\boldsymbol{q}},\ [\boldsymbol{\xi} - B\tilde{\boldsymbol{q}}] \in L_\infty^n . \tag{13.7}$$

Incorporating this information in the closed-loop system Equation (13.4), and knowing that $M(\boldsymbol{q}_d - \tilde{\boldsymbol{q}})^{-1}$ is bounded for all $\boldsymbol{q}_d, \tilde{\boldsymbol{q}} \in L_\infty^n$ and also that $C(\boldsymbol{q}_d - \tilde{\boldsymbol{q}},\dot{\boldsymbol{q}})\dot{\boldsymbol{q}}$ is bounded for all $\boldsymbol{q}_d, \tilde{\boldsymbol{q}}, \dot{\boldsymbol{q}} \in L_\infty^n$, it follows that the time derivative of the state vector is also bounded, *i.e.*

$$\dot{\boldsymbol{\xi}}, \dot{\tilde{\boldsymbol{q}}}, \ddot{\boldsymbol{q}} \in L_\infty^n, \tag{13.8}$$

and therefore,

$$\dot{\boldsymbol{\xi}} - B\dot{\tilde{\boldsymbol{q}}} \in L_\infty^n . \tag{13.9}$$

Using again the closed-loop Equation (13.4), we obtain the second time derivative of the state variables,

$$\begin{aligned}
\ddot{\boldsymbol{\xi}} &= -A\dot{\boldsymbol{\xi}} + AB\dot{\tilde{\boldsymbol{q}}} \\
\ddot{\tilde{\boldsymbol{q}}} &= -\ddot{\boldsymbol{q}} \\
\boldsymbol{q}^{(3)} &= -M(\boldsymbol{q})^{-1}\left[\frac{d}{dt}M(\boldsymbol{q})\right]M(\boldsymbol{q})^{-1}\left[K_p\tilde{\boldsymbol{q}} - K_v\left[\boldsymbol{\xi} - B\tilde{\boldsymbol{q}}\right] - C(\boldsymbol{q},\dot{\boldsymbol{q}})\dot{\boldsymbol{q}}\right] \\
&\quad + M(\boldsymbol{q})^{-1}\left[K_p\dot{\tilde{\boldsymbol{q}}} - K_v\left[\dot{\boldsymbol{\xi}} - B\dot{\tilde{\boldsymbol{q}}}\right] - \frac{d}{dt}\left[C(\boldsymbol{q},\dot{\boldsymbol{q}})\dot{\boldsymbol{q}}\right]\right]
\end{aligned}$$

where $\boldsymbol{q}^{(3)}$ denotes the third time derivative of the joint position $\boldsymbol{q}$ and we used

$$\frac{d}{dt}\left[M(\boldsymbol{q})^{-1}\right] = -M(\boldsymbol{q})^{-1}\left[\frac{d}{dt}M(\boldsymbol{q})\right]M(\boldsymbol{q})^{-1}.$$

In (13.7) and (13.8) we have already concluded that $\boldsymbol{\xi}, \tilde{\boldsymbol{q}}, \dot{\boldsymbol{q}}, \dot{\boldsymbol{\xi}}, \dot{\tilde{\boldsymbol{q}}}, \ddot{\boldsymbol{q}} \in L_\infty^n$ then, from the properties stated at the beginning of this analysis, we obtain

$$\ddot{\boldsymbol{\xi}}, \ddot{\tilde{\boldsymbol{q}}}, \boldsymbol{q}^{(3)} \in L_\infty^n, \tag{13.10}$$

and therefore,

$$\ddot{\boldsymbol{\xi}} - B\ddot{\tilde{\boldsymbol{q}}} \in L_\infty^n. \tag{13.11}$$

On the other hand, integrating both sides of (13.6) and using that $V(\boldsymbol{\xi}, \tilde{\boldsymbol{q}}, \dot{\tilde{\boldsymbol{q}}})$ is bounded along the trajectories, we obtain

$$[\boldsymbol{\xi} - B\tilde{\boldsymbol{q}}] \in L_2^n. \tag{13.12}$$

Considering (13.9), (13.12) and Lemma A.5, we obtain

$$\lim_{t\to\infty} [\boldsymbol{\xi}(t) - B\tilde{\boldsymbol{q}}(t)] = \boldsymbol{0}. \tag{13.13}$$

Next, we invoke Lemma A.6 with $\boldsymbol{f} = \boldsymbol{\xi} - B\tilde{\boldsymbol{q}}$. Using (13.13), (13.7), (13.9) and (13.11), we get from this lemma

$$\lim_{t\to\infty} \left[\dot{\boldsymbol{\xi}}(t) - B\dot{\tilde{\boldsymbol{q}}}(t)\right] = \boldsymbol{0}.$$

Consequently, using the closed-loop Equation (13.4) we get

$$\lim_{t\to\infty} -A[\boldsymbol{\xi}(t) - B\tilde{\boldsymbol{q}}(t)] + B\dot{\boldsymbol{q}} = \boldsymbol{0}.$$

From this expression and (13.13) we obtain

$$\lim_{t\to\infty} \dot{\boldsymbol{q}}(t) = \boldsymbol{0} \in \mathbb{R}^n. \tag{13.14}$$

Now, we show that $\lim_{t\to\infty} \tilde{\boldsymbol{q}}(t) = \boldsymbol{0} \in \mathbb{R}^n$. To that end, we consider again Lemma A.6 with $\boldsymbol{f} = \dot{\boldsymbol{q}}$. Incorporating (13.14), (13.7), (13.8) and (13.10) we get

$$\lim_{t\to\infty} \ddot{\boldsymbol{q}}(t) = \boldsymbol{0}.$$

Taking this into account in the closed-loop Equation (13.4) as well as (13.13) and (13.14), we get

$$\lim_{t\to\infty} M(\boldsymbol{q}_d - \tilde{\boldsymbol{q}}(t))^{-1}K_p\tilde{\boldsymbol{q}}(t) = \boldsymbol{0}.$$

So we conclude that

$$\lim_{t\to\infty} \tilde{\boldsymbol{q}}(t) = \boldsymbol{0} \in \mathbb{R}^n. \tag{13.15}$$

The last part of the proof, that is, the proof of $\lim_{t\to\infty} \boldsymbol{\xi}(t) = \mathbf{0}$ follows trivially from (13.13) and (13.15). Therefore, the origin is a globally attractive equilibrium point.

This completes the proof of global asymptotic stability of the origin of the closed-loop Equation (13.4).

We present next an example with the purpose of illustrating the performance of the Pelican robot under P"D" control with gravity compensation. As for all other examples on the Pelican robot, the results that we present are from laboratory experimentation.

Example 13.1. Consider the Pelican robot studied in Chapter 5, and depicted in Figure 5.2. The components of the vector of gravitational torques $\boldsymbol{g}(\boldsymbol{q})$ are given by

$$\begin{aligned} g_1(\boldsymbol{q}) &= (m_1 l_{c1} + m_2 l_1) g \sin(q_1) + m_2 l_{c2} g \sin(q_1 + q_2) \\ g_2(\boldsymbol{q}) &= m_2 l_{c2} g \sin(q_1 + q_2) \, . \end{aligned}$$

Consider the P"D" control law with gravity compensation on this robot for position control and where the design matrices K_p, K_v, A, B are taken diagonal and positive definite. In particular, pick

$$\begin{aligned} K_p &= \mathrm{diag}\{k_p\} = \mathrm{diag}\{30\} \quad [\mathrm{Nm/rad}] \, , \\ K_v &= \mathrm{diag}\{k_v\} = \mathrm{diag}\{7, \ 3\} \quad [\mathrm{Nm\ s/rad}] \, , \\ A &= \mathrm{diag}\{a_i\} = \mathrm{diag}\{30, \ 70\} \quad [1/\mathrm{s}] \, , \\ B &= \mathrm{diag}\{b_i\} = \mathrm{diag}\{30, \ 70\} \quad [1/\mathrm{s}] \, . \end{aligned}$$

The components of the control input $\boldsymbol{\tau}$ are given by

$$\begin{aligned} \tau_1 &= k_p \tilde{q}_1 - k_v \vartheta_1 + g_1(\boldsymbol{q}) \\ \tau_2 &= k_p \tilde{q}_2 - k_v \vartheta_2 + g_2(\boldsymbol{q}) \\ \dot{x}_1 &= -a_1 x_1 - a_1 b_1 q_1 \\ \dot{x}_2 &= -a_2 x_2 - a_2 b_2 q_2 \\ \vartheta_1 &= x_1 + b_1 q_1 \\ \vartheta_2 &= x_2 + b_2 q_2 \, . \end{aligned}$$

The initial conditions corresponding to the positions, velocities and states of the filters, are chosen as

$$\begin{aligned} q_1(0) &= 0, \qquad & q_2(0) &= 0 \\ \dot{q}_1(0) &= 0, \qquad & \dot{q}_2(0) &= 0 \\ x_1(0) &= 0, \qquad & x_2(0) &= 0 \, . \end{aligned}$$

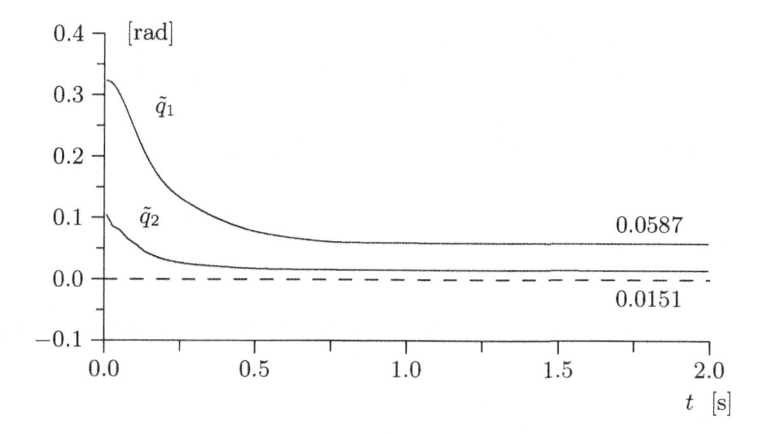

Figure 13.2. Graphs of position errors $\tilde{q}_1(t)$ and $\tilde{q}_2(t)$

The desired joint positions are chosen as

$$q_{d1} = \pi/10, \, q_{d2} = \pi/30 \quad [\text{rad}] \, .$$

In terms of the state vector of the closed-loop equation, the initial state is

$$\begin{bmatrix} \boldsymbol{\xi}(0) \\ \\ \tilde{\boldsymbol{q}}(0) \\ \\ \dot{\boldsymbol{q}}(0) \end{bmatrix} = \begin{bmatrix} b_1\pi/10 \\ b_2\pi/30 \\ \pi/10 \\ \pi/30 \\ 0 \\ 0 \end{bmatrix} = \begin{bmatrix} 9.423 \\ 7.329 \\ 0.3141 \\ 0.1047 \\ 0 \\ 0 \end{bmatrix} .$$

Figure 13.2 presents the experimental results and shows that the components of the position error $\tilde{\boldsymbol{q}}(t)$ tend asymptotically to a small nonzero constant. Although we expected that the error would tend to zero, the experimental behavior is mainly due to the presence of unmodeled friction at the joints. ◊

In a real implementation of a controller on an ordinary personal computer (as is the case of Example 13.1) typically the joint position $\boldsymbol{q}$ is sampled periodically by optical encoders and this is used to compute the joint velocity $\dot{\boldsymbol{q}}$. Indeed, if we denote by h the sampling period, the joint velocity at the instant kh is obtained as

$$\dot{\boldsymbol{q}}(kh) = \frac{\boldsymbol{q}(kh) - \boldsymbol{q}(kh-h)}{h},$$

that is, the differential operator $p = \frac{d}{dt}$ is replaced by $(1 - z^{-1})/h$, where z^{-1} is the delay operator that is, $z^{-1}\boldsymbol{q}(kh) = \boldsymbol{q}(kh - h)$. By the same argument, in the implementation of the P"D" control law with gravity compensation, (13.2)–(13.3), the variable $\boldsymbol{\vartheta}$ at instant kh may be computed as

$$\boldsymbol{\vartheta}(kh) = \frac{\boldsymbol{q}(kh) - \boldsymbol{q}(kh-h)}{h} + \frac{1}{2}\boldsymbol{\vartheta}(kh-h)$$

where we chose $A = \text{diag}\{a_i\} = \text{diag}\{h^{-1}\}$ and $B = \text{diag}\{b_i\} = \text{diag}\{2/h\}$.

13.2 P"D" Control with Desired Gravity Compensation

In this section we present a modification of PD control with desired gravity compensation, studied in Chapter 7, and whose characteristic is that it does not require the velocity term $\dot{\boldsymbol{q}}$ in its control law. The original references on this controller are cited at the end of the chapter.

This controller, that we call here P"D" control with desired gravity compensation, is described by

$$\boldsymbol{\tau} = K_p\tilde{\boldsymbol{q}} + K_v\left[\dot{\boldsymbol{q}}_d - \boldsymbol{\vartheta}\right] + \boldsymbol{g}(\boldsymbol{q}_d) \tag{13.16}$$

$$\dot{\boldsymbol{x}} = -A\boldsymbol{x} - AB\boldsymbol{q}$$

$$\boldsymbol{\vartheta} = \boldsymbol{x} + B\boldsymbol{q} \tag{13.17}$$

where $K_p, K_v \in \mathbb{R}^{n\times n}$ are diagonal positive definite matrices, $A = \text{diag}\{a_i\}$ and $B = \text{diag}\{b_i\}$ with a_i and b_i real strictly positive constants but otherwise arbitrary for all $i = 1, 2, \cdots, n$.

Figure 13.3 shows the block-diagram of the P"D" control with desired gravity compensation applied to robots. Notice that the measurement of the joint velocity $\dot{\boldsymbol{q}}$ is not required by the controller.

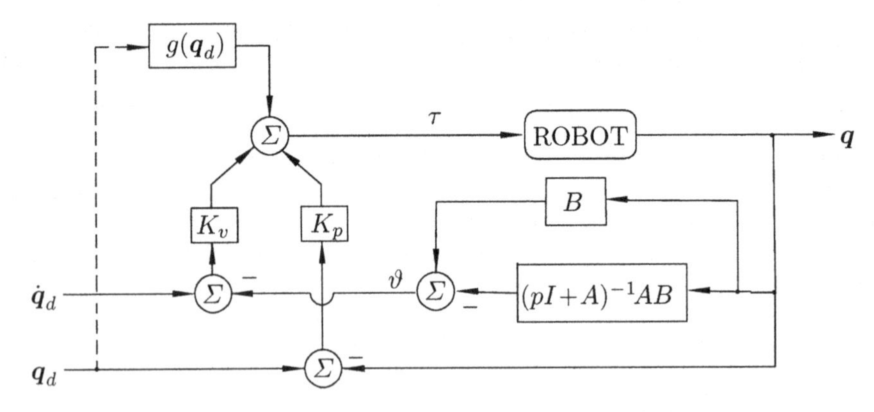

Figure 13.3. Block-diagram: P"D" control with desired gravity compensation

Comparing P"D" control with gravity compensation given by (13.2)–(13.3) with P"D" control with desired gravity compensation (13.16)–(13.17), we immediately notice the replacement of the term $\boldsymbol{g}(\boldsymbol{q})$ by the *feedforward* term $\boldsymbol{g}(\boldsymbol{q}_d)$.

The analysis of the control system in closed loop is similar to that from Section 13.1. The most noticeable difference is in the Lyapunov function considered for the proof of stability. Given the relative importance of the controller (13.16)–(13.17), we present next its complete study.

Define $\boldsymbol{\xi} = \boldsymbol{x} + B\boldsymbol{q}_d$. The equation that describes the behavior in closed loop is obtained by combining Equations (III.1) and (13.16)–(13.17), which may be expressed in terms of the state vector $\begin{bmatrix}\boldsymbol{\xi}^T & \tilde{\boldsymbol{q}}^T & \dot{\boldsymbol{q}}^T\end{bmatrix}^T$ as

$$\frac{d}{dt}\begin{bmatrix}\boldsymbol{\xi}\\ \tilde{\boldsymbol{q}}\\ \dot{\tilde{\boldsymbol{q}}}\end{bmatrix}=\begin{bmatrix}-A\boldsymbol{\xi}+AB\tilde{\boldsymbol{q}}+B\dot{\boldsymbol{q}}_d\\ \dot{\tilde{\boldsymbol{q}}}\\ \ddot{\boldsymbol{q}}_d-M(\boldsymbol{q})^{-1}[K_p\tilde{\boldsymbol{q}}+K_v[\dot{\boldsymbol{q}}_d-\boldsymbol{\xi}+B\tilde{\boldsymbol{q}}]+\boldsymbol{g}(\boldsymbol{q}_d)-C(\boldsymbol{q},\dot{\boldsymbol{q}})\dot{\boldsymbol{q}}-\boldsymbol{g}(\boldsymbol{q})]\end{bmatrix}$$

A sufficient condition for the origin $\begin{bmatrix}\boldsymbol{\xi}^T & \tilde{\boldsymbol{q}}^T & \dot{\tilde{\boldsymbol{q}}}^T\end{bmatrix}^T = \boldsymbol{0} \in \mathbb{R}^{3n}$ to be a unique equilibrium of the closed-loop equation is that the desired joint position $\boldsymbol{q}_d$ is a constant vector. In what follows of this section we assume that this is the case. Notice that in this scenario, the control law may be expressed by

$$\boldsymbol{\tau} = K_p\tilde{\boldsymbol{q}} - K_v\mathrm{diag}\left\{\frac{b_i p}{p+a_i}\right\}\boldsymbol{q} + \boldsymbol{g}(\boldsymbol{q}_d),$$

which is very close to PD with desired gravity compensation control law (8.1) when the desired position $\boldsymbol{q}_d$ is constant. The only difference is the substitution of the velocity term $\dot{\boldsymbol{q}}$ by

$$\boldsymbol{\vartheta} = \mathrm{diag}\left\{\frac{b_i p}{p+a_i}\right\}\boldsymbol{q},$$

thereby avoiding the use of velocity measurements $\dot{\boldsymbol{q}}(t)$ in the control law.

As we show below, if the matrix K_p is chosen so that

$$\lambda_{\min}\{K_p\} > k_g\,,$$

then the P"D" controller with desired gravity compensation verifies the position control objective, that is,

$$\lim_{t\to\infty}\boldsymbol{q}(t) = \boldsymbol{q}_d$$

for any constant vector $\boldsymbol{q}_d \in \mathbb{R}^n$.

Considering the desired position $\boldsymbol{q}_d$ to be constant, the closed-loop equation may then be written in terms of the new state vector $\begin{bmatrix}\boldsymbol{\xi}^T & \tilde{\boldsymbol{q}}^T & \dot{\boldsymbol{q}}^T\end{bmatrix}^T$ as

$$\frac{d}{dt}\begin{bmatrix}\boldsymbol{\xi}\\ \tilde{\boldsymbol{q}}\\ \dot{\boldsymbol{q}}\end{bmatrix}=\begin{bmatrix}-A\boldsymbol{\xi}+AB\tilde{\boldsymbol{q}}\\ -\dot{\boldsymbol{q}}\\ M(\boldsymbol{q})^{-1}\left[K_p\tilde{\boldsymbol{q}}-K_v[\boldsymbol{\xi}-B\tilde{\boldsymbol{q}}]+\boldsymbol{g}(\boldsymbol{q}_d)-C(\boldsymbol{q}_d-\tilde{\boldsymbol{q}},\dot{\boldsymbol{q}})\dot{\boldsymbol{q}}-\boldsymbol{g}(\boldsymbol{q}_d-\tilde{\boldsymbol{q}})\right]\end{bmatrix} \tag{13.18}$$

which, since $\boldsymbol{q}_d$ is constant, is an autonomous differential equation. Since the matrix K_p has been picked so that $\lambda_{\min}\{K_p\} > k_g$, then the origin $\begin{bmatrix}\boldsymbol{\xi}^T & \tilde{\boldsymbol{q}}^T & \dot{\boldsymbol{q}}^T\end{bmatrix}^T = \mathbf{0} \in \mathbb{R}^{3n}$ is the unique equilibrium of this equation (see the arguments in Section 8.2).

In order to study the stability of the origin, consider the Lyapunov function candidate

$$V(\boldsymbol{\xi},\tilde{\boldsymbol{q}},\dot{\boldsymbol{q}}) = \mathcal{K}(\boldsymbol{q}_d-\tilde{\boldsymbol{q}},\dot{\boldsymbol{q}}) + f(\tilde{\boldsymbol{q}}) + \frac{1}{2}(\boldsymbol{\xi}-B\tilde{\boldsymbol{q}})^T K_v B^{-1}(\boldsymbol{\xi}-B\tilde{\boldsymbol{q}}) \tag{13.19}$$

where

$$\mathcal{K}(\boldsymbol{q}_d-\tilde{\boldsymbol{q}},\dot{\boldsymbol{q}}) = \frac{1}{2}\dot{\boldsymbol{q}}^T M(\boldsymbol{q}_d-\tilde{\boldsymbol{q}})\dot{\boldsymbol{q}}$$

$$f(\tilde{\boldsymbol{q}}) = \mathcal{U}(\boldsymbol{q}_d-\tilde{\boldsymbol{q}}) - \mathcal{U}(\boldsymbol{q}_d) + \boldsymbol{g}(\boldsymbol{q}_d)^T\tilde{\boldsymbol{q}} + \frac{1}{2}\tilde{\boldsymbol{q}}^T K_p\tilde{\boldsymbol{q}}\,.$$

Notice first that the diagonal matrix K_vB^{-1} is positive definite. Since it has been assumed that $\lambda_{\min}\{K_p\} > k_g$, we have from Lemma 8.1 that $f(\tilde{\boldsymbol{q}})$ is a (globally) positive definite function of $\tilde{\boldsymbol{q}}$. Consequently, the function $V(\boldsymbol{\xi},\tilde{\boldsymbol{q}},\dot{\boldsymbol{q}})$ is also globally positive definite.

The time derivative of the Lyapunov function candidate yields

$$\begin{aligned}\dot{V}(\boldsymbol{\xi},\tilde{\boldsymbol{q}},\dot{\boldsymbol{q}}) &= \dot{\boldsymbol{q}}^T M(\boldsymbol{q})\ddot{\boldsymbol{q}} + \frac{1}{2}\dot{\boldsymbol{q}}^T\dot{M}(\boldsymbol{q})\dot{\boldsymbol{q}} - \dot{\tilde{\boldsymbol{q}}}^T\boldsymbol{g}(\boldsymbol{q}_d-\tilde{\boldsymbol{q}}) + \boldsymbol{g}(\boldsymbol{q}_d)^T\dot{\tilde{\boldsymbol{q}}}\\ &\quad + \tilde{\boldsymbol{q}}^T K_p\dot{\tilde{\boldsymbol{q}}} + [\boldsymbol{\xi}-B\tilde{\boldsymbol{q}}]^T K_v B^{-1}\left[\dot{\boldsymbol{\xi}}-B\dot{\tilde{\boldsymbol{q}}}\right].\end{aligned}$$

Using the closed-loop Equation (13.18) to solve for $\dot{\boldsymbol{\xi}}$, $\dot{\tilde{\boldsymbol{q}}}$ and $M(\boldsymbol{q})\ddot{\boldsymbol{q}}$, and canceling out some terms, we obtain

$$\begin{aligned}\dot{V}(\boldsymbol{\xi},\tilde{\boldsymbol{q}},\dot{\boldsymbol{q}}) &= -(\boldsymbol{\xi}-B\tilde{\boldsymbol{q}})^T K_v B^{-1} A(\boldsymbol{\xi}-B\tilde{\boldsymbol{q}})\\ &= -\begin{bmatrix}\boldsymbol{\xi}\\ \tilde{\boldsymbol{q}}\\ \dot{\boldsymbol{q}}\end{bmatrix}^T\begin{bmatrix}K_vB^{-1}A & -K_vA & 0\\ -K_vA & BK_vA & 0\\ 0 & 0 & 0\end{bmatrix}\begin{bmatrix}\boldsymbol{\xi}\\ \tilde{\boldsymbol{q}}\\ \dot{\boldsymbol{q}}\end{bmatrix}\end{aligned} \tag{13.20}$$

where we used (*cf.* Property 4.2)

$$\dot{\boldsymbol{q}}^T \left[\frac{1}{2} \dot{M}(\boldsymbol{q}) - C(\boldsymbol{q}, \dot{\boldsymbol{q}}) \right] \dot{\boldsymbol{q}} = 0 \,.$$

Clearly, the time derivative $\dot{V}(\boldsymbol{\xi}, \tilde{\boldsymbol{q}}, \dot{\boldsymbol{q}})$ of the Lyapunov function candidate is a globally semidefinite negative function. For this reason, according to the Theorem 2.3, the origin of the closed-loop Equation (13.18) is stable.

Since the closed-loop Equation (13.18) is autonomous, direct application of La Salle's Theorem 2.7 allows one to guarantee global asymptotic stability of the origin corresponding to the state space of the closed-loop system (*cf.* Problem 4 at the end of the chapter). Nevertheless, an alternative analysis, similar to that presented in Section 13.1, may also be carried out.

Since the origin $\begin{bmatrix} \boldsymbol{\xi}^T & \tilde{\boldsymbol{q}}^T & \dot{\boldsymbol{q}}^T \end{bmatrix}^T = \boldsymbol{0} \in \mathbb{R}^{3n}$ is a stable equilibrium, then if $\begin{bmatrix} \boldsymbol{\xi}(t)^T & \tilde{\boldsymbol{q}}(t)^T & \dot{\boldsymbol{q}}(t)^T \end{bmatrix}^T \to \boldsymbol{0} \in \mathbb{R}^{3n}$ when $t \to \infty$ (for all initial conditions), *i.e.* the equilibrium is globally attractive, then the origin is a globally asymptotically stable equilibrium. It is precisely this property that we show next.

In the development below we invoke further properties of the dynamic model of robot manipulators. Specifically, assuming that $\boldsymbol{q}, \dot{\boldsymbol{q}} \in L_\infty^n$ we have

- $M(\boldsymbol{q})^{-1}, \frac{d}{dt} M(\boldsymbol{q}) \in L_\infty^{n \times n}$
- $C(\boldsymbol{q}, \dot{\boldsymbol{q}})\dot{\boldsymbol{q}}, \boldsymbol{g}(\boldsymbol{q}), \frac{d}{dt}\boldsymbol{g}(\boldsymbol{q}) \in L_\infty^n$.

The latter follows from the regularity of the functions that define M, $\boldsymbol{g}$ and C. By the same reasoning, if moreover $\ddot{\boldsymbol{q}} \in L_\infty^n$ then

- $\frac{d}{dt} \left[C(\boldsymbol{q}, \dot{\boldsymbol{q}})\dot{\boldsymbol{q}} \right] \in L_\infty^n$.

The Lyapunov function $V(\boldsymbol{\xi}, \tilde{\boldsymbol{q}}, \dot{\boldsymbol{q}})$ given in (13.19) is positive definite and is composed of the sum of the following three non-negative terms

- $\frac{1}{2}\dot{\boldsymbol{q}}^T M(\boldsymbol{q})\dot{\boldsymbol{q}}$
- $\mathcal{U}(\boldsymbol{q}_d - \tilde{\boldsymbol{q}}) - \mathcal{U}(\boldsymbol{q}_d) + \boldsymbol{g}(\boldsymbol{q}_d)^T \tilde{\boldsymbol{q}} + \frac{1}{2}\tilde{\boldsymbol{q}}^T K_p \tilde{\boldsymbol{q}}$
- $\frac{1}{2}(\boldsymbol{\xi} - B\tilde{\boldsymbol{q}})^T K_v B^{-1} (\boldsymbol{\xi} - B\tilde{\boldsymbol{q}})$.

Since the time derivative $\dot{V}(\boldsymbol{\xi}, \tilde{\boldsymbol{q}}, \dot{\boldsymbol{q}})$, expressed in (13.6) is a negative semidefinite function, the Lyapunov function $V(\boldsymbol{\xi}, \tilde{\boldsymbol{q}}, \dot{\boldsymbol{q}})$ is bounded along trajectories. Therefore, the three non-negative listed terms above are also bounded along trajectories. Since, moreover, the potential energy $\mathcal{U}(\boldsymbol{q})$ of robots having only revolute joints is always bounded in its absolute value, it follows that

$$\dot{\boldsymbol{q}}, \tilde{\boldsymbol{q}}, \boldsymbol{\xi}, \boldsymbol{\xi} - B\tilde{\boldsymbol{q}} \in L_\infty^n \,. \tag{13.21}$$

Incorporating this information in the closed-loop Equation (13.18), and knowing that $M(\boldsymbol{q}_d - \tilde{\boldsymbol{q}})^{-1}$ and $\boldsymbol{g}(\boldsymbol{q}_d - \tilde{\boldsymbol{q}})$ are bounded for all $\boldsymbol{q}_d, \tilde{\boldsymbol{q}} \in L_\infty^n$ and also that $C(\boldsymbol{q}_d - \tilde{\boldsymbol{q}}, \dot{\boldsymbol{q}})\dot{\boldsymbol{q}}$ is bounded for all $\boldsymbol{q}_d, \tilde{\boldsymbol{q}}, \dot{\boldsymbol{q}} \in L_\infty^n$, it follows that the time derivative of the state vector is bounded, *i.e.*

$$\dot{\boldsymbol{\xi}},\ \dot{\tilde{\boldsymbol{q}}},\ \ddot{\boldsymbol{q}} \in L_\infty^n, \tag{13.22}$$

and therefore, it is also true that

$$\left[\dot{\boldsymbol{\xi}} - B\dot{\tilde{\boldsymbol{q}}}\right] \in L_\infty^n. \tag{13.23}$$

Using again the closed-loop Equation (13.18), we can compute the second time derivative of the variables state to obtain

$$\begin{aligned}
\ddot{\boldsymbol{\xi}} &= -A\dot{\boldsymbol{\xi}} + AB\dot{\tilde{\boldsymbol{q}}} \\
\ddot{\tilde{\boldsymbol{q}}} &= -\ddot{\boldsymbol{q}} \\
\boldsymbol{q}^{(3)} &= -M(\boldsymbol{q})^{-1}\left[\frac{d}{dt}M(\boldsymbol{q})\right]M(\boldsymbol{q})^{-1}\left[K_p\tilde{\boldsymbol{q}} - K_v\left[\boldsymbol{\xi} - B\tilde{\boldsymbol{q}}\right] - C(\boldsymbol{q},\dot{\boldsymbol{q}})\dot{\boldsymbol{q}}\right. \\
&\quad \left. +\boldsymbol{g}(\boldsymbol{q}_d) - \boldsymbol{g}(\boldsymbol{q})\right] \\
&\quad + M(\boldsymbol{q})^{-1}\left[K_p\dot{\tilde{\boldsymbol{q}}} - K_v\left(\dot{\boldsymbol{\xi}} - B\dot{\tilde{\boldsymbol{q}}}\right) - \frac{d}{dt}\left(C(\boldsymbol{q},\dot{\boldsymbol{q}})\dot{\boldsymbol{q}}\right) + \frac{d}{dt}\boldsymbol{g}(\boldsymbol{q})\right]
\end{aligned}$$

where $\boldsymbol{q}^{(3)}$ denotes the third time derivative of the joint position $\boldsymbol{q}$ and we used

$$\frac{d}{dt}\left[M(\boldsymbol{q})^{-1}\right] = -M(\boldsymbol{q})^{-1}\left[\frac{d}{dt}M(\boldsymbol{q})\right]M(\boldsymbol{q})^{-1}.$$

In (13.21) and (13.22) we concluded that $\boldsymbol{\xi}, \tilde{\boldsymbol{q}}, \dot{\boldsymbol{q}}, \dot{\boldsymbol{\xi}}, \dot{\tilde{\boldsymbol{q}}}, \ddot{\boldsymbol{q}} \in L_\infty^n$ then, from the properties stated at the beginning of this analysis, we obtain

$$\ddot{\boldsymbol{\xi}}, \ddot{\tilde{\boldsymbol{q}}}, \boldsymbol{q}^{(3)} \in L_\infty^n, \tag{13.24}$$

and therefore also

$$\ddot{\boldsymbol{\xi}} - B\ddot{\tilde{\boldsymbol{q}}} \in L_\infty^n. \tag{13.25}$$

On the other hand, from the time derivative $\dot{V}(\boldsymbol{\xi}, \tilde{\boldsymbol{q}}, \dot{\boldsymbol{q}})$, expressed in (13.20), we get

$$\boldsymbol{\xi} - B\tilde{\boldsymbol{q}} \in L_2^n. \tag{13.26}$$

Considering next (13.23), (13.26) and Lemma A.5, we conclude that

$$\lim_{t\to\infty} \boldsymbol{\xi}(t) - B\tilde{\boldsymbol{q}}(t) = \boldsymbol{0}. \tag{13.27}$$

Hence, using (13.27), (13.21), (13.23) and (13.25) together with Lemma A.6 we get

$$\lim_{t\to\infty} \dot{\boldsymbol{\xi}}(t) - B\dot{\tilde{\boldsymbol{q}}}(t) = \boldsymbol{0}$$

and consequently, taking $\dot{\boldsymbol{\xi}}$ and $\dot{\tilde{\boldsymbol{q}}}$ from the closed-loop Equation (13.18), we get

$$\lim_{t\to\infty} -A[\boldsymbol{\xi}(t) - B\tilde{\boldsymbol{q}}(t)] + B\dot{\boldsymbol{q}} = \mathbf{0}\,.$$

From this last expression and since we showed in (13.27) that $\lim_{t\to\infty} \boldsymbol{\xi}(t) - B\tilde{\boldsymbol{q}}(t) = \mathbf{0}$ it finally follows that

$$\lim_{t\to\infty} \dot{\boldsymbol{q}}(t) = \mathbf{0}\,. \tag{13.28}$$

We show next that $\lim_{t\to\infty} \tilde{\boldsymbol{q}}(t) = \mathbf{0} \in \mathbb{R}^n$. Using again Lemma A.6 with (13.28), (13.21), (13.22) and (13.24) we have

$$\lim_{t\to\infty} \ddot{\boldsymbol{q}}(t) = \mathbf{0}\,.$$

Taking this into account in the closed-loop Equation (13.18) as well as (13.27) and (13.28), we get

$$\lim_{t\to\infty} M(\boldsymbol{q}_d - \tilde{\boldsymbol{q}}(t))^{-1}\left[K_p\tilde{\boldsymbol{q}}(t) + \boldsymbol{g}(\boldsymbol{q}_d) - \boldsymbol{g}(\boldsymbol{q}_d - \tilde{\boldsymbol{q}}(t))\right] = \mathbf{0}$$

and therefore, since $\lambda_{\min}\{K_p\} > k_g$ we finally obtain from the methodology presented in Section 8.2,

$$\lim_{t\to\infty} \tilde{\boldsymbol{q}}(t) = \mathbf{0}\,. \tag{13.29}$$

The rest of the proof, that is that $\lim_{t\to\infty} \boldsymbol{\xi}(t) = \mathbf{0}$, follows directly from (13.27) and (13.29).

This completes the proof of global attractivity of the origin and, since we have already shown that the origin is Lyapunov stable, of global asymptotic stability of the origin of the closed-loop Equation (13.18).

We present next an example that demonstrates the performance that may be achieved with P"D" control with gravity compensation in particular, on the Pelican robot.

Example 13.2. Consider the Pelican robot presented in Chapter 5 and depicted in Figure 5.2. The components of the vector of gravitational torques $\boldsymbol{g}(\boldsymbol{q})$ are given by

$$\begin{aligned} g_1(\boldsymbol{q}) &= (m_1 l_{c1} + m_2 l_1) g \sin(q_1) + m_2 l_{c2} g \sin(q_1 + q_2) \\ g_2(\boldsymbol{q}) &= m_2 l_{c2} g \sin(q_1 + q_2)\,. \end{aligned}$$

According to Property 4.3, the constant k_g may be obtained as (see also Example 9.2):

$$
\begin{aligned}
k_g &= n\left(\max_{i,j,q}\left|\frac{\partial g_i(\boldsymbol{q})}{\partial q_j}\right|\right) \\
&= n((m_1 l_{c1} + m_2 l_1)g + m_2 l_{c2} g) \\
&= 23.94 \quad \left[\mathrm{kg\ m^2/s^2}\right].
\end{aligned}
$$

Consider the P"D" control with desired gravity compensation for this robot in position control. Let the design matrices $K_p, K_v A, B$ be diagonal and positive definite and satisfy

$$
\lambda_{\min}\{K_p\} > k_g .
$$

In particular, these matrices are taken to be

$$
\begin{aligned}
K_p &= \mathrm{diag}\{k_p\} = \mathrm{diag}\{30\} \quad [\mathrm{Nm/rad}], \\
K_v &= \mathrm{diag}\{k_v\} = \mathrm{diag}\{7,\ 3\} \quad [\mathrm{Nm\ s/rad}], \\
A &= \mathrm{diag}\{a_i\} = \mathrm{diag}\{30,\ 70\} \quad [1/\mathrm{s}], \\
B &= \mathrm{diag}\{b_i\} = \mathrm{diag}\{30,\ 70\} \quad [1/\mathrm{s}].
\end{aligned}
$$

The components of the control input $\boldsymbol{\tau}$ are given by

$$
\begin{aligned}
\tau_1 &= k_p\tilde{q}_1 - k_v\vartheta_1 + g_1(\boldsymbol{q}_d) \\
\tau_2 &= k_p\tilde{q}_2 - k_v\vartheta_2 + g_2(\boldsymbol{q}_d) \\
\dot{x}_1 &= -a_1 x_1 - a_1 b_1 q_1 \\
\dot{x}_2 &= -a_2 x_2 - a_2 b_2 q_2 \\
\vartheta_1 &= x_1 + b_1 q_1 \\
\vartheta_2 &= x_2 + b_2 q_2 .
\end{aligned}
$$

The initial conditions corresponding to the positions, velocities and states of the filters, are chosen as

$$
\begin{aligned}
q_1(0) &= 0, \qquad & q_2(0) &= 0 \\
\dot{q}_1(0) &= 0, \qquad & \dot{q}_2(0) &= 0 \\
x_1(0) &= 0, \qquad & x_2(0) &= 0 .
\end{aligned}
$$

The desired joint positions are

$$
q_{d1} = \pi/10, \qquad q_{d2} = \pi/30 \quad [\mathrm{rad}].
$$

In terms of the state vector of the closed-loop equation, the initial state is

$$
\begin{bmatrix} \boldsymbol{\xi}(0) \\ \\ \tilde{\boldsymbol{q}}(0) \\ \\ \dot{\boldsymbol{q}}(0) \end{bmatrix} = \begin{bmatrix} b_1\pi/10 \\ b_2\pi/30 \\ \pi/10 \\ \pi/30 \\ 0 \\ 0 \end{bmatrix} = \begin{bmatrix} 9.423 \\ 7.329 \\ 0.3141 \\ 0.1047 \\ 0 \\ 0 \end{bmatrix}.
$$

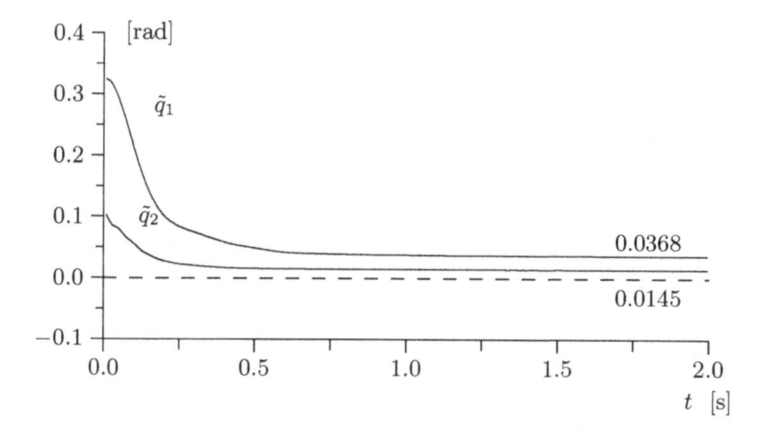

Figure 13.4. Graphs of position errors $\tilde{q}_1(t)$ and $\tilde{q}_2(t)$

Figure 13.4 shows the experimental results; again, as in the previous controller, it shows that the components of the position error $\tilde{\boldsymbol{q}}(t)$ tend asymptotically to a small constant nonzero value due, mainly, to the friction effects in the prototype. ♢

13.3 Conclusions

We may summarize the material of this chapter in the following remarks.

Consider the P"D" control with gravity compensation for n-DOF robots. Assume that the desired position $\boldsymbol{q}_d$ is constant.

- If the matrices K_p, K_v, A and B of the controller P"D" with gravity compensation are diagonal positive definite, then the origin of the closed-loop equation expressed in terms of the state vector $\begin{bmatrix} \boldsymbol{\xi}^T & \tilde{\boldsymbol{q}}^T & \dot{\boldsymbol{q}}^T \end{bmatrix}^T$, is a globally asymptotically stable equilibrium. Consequently, for any initial condition $\boldsymbol{q}(0), \dot{\boldsymbol{q}}(0) \in \mathbb{R}^n$, we have $\lim_{t\to\infty} \tilde{\boldsymbol{q}}(t) = \boldsymbol{0} \in \mathbb{R}^n$.

Consider the P"D" control with desired gravity compensation for n-DOF robots. Assume that the desired position $\boldsymbol{q}_d$ is constant.

- If the matrices K_p, K_v, A and B of the controller P"D" with desired gravity compensation are taken diagonal positive definite, and such that $\lambda_{\min}\{K_p\} > k_g$, then the origin of the closed-loop equation, expressed in terms of the state vector $\begin{bmatrix} \boldsymbol{\xi}^T & \tilde{\boldsymbol{q}}^T & \dot{\boldsymbol{q}}^T \end{bmatrix}^T$ is globally asymptotically stable. In particular, for any initial condition $\boldsymbol{q}(0), \dot{\boldsymbol{q}}(0) \in \mathbb{R}^n$, we have $\lim_{t\to\infty} \tilde{\boldsymbol{q}}(t) = \boldsymbol{0} \in \mathbb{R}^n$.

Bibliography

Studies of motion control for robot manipulators without the requirement of velocity measurements, started at the beginning of the 1990s. Some of the early related references are the following:

- Nicosia S., Tomei P., 1990, *"Robot control by using joint position measurements"*, IEEE Transactions on Automatic Control, Vol. 35, No. 9, September.
- Berghuis H., Löhnberg P., Nijmeijer H., 1991, *"Tracking control of robots using only position measurements"*, Proceedings of IEEE Conference on Decision and Control, Brighton, England, December, pp. 1049–1050.
- Canudas C., Fixot N., 1991, *"Robot control via estimated state feedback"*, IEEE Transactions on Automatic Control, Vol. 36, No. 12, December.
- Canudas C., Fixot N., Åström K. J., 1992, *"Trajectory tracking in robot manipulators via nonlinear estimated state feedback"*, IEEE Transactions on Robotics and Automation, Vol. 8, No. 1, February.
- Ailon A., Ortega R., 1993, *"An observer-based set-point controller for robot manipulators with flexible joints"*, Systems and Control Letters, Vol. 21, October, pp. 329–335.

The motion control problem for a time-varying trajectory $\boldsymbol{q}_d(t)$ without velocity measurements, with a rigorous proof of global asymptotic stability of the origin of the closed-loop system, was first solved for one-degree-of-freedom robots (including a term that is quadratic in the velocities) in

- Loría A., 1996, *"Global tracking control of one degree of freedom Euler-Lagrange systems without velocity measurements"*, European Journal of Control, Vol. 2, No. 2, June.

This result was extended to the case of n-DOF robots in

- Zergeroglu E., Dawson, D. M., Queiroz M. S. de, Krstić M., 2000, *"On global output feedback tracking control of robot manipulators"*, in Proceedings of Conferenece on Decision and Control, Sydney, Australia, pp. 5073–5078.

The controller called here, P"D" with gravity compensation and characterized by Equations (13.2)–(13.3) was independently proposed in

- Kelly R., 1993, *"A simple set-point robot controller by using only position measurements"*, 12th IFAC World Congress, Vol. 6, Sydney, Australia, July, pp. 173–176.
- Berghuis H., Nijmeijer H., 1993, *"Global regulation of robots using only position measurements"*, Systems and Control Letters, Vol. 21, October, pp. 289–293.

The controller called here, P"D" with desired gravity compensation and characterized by Equations (13.16)–(13.17) was independently proposed in the latter two references; the formal proof of global asymptotic stability was presented in the second.

Problems

1. Consider the following variant of the controller P"D" with gravity compensation[1]:

$$\begin{aligned} \boldsymbol{\tau} &= K_p\tilde{\boldsymbol{q}} + K_v\boldsymbol{\vartheta} + \boldsymbol{g}(\boldsymbol{q}) \\ \dot{\boldsymbol{x}} &= -A\boldsymbol{x} - AB\tilde{\boldsymbol{q}} \\ \boldsymbol{\vartheta} &= \boldsymbol{x} + B\tilde{\boldsymbol{q}} \end{aligned}$$

where $K_p, K_v \in \mathbb{R}^{n\times n}$ are diagonal positive definite matrices, $A = \mathrm{diag}\{a_i\}$ and $B = \mathrm{diag}\{b_i\}$ with a_i, b_i real strictly positive numbers. Assume that the desired joint position $\boldsymbol{q}_d \in \mathbb{R}^n$ is constant.

   a) Obtain the closed-loop equation expressed in terms of the state vector $\begin{bmatrix}\boldsymbol{x}^T & \tilde{\boldsymbol{q}}^T & \dot{\boldsymbol{q}}^T\end{bmatrix}^T$.

   b) Verify that the vector

$$\begin{bmatrix}\boldsymbol{x} \\ \tilde{\boldsymbol{q}} \\ \dot{\boldsymbol{q}}\end{bmatrix} = \begin{bmatrix}\boldsymbol{0} \\ \boldsymbol{0} \\ \boldsymbol{0}\end{bmatrix} \in \mathbb{R}^{3n}$$

   is the unique equilibrium of the closed-loop equation.

   c) Show that the origin of the closed-loop equation is a stable equilibrium point.

   Hint: Use the following Lyapunov function candidate[2]:

$$\begin{aligned} V(\boldsymbol{x}, \tilde{\boldsymbol{q}}, \dot{\boldsymbol{q}}) = \frac{1}{2}\dot{\boldsymbol{q}}^T M(\boldsymbol{q})\dot{\boldsymbol{q}} &+ \frac{1}{2}\tilde{\boldsymbol{q}}^T K_p\tilde{\boldsymbol{q}} \\ &+ \frac{1}{2}\left(\boldsymbol{x} + B\tilde{\boldsymbol{q}}\right)^T K_v B^{-1}\left(\boldsymbol{x} + B\tilde{\boldsymbol{q}}\right) . \end{aligned}$$

2. Consider the model of robots with elastic joints (3.27) and (3.28),

[1] This controller was analyzed in Berghuis H., Nijmeijer H., 1993, *"Global regulation of robots using only position measurements"*, Systems and Control Letters, Vol. 21, October, pp. 289–293.

[2] By virtue of La Salle's Theorem it may also be proved that the origin is globally asymptotically stable.

$$
\begin{aligned}
M(\boldsymbol{q})\ddot{\boldsymbol{q}} + C(\boldsymbol{q},\dot{\boldsymbol{q}})\dot{\boldsymbol{q}} + \boldsymbol{g}(\boldsymbol{q}) + K(\boldsymbol{q}-\boldsymbol{\theta}) &= \boldsymbol{0} \\
J\ddot{\boldsymbol{\theta}} - K(\boldsymbol{q}-\boldsymbol{\theta}) &= \boldsymbol{\tau}\,.
\end{aligned}
$$

It is assumed that only the vector of positions $\boldsymbol{\theta}$ of motors axes, but not the velocities vector $\dot{\boldsymbol{\theta}}$, is measured. We require that $\boldsymbol{q}(t) \to \boldsymbol{q}_d$, where $\boldsymbol{q}_d$ is constant.

A variant of the P"D" control with desired gravity compensation is[3]

$$
\begin{aligned}
\boldsymbol{\tau} &= K_p\tilde{\boldsymbol{\theta}} - K_v\boldsymbol{\vartheta} + \boldsymbol{g}(\boldsymbol{q}_d) \\
\dot{\boldsymbol{x}} &= -A\boldsymbol{x} - AB\boldsymbol{\theta} \\
\boldsymbol{\vartheta} &= \boldsymbol{x} + B\boldsymbol{\theta}
\end{aligned}
$$

where

$$
\begin{aligned}
\tilde{\boldsymbol{\theta}} &= \boldsymbol{q}_d - \boldsymbol{\theta} + K^{-1}\boldsymbol{g}(\boldsymbol{q}_d) \\
\tilde{\boldsymbol{q}} &= \boldsymbol{q}_d - \boldsymbol{q}
\end{aligned}
$$

and $K_p, K_v, A, B \in \mathbb{R}^{n\times n}$ are diagonal positive definite matrices.

a) Verify that the closed-loop equation in terms of the state vector $\begin{bmatrix}\boldsymbol{\xi}^T & \tilde{\boldsymbol{q}}^T & \tilde{\boldsymbol{\theta}}^T & \dot{\boldsymbol{q}}^T & \dot{\boldsymbol{\theta}}^T\end{bmatrix}^T$ may be written as

$$
\frac{d}{dt}\begin{bmatrix}\boldsymbol{\xi} \\ \tilde{\boldsymbol{q}} \\ \tilde{\boldsymbol{\theta}} \\ \dot{\boldsymbol{q}} \\ \dot{\boldsymbol{\theta}}\end{bmatrix} = \begin{bmatrix} -A\boldsymbol{\xi} + AB\tilde{\boldsymbol{\theta}} \\ -\dot{\boldsymbol{q}} \\ -\dot{\boldsymbol{\theta}} \\ M(\boldsymbol{q})^{-1}\left[-K(\tilde{\boldsymbol{\theta}}-\tilde{\boldsymbol{q}}) + \boldsymbol{g}(\boldsymbol{q}_d) - C(\boldsymbol{q},\dot{\boldsymbol{q}})\dot{\boldsymbol{q}} - \boldsymbol{g}(\boldsymbol{q})\right] \\ J^{-1}\left[K_p\tilde{\boldsymbol{\theta}} - K_v(\boldsymbol{\xi} - B\tilde{\boldsymbol{\theta}}) + K(\tilde{\boldsymbol{\theta}}-\tilde{\boldsymbol{q}})\right] \end{bmatrix}
$$

where $\boldsymbol{\xi} = \boldsymbol{x} + B\left[\boldsymbol{q}_d + K^{-1}\boldsymbol{g}(\boldsymbol{q}_d)\right]$.

b) Verify that the origin is an equilibrium of the closed-loop equation.

c) Show that if $\lambda_{\min}\{K_p\} > k_g$ and $\lambda_{\min}\{K\} > k_g$, then the origin is a stable equilibrium point.

Hint: Use the following Lyapunov function and La Salle's Theorem 2.7.

[3] This controller was proposed and analyzed in Kelly R., Ortega R., Ailon A., Loria A., 1994, *"Global regulation of flexible joint robots using approximate differentiation"*, IEEE Transactions on Automatic Control, Vol. 39, No. 6, June, pp. 1222–1224.

$$V(\tilde{\boldsymbol{q}}, \tilde{\boldsymbol{\theta}}, \dot{\boldsymbol{q}}, \dot{\boldsymbol{\theta}}) = V_1(\tilde{\boldsymbol{q}}, \tilde{\boldsymbol{\theta}}, \dot{\boldsymbol{q}}, \dot{\boldsymbol{\theta}}) + \frac{1}{2}\tilde{\boldsymbol{q}}^T K \tilde{\boldsymbol{q}} + V_2(\tilde{\boldsymbol{q}}) + \frac{1}{2}\left[\boldsymbol{\xi} - B\tilde{\boldsymbol{\theta}}\right]^T K_v B^{-1}\left[\boldsymbol{\xi} - B\tilde{\boldsymbol{\theta}}\right]$$

where

$$V_1(\tilde{\boldsymbol{q}}, \tilde{\boldsymbol{\theta}}, \dot{\boldsymbol{q}}, \dot{\boldsymbol{\theta}}) = \frac{1}{2}\dot{\boldsymbol{q}}^T M(\boldsymbol{q})\dot{\boldsymbol{q}} + \frac{1}{2}\dot{\boldsymbol{\theta}}^T J\dot{\boldsymbol{\theta}} + \frac{1}{2}\tilde{\boldsymbol{\theta}}^T K_p\tilde{\boldsymbol{\theta}} + \frac{1}{2}\tilde{\boldsymbol{\theta}}^T K\tilde{\boldsymbol{\theta}} - \tilde{\boldsymbol{\theta}}^T K\tilde{\boldsymbol{q}}$$

$$V_2(\tilde{\boldsymbol{q}}) = \mathcal{U}(\boldsymbol{q}_d - \tilde{\boldsymbol{q}}) - \mathcal{U}(\boldsymbol{q}_d) + \tilde{\boldsymbol{q}}^T \boldsymbol{g}(\boldsymbol{q}_d)$$

and verify that

$$\dot{V}(\tilde{\boldsymbol{q}}, \tilde{\boldsymbol{\theta}}, \dot{\boldsymbol{q}}, \dot{\boldsymbol{\theta}}) = -\left[\boldsymbol{\xi} - B\tilde{\boldsymbol{\theta}}\right]^T K_v B^{-1} A\left[\boldsymbol{\xi} - B\tilde{\boldsymbol{\theta}}\right].$$

3. Use La Salle's Theorem 2.7 to show global asymptotic stability of the origin of the closed-loop equations corresponding to the P"D" controller with gravity compensation, *i.e.* Equation (13.4).
4. Use La Salle's Theorem 2.7 to show global asymptotic stability of the origin of the closed-loop equations corresponding to the P"D" controller with desired gravity compensation, *i.e.* Equation (13.18).

14

Introduction to Adaptive Robot Control

Up to this chapter we have studied several control techniques which achieve the objective of position and motion control of manipulators. The standing implicit assumptions in the preceding chapters are that:

- The model is accurately known, *i.e.* either all the nonlinearities involved are known or they are negligible.
- The constant physical parameters such as link inertias, masses, lengths to the centers of mass and even the masses of the diverse objects which may be handled by the end-effector of the robot, are accurately known.

Obviously, while these considerations allow one to prove certain stability and convergence properties for the controllers studied in previous chapters, they must be taken with care. In robot control practice, either of these assumptions or both, may not hold. For instance, we may be neglecting considerable joint elasticity, friction or, even if we think we know accurately the masses and inertias of the robot, we cannot estimate the mass of the objects carried by the end-effector, which depend on the task accomplished.

Two general techniques in control theory and practice deal with these phenomena, respectively: *robust* control and *adaptive* control. Roughly, the first aims at controlling, with a small error, a *class* of robot manipulators with the same robust controller. That is, given a robot manipulator model, one designs a control law which achieves the motion control objective, with a small error, for the given model but to which is added a *known* nonlinearity.

Adaptive control is a design approach tailored for high performance applications in control systems with uncertainty in the parameters. That is, uncertainty in the dynamic system is assumed to be characterized by a set of unknown *constant* parameters. However, the design of adaptive controllers requires the precise knowledge of the structure of the system being controlled.

Certainly one may consider other variants such as adaptive control for systems with time-varying parameters, or *robust adaptive* control for systems with structural and parameter uncertainty.

In this and the following chapters we concentrate specifically on adaptive control of robot manipulators with *constant* parameters and for which we assume that we have no structural uncertainties. In this chapter we present an introduction to adaptive control of manipulators. In subsequent chapters we describe and analyze two adaptive controllers for robots. They correspond to the adaptive versions of

- PD control with adaptive desired gravity compensation,
- PD control with adaptive compensation.

14.1 Parameterization of the Dynamic Model

The dynamic model of robot manipulators[1], as we know, is given by Lagrange's equations, which we repeat here in their compact form:

$$M(\boldsymbol{q})\ddot{\boldsymbol{q}} + C(\boldsymbol{q}, \dot{\boldsymbol{q}})\dot{\boldsymbol{q}} + \boldsymbol{g}(\boldsymbol{q}) = \boldsymbol{\tau}\,. \tag{14.1}$$

In previous chapters we have not emphasized the fact that the elements of the inertia matrix $M(\boldsymbol{q})$, the centrifugal and Coriolis forces matrix $C(\boldsymbol{q}, \dot{\boldsymbol{q}})$ and the vector of gravitational torques $\boldsymbol{g}(\boldsymbol{q})$, depend not only on the geometry of the corresponding robot but also on the numerical values of diverse parameters such as masses, inertias and distances to centers of mass.

The scenario in which these parameters and the geometry of the robot are exactly known is called in the context of adaptive control, *the ideal case.* A more realistic scenario is usually that in which the numerical values of some parameters of the robot are unknown. Such is the case, for instance, when the object manipulated by the end-effector of the robot (which may be considered as part of the last link) is of uncertain mass and/or inertia. The consequence in this situation cannot be overestimated; due to the uncertainty in some of the parameters of the robot model it is impossible to use the model-based control laws from any of the previous chapters since they rely on an accurate knowledge of the dynamic model. The adaptive controllers are useful precisely in this *more realistic* case.

To emphasize the dependence of the dynamic model on the dynamic parameters, from now on we write the dynamic model (14.1) explicitly as a function of the vector of unknown dynamic parameters, $\boldsymbol{\theta}$, that is[2],

[1] Under the ideal conditions of rigid links, no elasticity at joints, no friction and having actuators with negligible dynamics.

[2] In this textbook we have used $\boldsymbol{\theta}$ to denote the joint positions of the motor shafts for models of robots with elastic joints. With an abuse of notation, in this and the

$$M(\boldsymbol{q},\boldsymbol{\theta})\ddot{\boldsymbol{q}} + C(\boldsymbol{q},\dot{\boldsymbol{q}},\boldsymbol{\theta})\dot{\boldsymbol{q}} + \boldsymbol{g}(\boldsymbol{q},\boldsymbol{\theta}) = \boldsymbol{\tau}\,. \tag{14.2}$$

The vector of parameters $\boldsymbol{\theta}$ may be of any dimension, that is, it does not depend in any specific way on the number of degrees of freedom or on whether the robot has revolute or prismatic joints *etc.* Notwithstanding, an upper-bound on the dimension is determined by the number of degrees of freedom. Therefore, we simply say that $\boldsymbol{\theta} \in \mathbb{R}^m$ where m is some known constant. It is also important to stress that the dynamic parameters, denoted here by $\boldsymbol{\theta}$, *do not* necessarily correspond to the individual physical parameters of the robot, as is illustrated in the following example.

Example 14.1. Consider the example of an ideal pendulum with its mass m concentrated at the tip, at a distance l from its axis of rotation. Its dynamic model is given by

$$ml^2\ddot{q} + mgl\sin(q) = \tau \tag{14.3}$$

hence, compared to (14.2) we identify $M(q,\boldsymbol{\theta}) = ml^2$, $g(q,\boldsymbol{\theta}) = mgl\sin(q)$. Hence, assuming that both the mass m and the length from the joint axis to the center of mass l, are unknown, we identify the vector of dynamic parameters as

$$\boldsymbol{\theta} = \begin{bmatrix} ml^2 \\ mgl \end{bmatrix},$$

which is, strictly speaking, a nonlinear vectorial function of the physical parameters m and l, since $\boldsymbol{\theta}$ depends on products of them. ♢

Note that here, the number of dynamic parameters coincides with the number of physical parameters, however, this is in general not the case as is clear from the examples below.

14.1.1 Linearity in the Dynamic Parameters

Example 14.1 also shows that the dynamic model (14.3) is linear in the parameters $\boldsymbol{\theta}$. To see this more clearly, notice that we may write

$$\begin{aligned} ml^2\ddot{q} + mgl\sin(q) &= [\ddot{q} \quad \sin(q)] \begin{bmatrix} ml^2 \\ mgl \end{bmatrix} \\ &=: \Phi(q,\ddot{q})\boldsymbol{\theta}\,. \end{aligned}$$

---

other chapters on adaptive control, the symbol $\boldsymbol{\theta}$ denotes the vector of dynamic parameters.

That is, the dynamic model (14.3) with zero input ($\tau = 0$), can be rewritten as the product of a vector function $\boldsymbol{\Phi}$ which contains nonlinear terms of the state (the generalized coordinates and its derivatives) and the vector of dynamic parameters, $\boldsymbol{\theta}$.

This property is commonly known as "linearity in the parameters" or "linear parameterization". It is a property possessed by many nonlinear systems and, in particular, by a fairly large class of robot manipulators. It is also our standing hypothesis for the subsequent chapters hence, we enunciate it formally below.

Property 14.1. Linearity in the dynamic parameters.

For the matrices $M(\boldsymbol{q}, \boldsymbol{\theta})$, $C(\boldsymbol{q}, \dot{\boldsymbol{q}}, \boldsymbol{\theta})$ and the vector $\boldsymbol{g}(\boldsymbol{q}, \boldsymbol{\theta})$ from the dynamic model (14.2), we have the following.

1. For all $\boldsymbol{u}$, $\boldsymbol{v}$, $\boldsymbol{w} \in \mathbb{R}^n$ it holds that

$$M(\boldsymbol{q}, \boldsymbol{\theta})\boldsymbol{u} + C(\boldsymbol{q}, \boldsymbol{w}, \boldsymbol{\theta})\boldsymbol{v} + \boldsymbol{g}(\boldsymbol{q}, \boldsymbol{\theta}) = \boldsymbol{\Phi}(\boldsymbol{q}, \boldsymbol{u}, \boldsymbol{v}, \boldsymbol{w})\boldsymbol{\theta} + \boldsymbol{\kappa}(\boldsymbol{q}, \boldsymbol{u}, \boldsymbol{v}, \boldsymbol{w}) \tag{14.4}$$

where $\boldsymbol{\kappa}(\boldsymbol{q}, \boldsymbol{u}, \boldsymbol{v}, \boldsymbol{w})$ is a vector of $n \times 1$, $\boldsymbol{\Phi}(\boldsymbol{q}, \boldsymbol{u}, \boldsymbol{v}, \boldsymbol{w})$ is a matrix of $n \times m$ and the vector $\boldsymbol{\theta} \in \mathbb{R}^m$ depends only on the dynamic parameters of the manipulator and its load.

2. Moreover[3], if $\boldsymbol{q}$, $\boldsymbol{u}$, $\boldsymbol{v}$, $\boldsymbol{w} \in L_\infty^n$ then $\boldsymbol{\Phi}(\boldsymbol{q}, \boldsymbol{u}, \boldsymbol{v}, \boldsymbol{w}) \in L_\infty^{n\times m}$.

It is worth remarking that one may always find a vector $\boldsymbol{\theta} \in \mathbb{R}^m$ for which $\boldsymbol{\kappa}(\boldsymbol{q}, \boldsymbol{u}, \boldsymbol{v}, \boldsymbol{w}) \equiv \boldsymbol{0} \in \mathbb{R}^n$. With this under consideration, setting $\boldsymbol{u} = \ddot{\boldsymbol{q}}$, $\boldsymbol{v} = \boldsymbol{w} = \dot{\boldsymbol{q}}$, on occasions it appears useful to rewrite Equation (14.4) in the simplified form

$$Y(\boldsymbol{q}, \dot{\boldsymbol{q}}, \ddot{\boldsymbol{q}})\boldsymbol{\theta} = M(\boldsymbol{q}, \boldsymbol{\theta})\ddot{\boldsymbol{q}} + C(\boldsymbol{q}, \dot{\boldsymbol{q}}, \boldsymbol{\theta})\dot{\boldsymbol{q}} + \boldsymbol{g}(\boldsymbol{q}, \boldsymbol{\theta}) \tag{14.5}$$

where $Y(\boldsymbol{q}, \dot{\boldsymbol{q}}, \ddot{\boldsymbol{q}}) = \boldsymbol{\Phi}(\boldsymbol{q}, \ddot{\boldsymbol{q}}, \dot{\boldsymbol{q}}, \dot{\boldsymbol{q}})$ is a matrix of dimension $n \times m$ and $\boldsymbol{\theta}$ is a vector of dimension $m \times 1$ which contains m constants that depend on the dynamic parameters. The constant n is clearly the number of DOF and m depends on the selection of the dynamic parameters of the robot.

Notice that Property 14.1 is stated in fair generality, *i.e.* it is not assumed, as for many other properties stated in Chapter 4, that the robot must have only revolute joints.

It is also important to underline, and it must be clear from Example 14.1, that the dynamic model of the robot is not necessarily linear in terms of the

[3] We remind the student reader that the notation L_∞ is described in detail in Appendix A which is left as self-study.

masses, inertias and distances of the centers of mass of the links but rather, it is linear in terms of the dynamic parameters $\boldsymbol{\theta}$ which in general, are *nonlinear functions* of the physical parameters. Therefore, given a selection of masses, inertias and distances to the centers of mass (of all the links), called here 'parameters of interest', the dynamic parameters are obtained from the robot model according to (14.4).

In general, the relation between the parameters of interest and the dynamic parameters is not available in a simple manner, but by developing (14.4) explicitly and using the fact that the matrix $\Phi(\boldsymbol{q}, \boldsymbol{u}, \boldsymbol{v}, \boldsymbol{w})$ as well as the vector $\boldsymbol{\kappa}(\boldsymbol{q}, \boldsymbol{u}, \boldsymbol{v}, \boldsymbol{w})$, do not depend on the dynamic parameters $\boldsymbol{\theta}$. This methodology is illustrated below through several examples; the procedure to determine the dynamic parameters may be too elaborate for robots with a large number of degrees of freedom. However, procedures to characterize the dynamic parameters are available.

Example 14.2. The right-hand side of the dynamic model of the device studied in Example 3.2, that is, of Equation (3.5),

$$m_2 l_2^2 \cos^2(\varphi)\ddot{q} = \tau$$

may be expressed in the form (14.5) where

$$\begin{aligned} Y(q, \dot{q}, \ddot{q}) &= l_2^2 \cos^2(\varphi)\ddot{q} \\ \theta &= m_2 \,. \end{aligned}$$

The following example shows that, as expected, the model of the Pelican prototype of Chapter 5 satisfies the linear parameterization property. This is used extensively later to design adaptive control schemes for this prototype in succeeding examples.

Example 14.3. Consider the Pelican manipulator moving on a vertical plane under the action of gravity as depicted in Figure 5.2. For simplicity, the manipulator is assumed to have two rigid links of unitary length ($l_1 = l_2 = 1$) and masses m_1 and m_2 concentrated at the ends of the links ($l_{c1} = l_{c2} = 1$, and $I_1 = I_2 = 0$). The dynamic model associated with the manipulator was obtained in Chapter 5 and is described by Equations (5.3) and (5.4):

$$\begin{aligned} \tau_1 = {} & [[m_1 + m_2] + m_2 + 2m_2 \cos(q_2)]\, \ddot{q}_1 \\ & + [m_2 + m_2 \cos(q_2)]\, \ddot{q}_2 \\ & - 2m_2 \sin(q_2)\dot{q}_1\dot{q}_2 - m_2 \sin(q_2)\dot{q}_2^2 \\ & + [m_1 + m_2]g \sin(q_1) \\ & + m_2 g \sin(q_1 + q_2) \end{aligned} \tag{14.6}$$

$$\tau_2 = [m_2 + m_2 \cos(q_2)]\, \ddot{q}_1 + m_2 \ddot{q}_2 \\ + m_2 \sin(q_2)\dot{q}_1^2 + m_2 g \sin(q_1 + q_2)\,. \tag{14.7}$$

The dynamic parameters in the model are the masses m_1 and m_2. Define the vector θ of dynamic parameters as $\boldsymbol{\theta} = [m_1 \;\; m_2]^T$.

The set of dynamic Equations (14.6) and (14.7) may be rewritten in linear terms of $\boldsymbol{\theta}$, that is, in the form (14.5):

$$Y(\boldsymbol{q}, \dot{\boldsymbol{q}}, \ddot{\boldsymbol{q}})\boldsymbol{\theta} = \boldsymbol{\tau}\,,$$

where $Y(\boldsymbol{q}, \dot{\boldsymbol{q}}, \ddot{\boldsymbol{q}})$ is the matrix of dimension 2×2:

$$Y(\boldsymbol{q}, \dot{\boldsymbol{q}}, \ddot{\boldsymbol{q}}) = \begin{bmatrix} Y_{11}(\boldsymbol{q}, \dot{\boldsymbol{q}}, \ddot{\boldsymbol{q}}) & Y_{12}(\boldsymbol{q}, \dot{\boldsymbol{q}}, \ddot{\boldsymbol{q}}) \\ Y_{21}(\boldsymbol{q}, \dot{\boldsymbol{q}}, \ddot{\boldsymbol{q}}) & Y_{22}(\boldsymbol{q}, \dot{\boldsymbol{q}}, \ddot{\boldsymbol{q}}) \end{bmatrix}$$

with

$$\begin{aligned} Y_{11}(\boldsymbol{q}, \dot{\boldsymbol{q}}, \ddot{\boldsymbol{q}}) &= \ddot{q}_1 + gS_1 \\ Y_{12}(\boldsymbol{q}, \dot{\boldsymbol{q}}, \ddot{\boldsymbol{q}}) &= 2\ddot{q}_1 + \ddot{q}_2 + C_2(2\ddot{q}_1 + \ddot{q}_2) \\ &\quad - S_2\dot{q}_2^2 - 2S_2\dot{q}_1\dot{q}_2 + gS_{12} + gS_1 \\ Y_{21}(\boldsymbol{q}, \dot{\boldsymbol{q}}, \ddot{\boldsymbol{q}}) &= 0 \\ Y_{22}(\boldsymbol{q}, \dot{\boldsymbol{q}}, \ddot{\boldsymbol{q}}) &= C_2\ddot{q}_1 + \ddot{q}_1 + \ddot{q}_2 + S_2\dot{q}_1^2 + gS_{12}\,. \end{aligned}$$

where $C_i = \cos(q_i)$, $S_i = \sin(q_i)$, $C_{12} = \cos(q_1 + q_2)$ and $S_{12} = \sin(q_1 + q_2)$. ♢

From Property 14.1, one can also show that the dynamic model of robots with (linear) actuators described by (3.33) also satisfies a linearity relation in terms of the dynamic parameters of the robot as well as in terms of the actuator constants. Specifically, for all $\boldsymbol{q}, \dot{\boldsymbol{q}}, \ddot{\boldsymbol{q}} \in \mathbb{R}^n$,

$$\begin{aligned} \Omega(\boldsymbol{q}, \dot{\boldsymbol{q}}, \ddot{\boldsymbol{q}})\boldsymbol{\theta} &= K^{-1}\left[R\, M(\boldsymbol{q}) + J\right]\ddot{\boldsymbol{q}} + K^{-1}R\, C(\boldsymbol{q}, \dot{\boldsymbol{q}})\dot{\boldsymbol{q}} \\ &\quad + K^{-1}R\, \boldsymbol{g}(\boldsymbol{q}) + K^{-1}R\boldsymbol{f}(\dot{\boldsymbol{q}}) + K^{-1}B\dot{\boldsymbol{q}} \\ &= \boldsymbol{v} \end{aligned} \tag{14.8}$$

where $\Omega(\boldsymbol{q}, \dot{\boldsymbol{q}}, \ddot{\boldsymbol{q}})$ is a matrix of dimension $n \times m$ and $\boldsymbol{\theta}$ is a vector of dimension $m \times 1$ that contains m constants that depend on the dynamic parameters of the robot and on those of the actuators.

Example 14.4. Consider the pendulum depicted in Figure 3.13 and whose dynamic model is derived in Example 3.8, that is,

$$\left[J_m + \frac{J_L}{r^2}\right]\ddot{q} + \left[f_m + \frac{f_L}{r^2} + \frac{K_a K_b}{R_a}\right]\dot{q} + \frac{k_L}{r^2}\sin(q) = \frac{K_a}{rR_a}v\,.$$

The dynamic equation may be written in linear terms of the vector $\boldsymbol{\theta}$, that is, in the form (14.8),

$$\Omega(q, \dot{q}, \ddot{q})\boldsymbol{\theta} = v$$

where

$$\Omega(q, \dot{q}, \ddot{q}) = [\ddot{q} \quad \dot{q} \quad \sin(q)]$$

$$\boldsymbol{\theta} = \frac{rR_a}{K_a}\left[J_m + \frac{J_L}{r^2} \qquad f_m + \frac{K_a K_b}{R_a} + \frac{f_L}{r^2} \qquad \frac{k_L}{r^2}\right]^T .$$

♦

14.1.2 The Nominal Model

We remark that for any given robot the vector of dynamic parameters $\boldsymbol{\theta}$ is not unique since it depends on how the parameters of interest are chosen. In the context of adaptive control, the parameters of interest are those whose numerical values are unknown. Usually, these are the mass, the inertia and the physical location of the center of mass of the last link of the robot.

For instance, as mentioned and illustrated through examples above, the vector $\boldsymbol{\kappa}(\boldsymbol{q}, \boldsymbol{u}, \boldsymbol{v}, \boldsymbol{w})$ and the matrix $\Phi(\boldsymbol{q}, \boldsymbol{u}, \boldsymbol{v}, \boldsymbol{w})$ are obtained from knowledge of the dynamic model of the robot under study, as well as from the vector $\boldsymbol{\theta} \in \mathbb{R}^m$ formed by the selection of the m dynamic parameters of interest. Naturally, it is always possible to choose a vector of dynamic parameters $\boldsymbol{\theta}$ for which (14.4) holds with $\boldsymbol{\kappa}(\boldsymbol{q}, \boldsymbol{u}, \boldsymbol{v}, \boldsymbol{w}) = \mathbf{0} \in \mathbb{R}^n$.

However, on certain occasions it may also appear useful to separate from the dynamics (14.2), those terms (if any) which involve *known* dynamic parameters or simply, that are *independent* of these. In such case, the parameterization (14.4) may be expressed as

$$\begin{aligned} M(\boldsymbol{q}, \boldsymbol{\theta})\boldsymbol{u} + C(\boldsymbol{q}, \boldsymbol{w}, \boldsymbol{\theta})\boldsymbol{v} + \boldsymbol{g}(\boldsymbol{q}, \boldsymbol{\theta}) = \\ \Phi(\boldsymbol{q}, \boldsymbol{u}, \boldsymbol{v}, \boldsymbol{w})\boldsymbol{\theta} + M_0(\boldsymbol{q})\boldsymbol{u} + C_0(\boldsymbol{q}, \boldsymbol{w})\boldsymbol{v} + \boldsymbol{g}_0(\boldsymbol{q}), \end{aligned} \tag{14.9}$$

where we may identify the *nominal model* or nominal part of the model,

$$\boldsymbol{\kappa}(\boldsymbol{q}, \boldsymbol{u}, \boldsymbol{v}, \boldsymbol{w}) = M_0(\boldsymbol{q})\boldsymbol{u} + C_0(\boldsymbol{q}, \boldsymbol{w})\boldsymbol{v} + \boldsymbol{g}_0(\boldsymbol{q}) .$$

That is, the matrices $M_0(\boldsymbol{q})$, $C_0(\boldsymbol{q}, \boldsymbol{w})$ and the vector $\boldsymbol{g}_0(\boldsymbol{q})$ represent respectively, parts of the matrices $M(\boldsymbol{q})$, $C(\boldsymbol{q}, \dot{\boldsymbol{q}})$ and of the vector $\boldsymbol{g}(\boldsymbol{q})$ that do not depend on the vector of *unknown* dynamic parameters $\boldsymbol{\theta}$.

According with the parameterization (14.9), given a vector $\hat{\boldsymbol{\theta}} \in \mathbb{R}^m$, the expression $\Phi(\boldsymbol{q}, \boldsymbol{u}, \boldsymbol{v}, \boldsymbol{w})\hat{\boldsymbol{\theta}}$ corresponds to

$$\boldsymbol{\Phi}(\boldsymbol{q},\boldsymbol{u},\boldsymbol{v},\boldsymbol{w})\hat{\boldsymbol{\theta}} = \\ M(\boldsymbol{q},\hat{\boldsymbol{\theta}})\boldsymbol{u}+C(\boldsymbol{q},\boldsymbol{w},\hat{\boldsymbol{\theta}})\boldsymbol{v}+\boldsymbol{g}(\boldsymbol{q},\hat{\boldsymbol{\theta}})-M_0(\boldsymbol{q})\boldsymbol{u}-C_0(\boldsymbol{q},\boldsymbol{w})\boldsymbol{v}-\boldsymbol{g}_0(\boldsymbol{q}). \tag{14.10}$$

A particular case of parameterization (14.9) is when $\boldsymbol{u} = \boldsymbol{v} = \boldsymbol{w} = \boldsymbol{0} \in \mathbb{R}^n$. In this scenario we have the following parameterization of the vector of gravitational torques:

$$\boldsymbol{g}(\boldsymbol{q},\boldsymbol{\theta}) = \boldsymbol{\Phi}(\boldsymbol{q},\boldsymbol{0},\boldsymbol{0},\boldsymbol{0})\boldsymbol{\theta} + \boldsymbol{g}_0(\boldsymbol{q})\,.$$

The following example is presented with the purpose of illustrating these ideas.

Example 14.5. Consider the model of a pendulum of mass m, inertia J with respect to the axis of rotation, and distance l from the axis of rotation to the center of mass. The torque τ is applied at the axis of rotation, that is,

$$J\ddot{q} + mgl\,\sin(q) = \tau\,.$$

We clearly identify $M(q) = J$, $C(q,\dot{q}) = 0$ and $g(q) = mgl\,\sin(q)$.

Consider as parameters of interest the mass m and the inertia J. The parameterization (14.9) in this scenario is

$$\begin{aligned} &M(q,\boldsymbol{\theta})u + C(q,w,\boldsymbol{\theta})v + g(q,\boldsymbol{\theta}) \\ &\quad = Ju + mgl\,\sin(q) \\ &\quad = \Phi(q,u,v,w)\boldsymbol{\theta} + M_0(q)u + C_0(q,w)v + g_0(q) \end{aligned} \tag{14.11}$$

where

$$\Phi(q,u,v,w) = \begin{bmatrix} u & gl\sin(q) \end{bmatrix} \tag{14.12}$$

$$\boldsymbol{\theta} = \begin{bmatrix} J \\ m \end{bmatrix}$$

$$M_0(q) = C_0(q,w) = g_0(q) = 0\,. \tag{14.13}$$

The reader may appreciate that particularly for this example, the dynamic parameters coincide precisely with the parameters of interest, m and J.

On the other hand, defining $\hat{\boldsymbol{\theta}} \in \mathbb{R}^2$ as

$$\hat{\boldsymbol{\theta}} = \begin{bmatrix} \hat{J} \\ \hat{m} \end{bmatrix},$$

the expression (14.10) becomes

$$\begin{aligned} &M(q,\hat{\boldsymbol{\theta}})u + C(q,w,\hat{\boldsymbol{\theta}})v + g(q,\hat{\boldsymbol{\theta}}) \\ &\quad = \hat{J}u + \hat{m}gl\,\sin(q) \\ &\quad = \Phi(q,u,v,w)\hat{\boldsymbol{\theta}} + M_0(q)u + C_0(q,w)v + g_0(q) \end{aligned}$$

where $\Phi(q,u,v,w)$, $M_0(q)$, $C_0(q,w)$ and $g_0(q)$ are exactly the same as in (14.12) and (14.13) respectively.

Assume now that the unique parameter of interest is the inertia J. The expression (14.9) is given again by (14.11) but now

$$\Phi(q,u,v,w) = u \tag{14.14}$$
$$\theta = J$$
$$M_0(q) = 0$$
$$C_0(q,w) = 0$$
$$g_0(q) = mgl\,\sin(q)\,. \tag{14.15}$$

On the other hand, defining $\hat{\theta} = \hat{J}$, the expression (14.10) becomes

$$\begin{aligned} & M(q,\hat{\theta})u + C(q,w,\hat{\theta})v + g(q,\hat{\theta}) \\ & = \hat{J}u + mgl\,\sin(q) \\ & = \Phi(q,u,v,w)\hat{\theta} + M_0(q)u + C_0(q,w)v + g_0(q), \end{aligned}$$

where $\Phi(q,u,v,w)$, $M_0(q)$, $C_0(q,w)$ and $g_0(q)$ are exactly (14.14)–(14.15). ♢

We present next an example of a planar 2-DOF robot. This robot is used in succeeding chapters with the aim of illustrating different adaptive controllers.

Example 14.6. Consider the planar manipulator having two DOF illustrated in Figure 14.1 and whose dynamic model was obtained in Example 3.3.

The dynamic model of the considered 2-DOF planar manipulator is given by (3.8)–(3.9), and may be written as

$$\theta_1\ddot{q}_1 + (\theta_3 C_{21} + \theta_4 S_{21})\,\ddot{q}_2 - \theta_3 S_{21}\dot{q}_2^2 + \theta_4 C_{21}\dot{q}_2^2 = \tau_1 \tag{14.16}$$

$$(\theta_3 C_{21} + \theta_4 S_{21})\,\ddot{q}_1 + \theta_2\ddot{q}_2 + \theta_3 S_{21}\dot{q}_1^2 - \theta_4 C_{21}\dot{q}_1^2 = \tau_2 \tag{14.17}$$

where

$$\begin{aligned} \theta_1 &= m_1 l_{c1}^2 + m_2 l_1^2 + I_1 \\ \theta_2 &= m_2 l_{c2}^2 + I_2 \\ \theta_3 &= m_2 l_1 l_{c2}\,\cos(\delta) \\ \theta_4 &= m_2 l_1 l_{c2}\,\sin(\delta)\,. \end{aligned}$$

It is easy to see that the previous model may be written in the standard form of the robot model (14.1) where

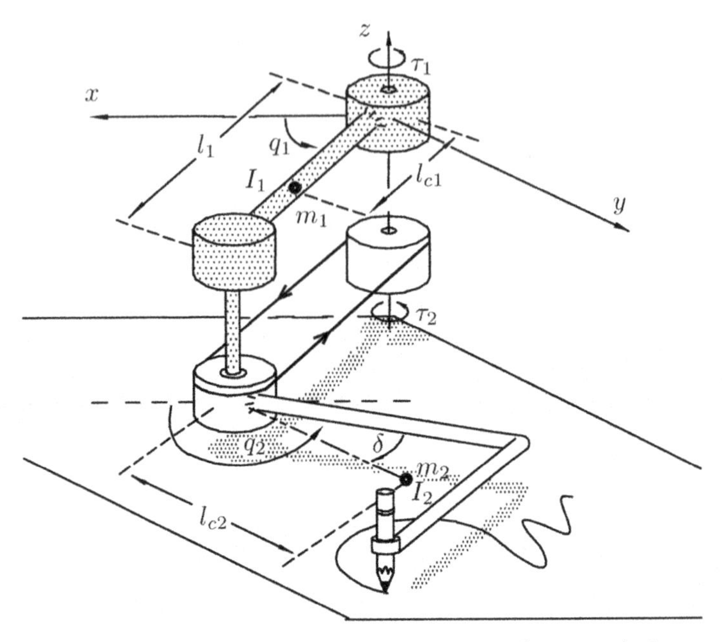

Figure 14.1. Planar 2-DOF manipulator on a horizontal plane

$$M(\boldsymbol{q}, \boldsymbol{\theta}) = \begin{bmatrix} \theta_1 & \theta_3 C_{21} + \theta_4 S_{21} \\ \theta_3 C_{21} + \theta_4 S_{21} & \theta_2 \end{bmatrix}$$

$$C(\boldsymbol{q}, \dot{\boldsymbol{q}}, \boldsymbol{\theta}) = \begin{bmatrix} 0 & (\theta_4 C_{21} - \theta_3 S_{21})\, \dot{q}_2 \\ (\theta_3 S_{21} - \theta_4 C_{21})\, \dot{q}_1 & 0 \end{bmatrix}$$

$$\boldsymbol{g}(\boldsymbol{q}, \boldsymbol{\theta}) = \boldsymbol{0}\,.$$

The dynamic model of the robot has been written in terms of the components θ_1, θ_2, θ_3 and θ_4 of the vector of unknown dynamic parameters. As mentioned above, these depend on the physical characteristics of the manipulator such as the masses and inertias of its links. The vector of dynamic parameters $\boldsymbol{\theta}$ is given directly by

$$\boldsymbol{\theta} = \begin{bmatrix} \theta_1 \\ \theta_2 \\ \theta_3 \\ \theta_4 \end{bmatrix} \in \mathbb{R}^4\,.$$

Next, define the vectors

$$\boldsymbol{u} = \begin{bmatrix} u_1 \\ u_2 \end{bmatrix}, \qquad \boldsymbol{v} = \begin{bmatrix} v_1 \\ v_2 \end{bmatrix}, \qquad \boldsymbol{w} = \begin{bmatrix} w_1 \\ w_2 \end{bmatrix}.$$

The parameterization (14.9) in this example yields

$$
\begin{aligned}
M(\boldsymbol{q},\boldsymbol{\theta})\boldsymbol{u} &+ C(\boldsymbol{q},\boldsymbol{w},\boldsymbol{\theta})\boldsymbol{v} + \boldsymbol{g}(\boldsymbol{q},\boldsymbol{\theta}) \\
&= \begin{bmatrix} \theta_1 & \theta_3 C_{21} + \theta_4 S_{21} \\ \theta_3 C_{21} + \theta_4 S_{21} & \theta_2 \end{bmatrix} \boldsymbol{u} \\
&\quad + \begin{bmatrix} 0 & (\theta_4 C_{21} - \theta_3 S_{21})\, w_2 \\ (\theta_3 S_{21} - \theta_4 C_{21})\, w_1 & 0 \end{bmatrix} \boldsymbol{v} \\
&= \begin{bmatrix} \Phi_{11} & \Phi_{12} & \Phi_{13} & \Phi_{14} \\ \Phi_{21} & \Phi_{22} & \Phi_{23} & \Phi_{24} \end{bmatrix} \begin{bmatrix} \theta_1 \\ \theta_2 \\ \theta_3 \\ \theta_4 \end{bmatrix} \\
&\quad + M_0(\boldsymbol{q})\boldsymbol{u} + C_0(\boldsymbol{q},\boldsymbol{w})\boldsymbol{v} + \boldsymbol{g}_0(\boldsymbol{q}).
\end{aligned}
$$

After spelling out $M(\boldsymbol{q},\boldsymbol{\theta})\boldsymbol{u} + C(\boldsymbol{q},\boldsymbol{w},\boldsymbol{\theta})\boldsymbol{v} + \boldsymbol{g}(\boldsymbol{q},\boldsymbol{\theta})$, it may be verified that

$$
\begin{aligned}
\Phi_{11} &= u_1 \\
\Phi_{12} &= 0 \\
\Phi_{13} &= C_{21} u_2 - S_{21} w_2 v_2 \\
\Phi_{14} &= S_{21} u_2 + C_{21} w_2 v_2 \\
\Phi_{21} &= 0 \\
\Phi_{22} &= u_2 \\
\Phi_{23} &= C_{21} u_1 + S_{21} w_1 v_1 \\
\Phi_{24} &= S_{21} u_1 - C_{21} w_1 v_1 \\
M_0(\boldsymbol{q}) &= 0 \in \mathbb{R}^{2\times 2} \\
C_0(\boldsymbol{q},\boldsymbol{w}) &= 0 \in \mathbb{R}^{2\times 2} \\
\boldsymbol{g}_0(\boldsymbol{q}) &= \boldsymbol{0} \in \mathbb{R}^{2}\,.
\end{aligned}
$$

Finally, we present an example of the Pelican robot, with which the reader must already be familiar.

Example 14.7. Consider the Pelican robot presented in Chapter 5, and shown in Figure 5.2. Its dynamic model is repeated here for convenience:

$$
\underbrace{\begin{bmatrix} M_{11}(\boldsymbol{q}) & M_{12}(\boldsymbol{q}) \\ M_{21}(\boldsymbol{q}) & M_{22}(\boldsymbol{q}) \end{bmatrix}}_{M(\boldsymbol{q})} \ddot{\boldsymbol{q}} + \underbrace{\begin{bmatrix} C_{11}(\boldsymbol{q},\dot{\boldsymbol{q}}) & C_{12}(\boldsymbol{q},\dot{\boldsymbol{q}}) \\ C_{21}(\boldsymbol{q},\dot{\boldsymbol{q}}) & C_{22}(\boldsymbol{q},\dot{\boldsymbol{q}}) \end{bmatrix}}_{C(\boldsymbol{q},\dot{\boldsymbol{q}})} \dot{\boldsymbol{q}} + \underbrace{\begin{bmatrix} g_1(\boldsymbol{q}) \\ g_2(\boldsymbol{q}) \end{bmatrix}}_{\boldsymbol{g}(\boldsymbol{q})} = \boldsymbol{\tau}
$$

where

$$
\begin{aligned}
M_{11}(\boldsymbol{q}) &= m_1 l_{c1}^2 + m_2 \left[l_1^2 + l_{c2}^2 + 2 l_1 l_{c2} \cos(q_2) \right] + I_1 + I_2 \\
M_{12}(\boldsymbol{q}) &= m_2 \left[l_{c2}^2 + l_1 l_{c2} \cos(q_2) \right] + I_2 \\
M_{21}(\boldsymbol{q}) &= m_2 \left[l_{c2}^2 + l_1 l_{c2} \cos(q_2) \right] + I_2 \\
M_{22}(\boldsymbol{q}) &= m_2 l_{c2}^2 + I_2
\end{aligned}
$$

$$
\begin{aligned}
C_{11}(\boldsymbol{q}, \dot{\boldsymbol{q}}) &= -m_2 l_1 l_{c2} \sin(q_2) \dot{q}_2 \\
C_{12}(\boldsymbol{q}, \dot{\boldsymbol{q}}) &= -m_2 l_1 l_{c2} \sin(q_2) \left[\dot{q}_1 + \dot{q}_2 \right] \\
C_{21}(\boldsymbol{q}, \dot{\boldsymbol{q}}) &= m_2 l_1 l_{c2} \sin(q_2) \dot{q}_1 \\
C_{22}(\boldsymbol{q}, \dot{\boldsymbol{q}}) &= 0
\end{aligned}
$$

$$
\begin{aligned}
g_1(\boldsymbol{q}) &= \left[m_1 l_{c1} + m_2 l_1 \right] g \sin(q_1) + m_2 l_{c2} g \sin(q_1 + q_2) \\
g_2(\boldsymbol{q}) &= m_2 l_{c2} g \sin(q_1 + q_2) .
\end{aligned}
$$

For this example we have selected as parameters of interest, the mass m_2, the inertia I_2 and the location of the center of mass of the second link, l_{c2}. In contrast to the previous example where the dynamic model (14.16)–(14.17) was written directly in terms of the dynamic parameters, here it is necessary to determine the latter as functions of the parameters of interest.

To that end, define first the vectors

$$
\boldsymbol{u} = \begin{bmatrix} u_1 \\ u_2 \end{bmatrix}, \quad \boldsymbol{v} = \begin{bmatrix} v_1 \\ v_2 \end{bmatrix}, \quad \boldsymbol{w} = \begin{bmatrix} w_1 \\ w_2 \end{bmatrix}.
$$

The development of the parameterization (14.9) in this example leads to

$$
\begin{aligned}
& M(\boldsymbol{q}, \boldsymbol{\theta})\boldsymbol{u} + C(\boldsymbol{q}, \boldsymbol{w}, \boldsymbol{\theta})\boldsymbol{v} + \boldsymbol{g}(\boldsymbol{q}, \boldsymbol{\theta}) = \\
& \qquad \begin{bmatrix} \Phi_{11} & \Phi_{12} & \Phi_{13} \\ \Phi_{21} & \Phi_{22} & \Phi_{23} \end{bmatrix} \begin{bmatrix} \theta_1 \\ \theta_2 \\ \theta_3 \end{bmatrix} + M_0(\boldsymbol{q})\boldsymbol{u} + C_0(\boldsymbol{q}, \boldsymbol{w})\boldsymbol{v} + \boldsymbol{g}_0(\boldsymbol{q}),
\end{aligned}
$$

where

$$
\begin{aligned}
\Phi_{11} &= l_1^2 u_1 + l_1 g \sin(q_1) \\
\Phi_{12} &= 2 l_1 \cos(q_2) u_1 + l_1 \cos(q_2) u_2 - l_1 \sin(q_2) w_2 v_1 \\
&\qquad - l_1 \sin(q_2) [w_1 + w_2] v_2 + g \sin(q_1 + q_2) \\
\Phi_{13} &= u_1 + u_2 \\
\Phi_{21} &= 0 \\
\Phi_{22} &= l_1 \cos(q_2) u_1 + l_1 \sin(q_2) w_1 v_1 + g \sin(q_1 + q_2) \\
\Phi_{23} &= u_1 + u_2 \\
\boldsymbol{\theta} &= \begin{bmatrix} \theta_1 \\ \theta_2 \\ \theta_3 \end{bmatrix} = \begin{bmatrix} m_2 \\ m_2 l_{c2} \\ m_2 l_{c2}^2 + I_2 \end{bmatrix}
\end{aligned}
$$

$$
\begin{aligned}
M_0(\boldsymbol{q}) &= \begin{bmatrix} m_1 l_{c1}^2 + I_1 & 0 \\ 0 & 0 \end{bmatrix} \\
C_0(\boldsymbol{q}, \boldsymbol{w}) &= \begin{bmatrix} 0 & 0 \\ 0 & 0 \end{bmatrix} \\
\boldsymbol{g}_0(\boldsymbol{q}) &= \begin{bmatrix} m_1 l_{c1} g \sin(q_1) \\ 0 \end{bmatrix} .
\end{aligned}
$$

Notice that effectively, the vector of dynamic parameters $\boldsymbol{\theta}$ depends exclusively on the parameters of interest m_2, I_2 and l_{c2}. ♢

14.2 The Adaptive Robot Control Problem

We have presented and discussed so far the fundamental property of linear parameterization of robot manipulators. All the adaptive controllers that we study in the following chapters rely on the assumption that this property holds.

Also, it is assumed that uncertainty in the model of the manipulator consists only of the lack of knowledge of the numerical values of the elements of $\boldsymbol{\theta}$. Hence, the structural form of the model of the manipulator is assumed to be exactly known, that is, the matrices $\Phi(\boldsymbol{q}, \boldsymbol{u}, \boldsymbol{v}, \boldsymbol{w})$, $M_0(\boldsymbol{q})$, $C_0(\boldsymbol{q}, \boldsymbol{w})$ and the vector $\boldsymbol{g}_0(\boldsymbol{q})$ are assumed to be known.

Formally, the control problem that we address in this text may be stated in the following terms. Consider the dynamic equation of n-DOF robots (14.2) taking into account the linear parameterization (14.9) that is,

$$
M(\boldsymbol{q}, \boldsymbol{\theta})\ddot{\boldsymbol{q}} + C(\boldsymbol{q}, \dot{\boldsymbol{q}}, \boldsymbol{\theta})\dot{\boldsymbol{q}} + \boldsymbol{g}(\boldsymbol{q}, \boldsymbol{\theta}) = \boldsymbol{\tau}
$$

or equivalently,

$$
\Phi(\boldsymbol{q}, \ddot{\boldsymbol{q}}, \dot{\boldsymbol{q}}, \dot{\boldsymbol{q}})\boldsymbol{\theta} + M_0(\boldsymbol{q})\ddot{\boldsymbol{q}} + C_0(\boldsymbol{q}, \dot{\boldsymbol{q}})\dot{\boldsymbol{q}} + \boldsymbol{g}_0(\boldsymbol{q}) = \boldsymbol{\tau} .
$$

Assume that the matrices $\Phi(\boldsymbol{q}, \ddot{\boldsymbol{q}}, \dot{\boldsymbol{q}}, \dot{\boldsymbol{q}}) \in \mathbb{R}^{n\times m}$, $M_0(\boldsymbol{q}), C_0(\boldsymbol{q}, \dot{\boldsymbol{q}}) \in \mathbb{R}^{n\times n}$ and the vector $\boldsymbol{g}_0(\boldsymbol{q}) \in \mathbb{R}^n$ are **known** but that the constant vector of dynamic parameters (which includes, for instance, inertias and masses) $\boldsymbol{\theta} \in \mathbb{R}^m$ is **unknown**[4]. Given a set of vectorial bounded functions $\boldsymbol{q}_d$, $\dot{\boldsymbol{q}}_d$ and $\ddot{\boldsymbol{q}}_d$, referred to as desired joint positions, velocities and accelerations, we seek to design controllers that achieve the position or motion control objectives. The solutions given in this textbook to this problem consist of the so-called *adaptive* controllers.

[4] By '$\Phi(q, \ddot{q}, \dot{q}, \dot{q})$ and $C_0(q, \dot{q})$ known' we understand that $\Phi(q, u, v, w)$ and $C_0(q, w)$ are known respectively. By '$\boldsymbol{\theta} \in \mathbb{R}^m$ unknown' we mean that the numerical values of its m components $\theta_1, \theta_2, \cdots, \theta_m$ are unknown.

We present next an example with the purpose of illustrating the control problem formulated above.

Example 14.8. Consider again the model of a pendulum of mass m, inertia J with respect to the axis of rotation, and distance l from the axis of rotation to its center of mass. The torque τ is applied at the axis of rotation, that is,

$$J\ddot{q} + mgl\,\sin(q) = \tau\,.$$

We clearly identify $M(q) = J$, $C(q,\dot{q}) = 0$ and $g(q) = mgl\,\sin(q)$.

Consider as parameter of interest, the inertia J. The model of the pendulum may be written in the generic form (14.9)

$$\begin{aligned} M(q,\theta)u + C(q,w,\theta)v + g(q,\theta) & \\ &= Ju + mgl\,\sin(q) \\ &= \Phi(q,u,v,w)\theta + M_0(q)u + C_0(q,w)v + g_0(q), \end{aligned}$$

where

$$\begin{aligned} \Phi(q,u,v,w) &= u \\ \theta &= J \\ M_0(q) &= 0 \\ C_0(q,w) &= 0 \\ g_0(q) &= mgl\,\sin(q)\,. \end{aligned}$$

Assume that the values of the mass m, the distance l and the gravity acceleration g are known but that the value of the inertia $\theta = J$ is unknown (yet constant). The control problem consists in designing a controller that is capable of achieving the motion control objective

$$\lim_{t\to\infty} \tilde{q}(t) = 0 \in \mathbb{R}$$

for any desired joint trajectory $q_d(t)$ (with bounded first and second time derivatives). The reader may notice that this problem formulation has not been addressed by any of the controllers presented in previous chapters. ♢

It is important to stress that the lack of knowledge of the vector of dynamic parameters of the robot, $\boldsymbol{\theta}$ and consequently, the uncertainty in its dynamic model make impossible the use of controllers which rely on accurate knowledge of the robot model, such as those studied in the chapters of Part II of this textbook. This has been the main reason that motivates the presentation of adaptive controllers in this part of the text. Certainly, if by any other means it is possible to determine the dynamic parameters, the use of an adaptive controller is unnecessary.

Another important observation about the control problem formulated above is the following. We have said explicitly that the vector of dynamic parameters $\boldsymbol{\theta} \in \mathbb{R}^m$ is assumed unknown but constant. This means precisely that the components of this vector do not vary as functions of time. Consequently, in the case where the parametric uncertainty comes from the mass or the inertia corresponding to the manipulated load by the robot[5], this must always be the same object, and therefore, it may not be latched or changed. Obviously this is a serious restriction from a practical viewpoint but it is necessary for the stability analysis of any adaptive controller if one is interested in guaranteeing achievement of the motion or position control objectives. As a matter of fact, the previous remarks also apply universally to all controllers that have been studied in previous chapters of this textbook. The reader should not be surprised by this fact since in the stability analyses the dynamic model of robot manipulators (*including* the manipulated object) is given by

$$M(\boldsymbol{q})\ddot{\boldsymbol{q}} + C(\boldsymbol{q}, \dot{\boldsymbol{q}})\dot{\boldsymbol{q}} + \boldsymbol{g}(\boldsymbol{q}) = \boldsymbol{\tau}$$

where we have implicitly involved the hypothesis that its parameters are *constant.* Naturally, in the case of model-based controllers for robots, these constant parameters must in addition, be known. In the scenario where the parameters vary with time then this variation must be known exactly.

14.3 Parameterization of the Adaptive Controller

The control laws to solve the position and motion control problems for robot manipulators may be written in the functional form

$$\boldsymbol{\tau} = \boldsymbol{\tau}(\boldsymbol{q}, \dot{\boldsymbol{q}}, \boldsymbol{q}_d, \dot{\boldsymbol{q}}_d, \ddot{\boldsymbol{q}}_d, M(\boldsymbol{q}), C(\boldsymbol{q}, \dot{\boldsymbol{q}}), \boldsymbol{g}(\boldsymbol{q})) . \tag{14.18}$$

In general, these control laws are formed by the sum of two terms; the first, which does not depend explicitly on the dynamic model of the robot to be controlled, and a second one which does. Therefore, giving a little 'more' structure to (14.18), we may write that most of the control laws have the form

$$\boldsymbol{\tau} = \boldsymbol{\tau}_1(\boldsymbol{q}, \dot{\boldsymbol{q}}, \boldsymbol{q}_d, \dot{\boldsymbol{q}}_d, \ddot{\boldsymbol{q}}_d) + M(\boldsymbol{q})\boldsymbol{u} + C(\boldsymbol{q}, \boldsymbol{w})\boldsymbol{v} + \boldsymbol{g}(\boldsymbol{q}),$$

where the vectors $\boldsymbol{u}, \boldsymbol{v}, \boldsymbol{w} \in \mathbb{R}^n$ depend in general on the positions $\boldsymbol{q}$, velocities $\dot{\boldsymbol{q}}$ and on the desired trajectory and its derivatives, $\boldsymbol{q}_d, \dot{\boldsymbol{q}}_d$ and $\ddot{\boldsymbol{q}}_d$. The term

[5] The manipulated object (load) may be considered as part of the last link of the robot.

$\boldsymbol{\tau}_1(\boldsymbol{q}, \dot{\boldsymbol{q}}, \boldsymbol{q}_d, \dot{\boldsymbol{q}}_d, \ddot{\boldsymbol{q}}_d)$, which does not depend on the dynamic model, usually corresponds to linear control terms of PD type, *i.e.*

$$\boldsymbol{\tau}_1(\boldsymbol{q}, \dot{\boldsymbol{q}}, \boldsymbol{q}_d, \dot{\boldsymbol{q}}_d, \ddot{\boldsymbol{q}}_d) = K_p[\boldsymbol{q}_d - \boldsymbol{q}] + K_v[\dot{\boldsymbol{q}}_d - \dot{\boldsymbol{q}}]$$

where K_p and K_v are gain matrices of position and velocity (or derivative gain) respectively.

Certainly, the structure of some position control laws do not depend on the dynamic model of the robot to be controlled; *e.g.* such is the case for PD and PID control laws. Other control laws require only part of the dynamic model of the robot; *e.g.* PD control with gravity compensation.

In general an adaptive controller is formed of two main parts:

- control law or controller;
- adaptive (update) law.

At this point it is worth remarking that we have not spoken of any particular adaptive controller to solve a given control problem. Indeed, there may exist many control and adaptive laws that allow one to solve a specific control problem. However, in general the control law is an algebraic equation that calculates the control action and which may be written in the generic form

$$\boldsymbol{\tau} = \boldsymbol{\tau}_1(\boldsymbol{q}, \dot{\boldsymbol{q}}, \boldsymbol{q}_d, \dot{\boldsymbol{q}}_d, \ddot{\boldsymbol{q}}_d) + M(\boldsymbol{q}, \hat{\boldsymbol{\theta}})\boldsymbol{u} + C(\boldsymbol{q}, \boldsymbol{w}, \hat{\boldsymbol{\theta}})\boldsymbol{v} + \boldsymbol{g}(\boldsymbol{q}, \hat{\boldsymbol{\theta}}) \tag{14.19}$$

where in general, the vectors $\boldsymbol{u}, \boldsymbol{v}, \boldsymbol{w} \in \mathbb{R}^n$ depend on the positions $\boldsymbol{q}$ and velocities $\dot{\boldsymbol{q}}$ as well as on the desired trajectory $\boldsymbol{q}_d$, and its derivatives $\dot{\boldsymbol{q}}_d$ and $\ddot{\boldsymbol{q}}_d$. The vector $\hat{\boldsymbol{\theta}} \in \mathbb{R}^m$ is referred to as the vector of adaptive parameters even though it actually corresponds to the vectorial function of time $\hat{\boldsymbol{\theta}}(t)$, which is such that (14.10) holds for all $t \geq 0$. It is important to mention that on some occasions, the control law may be a dynamic equation and not just 'algebraic'.

Typically, the control law (14.19) is chosen so that when substituting the vector of adaptive parameters $\hat{\boldsymbol{\theta}}$ by the vector of dynamic parameters $\boldsymbol{\theta}$ (which yields a nonadaptive controller), the resulting closed-loop system meets the control objective. As a matter of fact, in the case of control of robot manipulators nonadaptive control strategies that do not guarantee global asymptotic stability of the origin $\begin{bmatrix}\tilde{\boldsymbol{q}}^T & \dot{\tilde{\boldsymbol{q}}}^T\end{bmatrix}^T = \boldsymbol{0} \in \mathbb{R}^{2n}$ or $\begin{bmatrix}\tilde{\boldsymbol{q}}^T & \dot{\boldsymbol{q}}^T\end{bmatrix}^T = \boldsymbol{0} \in \mathbb{R}^{2n}$ for the case when $\boldsymbol{q}_d(t)$ is constant, are not candidates for adaptive versions, at least not with the standard design tools.

The adaptive law allows one to determine $\hat{\boldsymbol{\theta}}(t)$ and in general, may be written as a differential equation of $\hat{\boldsymbol{\theta}}$. An adaptive law commonly used in continuous adaptive systems is the so-called integral law or gradient type

$$\hat{\boldsymbol{\theta}}(t) = \Gamma \int_0^t \boldsymbol{\psi}\left(s, \boldsymbol{q}, \dot{\boldsymbol{q}}, \ddot{q}, \boldsymbol{q}_d, \dot{\boldsymbol{q}}_d, \ddot{\boldsymbol{q}}_d\right)\, ds + \hat{\boldsymbol{\theta}}(0) \tag{14.20}$$

where[6] $\Gamma = \Gamma^T \in \mathbb{R}^{m\times m}$ and $\hat{\boldsymbol{\theta}}(0) \in \mathbb{R}^m$ are design parameters while $\boldsymbol{\psi}$ is a vectorial function to be determined, of dimension m.

The symmetric matrix Γ is usually diagonal and positive definite and is called 'adaptive gain'. The "magnitude" of the adaptive gain Γ is related proportionally to the "rapidity of adaptation" of the control system vis-a-vis the parametric uncertainty of the dynamic model. The design procedures for adaptive controllers that use integral adaptive laws (14.20) in general, do not provide any guidelines to determine specifically the adaptive gain Γ. In practice one simply applies 'experience' to a trial-and-error approach until satisfactory behavior of the control system is obtained and usually, the adaptive gain is initially chosen to be "small".

On the other hand, $\hat{\boldsymbol{\theta}}(0)$ is an arbitrary vector even though in practice, we choose it as the best approximation available to the unknown vector of dynamic parameters, $\boldsymbol{\theta}$.

Figure 14.2 shows a block-diagram of the adaptive control of a robot.

An equivalent representation of the adaptive law is obtained by differentiating (14.20) with respect to time, that is,

$$\dot{\hat{\boldsymbol{\theta}}}(t) = \Gamma \boldsymbol{\psi}\left(s, \boldsymbol{q}, \dot{\boldsymbol{q}}, \ddot{\boldsymbol{q}}, \boldsymbol{q}_d, \dot{\boldsymbol{q}}_d, \ddot{\boldsymbol{q}}_d\right) . \tag{14.21}$$

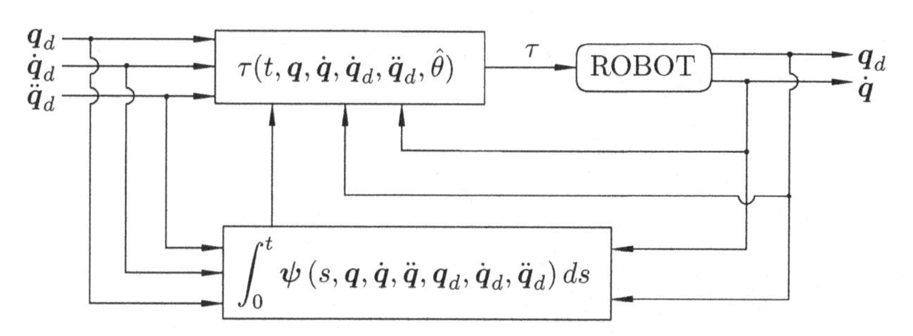

Figure 14.2. Block-diagram: generic adaptive control of robots

It is desirable, from a practical viewpoint, that the control law (14.19) as well as the adaptive law (14.20) or (14.21), do not depend explicitly on the joint acceleration $\ddot{\boldsymbol{q}}$.

14.3.1 Stability and Convergence of Adaptive Control Systems

An important topic in adaptive control systems is parametric convergence. The concept of parametric convergence refers to the asymptotic properties of

[6] In (14.20) as in other integrals, we avoid the cumbersome notation $\boldsymbol{\psi}(t, \boldsymbol{q}(t), \dot{\boldsymbol{q}}(t), \ddot{\boldsymbol{q}}(t), \boldsymbol{q}_d(t), \dot{\boldsymbol{q}}_d(t), \ddot{\boldsymbol{q}}_d(t))$.

the vector of adaptive parameters $\hat{\boldsymbol{\theta}}$. For a given adaptive system, if the limit of $\hat{\boldsymbol{\theta}}(t)$ when $t \to \infty$ exists and is such that

$$\lim_{t \to \infty} \hat{\boldsymbol{\theta}}(t) = \boldsymbol{\theta},$$

then we say that the adaptive system guarantees parametric convergence. As a matter of fact, parametric convergence is not an intrinsic characteristic of an adaptive controller. The latter depends, of course, on the adaptive controller itself but also on the behavior of some functions which may be internal or eventually external to the system in closed loop.

The study of the conditions to obtain parametric convergence in adaptive control systems is in general elaborate and requires additional mathematical tools to those presented in this text. For this reason, this topic is excluded.

The methodology of stability analysis for adaptive control systems for robot manipulators that is treated in this textbook is based on Lyapunov stability theory, following the guidelines of their nonadaptive counterparts. The main difference with respect to the analyses presented before is the inclusion of the parametric errors vector $\tilde{\boldsymbol{\theta}} \in \mathbb{R}^m$ defined as

$$\tilde{\boldsymbol{\theta}} = \hat{\boldsymbol{\theta}} - \boldsymbol{\theta}$$

in the closed loop equation's state vector.

The dynamic equations that characterize adaptive control systems in closed loop have the general form

$$\frac{d}{dt}\begin{bmatrix} \tilde{\boldsymbol{q}} \\ \dot{\tilde{\boldsymbol{q}}} \\ \tilde{\boldsymbol{\theta}} \end{bmatrix} = \boldsymbol{f}\left(t, \boldsymbol{q}, \dot{\boldsymbol{q}}, \boldsymbol{q}_d, \dot{\boldsymbol{q}}_d, \ddot{\boldsymbol{q}}_d, \tilde{\boldsymbol{\theta}}\right),$$

for which the origin is an equilibrium point. In general, unless we make appropriate hypotheses on the reference trajectories, the origin in adaptive control systems is *not* the only equilibrium point; as a matter of fact, it is not even an isolated equilibrium! The study of such systems is beyond the scope of the present text. For this reason, we do not study asymptotic stability (neither local nor global) but only stability and convergence of the position errors. That is, we show by other arguments, achievement of the control objective

$$\lim_{t \to \infty} \tilde{\boldsymbol{q}}(t) = \mathbf{0}.$$

We wish to emphasize the significance of the last phrase. Notice that we are claiming that even though we do not study and in general do *not* guarantee parameter convergence (to their true values) for any of the adaptive controllers studied in this text, we are implicitly saying that one can still achieve the motion control objective. This, in the presence of multiple equilibria and parameter uncertainty.

That one can achieve the control objective under parameter uncertainty is a fundamental truth that holds for many nonlinear systems and is commonly known as certainty equivalence.

Bibliography

The first adaptive control system with a rigorous proof of stability for the problem of motion control of robots, as far as we know, was reported in

- Craig J., Hsu P., Sastry S., 1986, *"Adaptive control of mechanical manipulators"*, Proceedings of the 1986 IEEE International Conference on Robotics and Automation, San Francisco, CA., April, pp. 190–195. Also reported in The International Journal of Robotics Research, Vol. 6, No. 2, Summer 1987, pp. 16–28.

A key step in the study of this controller, and by the way, also in that of the succeeding controllers in the literature, was the use of the linear-parameterization property of the robot model (see Property 14.1 above). This first adaptive controller needed *a priori* knowledge of bounds on the dynamic parameters as well as the measurement of the vector of joint accelerations $\ddot{q}$. After this first adaptive controller a series of adaptive controllers that did not need knowledge of the bounds on the parameters nor the measurement of joint accelerations were developed. A list containing some of the most relevant related references is presented next.

- Middleton R. H., Goodwin G. C., 1986. *"Adaptive computed torque control for rigid link manipulators"*, Proceedings of the 25th Conference on Decision and Control, Athens, Greece, December, pp. 68–73. Also reported in Systems and Control Letters, Vol. 10, pp. 9–16, 1988.
- Slotine J. J., Li W., 1987, *"On the adaptive control of robot manipulators"*, The International Journal of Robotics Research, Vol. 6, No. 3, pp. 49–59.
- Sadegh N., Horowitz R., 1987. *"Stability analysis of an adaptive controller for robotic manipulators"*, Proceedings of the 1987 IEEE International Conference on Robotics and Automation, Raleigh NC., April, pp. 1223–1229.
- Bayard D., Wen J. T., 1988. *"New class of control law for robotic manipulators. Part 2: Adaptive case"*, International Journal of Control, Vol. 47, No. 5, pp. 1387–1406.
- Slotine J. J., Li W., 1988, *"Adaptive manipulator control: A case study"*, IEEE Transactions on Automatic Control, Vol. 33, No. 11, November, pp. 995–1003.

- Kelly R., Carelli R., Ortega R., 1989, *"Adaptive motion control design to robot manipulators: An input–output approach"*, International Journal of Control, Vol. 50, No. 6, September, pp. 2563–2581.
- Landau I. D., Horowitz R., 1989, *"Applications of the passivity approach to the stability analysis of adaptive controllers for robot manipulators"*, International Journal of Adaptive Control and Signal Processing, Vol. 3, pp. 23–38.
- Sadegh N., Horowitz R., 1990, *"An exponential stable adaptive control law for robot manipulators"*, IEEE Transactions on Robotics and Automation, Vol. 6, No. 4, August, pp. 491–496.
- Kelly R., 1990, *"Adaptive computed torque plus compensation control for robot manipulators"*, Mechanism and Machine Theory, Vol. 25, No. 2, pp. 161–165.
- Johansson R., 1990, *"Adaptive control of manipulator motion"*, IEEE Transactions on Robotics and Automation, Vol. 6, No. 4, August, pp. 483–490.
- Lozano R., Canudas C., 1990, *"Passivity based adaptive control for mechanical manipulators using LS–type estimation"*, IEEE Transactions on Automatic Control, Vol. 25, No. 12, December, pp. 1363–1365.
- Lozano R., Brogliato B., 1992, *"Adaptive control of robot manipulators with flexible joints"*, IEEE Transactions on Automatic Control, Vol, 37, No. 2, February, pp. 174–181.
- Canudas C., Fixot N., 1992, *"Adaptive control of robot manipulators via velocity estimated feedback"*, IEEE Transactions on Automatic Control, Vol. 37, No. 8, August, pp. 1234–1237.
- Hsu L., Lizarralde F., 1993, *"Variable structure adaptive tracking control of robot manipulators without velocity measurement"*, 12th IFAC World Congress, Sydney, Australia, July, Vol. 1, pp. 145–148.
- Yu T., Arteaga A., 1994, *"Adaptive control of robots manipulators based on passivity"*, IEEE Transactions on Automatic Control, Vol. 39, No. 9, September, pp. 1871–1875.

An excellent introductory tutorial to adaptive motion control of robot manipulators is presented in

- Ortega R., Spong M., 1989. *"Adaptive motion control of rigid robots: A tutorial"*, Automatica, Vol. 25, No. 6, pp. 877–888.

Nowadays, we also count on several textbooks that are devoted in part to the study of adaptive controllers for robot manipulators. We cite among these:

- Craig J., 1988, *"Adaptive control of mechanical manipulators"*, Addison–Wesley Pub. Co.

- Spong M., Vidyasagar M., 1989, *"Robot dynamics and control"*, John Wiley and Sons.
- Stoten D. P., 1990, *"Model reference adaptive control of manipulators"*, John Wiley and Sons.
- Slotine J. J., Li W., 1991, *"Applied nonlinear control"*, Prentice-Hall.
- Lewis F. L., Abdallah C. T., Dawson D. M., 1993, *"Control of robot manipulators"*, Macmillan Pub. Co.
- Arimoto S., 1996, *"Control theory of non-linear mechanical systems"*, Oxford University Press, New York.

A detailed description of the basic concepts of adaptive control systems may be found in following texts.

- Anderson B. D. O., Bitmead R. R., Johnson C. R., Kokotović P., Kosut R., Mareels I. M. Y., Praly L., Riedle B. D., 1986, *"Stability of adaptive systems: Passivity and averaging analysis"*, The MIT Press, Cambridge, MA.
- Sastry S., Bodson M., 1989, *"Adaptive control–stability, convergence and robustness"*, Prentice-Hall.
- Narendra K., Annaswamy A., 1989, *"Stable adaptive systems"*, Prentice-Hall.
- Åström K. J., Wittenmark B., 1995, Second Edition, *"Adaptive control"*, Addison–Wesley Pub. Co.
- Kristić M., Kanellakopoulos I., Kokotović P., 1995, *"Nonlinear and adaptive control design"*, John Wiley and Sons, Inc.
- Marino R., Tomei P., 1995, *"Nonlinear control design"*, Prentice-Hall.
- Khalil H., 1996, *"Nonlinear systems"*, Second Edition, Prentice-Hall.
- Landau I. D., Lozano R., M'Saad M., 1998, "Adaptive control", Springer-Verlag: London.

The following references present the analysis and experimentation of various adaptive controllers for robots

- de Jager B., 1992, *"Practical evaluation of robust control for a class of nonlinear mechanical dynamic systems"*, PhD. thesis, Eindhoven University of Technology, The Netherlands, November.
- Whitcomb L. L., Rizzi A., Koditschek D. E., 1993, *"Comparative experiments with a new adaptive controller for robot arms"*, IEEE Transactions on Robotics and Automation, Vol. 9, No. 1, February.
- Berghuis H., 1993, *"Model-based robot control: from theory to practice"*, PhD. thesis, University of Twente, The Netherlands, June.

In recent years a promising approach appeared for robot control which is called 'learning'. This approach is of special interest in the case of parametric uncertainty in the model of the robot and when the specified motion is periodic. The interested reader is invited to see

- Arimoto S., 1990, " *Learning control theory for robotic motion*", International Journal of Adaptive Control and Signal Processing, Vol. 4, No. 6, pp. 543–564.
- Massner W., Horowitz R., Kao W., Boals M., 1991, *"A new adaptive learning rule"*, IEEE Transactions on Automatic Control, Vol. 36, No. 2, February, pp. 188–197.
- Arimoto S., Naniwa T., Parra–Vega V., Whitcomb L. L., 1995, *"A class of quasi-natural potentials for robot servo loops and its role in adaptive learning controls"*, Intelligent and Soft Computing, Vol. 1, No. 1, pp. 85–98.

Property 14.1 on the linearity of the robots dynamic model in the dynamic parameters has been reported in

- Khosla P., Kanade T., 1985, *"Parameter identification of robot dynamics"*, Proceedings 24th IEEE Conference on Decision and Control, Fort Lauderdale FL, December.
- Spong M., Vidyasagar M., 1989, *"Robot dynamics and control"*, John Wiley and Sons.
- Whitcomb L. L., Rizzi A., Koditschek D. E., 1991, *"Comparative experiments with a new adaptive controller for robot arms"*, Center for Systems Science, Dept. of Electrical Engineering, Yale University, Technical Report TR9101, February.

To the best of the authors' knowledge, the only rigorous proof of global *uniform* asymptotic stability for adaptive motion control of robot manipulators, *i.e.* including a proof of uniform global asymptotic convergence of the parameters, is given in

- A. Loría, R. Kelly and A. Teel, 2003, "Uniform parametric convergence in the adaptive control of manipulators: a case restudied", in Proceedings of International Conference on Robotics and Automation, Taipei, Taiwan, pp. 1062–1067.

Problems

1. Consider the simplified Cartesian mechanical device of Figure 14.3.
   Express the dynamic model in the form
   $$M(\boldsymbol{q})\ddot{\boldsymbol{q}} + C(\boldsymbol{q},\dot{\boldsymbol{q}})\dot{\boldsymbol{q}} + \boldsymbol{g}(\boldsymbol{q}) = Y(\boldsymbol{q},\dot{\boldsymbol{q}},\ddot{\boldsymbol{q}})\boldsymbol{\theta}$$
   where $\boldsymbol{\theta} = [m_1 + m_2 \;\; m_2]^T$.

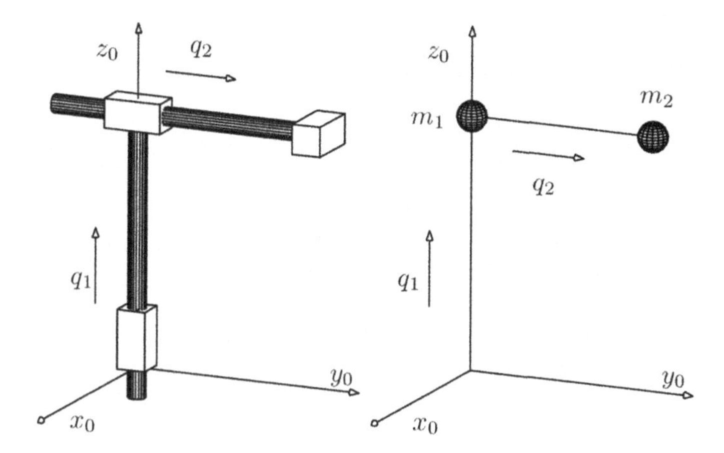

Figure 14.3. Problem 2. Cartesian robot.

15

PD Control with Adaptive Desired Gravity Compensation

It must be clear at this point that position control – regulation – is one of the simplest control objectives that may be formulated for robot manipulators. In spite of this apparent simplicity, the controllers which may achieve it globally require, in general, knowledge of at least the vector of gravitational torques $\boldsymbol{g}(\boldsymbol{q})$ of the dynamic robot model in question. Among the simplest controllers we have the following:

- PD control with gravity compensation;
- PD control with desired gravity compensation.

The first satisfies the position control objective globally with a trivial choice of the design parameter (*cf.* Chapter 7) while the second, even though it also achieves the position control objective globally, it requires a particular choice of design parameters (*cf.* Chapter 8). Nevertheless, the second controller is more attractive from a practical viewpoint due to its relative simplicity. As mentioned above, a common feature of both controllers is the use of the vector of gravitational torques $\boldsymbol{g}(\boldsymbol{q})$. The knowledge of this vector must be complete in the sense that both the structure of $\boldsymbol{g}(\boldsymbol{q})$ and the numerical values of the dynamic parameters must be known. Naturally in the case in which one of them is unknown, the previous control schemes may not be implemented.

In this chapter we study an adaptive control that is capable of satisfying the position control objective globally without requiring exact knowledge of the numerical values involved in the dynamic model of the robot to be controlled. We consider the scenario where all the joints of the robot are revolute. Specifically, we study the adaptive version of PD control with desired gravity compensation.

The material presented in this chapter has been taken from the corresponding references cited at the end of the chapter.

15.1 The Control and Adaptive Laws

We start by recalling the PD control law with desired gravity compensation given by (8.1), and which we repeat below for convenience,

$$\boldsymbol{\tau} = K_p\tilde{\boldsymbol{q}} + K_v\dot{\tilde{\boldsymbol{q}}} + \boldsymbol{g}(\boldsymbol{q}_d)\,.$$

We also recall that $K_p, K_v \in \mathbb{R}^{n\times n}$ are symmetric positive definite matrices chosen by the designer. As is customary, the position error is denoted by $\tilde{\boldsymbol{q}} = \boldsymbol{q}_d - \boldsymbol{q}$, while $\boldsymbol{q}_d \in \mathbb{R}^n$ stands for the desired joint position. In this chapter we assume that the desired joint position $\boldsymbol{q}_d$ is constant and therefore the control law covers the form

$$\boldsymbol{\tau} = K_p\tilde{\boldsymbol{q}} - K_v\dot{\boldsymbol{q}} + \boldsymbol{g}(\boldsymbol{q}_d)\,. \tag{15.1}$$

The practical convenience of this control law with respect to that of PD control with gravity compensation, given by (7.1), is evident. Indeed, the vector $\boldsymbol{g}(\boldsymbol{q}_d)$ used in the control law (15.1) depends on $\boldsymbol{q}_d$ and not on $\boldsymbol{q}$ therefore, it may be evaluated "off-line" once $\boldsymbol{q}_d$ is defined. In other words, it is unnecessary to compute $\boldsymbol{g}(\boldsymbol{q})$ in real time.

For further development it is also worth recalling Property 14.1, which establishes that the dynamic model of a n-DOF robot (with a manipulated load included) may be written using the parameterization (14.9), *i.e.* as

$$\begin{aligned} M(\boldsymbol{q},\boldsymbol{\theta})\boldsymbol{u} + C(\boldsymbol{q},\boldsymbol{w},\boldsymbol{\theta})\boldsymbol{v} + \boldsymbol{g}(\boldsymbol{q},\boldsymbol{\theta}) &= \\ \Phi(\boldsymbol{q},\boldsymbol{u},\boldsymbol{v},\boldsymbol{w})\boldsymbol{\theta} + M_0(\boldsymbol{q})\boldsymbol{u} + C_0(\boldsymbol{q},\boldsymbol{w})\boldsymbol{v} + \boldsymbol{g}_0(\boldsymbol{q}), \end{aligned} \tag{15.2}$$

where $\Phi(\boldsymbol{q},\boldsymbol{u},\boldsymbol{v},\boldsymbol{w}) \in \mathbb{R}^{n\times m}$, $M_0(\boldsymbol{q}) \in \mathbb{R}^{n\times n}$, $C_0(\boldsymbol{q},\boldsymbol{w}) \in \mathbb{R}^{n\times n}$, $\boldsymbol{g}_0(\boldsymbol{q}) \in \mathbb{R}^n$ and $\boldsymbol{\theta} \in \mathbb{R}^m$. The vector $\boldsymbol{\theta}$, known by the name vector of dynamic parameters, contains elements that depend precisely on physical parameters such as masses and inertias of the links of the manipulator and on the load. The matrices $M_0(\boldsymbol{q})$, $C_0(\boldsymbol{q},\boldsymbol{w})$ and the vector $\boldsymbol{g}_0(\boldsymbol{q})$ represent parts of the matrices $M(\boldsymbol{q})$, $C(\boldsymbol{q},\dot{\boldsymbol{q}})$ and of the vector $\boldsymbol{g}(\boldsymbol{q})$ that do not depend on the vector of (unknown) dynamic parameters $\boldsymbol{\theta}$.

By virtue of the previous fact, notice that the following expression is valid for all $\boldsymbol{x} \in \mathbb{R}^n$

$$\boldsymbol{g}(\boldsymbol{x},\boldsymbol{\theta}) = \Phi(\boldsymbol{x},\boldsymbol{0},\boldsymbol{0},\boldsymbol{0})\boldsymbol{\theta} + \boldsymbol{g}_0(\boldsymbol{x}), \tag{15.3}$$

where we set

$$\begin{aligned} \boldsymbol{u} &= \boldsymbol{0} \\ \boldsymbol{v} &= \boldsymbol{0} \\ \boldsymbol{w} &= \boldsymbol{0}\,. \end{aligned}$$

On the other hand, using (14.10), we conclude that for any vector $\hat{\boldsymbol{\theta}} \in \mathbb{R}^m$ and $\boldsymbol{x} \in \mathbb{R}^n$

$$g(\boldsymbol{x}, \hat{\boldsymbol{\theta}}) = \Phi(\boldsymbol{x}, \mathbf{0}, \mathbf{0}, \mathbf{0})\hat{\boldsymbol{\theta}} + \boldsymbol{g}_0(\boldsymbol{x})\,. \tag{15.4}$$

For notational simplicity, in the sequel we use the following abbreviation

$$\Phi_g(\boldsymbol{x}) = \Phi(\boldsymbol{x}, \mathbf{0}, \mathbf{0}, \mathbf{0})\,. \tag{15.5}$$

Considering (15.3) with $\boldsymbol{x} = \boldsymbol{q}_d$, the PD control law with desired gravity compensation, (15.1), may also be written as

$$\boldsymbol{\tau} = K_p\tilde{\boldsymbol{q}} - K_v\dot{\boldsymbol{q}} + \Phi_g(\boldsymbol{q}_d)\boldsymbol{\theta} + \boldsymbol{g}_0(\boldsymbol{q}_d)\,. \tag{15.6}$$

It is important to emphasize that in the implementation of the PD control law with desired gravity compensation, (15.1) or, equivalently (15.6), knowledge of the dynamic parameters $\boldsymbol{\theta}$ of the robot (including the manipulated load) is required.

In the sequel, we assume that the vector $\boldsymbol{\theta} \in \mathbb{R}^m$ of dynamic parameters is unknown but constant. Obviously, in this scenario, PD control with desired gravity compensation may not be used for robot control. Nevertheless, we assume that the unknown dynamic parameters $\boldsymbol{\theta}$ lay in a known region $\Omega \subset \mathbb{R}^m$ of the space $\mathbb{R}^m$. In other words, even though the vector $\boldsymbol{\theta}$ is supposed to be unknown, we assume that the set Ω in which $\boldsymbol{\theta}$ lays is known. The set Ω may be arbitrarily "large" but has to be bounded. In practice, the set Ω may be determined from upper and lower-bounds on the dynamic parameters which, as has been mentioned, are functions of the masses, inertias and location of the centers of mass of the links.

The solution that we consider in this chapter to the position control problem formulated above consists in the so-called adaptive version of PD control with desired gravity compensation, that is, PD control with adaptive desired gravity compensation.

The structure of the motion adaptive control schemes for robot manipulators that are studied in this text are defined by means of a control law like (14.19) and an adaptive law like (14.20). In the particular case of position control these control laws take the form

$$\boldsymbol{\tau} = \boldsymbol{\tau}\left(t, \boldsymbol{q}, \dot{\boldsymbol{q}}, \boldsymbol{q}_d, \hat{\boldsymbol{\theta}}\right) \tag{15.7}$$

$$\hat{\boldsymbol{\theta}}(t) = \Gamma \int_0^t \boldsymbol{\psi}\left(t, \boldsymbol{q}, \dot{\boldsymbol{q}}, \boldsymbol{q}_d\right)\, dt + \hat{\boldsymbol{\theta}}(0), \tag{15.8}$$

where $\Gamma = \Gamma^T \in \mathbb{R}^{m \times m}$ (adaptation gain) and $\hat{\boldsymbol{\theta}}(0) \in \mathbb{R}^m$ are design parameters while $\boldsymbol{\psi}$ is a vectorial function to be determined, and has dimension m.

The PD control with adaptive desired gravity compensation is described in (15.7)–(15.8) where

$$
\begin{aligned}
\tau &= K_p\tilde{q} - K_v\dot{q} + g(q_d, \hat{\theta}) && (15.9)\\
&= K_p\tilde{q} - K_v\dot{q} + \Phi_g(q_d)\hat{\theta} + g_0(q_d), && (15.10)
\end{aligned}
$$

and

$$
\hat{\theta}(t) = \Gamma \Phi_g(q_d)^T \int_0^t \left[\frac{\varepsilon_0}{1+\|\tilde{q}\|}\tilde{q} - \dot{q}\right] ds + \hat{\theta}(0), \qquad (15.11)
$$

where $K_p, K_v \in \mathbb{R}^{n\times n}$ and $\Gamma \in \mathbb{R}^{m\times m}$ are symmetric positive definite design matrices and ε_0 is a positive constant that satisfies conditions that are given later on. The pass from (15.9) to (15.10) was made by using (15.4) with $x = q_d$.

Notice that the control law (15.10) does not depend on the dynamic parameters θ but on the so-called adaptive parameters $\hat{\theta}$ that in their turn, are obtained from the adaptive law (15.11) which of course, does not depend either on θ.

Among the design parameters of the adaptive controller formed by Equations (15.10)–(15.11), only the matrix K_p and the real positive constant ε_0 must be chosen carefully. To that end, we start by defining $\lambda_{\text{Max}}\{M\}$, k_{C1} and k_g as

- $\lambda_{\text{Max}}\{M(q,\theta)\} \le \lambda_{\text{Max}}\{M\} \qquad \forall\, q \in \mathbb{R}^n, \quad \theta \in \Omega$
- $\|C(q,\dot{q},\theta)\| \le k_{C1}\,\|\dot{q}\| \qquad \forall\, q,\dot{q} \in \mathbb{R}^n, \quad \theta \in \Omega$
- $\|g(x,\theta) - g(y,\theta)\| \le k_g\,\|x - y\| \qquad \forall\, x,y \in \mathbb{R}^n, \quad \theta \in \Omega.$

Notice that these conditions are compatible with those established in Chapter 4. The constants $\lambda_{\text{Max}}\{M\}$, k_{C1} and k_g are considered known. Naturally, to obtain them it is necessary to know explicitly the matrices $M(q,\theta)$, $C(q,\dot{q},\theta)$ and of the vector $g(q,\theta)$, as well as of the set Ω, but one does not require to know the exact vector of dynamic parameters θ.

The symmetric positive definite matrix K_p and the positive constant ε_0 are chosen so that the following design conditions be verified.

C.1) $\lambda_{\min}\{K_p\} > k_g$,

C.2) $\sqrt{\dfrac{2\lambda_{\min}\{K_p\}}{\varepsilon_2\lambda_{\text{Max}}\{M\}}} > \varepsilon_0$,

C.3) $\dfrac{2\lambda_{\min}\{K_v\}[\lambda_{\min}\{K_p\} - k_g]}{\lambda^2_{\text{Max}}\{K_v\}} > \varepsilon_0$,

C.4) $\dfrac{\lambda_{\min}\{K_v\}}{2\,[k_{C1} + 2\lambda_{\text{Max}}\{M\}]} > \varepsilon_0$

where ε_2 is defined so that

$$
\varepsilon_2 = \frac{2\varepsilon_1}{\varepsilon_1 - 2} \qquad (15.12)
$$

and ε_1 satisfies the inequality

$$\frac{2\lambda_{\min}\{K_p\}}{k_g} > \varepsilon_1 > 2\,. \tag{15.13}$$

It is important to underline that once the matrix K_p is fixed in accordance with condition C.1 and the matrix K_v has been chosen arbitrarily but of course, symmetric positive definite, then it is always possible to find a set of strictly positive values for ε_0 for which the conditions C.2–C.4 are also verified.

Before proceeding to derive the closed-loop equation we define the parameter errors vector $\tilde{\boldsymbol{\theta}} \in \mathbb{R}^m$ as

$$\tilde{\boldsymbol{\theta}} = \hat{\boldsymbol{\theta}} - \boldsymbol{\theta}\,. \tag{15.14}$$

The parametric errors vector $\tilde{\boldsymbol{\theta}}$ is unknown since this is obtained as a function of the vector of dynamic parameters $\boldsymbol{\theta}$ which is assumed to be unknown. Nevertheless, the parametric error $\tilde{\boldsymbol{\theta}}$ is introduced here only for analytical purposes and evidently, it is not used by the controller.

From the definition of the parametric errors vector $\tilde{\boldsymbol{\theta}}$ in (15.14), it may be verified that

$$\begin{aligned}\Phi_g(\boldsymbol{q}_d)\hat{\boldsymbol{\theta}} &= \Phi_g(\boldsymbol{q}_d)\tilde{\boldsymbol{\theta}} + \Phi_g(\boldsymbol{q}_d)\boldsymbol{\theta} \\ &= \Phi_g(\boldsymbol{q}_d)\tilde{\boldsymbol{\theta}} + \boldsymbol{g}(\boldsymbol{q}_d, \boldsymbol{\theta}) - \boldsymbol{g}_0(\boldsymbol{q}_d),\end{aligned}$$

where we used (15.3) with $\boldsymbol{x} = \boldsymbol{q}_d$.

Using the expression above, the control law (15.10) may be written as

$$\boldsymbol{\tau} = K_p\tilde{\boldsymbol{q}} - K_v\dot{\boldsymbol{q}} + \Phi_g(\boldsymbol{q}_d)\tilde{\boldsymbol{\theta}} + \boldsymbol{g}(\boldsymbol{q}_d, \boldsymbol{\theta})\,.$$

Using the control law as written above and substituting the control action $\boldsymbol{\tau}$ in the equation of the robot model (14.2), we obtain

$$M(\boldsymbol{q}, \boldsymbol{\theta})\ddot{\boldsymbol{q}} + C(\boldsymbol{q}, \dot{\boldsymbol{q}}, \boldsymbol{\theta})\dot{\boldsymbol{q}} = K_p\tilde{\boldsymbol{q}} - K_v\dot{\boldsymbol{q}} + \Phi_g(\boldsymbol{q}_d)\tilde{\boldsymbol{\theta}} + \boldsymbol{g}(\boldsymbol{q}_d, \boldsymbol{\theta}) - \boldsymbol{g}(\boldsymbol{q}, \boldsymbol{\theta})\,. \tag{15.15}$$

On the other hand, since the vector of dynamic parameters $\boldsymbol{\theta}$ has been assumed to be constant, its time derivative is zero, $\dot{\boldsymbol{\theta}} = \boldsymbol{0} \in \mathbb{R}^m$. Therefore, taking the derivative with respect to time of the parametric errors vector $\tilde{\boldsymbol{\theta}}$, defined in (15.14), we obtain $\dot{\tilde{\boldsymbol{\theta}}} = \dot{\hat{\boldsymbol{\theta}}}$. In its turn, the time derivative of the vector of adaptive parameters $\hat{\boldsymbol{\theta}}$ is obtained by derivating with respect to time the adaptive law (15.11). Using these arguments we finally get

$$\dot{\tilde{\boldsymbol{\theta}}} = \Gamma\Phi_g(\boldsymbol{q}_d)^T\left[\frac{\varepsilon_0}{1 + \|\tilde{\boldsymbol{q}}\|}\tilde{\boldsymbol{q}} - \dot{\boldsymbol{q}}\right]. \tag{15.16}$$

From all the above conclude that the closed-loop equation is formed by Equations (15.15) and (15.16) and it may be written as

$$\frac{d}{dt}\begin{bmatrix}\tilde{\boldsymbol{q}}\\ \dot{\boldsymbol{q}}\\ \tilde{\boldsymbol{\theta}}\end{bmatrix} = \begin{bmatrix} -\dot{\boldsymbol{q}} \\ M(\boldsymbol{q},\boldsymbol{\theta})^{-1}\left[K_p\tilde{\boldsymbol{q}} - K_v\dot{\boldsymbol{q}} + \Phi_g(\boldsymbol{q}_d)\tilde{\boldsymbol{\theta}} - C(\boldsymbol{q},\dot{\boldsymbol{q}},\boldsymbol{\theta})\dot{\boldsymbol{q}} + \boldsymbol{g}(\boldsymbol{q}_d,\boldsymbol{\theta}) - \boldsymbol{g}(\boldsymbol{q},\boldsymbol{\theta})\right] \\ \Gamma\Phi_g(\boldsymbol{q}_d)^T\left[\frac{\varepsilon_0}{1+\|\tilde{\boldsymbol{q}}\|}\tilde{\boldsymbol{q}} - \dot{\boldsymbol{q}}\right]\end{bmatrix} \tag{15.17}$$

Notice that this is a set of autonomous differential equations with state $\left[\tilde{\boldsymbol{q}}^T\ \dot{\boldsymbol{q}}^T\ \tilde{\boldsymbol{\theta}}^T\right]^T$ and the origin of the state space, *i.e.*

$$\begin{bmatrix}\tilde{\boldsymbol{q}}\\ \dot{\boldsymbol{q}}\\ \tilde{\boldsymbol{\theta}}\end{bmatrix} = \boldsymbol{0} \in \mathbb{R}^{2n+m},$$

is an equilibrium point of (15.17).

15.2 Stability Analysis

The stability analysis of the origin of the state space of closed-loop equation follows along the guidelines of Section 8.4. Consider the following extension of the Lyapunov function candidate (8.23) with the additional term $\frac{1}{2}\tilde{\boldsymbol{\theta}}^T\Gamma^{-1}\tilde{\boldsymbol{\theta}}$, *i.e.*

$$V(\tilde{\boldsymbol{q}},\dot{\boldsymbol{q}},\tilde{\boldsymbol{\theta}}) = \frac{1}{2}\begin{bmatrix}\tilde{\boldsymbol{q}}\\ \dot{\boldsymbol{q}}\\ \tilde{\boldsymbol{\theta}}\end{bmatrix}^T \overbrace{\begin{bmatrix}\frac{2}{\varepsilon_2}K_p & -\frac{\varepsilon_0}{1+\|\tilde{\boldsymbol{q}}\|}M(\boldsymbol{q},\boldsymbol{\theta}) & 0\\ -\frac{\varepsilon_0}{1+\|\tilde{\boldsymbol{q}}\|}M(\boldsymbol{q},\boldsymbol{\theta}) & M(\boldsymbol{q},\boldsymbol{\theta}) & 0\\ 0 & 0 & \Gamma^{-1}\end{bmatrix}}^{P}\begin{bmatrix}\tilde{\boldsymbol{q}}\\ \dot{\boldsymbol{q}}\\ \tilde{\boldsymbol{\theta}}\end{bmatrix}$$
$$\underbrace{+\,\mathcal{U}(\boldsymbol{q},\boldsymbol{\theta}) - \mathcal{U}(\boldsymbol{q}_d,\boldsymbol{\theta}) + \boldsymbol{g}(\boldsymbol{q}_d,\boldsymbol{\theta})^T\tilde{\boldsymbol{q}} + \frac{1}{\varepsilon_1}\tilde{\boldsymbol{q}}^TK_p\tilde{\boldsymbol{q}}}_{f(\tilde{\boldsymbol{q}})}$$

$$
\begin{aligned}
&= \frac{1}{2}\dot{\boldsymbol{q}}^T M(\boldsymbol{q},\boldsymbol{\theta})\dot{\boldsymbol{q}} + \mathcal{U}(\boldsymbol{q},\boldsymbol{\theta}) - \mathcal{U}(\boldsymbol{q}_d,\boldsymbol{\theta}) + \boldsymbol{g}(\boldsymbol{q}_d,\boldsymbol{\theta})^T\tilde{\boldsymbol{q}} \\
&\quad + \left(\frac{1}{\varepsilon_1} + \frac{1}{\varepsilon_2}\right)\tilde{\boldsymbol{q}}^T K_p \tilde{\boldsymbol{q}} - \frac{\varepsilon_0}{1+\|\tilde{\boldsymbol{q}}\|}\,\tilde{\boldsymbol{q}}^T M(\boldsymbol{q},\boldsymbol{\theta})\dot{\boldsymbol{q}} \\
&\quad + \frac{1}{2}\tilde{\boldsymbol{\theta}}^T \Gamma^{-1}\tilde{\boldsymbol{\theta}}\,,
\end{aligned}
\tag{15.18}
$$

where $f(\tilde{\boldsymbol{q}})$ is defined as in (8.18) and the constants $\varepsilon_0 > 0$, $\varepsilon_1 > 2$ and $\varepsilon_2 > 2$ are chosen so that

$$
\frac{2\lambda_{\min}\{K_p\}}{k_g} > \varepsilon_1 > 2 \tag{15.19}
$$

$$
\varepsilon_2 = \frac{2\varepsilon_1}{\varepsilon_1 - 2} \tag{15.20}
$$

$$
\sqrt{\frac{2\lambda_{\min}\{K_p\}}{\varepsilon_2\lambda_{\mathrm{Max}}\{M\}}} > \varepsilon_0 > 0\,. \tag{15.21}
$$

The condition (15.19) guarantees that $f(\tilde{\boldsymbol{q}})$ is a positive definite function (see Lemma 8.1), while (15.21) ensures that P is a positive definite matrix. Finally (15.20) implies that $\frac{1}{\varepsilon_1} + \frac{1}{\varepsilon_2} = \frac{1}{2}$. Notice that condition (15.21) corresponds exactly to condition C.2 which holds due to the hypothesis on the choice of ε_0.

Thus, to show that the Lyapunov function candidate $V(\tilde{\boldsymbol{q}},\dot{\boldsymbol{q}},\tilde{\boldsymbol{\theta}})$ is positive definite, we start by defining ε as

$$
\varepsilon = \varepsilon(\|\tilde{\boldsymbol{q}}\|) := \frac{\varepsilon_0}{1+\|\tilde{\boldsymbol{q}}\|}\,. \tag{15.22}
$$

Consequently, the inequality (15.21) implies that the matrix

$$
\frac{2}{\varepsilon_2}K_p - \left(\frac{\varepsilon_0}{1+\|\tilde{\boldsymbol{q}}\|}\right)^2 M(\boldsymbol{q},\boldsymbol{\theta}) = \frac{2}{\varepsilon_2}K_p - \varepsilon^2 M(\boldsymbol{q},\boldsymbol{\theta})
$$

is positive definite.

On the other hand, the Lyapunov function candidate (15.18) may be rewritten in the following manner:

$$
\begin{aligned}
V(\tilde{\boldsymbol{q}},\dot{\boldsymbol{q}},\tilde{\boldsymbol{\theta}}) &= \frac{1}{2}\left[-\dot{\boldsymbol{q}} + \varepsilon\tilde{\boldsymbol{q}}\right]^T M(\boldsymbol{q},\boldsymbol{\theta})\left[-\dot{\boldsymbol{q}} + \varepsilon\tilde{\boldsymbol{q}}\right] \\
&\quad + \frac{1}{2}\tilde{\boldsymbol{q}}^T\left[\frac{2}{\varepsilon_2}K_p - \varepsilon^2 M(\boldsymbol{q},\boldsymbol{\theta})\right]\tilde{\boldsymbol{q}} \\
&\quad + \frac{1}{2}\tilde{\boldsymbol{\theta}}^T\Gamma^{-1}\tilde{\boldsymbol{\theta}} \\
&\quad + \underbrace{\mathcal{U}(\boldsymbol{q},\boldsymbol{\theta}) - \mathcal{U}(\boldsymbol{q}_d,\boldsymbol{\theta}) + \boldsymbol{g}(\boldsymbol{q}_d,\boldsymbol{\theta})^T\tilde{\boldsymbol{q}} + \frac{1}{\varepsilon_1}\tilde{\boldsymbol{q}}^T K_p\tilde{\boldsymbol{q}}}_{f(\tilde{\boldsymbol{q}})}\,,
\end{aligned}
$$

which is a positive definite function since the matrices $M(\boldsymbol{q},\boldsymbol{\theta})$ and $\frac{2}{\varepsilon_2}K_p - \varepsilon^2 M(\boldsymbol{q},\boldsymbol{\theta})$ are positive definite and $f(\tilde{\boldsymbol{q}})$ is also a positive definite function (since $\lambda_{\min}\{K_p\} > k_g$ and from Lemma 8.1).

Now we proceed to compute the total time derivative of the Lyapunov function candidate (15.18). For notational simplicity, in the sequel we drop the argument $\boldsymbol{\theta}$ from the matrices $M(\boldsymbol{q},\boldsymbol{\theta})$, $C(\boldsymbol{q},\dot{\boldsymbol{q}},\boldsymbol{\theta})$, from the vectors $\boldsymbol{g}(\boldsymbol{q},\boldsymbol{\theta})$, $\boldsymbol{g}(\boldsymbol{q}_d,\boldsymbol{\theta})$ and from $\mathcal{U}(\boldsymbol{q},\boldsymbol{\theta})$ and $\mathcal{U}(\boldsymbol{q}_d,\boldsymbol{\theta})$. However, the reader should keep in mind that, strictly speaking, V depends on time since $\boldsymbol{\theta} = \hat{\boldsymbol{\theta}}(t) - \tilde{\boldsymbol{\theta}}$.

The time derivative of the Lyapunov function candidate (15.18) along the trajectories of the closed-loop Equation (15.17) becomes, after some simplifications,

$$\begin{aligned}\dot{V}(\tilde{\boldsymbol{q}},\dot{\boldsymbol{q}},\tilde{\boldsymbol{\theta}}) = {}& \dot{\boldsymbol{q}}^T\left[K_p\tilde{\boldsymbol{q}} - K_v\dot{\boldsymbol{q}} - C(\boldsymbol{q},\dot{\boldsymbol{q}})\dot{\boldsymbol{q}} + \boldsymbol{g}(\boldsymbol{q}_d) - \boldsymbol{g}(\boldsymbol{q})\right] + \frac{1}{2}\dot{\boldsymbol{q}}^T\dot{M}(\boldsymbol{q})\dot{\boldsymbol{q}} \\ & + \boldsymbol{g}(\boldsymbol{q})^T\dot{\boldsymbol{q}} - \boldsymbol{g}(\boldsymbol{q}_d)^T\dot{\boldsymbol{q}} - \tilde{\boldsymbol{q}}^T K_p\dot{\boldsymbol{q}} + \varepsilon\dot{\boldsymbol{q}}^T M(\boldsymbol{q})\dot{\boldsymbol{q}} - \varepsilon\tilde{\boldsymbol{q}}^T\dot{M}(\boldsymbol{q})\dot{\boldsymbol{q}} \\ & - \varepsilon\tilde{\boldsymbol{q}}^T\left[K_p\tilde{\boldsymbol{q}} - K_v\dot{\boldsymbol{q}} - C(\boldsymbol{q},\dot{\boldsymbol{q}})\dot{\boldsymbol{q}} + \boldsymbol{g}(\boldsymbol{q}_d) - \boldsymbol{g}(\boldsymbol{q})\right] \\ & - \dot{\varepsilon}\tilde{\boldsymbol{q}}^T M(\boldsymbol{q})\dot{\boldsymbol{q}},\end{aligned}$$

where we used $\boldsymbol{g}(\boldsymbol{q}) = \dfrac{\partial\mathcal{U}(\boldsymbol{q})}{\partial\boldsymbol{q}}$. After some further simplifications, the time derivative $\dot{V}(\tilde{\theta},\tilde{\boldsymbol{q}},\dot{\boldsymbol{q}})$ may be written as

$$\begin{aligned}\dot{V}(\tilde{\boldsymbol{q}},\dot{\boldsymbol{q}},\tilde{\boldsymbol{\theta}}) = {}& -\dot{\boldsymbol{q}}^T K_v\dot{\boldsymbol{q}} + \dot{\boldsymbol{q}}^T\left[\frac{1}{2}\dot{M}(\boldsymbol{q}) - C(\boldsymbol{q},\dot{\boldsymbol{q}})\right]\dot{\boldsymbol{q}} + \varepsilon\dot{\boldsymbol{q}}^T M(\boldsymbol{q})\dot{\boldsymbol{q}} \\ & - \varepsilon\tilde{\boldsymbol{q}}^T\left[\dot{M}(\boldsymbol{q}) - C(\boldsymbol{q},\dot{\boldsymbol{q}})\right]\dot{\boldsymbol{q}} - \varepsilon\tilde{\boldsymbol{q}}^T\left[K_p\tilde{\boldsymbol{q}} - K_v\dot{\boldsymbol{q}}\right] \\ & - \varepsilon\tilde{\boldsymbol{q}}^T\left[\boldsymbol{g}(\boldsymbol{q}_d) - \boldsymbol{g}(\boldsymbol{q})\right] - \dot{\varepsilon}\tilde{\boldsymbol{q}}^T M(\boldsymbol{q})\dot{\boldsymbol{q}}\,.\end{aligned}$$

Finally, considering Property 4.2, *i.e.* that the matrix $\frac{1}{2}\dot{M}(\boldsymbol{q}) - C(\boldsymbol{q},\dot{\boldsymbol{q}})$ is skew-symmetric and $\dot{M}(\boldsymbol{q}) = C(\boldsymbol{q},\dot{\boldsymbol{q}}) + C(\boldsymbol{q},\dot{\boldsymbol{q}})^T$, we get

$$\begin{aligned}\dot{V}(\tilde{\boldsymbol{q}},\dot{\boldsymbol{q}},\tilde{\boldsymbol{\theta}}) = {}& -\dot{\boldsymbol{q}}^T K_v\dot{\boldsymbol{q}} + \varepsilon\dot{\boldsymbol{q}}^T M(\boldsymbol{q})\dot{\boldsymbol{q}} - \varepsilon\tilde{\boldsymbol{q}}^T K_p\tilde{\boldsymbol{q}} + \varepsilon\tilde{\boldsymbol{q}}^T K_v\dot{\boldsymbol{q}} \\ & - \varepsilon\dot{\boldsymbol{q}}^T C(\boldsymbol{q},\dot{\boldsymbol{q}})\tilde{\boldsymbol{q}} - \varepsilon\tilde{\boldsymbol{q}}^T\left[\boldsymbol{g}(\boldsymbol{q}_d) - \boldsymbol{g}(\boldsymbol{q})\right] \\ & - \dot{\varepsilon}\tilde{\boldsymbol{q}}^T M(\boldsymbol{q})\dot{\boldsymbol{q}}\,.\end{aligned} \tag{15.23}$$

As we know now, to conclude stability by means of Lyapunov's direct method, it is sufficient to prove that $\dot{V}(\boldsymbol{0},\boldsymbol{0},\boldsymbol{0}) = 0$ and that $\dot{V}(\tilde{\boldsymbol{q}},\dot{\boldsymbol{q}},\tilde{\boldsymbol{\theta}}) \leq 0$ for all vectors $\begin{bmatrix}\tilde{\boldsymbol{q}}^T & \dot{\boldsymbol{q}}^T & \tilde{\boldsymbol{\theta}}^T\end{bmatrix}^T \neq \boldsymbol{0} \in \mathbb{R}^{2n+m}$. These conditions are verified for instance if $\dot{V}(\tilde{\boldsymbol{q}},\dot{\boldsymbol{q}},\tilde{\boldsymbol{\theta}})$ is negative semidefinite. Observe that at this moment, it is very difficult to ensure from (15.23), that $\dot{V}(\tilde{\boldsymbol{q}},\dot{\boldsymbol{q}},\tilde{\boldsymbol{\theta}})$ is a negative semidefinite function. With the aim of finding additional conditions on ε_0 so that $\dot{V}(\tilde{\boldsymbol{q}},\dot{\boldsymbol{q}},\tilde{\boldsymbol{\theta}})$ is negative semidefinite, we present next some upper-bounds over the following three terms:

- $-\varepsilon\dot{\boldsymbol{q}}^T C(\boldsymbol{q},\dot{\boldsymbol{q}})\tilde{\boldsymbol{q}}$
- $-\varepsilon\tilde{\boldsymbol{q}}^T\left[\boldsymbol{g}(\boldsymbol{q}_d)-\boldsymbol{g}(\boldsymbol{q})\right]$
- $-\dot{\varepsilon}\tilde{\boldsymbol{q}}^T M(\boldsymbol{q})\dot{\boldsymbol{q}}$.

First, with respect to $-\varepsilon\dot{\boldsymbol{q}}^T C(\boldsymbol{q},\dot{\boldsymbol{q}})\tilde{\boldsymbol{q}}$, we have

$$\begin{aligned}-\varepsilon\dot{\boldsymbol{q}}^T C(\boldsymbol{q},\dot{\boldsymbol{q}})\tilde{\boldsymbol{q}} &\leq \left|-\varepsilon\dot{\boldsymbol{q}}^T C(\boldsymbol{q},\dot{\boldsymbol{q}})\tilde{\boldsymbol{q}}\right| \\ &\leq \varepsilon\left\|\dot{\boldsymbol{q}}\right\|\left\|C(\boldsymbol{q},\dot{\boldsymbol{q}})\tilde{\boldsymbol{q}}\right\| \\ &\leq \varepsilon k_{C_1}\left\|\dot{\boldsymbol{q}}\right\|\left\|\dot{\boldsymbol{q}}\right\|\left\|\tilde{\boldsymbol{q}}\right\| \\ &\leq \varepsilon_0 k_{C_1}\left\|\dot{\boldsymbol{q}}\right\|^2 \end{aligned} \tag{15.24}$$

where we took into account Property 4.2, *i.e.* that $\|C(\boldsymbol{q},\boldsymbol{x})\boldsymbol{y}\| \leq k_{C_1}\|\boldsymbol{x}\|\|\boldsymbol{y}\|$, and the definition of ε in (15.22).

Next, concerning the term $-\varepsilon\tilde{\boldsymbol{q}}^T\left[\boldsymbol{g}(\boldsymbol{q}_d)-\boldsymbol{g}(\boldsymbol{q})\right]$, we have

$$\begin{aligned}-\varepsilon\tilde{\boldsymbol{q}}^T\left[\boldsymbol{g}(\boldsymbol{q}_d)-\boldsymbol{g}(\boldsymbol{q})\right] &\leq \left|-\varepsilon\tilde{\boldsymbol{q}}^T\left[\boldsymbol{g}(\boldsymbol{q}_d)-\boldsymbol{g}(\boldsymbol{q})\right]\right| \\ &\leq \varepsilon\left\|\tilde{\boldsymbol{q}}\right\|\left\|\boldsymbol{g}(\boldsymbol{q}_d)-\boldsymbol{g}(\boldsymbol{q})\right\| \\ &\leq \varepsilon k_g\left\|\tilde{\boldsymbol{q}}\right\|^2 \end{aligned} \tag{15.25}$$

where we used Property 4.3, *i.e.* that $\|\boldsymbol{g}(\boldsymbol{x})-\boldsymbol{g}(\boldsymbol{y})\| \leq k_g\|\boldsymbol{x}-\boldsymbol{y}\|$.

Finally, for the term $-\dot{\varepsilon}\tilde{\boldsymbol{q}}^T M(\boldsymbol{q})\dot{\boldsymbol{q}}$, we have

$$\begin{aligned}-\dot{\varepsilon}\tilde{\boldsymbol{q}}^T M(\boldsymbol{q})\dot{\boldsymbol{q}} &\leq \left|-\dot{\varepsilon}\tilde{\boldsymbol{q}}^T M(\boldsymbol{q})\dot{\boldsymbol{q}}\right| \\ &= \left|\frac{\varepsilon_0}{\|\tilde{\boldsymbol{q}}\|\left(1+\|\tilde{\boldsymbol{q}}\|\right)^2}\tilde{\boldsymbol{q}}^T\dot{\boldsymbol{q}}\tilde{\boldsymbol{q}}^T M(\boldsymbol{q})\dot{\boldsymbol{q}}\right| \\ &\leq \frac{\varepsilon_0}{\|\tilde{\boldsymbol{q}}\|\left(1+\|\tilde{\boldsymbol{q}}\|\right)^2}\left\|\tilde{\boldsymbol{q}}\right\|\left\|\dot{\boldsymbol{q}}\right\|\left\|\tilde{\boldsymbol{q}}\right\|\left\|M(\boldsymbol{q})\dot{\boldsymbol{q}}\right\| \\ &\leq \frac{\varepsilon_0}{1+\|\tilde{\boldsymbol{q}}\|}\left\|\dot{\boldsymbol{q}}\right\|^2\lambda_{\mathrm{Max}}\{M(\boldsymbol{q})\} \\ &\leq \varepsilon_0\lambda_{\mathrm{Max}}\{M\}\left\|\dot{\boldsymbol{q}}\right\|^2 \end{aligned} \tag{15.26}$$

where we considered again the definition of ε in (15.22) and Property 4.1, *i.e.* that $\lambda_{\mathrm{Max}}\{M\}\|\dot{\boldsymbol{q}}\| \geq \lambda_{\mathrm{Max}}\{M(\boldsymbol{q})\}\|\dot{\boldsymbol{q}}\| \geq \|M(\boldsymbol{q})\dot{\boldsymbol{q}}\|$.

From the inequalities (15.24), (15.25) and (15.26), it follows that the time derivative $\dot{V}(\tilde{\boldsymbol{q}},\dot{\boldsymbol{q}},\tilde{\boldsymbol{\theta}})$ in (15.23) reduces to

$$\begin{aligned}\dot{V}(\tilde{\boldsymbol{q}},\dot{\boldsymbol{q}},\tilde{\boldsymbol{\theta}}) \leq{}& -\dot{\boldsymbol{q}}^T K_v\dot{\boldsymbol{q}} + \varepsilon\dot{\boldsymbol{q}}^T M(\boldsymbol{q})\dot{\boldsymbol{q}} - \varepsilon\tilde{\boldsymbol{q}}^T K_p\tilde{\boldsymbol{q}} + \varepsilon\tilde{\boldsymbol{q}}^T K_v\dot{\boldsymbol{q}} \\ &+ \varepsilon_0 k_{C_1}\left\|\dot{\boldsymbol{q}}\right\|^2 + \varepsilon k_g\left\|\tilde{\boldsymbol{q}}\right\|^2 + \varepsilon_0\lambda_{\mathrm{Max}}\{M\}\left\|\dot{\boldsymbol{q}}\right\|^2 .\end{aligned}$$

which in turn may be rewritten as

$$\begin{aligned}\dot{V}(\tilde{\boldsymbol{q}},\dot{\boldsymbol{q}},\tilde{\boldsymbol{\theta}}) \leq{}& -\begin{bmatrix}\tilde{\boldsymbol{q}}\\ \dot{\boldsymbol{q}}\end{bmatrix}^T\begin{bmatrix}\varepsilon K_p & -\frac{\varepsilon}{2}K_v\\ -\frac{\varepsilon}{2}K_v & \frac{1}{2}K_v\end{bmatrix}\begin{bmatrix}\tilde{\boldsymbol{q}}\\ \dot{\boldsymbol{q}}\end{bmatrix} + \varepsilon k_g\left\|\tilde{\boldsymbol{q}}\right\|^2 \\ &-\frac{1}{2}\left[\lambda_{\min}\{K_v\} - 2\varepsilon_0(k_{C_1}+2\lambda_{\mathrm{Max}}\{M\})\right]\left\|\dot{\boldsymbol{q}}\right\|^2, \end{aligned} \tag{15.27}$$

where we used $-\dot{\boldsymbol{q}}^T K_v \dot{\boldsymbol{q}} \leq -\frac{1}{2}\dot{\boldsymbol{q}}^T K_v \dot{\boldsymbol{q}} - \frac{\lambda_{\min}\{K_v\}}{2}\|\dot{\boldsymbol{q}}\|^2$ and $\varepsilon \dot{\boldsymbol{q}}^T M(\boldsymbol{q})\dot{\boldsymbol{q}} \leq \varepsilon_0 \lambda_{\mathrm{Max}}\{M\}\|\dot{\boldsymbol{q}}\|^2$. Finally, from (15.27) we get

$$
\begin{aligned}
\dot{V}(\tilde{\boldsymbol{q}}, \dot{\boldsymbol{q}}, \tilde{\boldsymbol{\theta}}) \leq & -\varepsilon \begin{bmatrix} \|\tilde{\boldsymbol{q}}\| \\ \|\dot{\boldsymbol{q}}\| \end{bmatrix}^T \overbrace{\begin{bmatrix} \lambda_{\min}\{K_p\} - k_g & -\frac{1}{2}\lambda_{\mathrm{Max}}\{K_v\} \\ -\frac{1}{2}\lambda_{\mathrm{Max}}\{K_v\} & \frac{1}{2\varepsilon_0}\lambda_{\min}\{K_v\} \end{bmatrix}}^{Q} \begin{bmatrix} \|\tilde{\boldsymbol{q}}\| \\ \|\dot{\boldsymbol{q}}\| \end{bmatrix} \\
& -\frac{1}{2}\underbrace{\left[\lambda_{\min}\{K_v\} - 2\varepsilon_0(k_{C_1} + 2\lambda_{\mathrm{Max}}\{M\})\right]}_{\delta}\|\dot{\boldsymbol{q}}\|^2 . \qquad (15.28)
\end{aligned}
$$

From the inequality above, we may determine immediately the conditions for ε_0 to ensure that $\dot{V}(\tilde{\boldsymbol{q}}, \dot{\boldsymbol{q}}, \tilde{\boldsymbol{\theta}})$ is a negative semidefinite function. For this, we require first to guarantee that the matrix Q is positive definite and that $\delta > 0$. The matrix Q is positive definite if

$$
\begin{aligned}
\lambda_{\min}\{K_p\} &> k_g \qquad (15.29) \\
\frac{2\lambda_{\min}\{K_v\}(\lambda_{\min}\{K_p\} - k_g)}{\lambda^2_{\mathrm{Max}}\{K_v\}} &> \varepsilon_0
\end{aligned}
$$

while $\delta > 0$ if

$$
\frac{\lambda_{\min}\{K_v\}}{2(k_{C_1} + 2\lambda_{\mathrm{Max}}\{M\})} > \varepsilon_0 . \qquad (15.30)
$$

Observe that the three conditions (15.29)–(15.30) are satisfied since by hypothesis the matrix K_p and the constant ε_0 verify conditions C.1 and C.3–C.4 respectively. Therefore, the matrix Q is symmetric positive definite which means that $\lambda_{\min}\{Q\} > 0$.

Next, invoking the theorem of Rayleigh–Ritz (*cf.* page 24), we obtain

$$
\begin{aligned}
-\varepsilon \begin{bmatrix} \|\tilde{\boldsymbol{q}}\| \\ \|\dot{\boldsymbol{q}}\| \end{bmatrix}^T \overbrace{\begin{bmatrix} \lambda_{\min}\{K_p\} - k_g & -\frac{1}{2}\lambda_{\mathrm{Max}}\{K_v\} \\ -\frac{1}{2}\lambda_{\mathrm{Max}}\{K_v\} & \frac{1}{2\varepsilon_0}\lambda_{\min}\{K_v\} \end{bmatrix}}^{Q} \begin{bmatrix} \|\tilde{\boldsymbol{q}}\| \\ \|\dot{\boldsymbol{q}}\| \end{bmatrix} \leq \\
-\varepsilon\lambda_{\min}\{Q\}\left[\|\tilde{\boldsymbol{q}}\|^2 + \|\dot{\boldsymbol{q}}\|^2\right] .
\end{aligned}
$$

Incorporating this inequality in (15.28) and using the definition of ε we obtain

$$
\begin{aligned}
\dot{V}(\tilde{\boldsymbol{q}}, \dot{\boldsymbol{q}}, \tilde{\boldsymbol{\theta}}) &\leq -\frac{\varepsilon_0}{1 + \|\tilde{\boldsymbol{q}}\|}\lambda_{\min}\{Q\}\left[\|\tilde{\boldsymbol{q}}\|^2 + \|\dot{\boldsymbol{q}}\|^2\right] - \frac{\delta}{2}\|\dot{\boldsymbol{q}}\|^2 \\
&\leq -\varepsilon_0\lambda_{\min}\{Q\}\frac{\|\tilde{\boldsymbol{q}}\|^2}{1 + \|\tilde{\boldsymbol{q}}\|} - \frac{\delta}{2}\|\dot{\boldsymbol{q}}\|^2 . \qquad (15.31)
\end{aligned}
$$

Therefore, it appears that $\dot{V}(\tilde{\boldsymbol{q}}, \dot{\boldsymbol{q}}, \tilde{\boldsymbol{\theta}})$ expressed in (15.31), is a globally negative *semidefinite* function. Since moreover the Lyapunov function candidate (15.18) is globally positive definite, Theorem 2.3 allows one to guarantee that the origin of the state space of the closed-loop Equation (15.17) is stable and in particular that its solutions are bounded, that is,

$$\tilde{\boldsymbol{q}}, \dot{\boldsymbol{q}} \in L_\infty^n, \tag{15.32}$$

$$\tilde{\boldsymbol{\theta}} \in L_\infty^m.$$

Since $\dot{V}(\tilde{\boldsymbol{q}}, \dot{\boldsymbol{q}}, \tilde{\boldsymbol{\theta}})$ obtained in (15.31) is not negative definite we may not conclude yet that the origin is an asymptotically stable equilibrium point.

Hence, from the analysis presented so far it is not possible yet to conclude anything about the achievement of the position control objective. For this it is necessary to make some additional claims.

The idea consists in using Lemma A.5 (*cf.* page 392) which establishes that if a continuously differentiable function $\boldsymbol{f} : \mathbb{R}_+ \to \mathbb{R}^n$ satisfies $\boldsymbol{f} \in L_2^n$ and $\boldsymbol{f}, \dot{\boldsymbol{f}} \in L_\infty^n$ then $\lim_{t\to\infty} \boldsymbol{f}(t) = \boldsymbol{0} \in \mathbb{R}^n$.

Hence, if we wish to show that $\lim_{t\to\infty} \tilde{\boldsymbol{q}}(t) = \boldsymbol{0} \in \mathbb{R}^n$, and we know from (15.32) that $\tilde{\boldsymbol{q}} \in L_\infty^n$ and $\dot{\tilde{\boldsymbol{q}}} = -\dot{\boldsymbol{q}} \in L_\infty^n$, it is only left to prove that $\tilde{\boldsymbol{q}} \in L_2^n$, that is, to verify the existence of a finite positive constant k such that

$$k \geq \int_0^\infty \|\tilde{\boldsymbol{q}}(t)\|^2 \, dt.$$

This proof is developed below.

Since $\frac{\delta}{2}\|\dot{\boldsymbol{q}}\|^2 \geq 0$ for all $\dot{\boldsymbol{q}} \in \mathbb{R}^n$ then, from (15.31), the following inequality holds:

$$\frac{d}{dt}V(\tilde{\boldsymbol{q}}(t), \dot{\boldsymbol{q}}(t), \tilde{\boldsymbol{\theta}}(t)) \leq -\varepsilon_0 \lambda_{\min}\{Q\} \frac{\|\tilde{\boldsymbol{q}}(t)\|^2}{1 + \|\tilde{\boldsymbol{q}}(t)\|}. \tag{15.33}$$

The next step consists in integrating the inequality (15.33) from $t = 0$ to $t = \infty$, that is[1]

$$\int_{V_0}^{V_\infty} dV \leq -\varepsilon_0 \lambda_{\min}\{Q\} \int_0^\infty \frac{\|\tilde{\boldsymbol{q}}(t)\|^2}{1 + \|\tilde{\boldsymbol{q}}(t)\|} dt$$

where we defined $V_0 := V(0, \tilde{\boldsymbol{q}}(0), \dot{\boldsymbol{q}}(0), \tilde{\boldsymbol{\theta}}(0))$ and

[1] Recall that for functions $g(t)$ and $f(t)$ continuous in $a \leq t \leq b$, satisfying $g(t) \leq f(t)$ for all $a \leq t \leq b$, we have

$$\int_a^b g(t) \, dt \leq \int_a^b f(t) \, dt.$$

$$V_\infty := \lim_{t\to\infty} V(\tilde{\boldsymbol{q}}(t), \dot{\boldsymbol{q}}(t), \tilde{\boldsymbol{\theta}}(t)) .$$

The integral on the left-hand side of the inequality above may be trivially evaluated to obtain

$$V_\infty - V_0 \le -\varepsilon_0 \lambda_{\min}\{Q\} \int_0^\infty \frac{\|\tilde{\boldsymbol{q}}(t)\|^2}{1+\|\tilde{\boldsymbol{q}}(t)\|} dt ,$$

or in equivalent form

$$-V_0 \le -\varepsilon_0 \lambda_{\min}\{Q\} \int_0^\infty \frac{\|\tilde{\boldsymbol{q}}(t)\|^2}{1+\|\tilde{\boldsymbol{q}}(t)\|} dt - V_\infty . \tag{15.34}$$

Here it is worth recalling that the Lyapunov function candidate $V(\tilde{\boldsymbol{q}}, \dot{\tilde{\boldsymbol{q}}}, \tilde{\boldsymbol{\theta}})$ is positive definite, hence we may claim that $V_\infty \ge 0$ and therefore, from the inequality (15.34) we get

$$-V_0 \le -\varepsilon_0 \lambda_{\min}\{Q\} \int_0^\infty \frac{\|\tilde{\boldsymbol{q}}(t)\|^2}{1+\|\tilde{\boldsymbol{q}}(t)\|} dt .$$

From the latter expression it readily follows that

$$\frac{V_0}{\varepsilon_0 \lambda_{\min}\{Q\}} \ge \int_0^\infty \frac{\|\tilde{\boldsymbol{q}}(t)\|^2}{1+\|\tilde{\boldsymbol{q}}(t)\|} dt ,$$

where the left-hand side of the inequality above is constant, positive and bounded. This means that the position error $\tilde{\boldsymbol{q}}$ divided by $\sqrt{1+\|\tilde{\boldsymbol{q}}\|}$ belongs to the L_2^n space, *i.e.*

$$\frac{\tilde{\boldsymbol{q}}}{\sqrt{1+\|\tilde{\boldsymbol{q}}\|}} \in L_2^n . \tag{15.35}$$

Next, we use Lemma A.7. To that end, we express the position error $\tilde{\boldsymbol{q}}$ as the product of two functions in the following manner:

$$\tilde{\boldsymbol{q}} = \underbrace{\left[\sqrt{1+\|\tilde{\boldsymbol{q}}\|}\right]}_{h} \underbrace{\left[\frac{\tilde{\boldsymbol{q}}}{\sqrt{1+\|\tilde{\boldsymbol{q}}\|}}\right]}_{\boldsymbol{f}} .$$

As we showed in (15.32), the position error $\tilde{\boldsymbol{q}}$ belongs to the L_∞^n space and therefore, $\sqrt{1+\|\tilde{\boldsymbol{q}}\|} \in L_\infty$. On the other hand in (15.35) we concluded that the other factor belongs to the space L_2^n, hence $\tilde{\boldsymbol{q}}$ is the product of a bounded function times another which belongs to L_2^n. Using this and Lemma A.7 we obtain

$$\tilde{\boldsymbol{q}} \in L_2^n ,$$

which is what we wanted to prove.

Thus, from $\tilde{\boldsymbol{q}} \in L_2^n$, (15.32) and Lemma A.5 we conclude that the position error $\tilde{\boldsymbol{q}}$ tends asymptotically to the zero vector, *i.e.*

$$\lim_{t\to\infty} \tilde{\boldsymbol{q}}(t) = \boldsymbol{0} \in \mathbb{R}^n .$$

In words, the position control objective is achieved.

Invoking some additional arguments it may be verified that not only the position error $\tilde{\boldsymbol{q}}$ tends to zero asymptotically, but so does the velocity $\dot{\boldsymbol{q}}$. Nevertheless, these conclusions should not be extrapolated to the parametric errors $\tilde{\boldsymbol{\theta}}(t)$.

Thus, from *the* previous analysis we conclude that in general, the origin of the closed-loop Equation (15.17) may not be an asymptotically stable equilibrium point, not even locally. Nevertheless, as has been demonstrated the position control objective is guaranteed.

15.3 Examples

We present two examples that illustrate the application of PD control with adaptive desired gravity compensation.

Example 15.1. Consider the model of a pendulum of mass m, inertia J with respect to the axis of rotation, and distance l from the axis of rotation to the center of mass. A torque τ is applied at the axis of rotation, that is,

$$J\ddot{q} + mgl\,\sin(q) = \tau .$$

We clearly identify $M(q) = J$, $C(q,\dot{q}) = 0$ and $g(q) = mgl\,\sin(q)$.

In Example 14.8 we stated the following control problem. Assume that the values of mass m, distance l and gravity acceleration g, are known but that the value of the inertia J is unknown (but constant). The control problem consists now in designing a controller that is capable of satisfying the position control objective

$$\lim_{t\to\infty} q(t) = q_d \in \mathbb{R}$$

for any desired constant joint position q_d.

We may try to solve this control problem by means of PD control with adaptive desired gravity compensation. The parameter of interest that has been assumed unknown is the inertia J.

The parameterization corresponding to (15.2) is, in this example:

$$
\begin{aligned}
M(q, \theta)u + C(q, w, \theta)v + g(q, \theta) \\
&= Ju + mgl\, \sin(q) \\
&= \Phi(q, u, v, w)\theta + M_0(q)u + C_0(q, w)v + g_0(q),
\end{aligned}
$$

where

$$
\begin{aligned}
\Phi(q, u, v, w) &= u \\
\theta &= J \\
M_0(q) &= 0 \\
C_0(q, \dot{q}) &= 0 \\
g_0(q) &= mgl\, \sin(q)\,.
\end{aligned}
$$

Notice that according to the definition of $\Phi_g(\boldsymbol{x})$ we have

$$
\Phi_g(x) = \Phi(x, 0, 0, 0) = 0
$$

for all $x \in \mathbb{R}$.

Therefore, the adaptive control law given by Equations (15.10) and (15.11) becomes

$$
\begin{aligned}
\tau &= k_p\tilde{q} - k_v\dot{q} + \Phi_g(q_d)\hat{\theta} + g_0(q_d) \\
&= k_p\tilde{q} - k_v\dot{q} + mgl\, \sin(q_d)
\end{aligned}
$$

and

$$
\begin{aligned}
\hat{\theta}(t) &= \gamma\Phi_g(q_d)\int_0^t \left[\frac{\varepsilon_0}{1 + \|\tilde{q}\|}\tilde{q} - \dot{q}\right] ds + \hat{\theta}(0) \\
&= \hat{\theta}(0)\,.
\end{aligned}
$$

As the reader may notice not without surprise, the design of the PD controller with adaptive desired gravity compensation yields a non-adaptive controller (observe that the control law does not depend on the adaptive parameter $\hat{\theta}$ and consequently, there does not exist any adaptive law). Therefore, it simply corresponds to PD control with desired gravity compensation. This is because the parametric uncertainty in the model of the pendulum considers only the inertia J, otherwise the component $g(q) = mgl\, \sin(q)$ is completely known and therefore, the control problem that has been formulated may be solved directly for instance by the PD control law with desired gravity compensation, that is without appealing to any concept from adaptive control theory. Nevertheless, the control problem might not be solvable by PD control with desired gravity compensation if for example, the mass m were unknown. This interesting scenario is left as a problem at the end of the chapter.

Recall that condition C.1 establishes that the gain k_p must be larger than k_g; in this example, $k_g \geq mgl$. This is a sufficient condition to guarantee global asymptotic stability for the origin of a PD control with desired gravity compensation in closed loop with an ideal pendulum (see Chapter 8).

The moral of this example is significant: the application of adaptive controllers in the case of parametric uncertainty in the system must be carefully evaluated. As the control problem of this example shows adaptive control approaches are unnecessary in some cases. ◇

We present next the design of PD control with adaptive compensation for the Pelican robot presented in Chapter 5. The reader should notice that the resulting adaptive controller is more complex than in the previous example.

Example 15.2. Consider the Pelican robot studied in Chapter 5 and shown in Figure 5.2. Its dynamic model is recalled here for convenience:

$$\underbrace{\begin{bmatrix} M_{11}(\boldsymbol{q}) & M_{12}(\boldsymbol{q}) \\ M_{21}(\boldsymbol{q}) & M_{22}(\boldsymbol{q}) \end{bmatrix}}_{M(\boldsymbol{q})} \ddot{\boldsymbol{q}} + \underbrace{\begin{bmatrix} C_{11}(\boldsymbol{q},\dot{\boldsymbol{q}}) & C_{12}(\boldsymbol{q},\dot{\boldsymbol{q}}) \\ C_{21}(\boldsymbol{q},\dot{\boldsymbol{q}}) & C_{22}(\boldsymbol{q},\dot{\boldsymbol{q}}) \end{bmatrix}}_{C(\boldsymbol{q},\dot{\boldsymbol{q}})} \dot{\boldsymbol{q}} + \underbrace{\begin{bmatrix} g_1(\boldsymbol{q}) \\ g_2(\boldsymbol{q}) \end{bmatrix}}_{\boldsymbol{g}(\boldsymbol{q})} = \boldsymbol{\tau}$$

where

$$\begin{aligned}
M_{11}(\boldsymbol{q}) &= m_1 l_{c1}^2 + m_2 \left[l_1^2 + l_{c2}^2 + 2 l_1 l_{c2} \cos(q_2) \right] + I_1 + I_2 \\
M_{12}(\boldsymbol{q}) &= m_2 \left[l_{c2}^2 + l_1 l_{c2} \cos(q_2) \right] + I_2 \\
M_{21}(\boldsymbol{q}) &= m_2 \left[l_{c2}^2 + l_1 l_{c2} \cos(q_2) \right] + I_2 \\
M_{22}(\boldsymbol{q}) &= m_2 l_{c2}^2 + I_2
\end{aligned}$$

$$\begin{aligned}
C_{11}(\boldsymbol{q},\dot{\boldsymbol{q}}) &= -m_2 l_1 l_{c2} \sin(q_2) \dot{q}_2 \\
C_{12}(\boldsymbol{q},\dot{\boldsymbol{q}}) &= -m_2 l_1 l_{c2} \sin(q_2) \left[\dot{q}_1 + \dot{q}_2 \right] \\
C_{21}(\boldsymbol{q},\dot{\boldsymbol{q}}) &= m_2 l_1 l_{c2} \sin(q_2) \dot{q}_1 \\
C_{22}(\boldsymbol{q},\dot{\boldsymbol{q}}) &= 0
\end{aligned}$$

$$\begin{aligned}
g_1(\boldsymbol{q}) &= \left[m_1 l_{c1} + m_2 l_1 \right] g \sin(q_1) + m_2 l_{c2} g \sin(q_1 + q_2) \\
g_2(\boldsymbol{q}) &= m_2 l_{c2} g \sin(q_1 + q_2) \,.
\end{aligned}$$

For this example we consider parametric uncertainty in the mass m_2, the inertia I_2 and in the location of the center of mass l_{c2} of the second link; that is, the numerical values of these constants are not known exactly. Nevertheless, we assume we know upper-bounds on these constants, and they are denoted by $\overline{m_2}$, $\overline{I_2}$ and $\overline{l_{c2}}$ respectively, that is,

$$m_2 \leq \overline{m_2}; \qquad I_2 \leq \overline{I_2}; \qquad l_{c2} \leq \overline{l_{c2}}.$$

The control problem consists in driving asymptotically to zero the position error $\tilde{\boldsymbol{q}}(t)$ for any constant vector of desired joint positions $\boldsymbol{q}_d(t)$. Notice that in view of the supposed parametric uncertainty, the solution of the control problem is not trivial. In particular, the lack of knowledge of m_2 and l_{c2} has a direct impact on the uncertainty in the vector of gravitational torques $\boldsymbol{g}(\boldsymbol{q})$. The solution that we give below is based on PD control with adaptive desired gravity compensation.

The robot considered here, including parametric uncertainty, was analyzed in Example 14.7 where we used the (unknown) dynamic parameters vector $\boldsymbol{\theta} \in \mathbb{R}^3$,

$$\boldsymbol{\theta} = \begin{bmatrix} \theta_1 \\ \theta_2 \\ \theta_3 \end{bmatrix} = \begin{bmatrix} m_2 \\ m_2 l_{c2} \\ m_2 l_{c2}^2 + I_2 \end{bmatrix}.$$

The structure of the PD control law with adaptive desired gravity compensation is given by (15.10)–(15.11), *i.e.*

$$\boldsymbol{\tau} = K_p \tilde{\boldsymbol{q}} - K_v \dot{\boldsymbol{q}} + \Phi_g(\boldsymbol{q}_d)\hat{\boldsymbol{\theta}} + \boldsymbol{g}_0(\boldsymbol{q}_d)$$

$$\hat{\boldsymbol{\theta}}(t) = \Gamma \Phi_g(\boldsymbol{q}_d)^T \int_0^t \left[\frac{\varepsilon_0}{1 + \|\tilde{\boldsymbol{q}}\|} \tilde{\boldsymbol{q}} - \dot{\boldsymbol{q}} \right] ds + \hat{\boldsymbol{\theta}}(0)$$

where $K_p, K_v \in \mathbb{R}^{n \times n}$ and $\Gamma \in \mathbb{R}^{m \times m}$ are symmetric positive definite design matrices and ε_0 is a positive constant, which must be chosen appropriately. The vector $\boldsymbol{g}_0(\boldsymbol{q})$ was obtained previously for the robot considered here, in Example 14.7, as

$$\boldsymbol{g}_0(\boldsymbol{q}_d) = \begin{bmatrix} m_1 l_{c1} g \sin(q_{d1}) \\ 0 \end{bmatrix}.$$

In Example 14.7 we determined $\Phi(\boldsymbol{q}, \boldsymbol{u}, \boldsymbol{v}, \boldsymbol{w})$. Therefore, the matrix $\Phi_g(\boldsymbol{q}_d)$ follows from (15.5) as

$$\begin{aligned} \Phi_g(\boldsymbol{q}_d) &= \Phi(\boldsymbol{q}_d, \boldsymbol{0}, \boldsymbol{0}, \boldsymbol{0}) \\ &= \begin{bmatrix} l_1 g \sin(q_{d1}) & g \sin(q_{d1} + q_{d2}) & 0 \\ 0 & g \sin(q_{d1} + q_{d2}) & 0 \end{bmatrix}. \end{aligned}$$

Once the structure of the controller has been defined, we proceed to determine its parameters. For this, we see that we need to compute the matrices K_p and K_v, as well as the constant ε_0 in accordance with conditions C.1 through C.4 (*cf.* page 340). To that end, we first need to determine the numerical values of the constants $\lambda_{\mathrm{Max}}\{M\}$, k_{C1} and k_g which must satisfy

$$\bullet \lambda_{\text{Max}}\{M(\boldsymbol{q},\boldsymbol{\theta})\} \leq \lambda_{\text{Max}}\{M\} \qquad \forall\, \boldsymbol{q} \in \mathbb{R}^n, \quad \boldsymbol{\theta} \in \Omega$$

$$\bullet \left\|C(\boldsymbol{q},\dot{\boldsymbol{q}},\boldsymbol{\theta})\right\| \leq k_{C1} \left\|\dot{\boldsymbol{q}}\right\| \qquad \forall\, \boldsymbol{q},\dot{\boldsymbol{q}} \in \mathbb{R}^n, \quad \boldsymbol{\theta} \in \Omega$$

$$\bullet \left\|\boldsymbol{g}(\boldsymbol{x},\boldsymbol{\theta}) - \boldsymbol{g}(\boldsymbol{y},\boldsymbol{\theta})\right\| \leq k_g \left\|\boldsymbol{x} - \boldsymbol{y}\right\| \ \forall\, \boldsymbol{x},\boldsymbol{y} \in \mathbb{R}^n, \quad \boldsymbol{\theta} \in \Omega.$$

Therefore, it appears necessary to characterize the set $\Omega \subset \mathbb{R}^3$ to which belongs the vector of unknown dynamic parameters $\boldsymbol{\theta}$. This can be done by using the upper-bounds $\overline{m_2}$, $\overline{I_2}$ and $\overline{l_{c2}}$ which are assumed to be known. The set Ω is then given by

$$\Omega = \left\{ \begin{bmatrix} x_1 \\ x_2 \\ x_3 \end{bmatrix} \in \mathbb{R}^3 : |x_1| \leq \overline{m_2}; |x_2| \leq \overline{m_2}\overline{l_{c2}}; |x_3| \leq \overline{m_2}\overline{l_{c2}}^2 + \overline{I_2} \right\}.$$

Expressions for the constants $\lambda_{\text{Max}}\{M\}$, k_{C1} and k_g were obtained for the robot under study, in Chapter 5. In the case of parametric uncertainty considered here, such expressions are

$$\begin{aligned}
\lambda_{\text{Max}}\{M\} &\geq m_1 l_{c1}^2 + \overline{m_2}\left[l_1^2 + 2\overline{l_{c2}}^2 + 3 l_{c1}\overline{l_{c2}}\right] + I_1 + \overline{I_2} \\
k_{C1} &\geq n^2 \overline{m_2} l_1 \overline{l_{c2}} \\
k_g &\geq n\left[m_1 l_{c1} + \overline{m_2} l_1 + \overline{m_2}\overline{l_{c2}}\right] g\,.
\end{aligned}$$

Considering the numerical values shown in Table 5.1 of Chapter 5, and fixing the following values for the bounds,

$$\begin{aligned}
\overline{m_2} &= 2.898 && [\text{kg}] \\
\overline{I_2} &= 0.0125 && [\text{kg m}^2] \\
\overline{l_{c2}} &= 0.02862 && [\text{m}],
\end{aligned}$$

we finally obtain the values:

$$\begin{aligned}
\lambda_{\text{Max}}\{M\} &= 0.475 && [\text{kg m}^2] \\
k_{C1} &= 0.086 && [\text{kg m}^2] \\
k_g &= 28.99 && [\text{kg m}^2/\text{s}^2]\,.
\end{aligned}$$

The next step consists in using the previous information together with conditions C.1 through C.4 (*cf.* page 340) to calculate the matrices K_p, K_v and the constants ε_0 and ε_2. As a matter of fact we may simply choose K_p so as to satisfy condition C.1, any positive definite matrix K_v and any constant ε_2 strictly larger than two. Finally, using conditions C.2 through C.4 we obtain ε_0. The choice of the latter is detailed below.

Condition C.1 establishes the inequality

$$\lambda_{\min}\{K_p\} > k_g$$

where $k_g = 28.99$. Hence the matrix $K_p = \text{diag}\{k_p\} = \text{diag}\{30\}$ satisfies such a condition.

The matrix K_v is chosen arbitrarily but of course it must be symmetric positive definite. For instance, we may fix it at $K_v = \text{diag}\{k_v\} = \text{diag}\{7,\ 3\}$.

Next, we choose ε_1 in accordance with the inequality (15.13), that is,

$$\frac{2\lambda_{\min}\{K_p\}}{k_g} > \varepsilon_1 > 2,$$

where for instance, $\varepsilon_1 = 2.01$ is an appropriate value. The constant ε_2 is determined from the definition (15.12), that is,

$$\varepsilon_2 = \frac{2\varepsilon_1}{\varepsilon_1 - 2},$$

so we get $\varepsilon_2 = 402$.

Using all this information, we immediately verify that

$$\sqrt{\frac{2\lambda_{\min}\{K_p\}}{\varepsilon_2 \lambda_{\mathrm{Max}}\{M\}}} = 0.561$$

$$\frac{2\lambda_{\min}\{K_v\}[\lambda_{\min}\{K_p\} - k_g]}{\lambda_{\mathrm{Max}}^2\{K_v\}} = 0.124$$

$$\frac{\lambda_{\min}\{K_v\}}{2\left[k_{C1} + 2\lambda_{\mathrm{Max}}\{M\}\right]} = 1.448$$

According to conditions C.2 through C.4, the positive constant ε_0 must be strictly smaller than the previous quantities. Therefore, we choose $\varepsilon_0 = 0.12$.

The adaptation gains matrix Γ must be symmetric positive definite. A choice is $\Gamma = \text{diag}\{\gamma_1,\ \gamma_2\} = \text{diag}\{500,\ 10\}$. The vector of initial adaptive parameters is arbitrary, and here it is taken to be zero: $\hat{\boldsymbol{\theta}}(0) = \mathbf{0}$.

In summary, the control law may be written as

$$\begin{aligned}
\tau_1 &= k_p\tilde{q}_1 - k_v\dot{q}_1 + l_1 g \sin(q_{d1})\hat{\theta}_1 + g \sin(q_{d1} + q_{d2})\hat{\theta}_2 \\
&\quad + m_1 l_{c1} g \sin(q_{d1}) \\
\tau_2 &= k_p\tilde{q}_2 - k_v\dot{q}_2 + g \sin(q_{d1} + q_{d2})\hat{\theta}_2 .
\end{aligned}$$

Notice that the control law does not depend on the adaptive parameter $\hat{\theta}_3$. This is because the vector of gravitational torques $\boldsymbol{g}(\boldsymbol{q})$ does not depend explicitly on the dynamic parameter $\theta_3 = m_2 l_{c2}^2 + I_2$. Consequently, the adaptive law has only the following two components instead of three:

$$\hat{\theta}_1(t) = \gamma_1 l_1 g \sin(q_{d1}) \int_0^t \left[\frac{\varepsilon_0}{1+\|\tilde{\boldsymbol{q}}\|}\tilde{q}_1 - \dot{q}_1\right] ds + \hat{\theta}_1(0)$$

$$\begin{aligned}\hat{\theta}_2(t) &= \gamma_2 g \sin(q_{d1}+q_{d2}) \int_0^t \left[\frac{\varepsilon_0}{1+\|\tilde{\boldsymbol{q}}\|}\tilde{q}_1 - \dot{q}_1\right] ds \\ &+ \gamma_2 g \sin(q_{d1}+q_{d2}) \int_0^t \left[\frac{\varepsilon_0}{1+\|\tilde{\boldsymbol{q}}\|}\tilde{q}_2 - \dot{q}_2\right] ds + \hat{\theta}_2(0)\,.\end{aligned}$$

We describe next the laboratory experimental results. The initial conditions corresponding to the positions and velocities, are chosen as

$$\begin{aligned} q_1(0) &= 0, & q_2(0) &= 0 \\ \dot{q}_1(0) &= 0, & \dot{q}_2(0) &= 0\,. \end{aligned}$$

The desired joint positions are chosen as

$$q_{d1} = \pi/10, \qquad q_{d2} = \pi/30 \quad [\text{rad}]\,.$$

In terms of the state vector of the closed-loop equation, the initial state is

$$\begin{bmatrix} \tilde{\boldsymbol{q}}(0) \\ \\ \dot{\boldsymbol{q}}(0) \end{bmatrix} = \begin{bmatrix} \pi/10 \\ \pi/30 \\ 0 \\ 0 \end{bmatrix} = \begin{bmatrix} 0.3141 \\ 0.1047 \\ 0 \\ 0 \end{bmatrix} \quad [\text{rad}]\,.$$

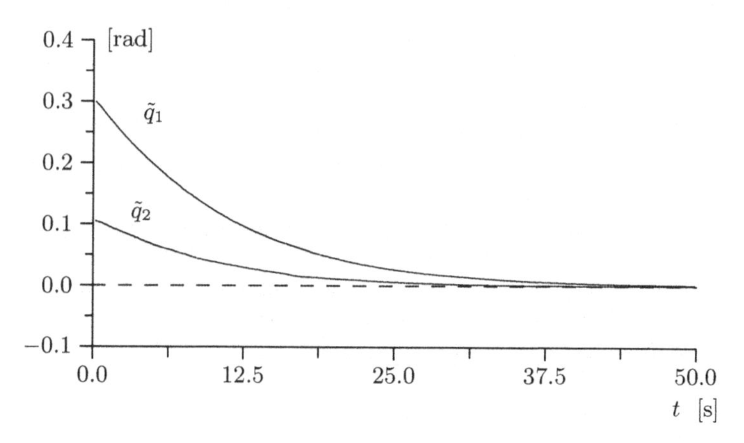

Figure 15.1. Graphs of position errors $\tilde{q}_1$ and $\tilde{q}_2$

Figures 15.1 and 15.2 present the experimental results. In particular, Figure 15.1 shows that the components of the position error $\tilde{\boldsymbol{q}}(t)$ tend asymptotically to zero in spite of the non-modeled friction phenomenon. The evolution in time of the adaptive parameters is shown

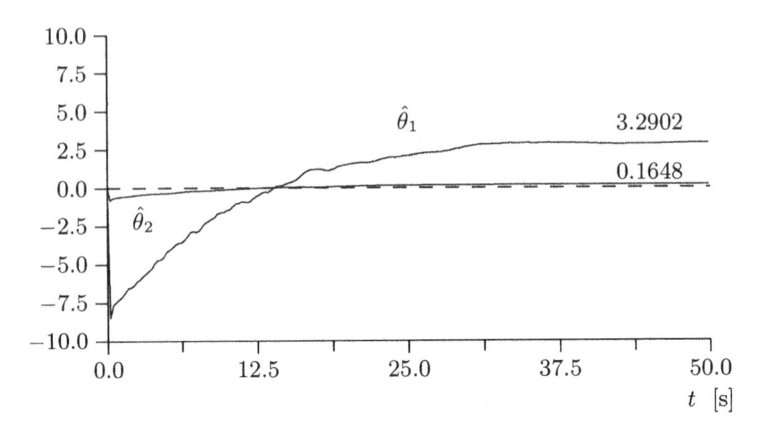

Figure 15.2. Graphs of adaptive parameters $\hat{\theta}_1$ and $\hat{\theta}_2$

in Figure 15.2 where we appreciate that both parameters tend to values which are relatively near the unknown values of θ_1 and θ_2, *i.e.*

$$\lim_{t\to\infty}\begin{bmatrix}\hat{\theta}_1(t)\\ \hat{\theta}_2(t)\end{bmatrix}=\begin{bmatrix}3.2902\\ 0.1648\end{bmatrix}\approx\begin{bmatrix}\theta_1\\ \theta_2\end{bmatrix}=\begin{bmatrix}m_2\\ m_2 l_{c2}\end{bmatrix}=\begin{bmatrix}2.0458\\ 0.047\end{bmatrix}.$$

As mentioned in Chapter 14 the latter phenomenon, *i.e.* that $\hat{\boldsymbol{\theta}}(t)\to\boldsymbol{\theta}$ as $t\to\infty$ is called parametric convergence and the proof of this property relies on a property called *persistency of excitation.* Verifying this property in applications is in general a difficult task and as a matter of fact, often in complex (nonlinear) adaptive control systems it may be expected that parameters do not converge to their true values.

Similarly as for PID control, it may be appreciated from Figure 15.1 that the temporal evolution of the position errors is slow. Note that the timescale spans 50 s. Hence, as for the case of PID control the transient response here is slower than that under PD control with gravity compensation (see Figure 7.3) or PD control with desired gravity compensation (see Figure 8.4). As before, if instead of limiting the value of ε_0 we use the same gains as for the latter controllers, the performance is improved, as can be appreciated from Figure 15.3. For this, we set the gains to

$$K_p=\begin{bmatrix}30 & 0\\ 0 & 30\end{bmatrix}\quad[\mathrm{Nm/rad}],$$

$$K_v=\begin{bmatrix}7 & 0\\ 0 & 3\end{bmatrix}\quad[\mathrm{Nm\,s/rad}],$$

$$\Gamma=\begin{bmatrix}500 & 0\\ 0 & 10\end{bmatrix}\quad[\mathrm{Nm/rad\,s}],$$

and $\varepsilon_0 = 5$, *i.e.* K_p and K_v have the same values as for the PD controllers.

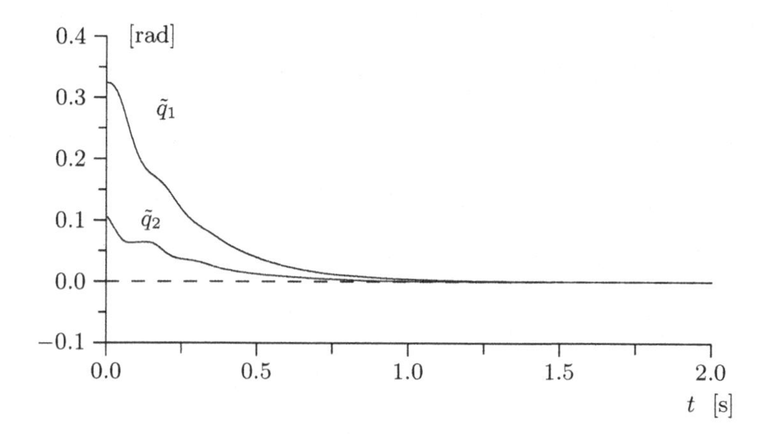

Figure 15.3. Graphs of position errors $\tilde{q}_1$ and $\tilde{q}_2$

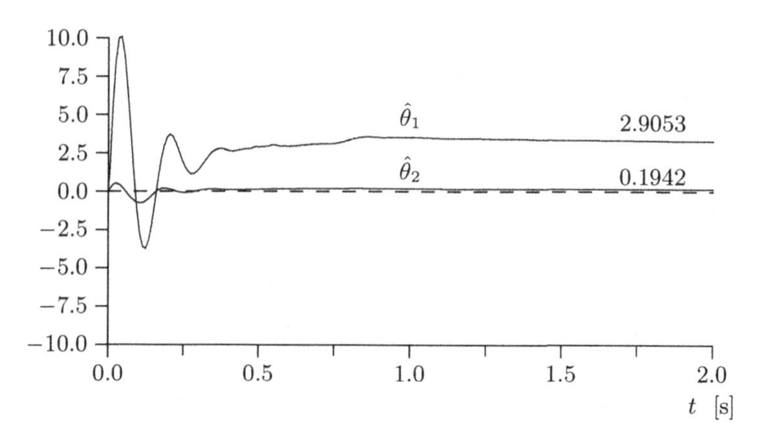

Figure 15.4. Graphs of adaptive parameters $\hat{\theta}_1$ and $\hat{\theta}_2$

15.4 Conclusions

We may draw the following conclusions from the analysis presented in this chapter.

Consider the PD control law with adaptive desired compensation gravity of robots with n revolute rigid joints.

- Assume that the desired position $\boldsymbol{q}_d$ is constant.
- Assume that the symmetric matrices K_p and K_v of the control law satisfy
  - $\lambda_{\rm Max}\{K_p\} \geq \lambda_{\rm min}\{K_p\} > k_g$
  - $K_v > 0$.

  Choose the constants ε_1 and ε_2 in accordance with

$$\frac{2\lambda_{\rm min}\{K_p\}}{k_g} > \varepsilon_1 > 2,$$

$$\varepsilon_2 = \frac{2\varepsilon_1}{\varepsilon_1 - 2}.$$

  If the symmetric matrix Γ and the constant ε_0 from the adaptive law satisfy
  - $\Gamma > 0$
  - $\sqrt{\dfrac{2\lambda_{\rm min}\{K_p\}}{\varepsilon_2\lambda_{\rm Max}\{M\}}} > \varepsilon_0$
  - $\dfrac{2\lambda_{\rm min}\{K_v\}[\lambda_{\rm min}\{K_p\} - k_g]}{\lambda^2_{\rm Max}\{K_v\}} > \varepsilon_0$
  - $\dfrac{\lambda_{\rm min}\{K_v\}}{2\left[k_{C1} + 2\lambda_{\rm Max}\{M\}\right]} > \varepsilon_0,$

  then the origin of the closed-loop equation expressed in terms of the state vector $\begin{bmatrix}\tilde{\boldsymbol{q}}^T & \dot{\boldsymbol{q}}^T & \tilde{\boldsymbol{\theta}}^T\end{bmatrix}^T$, is a stable equilibrium. Moreover, the position control objective is achieved globally. In particular we have $\lim_{t\to\infty}\tilde{\boldsymbol{q}}(t) = \boldsymbol{0}$ for all initial condition $\begin{bmatrix}\tilde{\boldsymbol{q}}(0)^T & \dot{\boldsymbol{q}}(0)^T & \tilde{\boldsymbol{\theta}}(0)^T\end{bmatrix}^T \in \mathbb{R}^{2n+m}$.

Bibliography

The material of this chapter has been adapted from

- Kelly R., 1993, *"Comments on 'Adaptive PD controller for robot manipulators' "*, IEEE Transactions on Robotics and Automation, Vol. 9, No. 1, p. 117–119.

The Lyapunov function (15.18) follows the ideas reported in

- Whitcomb L. L., Rizzi A., Koditschek D. E., 1993, *"Comparative experiments with a new adaptive controller for robot arms"*, IEEE Transactions on Robotics and Automation, Vol. 9, No. 1, p. 59–70.

An adaptive version of PD control with gravity compensation has been presented in

- Tomei P., 1991, *"Adaptive PD controller for robot manipulators"*, IEEE Transactions on Robotics and Automation, Vol. 7, No. 4, p. 565–570.

Problems

1. Consider the Example 15.1. In this example we assumed uncertainty in the inertia J. Obtain explicitly the control and adaptive laws corresponding to PD control with adaptive desired compensation assuming now that the uncertainty is on the mass m.
2. Show that the PD control law with adaptive desired gravity compensation, given by (15.10)–(15.11), may be written as a controller of type PID with "normalized" integral action, that is,

$$\boldsymbol{\tau} = K_P\tilde{\boldsymbol{q}} - K_v\dot{\boldsymbol{q}} + K_i \int_0^t \frac{\tilde{\boldsymbol{q}}}{1 + \|\tilde{\boldsymbol{q}}\|}\, ds$$

where we defined

$$\begin{aligned} K_P &= K_p + \Phi_g(\boldsymbol{q}_d)\Gamma\Phi_g(\boldsymbol{q}_d)^T \\ K_i &= \varepsilon_0\Phi_g(\boldsymbol{q}_d)\Gamma\Phi_g(\boldsymbol{q}_d)^T \\ \Phi_g(\boldsymbol{q}_d)\hat{\boldsymbol{\theta}}(0) &= -\boldsymbol{g}_0(\boldsymbol{q}_d)\,. \end{aligned}$$

16

PD Control with Adaptive Compensation

As mentioned in Chapter 11, in 1987 an adaptive control system – control law and adaptive law – was proposed to solve the motion control problem for robot manipulators under parameter uncertainty and, since then, this control scheme has become increasingly popular in the study of robot control. This is the so-called adaptive controller of Slotine and Li. In Chapter 11 we present the 'non-adaptive' version of this controller, which we have called, PD control with compensation. In the present chapter we study the same control law in its original form, *i.e.* with adaptation. As usual, related references are cited at the end of the chapter.

In Chapter 11 we showed that in the scenario that the dynamic robot model is exactly known, that is, both its structure and its dynamic parameters are known, this control law may be used to achieve the motion control objective, globally; moreover, with a trivial choice of design parameters.

In this chapter we consider the case where the dynamic parameters are unknown but constant.

16.1 The Control and Adaptive Laws

First, it is worth recalling that the PD control law with compensation is given by (11.1), *i.e.*

$$\boldsymbol{\tau} = K_p\tilde{\boldsymbol{q}} + K_v\dot{\tilde{\boldsymbol{q}}} + M(\boldsymbol{q})\left[\ddot{\boldsymbol{q}}_d + \Lambda\dot{\tilde{\boldsymbol{q}}}\right] + C(\boldsymbol{q},\dot{\boldsymbol{q}})\left[\dot{\boldsymbol{q}}_d + \Lambda\tilde{\boldsymbol{q}}\right] + \boldsymbol{g}(\boldsymbol{q}), \tag{16.1}$$

where $K_p, K_v \in \mathbb{R}^{n\times n}$ are symmetric positive definite design matrices, $\tilde{\boldsymbol{q}} = \boldsymbol{q}_d - \boldsymbol{q}$ denotes the position error, and Λ is defined as

$$\Lambda = K_v^{-1}K_p\,.$$

Notice that Λ is the product of two symmetric positive definite matrices. Even though it is not necessarily symmetric nor positive definite, it is always nonsingular. This property of Λ is used below.

Next, it is worth recalling Property 14.1, which establishes that the dynamic model of an n-DOF robot (with manipulated load included) may be written according with the parameterization (14.9), *i.e.*

$$\begin{aligned} &M(\boldsymbol{q},\boldsymbol{\theta})\boldsymbol{u} + C(\boldsymbol{q},\boldsymbol{w},\boldsymbol{\theta})\boldsymbol{v} + \boldsymbol{g}(\boldsymbol{q},\boldsymbol{\theta}) \\ &\quad = \Phi(\boldsymbol{q},\boldsymbol{u},\boldsymbol{v},\boldsymbol{w})\boldsymbol{\theta} + M_0(\boldsymbol{q})\boldsymbol{u} + C_0(\boldsymbol{q},\boldsymbol{w})\boldsymbol{v} + \boldsymbol{g}_0(\boldsymbol{q}) \end{aligned}$$

where $\Phi(\boldsymbol{q},\boldsymbol{u},\boldsymbol{v},\boldsymbol{w}) \in \mathbb{R}^{n\times m}$, $M_0(\boldsymbol{q}) \in \mathbb{R}^{n\times n}$, $C_0(\boldsymbol{q},\boldsymbol{w}) \in \mathbb{R}^{n\times n}$, $\boldsymbol{g}_0(\boldsymbol{q}) \in \mathbb{R}^n$ and $\boldsymbol{\theta} \in \mathbb{R}^m$. The vector $\boldsymbol{\theta}$, referred to as the vector of dynamic parameters, contains elements that depend precisely on the dynamic parameters of the manipulator and on the manipulated load. The matrices $M_0(\boldsymbol{q})$, $C_0(\boldsymbol{q},\boldsymbol{w})$ and the vector $\boldsymbol{g}_0(\boldsymbol{q})$ represent parts of the matrices $M(\boldsymbol{q})$, $C(\boldsymbol{q},\dot{\boldsymbol{q}})$ and of the vector $\boldsymbol{g}(\boldsymbol{q})$ that do not depend on the vector of dynamic parameters $\boldsymbol{\theta}$ respectively.

By virtue of the previous fact, notice that the following holds:

$$\begin{aligned} &M(\boldsymbol{q},\boldsymbol{\theta})\left[\ddot{\boldsymbol{q}}_d + \Lambda\dot{\tilde{\boldsymbol{q}}}\right] + C(\boldsymbol{q},\dot{\boldsymbol{q}},\boldsymbol{\theta})\left[\dot{\boldsymbol{q}}_d + \Lambda\tilde{\boldsymbol{q}}\right] + \boldsymbol{g}(\boldsymbol{q},\boldsymbol{\theta}) \\ &\quad = \Phi(\boldsymbol{q},\ddot{\boldsymbol{q}}_d + \Lambda\dot{\tilde{\boldsymbol{q}}},\dot{\boldsymbol{q}}_d + \Lambda\tilde{\boldsymbol{q}},\dot{\boldsymbol{q}})\boldsymbol{\theta} + M_0(\boldsymbol{q})\left[\ddot{\boldsymbol{q}}_d + \Lambda\dot{\tilde{\boldsymbol{q}}}\right] \\ &\qquad + C_0(\boldsymbol{q},\dot{\boldsymbol{q}})\left[\dot{\boldsymbol{q}}_d + \Lambda\tilde{\boldsymbol{q}}\right] + \boldsymbol{g}_0(\boldsymbol{q}), \end{aligned} \tag{16.2}$$

where we defined

$$\begin{aligned} \boldsymbol{u} &= \ddot{\boldsymbol{q}}_d + \Lambda\dot{\tilde{\boldsymbol{q}}} \\ \boldsymbol{v} &= \dot{\boldsymbol{q}}_d + \Lambda\tilde{\boldsymbol{q}} \\ \boldsymbol{w} &= \dot{\boldsymbol{q}}\,. \end{aligned}$$

On the other hand, from (14.10) we conclude that for any vector $\hat{\boldsymbol{\theta}} \in \mathbb{R}^m$,

$$\begin{aligned} &M(\boldsymbol{q},\hat{\boldsymbol{\theta}})\left[\ddot{\boldsymbol{q}}_d + \Lambda\dot{\tilde{\boldsymbol{q}}}\right] + C(\boldsymbol{q},\dot{\boldsymbol{q}},\hat{\boldsymbol{\theta}})\left[\dot{\boldsymbol{q}}_d + \Lambda\tilde{\boldsymbol{q}}\right] + \boldsymbol{g}(\boldsymbol{q},\hat{\boldsymbol{\theta}}) \\ &\quad = \Phi(\boldsymbol{q},\ddot{\boldsymbol{q}}_d + \Lambda\dot{\tilde{\boldsymbol{q}}},\dot{\boldsymbol{q}}_d + \Lambda\tilde{\boldsymbol{q}},\dot{\boldsymbol{q}})\hat{\boldsymbol{\theta}} + M_0(\boldsymbol{q})\left[\ddot{\boldsymbol{q}}_d + \Lambda\dot{\tilde{\boldsymbol{q}}}\right] \\ &\qquad + C_0(\boldsymbol{q},\dot{\boldsymbol{q}})\left[\dot{\boldsymbol{q}}_d + \Lambda\tilde{\boldsymbol{q}}\right] + \boldsymbol{g}_0(\boldsymbol{q})\,. \end{aligned} \tag{16.3}$$

For notational simplicity, in the sequel we use the abbreviation

$$\Phi = \Phi(\boldsymbol{q},\ddot{\boldsymbol{q}}_d + \Lambda\dot{\tilde{\boldsymbol{q}}},\dot{\boldsymbol{q}}_d + \Lambda\tilde{\boldsymbol{q}},\dot{\boldsymbol{q}})\,.$$

Considering (16.2), the PD control law with compensation, (16.1) may also be written as

$$\tau = K_p\tilde{q} + K_v\dot{\tilde{q}} + \Phi\theta + M_0(q)\left[\ddot{q}_d + \Lambda\dot{\tilde{q}}\right] + C_0(q,\dot{q})\left[\dot{q}_d + \Lambda\tilde{q}\right] + g_0(q)\,. \quad (16.4)$$

It is important to emphasize that the realization of the PD control law with compensation, (16.1), or equivalently (16.4), requires knowledge of the dynamic parameters of the robot, including the manipulated load, *i.e.* θ. In the sequel, we assume that the vector $\theta \in \mathbb{R}^m$ of dynamic parameters is unknown but constant. Of course, in this scenario, the control law (16.4) may not be implemented. Therefore, the solution considered in this chapter for the formulated control problem consists in applying PD control with adaptive compensation.

As is explained in Chapter 14, the structure of the adaptive controllers for motion control of robot manipulators that are studied in this text are of the form (14.19) with an adaptive law (14.20), *i.e.*[1]

$$\tau = \tau\left(t, q, \dot{q}, q_d, \dot{q}_d, \ddot{q}_d, \hat{\theta}\right) \quad (16.5)$$

$$\hat{\theta}(t) = \Gamma \int_0^t \psi\left(s, q, \dot{q}, q_d, \dot{q}_d, \ddot{q}_d\right)\, ds + \hat{\theta}(0) \quad (16.6)$$

where $\Gamma = \Gamma^T \in \mathbb{R}^{m\times m}$ (adaptive gain) and $\hat{\theta}(0) \in \mathbb{R}^m$ are design parameters while ψ is a vectorial function of dimension m to be determined.

The PD control law with adaptive compensation is given by (16.5)–(16.6) where

$$\begin{aligned}\tau &= K_p\tilde{q} + K_v\dot{\tilde{q}} + M(q,\hat{\theta})\left[\ddot{q}_d + \Lambda\dot{\tilde{q}}\right] + C(q,\dot{q},\hat{\theta})\left[\dot{q}_d + \Lambda\tilde{q}\right] \\ &\quad + g(q,\hat{\theta}) & (16.7)\\ &= K_p\tilde{q} + K_v\dot{\tilde{q}} + \Phi\hat{\theta} + M_0(q)\left[\ddot{q}_d + \Lambda\dot{\tilde{q}}\right] + C_0(q,\dot{q})\left[\dot{q}_d + \Lambda\tilde{q}\right] \\ &\quad + g_0(q), & (16.8)\end{aligned}$$

and

$$\hat{\theta}(t) = \Gamma \int_0^t \Phi^T\left[\dot{\tilde{q}} + \Lambda\tilde{q}\right]\, ds + \hat{\theta}(0), \quad (16.9)$$

where $K_p, K_v \in \mathbb{R}^{n\times n}$ and $\Gamma \in \mathbb{R}^{m\times m}$ are symmetric positive definite design matrices. The pass from (16.7) to (16.8) follows by using (16.3). It is assumed that the centrifugal and Coriolis forces matrix $C(q,\dot{q},\theta)$ is chosen by means of the Christoffel symbols (*cf.* Equation 3.21).

Notice that the control law (16.8) does not depend on the dynamic parameters θ but on the adaptive parameters $\hat{\theta}$, which in turn, are obtained from the adaptive law (16.9), which of course, does not depend on θ either.

[1] In (16.6) as in other integrals throughout the chapter, we avoid the correct but cumbersome notation $\psi\left(s, q(s), \dot{q}(s), q_d(s), \dot{q}_d(s), \ddot{q}_d(s)\right)$.

Before proceeding to derive the closed-loop equation we first write the parametric errors vector $\tilde{\boldsymbol{\theta}} \in \mathbb{R}^m$ as

$$\tilde{\boldsymbol{\theta}} = \hat{\boldsymbol{\theta}} - \boldsymbol{\theta}. \tag{16.10}$$

The parametric errors vector $\tilde{\boldsymbol{\theta}}$ is unknown since it is a function of the vector of dynamic parameters $\boldsymbol{\theta}$ that has been assumed to be unknown. Nevertheless, the parametric error $\tilde{\boldsymbol{\theta}}$ is introduced only with analytic purposes, and it is not used by the controller.

From the definition of the parametric errors vector $\tilde{\boldsymbol{\theta}}$ in (16.10), it may be verified that

$$\begin{aligned}
\Phi\hat{\boldsymbol{\theta}} &= \Phi\tilde{\boldsymbol{\theta}} + \Phi\boldsymbol{\theta} \\
&= \Phi\tilde{\boldsymbol{\theta}} + M(\boldsymbol{q}, \boldsymbol{\theta})\left[\ddot{\boldsymbol{q}}_d + \Lambda\dot{\tilde{\boldsymbol{q}}}\right] + C(\boldsymbol{q}, \dot{\boldsymbol{q}}, \boldsymbol{\theta})\left[\dot{\boldsymbol{q}}_d + \Lambda\tilde{\boldsymbol{q}}\right] + \boldsymbol{g}(\boldsymbol{q}, \boldsymbol{\theta}) \\
&\quad -M_0(\boldsymbol{q})\left[\ddot{\boldsymbol{q}}_d + \Lambda\dot{\tilde{\boldsymbol{q}}}\right] - C_0(\boldsymbol{q}, \dot{\boldsymbol{q}})\left[\dot{\boldsymbol{q}}_d + \Lambda\tilde{\boldsymbol{q}}\right] - \boldsymbol{g}_0(\boldsymbol{q})
\end{aligned}$$

where we used (16.2).

Making use of this last expression, the control law (16.8) takes the form

$$\begin{aligned}
\boldsymbol{\tau} &= K_p\tilde{\boldsymbol{q}} + K_v\dot{\tilde{\boldsymbol{q}}} + \Phi\tilde{\boldsymbol{\theta}} \\
&\quad + M(\boldsymbol{q}, \boldsymbol{\theta})\left[\ddot{\boldsymbol{q}}_d + \Lambda\dot{\tilde{\boldsymbol{q}}}\right] + C(\boldsymbol{q}, \dot{\boldsymbol{q}}, \boldsymbol{\theta})\left[\dot{\boldsymbol{q}}_d + \Lambda\tilde{\boldsymbol{q}}\right] + \boldsymbol{g}(\boldsymbol{q}, \boldsymbol{\theta}).
\end{aligned}$$

Using the control law expressed above and substituting the control action $\boldsymbol{\tau}$ in the equation of the robot model (14.2), we get

$$M(\boldsymbol{q}, \boldsymbol{\theta})\left[\ddot{\tilde{\boldsymbol{q}}} + \Lambda\dot{\tilde{\boldsymbol{q}}}\right] + C(\boldsymbol{q}, \dot{\boldsymbol{q}}, \boldsymbol{\theta})\left[\dot{\tilde{\boldsymbol{q}}} + \Lambda\tilde{\boldsymbol{q}}\right] = -K_p\tilde{\boldsymbol{q}} - K_v\dot{\tilde{\boldsymbol{q}}} - \Phi\tilde{\boldsymbol{\theta}}. \tag{16.11}$$

On the other hand, since the vector of dynamic parameters $\boldsymbol{\theta}$ has been assumed constant, its time derivative is zero, that is $\dot{\boldsymbol{\theta}} = \boldsymbol{0} \in \mathbb{R}^m$. Therefore, the time derivative of the parametric errors vector $\tilde{\boldsymbol{\theta}}$ defined in (16.10), satisfies $\dot{\tilde{\boldsymbol{\theta}}} = \dot{\hat{\boldsymbol{\theta}}}$. In turn, the time derivative of the vector of adaptive parameters $\hat{\boldsymbol{\theta}}$ is obtained by differentiating with respect to time the adaptive law (16.9). Considering these facts we have

$$\dot{\tilde{\boldsymbol{\theta}}} = \Gamma\Phi^T\left[\dot{\tilde{\boldsymbol{q}}} + \Lambda\tilde{\boldsymbol{q}}\right]. \tag{16.12}$$

The closed-loop equation, which is formed of Equations (16.11) and (16.12), may be written as

$$\frac{d}{dt}\begin{bmatrix} \tilde{\boldsymbol{q}} \\ \dot{\tilde{\boldsymbol{q}}} \\ \tilde{\boldsymbol{\theta}} \end{bmatrix} = \begin{bmatrix} \dot{\tilde{\boldsymbol{q}}} \\ M(\boldsymbol{q}, \boldsymbol{\theta})^{-1}\left[-K_p\tilde{\boldsymbol{q}} - K_v\dot{\tilde{\boldsymbol{q}}} - \Phi\tilde{\boldsymbol{\theta}} - C(\boldsymbol{q}, \dot{\boldsymbol{q}}, \boldsymbol{\theta})\left[\dot{\tilde{\boldsymbol{q}}} + \Lambda\tilde{\boldsymbol{q}}\right]\right] - \Lambda\dot{\tilde{\boldsymbol{q}}} \\ \Gamma\Phi^T\left[\dot{\tilde{\boldsymbol{q}}} + \Lambda\tilde{\boldsymbol{q}}\right] \end{bmatrix}, \tag{16.13}$$

which is a nonautonomous differential equation and the origin of the state space, *i.e.*

$$\begin{bmatrix} \tilde{q} \\ \dot{\tilde{q}} \\ \tilde{\theta} \end{bmatrix} = \mathbf{0} \in \mathbb{R}^{2n+m},$$

is an equilibrium point.

16.2 Stability Analysis

The stability analysis of the origin of the state space for the closed-loop system is carried out using the Lyapunov function candidate

$$V(t, \tilde{q}, \dot{\tilde{q}}, \tilde{\theta}) = \frac{1}{2} \begin{bmatrix} \tilde{q} \\ \dot{\tilde{q}} \\ \tilde{\theta} \end{bmatrix}^T \begin{bmatrix} 2K_p + \Lambda^T M(q,\theta)\Lambda & \Lambda^T M(q,\theta) & 0 \\ M(q,\theta)\Lambda & M(q,\theta) & 0 \\ 0 & 0 & \Gamma^{-1} \end{bmatrix} \begin{bmatrix} \tilde{q} \\ \dot{\tilde{q}} \\ \tilde{\theta} \end{bmatrix}.$$

At first sight, it may not appear evident that the Lyapunov function candidate is positive definite, however, this may be clearer when rewriting it as

$$V(t, \tilde{q}, \dot{\tilde{q}}, \tilde{\theta}) = \frac{1}{2}\left[\dot{\tilde{q}} + \Lambda\tilde{q}\right]^T M(q,\theta)\left[\dot{\tilde{q}} + \Lambda\tilde{q}\right] + \tilde{q}^T K_p \tilde{q} + \frac{1}{2}\tilde{\theta}^T \Gamma^{-1}\tilde{\theta}. \quad (16.14)$$

It is interesting to remark that the function defined in (16.14) may be regarded as an extension of the Lyapunov function (11.3) used in the study of (nonadaptive) PD control with compensation. The only difference is the introduction of the term $\frac{1}{2}\tilde{\theta}^T\Gamma^{-1}\tilde{\theta}$ in the Lyapunov function candidate for the adaptive version.

The time derivative of the Lyapunov function candidate (16.14) becomes

$$\begin{aligned}\dot{V}(t, \tilde{q}, \dot{\tilde{q}}, \tilde{\theta}) = {} & \left[\dot{\tilde{q}} + \Lambda\tilde{q}\right]^T M(q,\theta)\left[\ddot{\tilde{q}} + \Lambda\dot{\tilde{q}}\right] + \frac{1}{2}\left[\dot{\tilde{q}} + \Lambda\tilde{q}\right]^T \dot{M}(q,\theta)\left[\dot{\tilde{q}} + \Lambda\tilde{q}\right] \\ & + 2\tilde{q}^T K_p \dot{\tilde{q}} + \tilde{\theta}^T \Gamma^{-1}\dot{\tilde{\theta}}.\end{aligned}$$

Solving for $M(q)\ddot{\tilde{q}}$ and $\dot{\tilde{\theta}}$ from the closed-loop Equation (16.13) and substituting in the previous equation, we obtain

$$\dot{V}(t, \tilde{q}, \dot{\tilde{q}}, \tilde{\theta}) = -\left[\dot{\tilde{q}} + \Lambda\tilde{q}\right]^T K_v \left[\dot{\tilde{q}} + \Lambda\tilde{q}\right] + 2\tilde{q}^T K_p \dot{\tilde{q}},$$

where we canceled the term

$$\left[\dot{\tilde{q}} + \Lambda\tilde{q}\right]^T \left[\frac{1}{2}\dot{M}(q,\theta) - C(q,\dot{q},\theta)\right] \left[\dot{\tilde{q}} + \Lambda\tilde{q}\right]$$

by virtue of Property 4.2. Now, using $K_p = K_v \Lambda$, the equation of $\dot{V}(t,\tilde{q},\dot{\tilde{q}},\tilde{\theta})$ reduces to

$$\begin{aligned}\dot{V}(t,\tilde{q},\dot{\tilde{q}},\tilde{\theta}) &= -\dot{\tilde{q}}^T K_v \dot{\tilde{q}} - \tilde{q}^T \Lambda^T K_v \Lambda \tilde{q} \\ &= -\begin{bmatrix} \tilde{q} \\ \dot{\tilde{q}} \\ \tilde{\theta} \end{bmatrix}^T \begin{bmatrix} \Lambda^T K_v \Lambda & 0 & 0 \\ 0 & K_v & 0 \\ 0 & 0 & 0 \end{bmatrix} \begin{bmatrix} \tilde{q} \\ \dot{\tilde{q}} \\ \tilde{\theta} \end{bmatrix}. \end{aligned} \tag{16.15}$$

Recalling that Λ is a nonsingular matrix while K_v is a symmetric positive definite matrix, it follows that the matrix $\Lambda^T K_v \Lambda$ is also symmetric positive definite (*cf.* Lemma 2.1). Therefore, it follows that $\dot{V}(t,\tilde{q},\dot{\tilde{q}},\tilde{\theta})$ expressed in (16.15) is a globally negative semidefinite function. Since moreover the Lyapunov function candidate (16.14) is globally positive definite, radially unbounded and decrescent, Theorem 2.3 guarantees that the origin of the closed-loop Equation (16.13) is uniformly stable and all the solutions are bounded, *i.e.*

$$\begin{aligned} \tilde{q}, \dot{\tilde{q}} &\in L_\infty^n, \\ \tilde{\theta} &\in L_\infty^m . \end{aligned} \tag{16.16}$$

Since $\dot{V}(t,\tilde{q},\dot{\tilde{q}},\tilde{\theta})$ obtained in (16.15) is not negative definite, we may not yet conclude that the origin is an asymptotically stable equilibrium point. On the other hand, La Salle's theorem may not be used either to show asymptotic stability since the closed-loop Equation (16.13) is nonautonomous since it depends implicitly on the functions $q_d(t)$ and $\hat{\theta}(t)$.

Thus, from the previous analysis, it is not possible yet to conclude anything about the achievement of the motion control objective. For this, it is necessary to present further arguments.

The idea consists in using Lemma A.5, which establishes the following. Consider a continuously differentiable function $f : \mathbb{R}_+ \to \mathbb{R}^n$ which satisfies

- $f \in L_2^n$
- $f, \dot{f} \in L_\infty^n$.

Then, the function f necessarily satisfies $\lim_{t\to\infty} f(t) = 0 \in \mathbb{R}^n$.

Therefore, if we wish to prove that $\lim_{t\to\infty} \tilde{q}(t) = 0 \in \mathbb{R}^n$, and we know from (16.16) and Theorem 2.3, that $\tilde{q}, \dot{\tilde{q}} \in L_\infty^n$, it is only left to prove that $\tilde{q} \in L_2^n$, that is, it is sufficient to show that there exists a finite positive constant k such that

$$k \geq \int_0^\infty \|\tilde{q}(t)\|^2 \, dt .$$

This proof is carried out below.

Since $\dot{\tilde{q}}^T K_v \dot{\tilde{q}} \geq 0$ for all $\dot{\tilde{q}} \in \mathbb{R}^n$, it follows from (16.15) that

$$\frac{d}{dt} V(t, \tilde{q}(t), \dot{\tilde{q}}(t), \tilde{\theta}(t)) \leq -\tilde{q}(t)^T \Lambda^T K_v \Lambda \tilde{q}(t) \leq -\lambda_{\min}\{\Lambda^T K_v \Lambda\} \|\tilde{q}(t)\|^2 \qquad (16.17)$$

where we used the fact that $\Lambda^T K_v \Lambda$ is a symmetric matrix and therefore (*cf.* Theorem of Rayleigh–Ritz, on page 24),

$$\tilde{q}^T \left[\Lambda^T K_v \Lambda\right] \tilde{q} \geq \lambda_{\min}\{\Lambda^T K_v \Lambda\} \|\tilde{q}\|^2$$

for all $\tilde{q} \in \mathbb{R}^n$. Notice that, moreover, $\lambda_{\min}\{\Lambda^T K_v \Lambda\} > 0$ since the matrix $\Lambda^T K_v \Lambda$ besides being symmetric, is also positive definite.

The next step consists in integrating the inequality (16.17) from $t = 0$ to $t = \infty$, that is[2]

$$\int_{V_0}^{V_\infty} dV \leq -\lambda_{\min}\{\Lambda^T K_v \Lambda\} \int_0^\infty \|\tilde{q}(t)\|^2 \, dt$$

where we defined $V_0 := (0, \tilde{q}(0), \dot{\tilde{q}}(0), \tilde{\theta}(0))$

$$V_\infty := \lim_{t\to\infty} V(t, \tilde{q}(t), \dot{q}(t), \tilde{\theta}(t)) .$$

The integral on the left-hand side of this inequality is calculated trivially to obtain

$$V_\infty - V_0 \leq -\lambda_{\min}\{\Lambda^T K_v \Lambda\} \int_0^\infty \|\tilde{q}(t)\|^2 \, dt,$$

or in equivalent form

$$-V_0 \leq -\lambda_{\min}\{\Lambda^T K_v \Lambda\} \int_0^\infty \|\tilde{q}(t)\|^2 \, dt - V_\infty . \qquad (16.18)$$

We recall that the Lyapunov function candidate $V(t, \tilde{q}, \dot{\tilde{q}}, \tilde{\theta})$ is positive definite, radially unbounded and decrescent and moreover, all the signals are bounded. Therefore, $\infty > V_\infty \geq 0$ and, from Inequality (16.18), we get

---

[2] Here we are using the following elementary facts: $\int_0^\infty \frac{dV}{dt} dt = \int_{V_0}^{V_\infty} dV$ and, that for functions $g(t)$ and $f(t)$ continuous in $a \leq t \leq b$, satisfying $g(t) \leq f(t)$ for all $a \leq t \leq b$, we have

$$\int_a^b g(t) \, dt \leq \int_a^b f(t) \, dt .$$

$$-V_0 \leq -\lambda_{\min}\{\Lambda^T K_v \Lambda\} \int_0^\infty \|\tilde{\boldsymbol{q}}(t)\|^2 \, dt\,.$$

From this expression we immediately conclude that

$$\frac{V_0}{\lambda_{\min}\{\Lambda^T K_v \Lambda\}} \geq \int_0^\infty \|\tilde{\boldsymbol{q}}(t)\|^2 \, dt$$

where the left-hand side of the inequality is finite positive and constant. This means that the position error $\tilde{\boldsymbol{q}}$ belongs to the L_2^n space which is precisely what we wanted to prove.

Thus, according to the arguments above, we may conclude now that the position error $\tilde{\boldsymbol{q}}$ tends asymptotically to the zero vector, *i.e.*

$$\lim_{t\to\infty} \tilde{\boldsymbol{q}}(t) = \mathbf{0} \in \mathbb{R}^n$$

or in words, that the motion control objective has been achieved.

Invoking further arguments one may also show that not only the position error $\tilde{\boldsymbol{q}}$ tends to zero asymptotically, but that so does the velocity error $\dot{\tilde{\boldsymbol{q}}}$. Nevertheless, these conclusions should not be extrapolated to the parametric error $\tilde{\boldsymbol{\theta}}$, unless the desired joint position $\boldsymbol{q}_d(t)$ satisfies some special properties that are not discussed here. From these comments we conclude that, *in general*, the origin of the closed-loop Equation (16.13) may not be asymptotically stable, not even locally. However, as has been demonstrated, the motion control objective is guaranteed.

16.3 Examples

Next, we present a series of examples which illustrate the application of PD control with adaptive compensation.

The first example is presented as a solution to the formulated control problem in Example 14.8.

Example 16.1. Consider the model of a pendulum of mass m, inertia J with respect to the axis of rotation and distance l from the axis of rotation to the center of mass. The torque τ is applied at the axis of rotation, that is,

$$J\ddot{q} + mgl\,\sin(q) = \tau\,.$$

We clearly identify $M(q) = J$, $C(q,\dot{q}) = 0$ and $g(q) = mgl\,\sin(q)$.

In Example 14.8 we formulated the following control problem. Assume that the values of the mass m, the distance l and the gravity acceleration g are known but that the value of the inertia J is unknown (yet constant). The control problem consists in designing a controller that is capable of satisfying the motion control objective

$$\lim_{t\to\infty} \tilde{q}(t) = 0 \in \mathbb{R}$$

for any desired joint position $q_d(t)$ (with bounded first and second time derivatives).

The solution to this control problem may be obtained via PD control with adaptive compensation. The parameter of interest that has been supposed unknown is the inertia J. The parameterization which corresponds to (16.2) is, in this example,

$$\begin{aligned} M(q,\theta)\left[\ddot{q}_d + \lambda\dot{\tilde{q}}\right] + C(q,\dot{q},\theta)\left[\dot{q}_d + \lambda\tilde{q}\right] + g(q,\theta) \\ = J\left[\ddot{q}_d + \lambda\dot{\tilde{q}}\right] + mgl\,\sin(q) \\ = \Phi\theta + M_0(q)\left[\ddot{q}_d + \lambda\dot{\tilde{q}}\right] + C_0(q,\dot{q})\left[\dot{q}_d + \lambda\tilde{q}\right] + g_0(q) \end{aligned}$$

where

$$\begin{aligned} \Phi &= \ddot{q}_d + \lambda\dot{\tilde{q}} \\ \theta &= J \\ M_0(q) &= 0 \\ C_0(q,\dot{q}) &= 0 \\ g_0(q) &= mgl\,\sin(q)\,. \end{aligned}$$

Therefore, the adaptive control system given by Equations (16.8) and (16.9) becomes

$$\begin{aligned} \tau &= k_p\tilde{q} + k_v\dot{\tilde{q}} + \Phi\hat{\theta} + g_0(q) \\ &= k_p\tilde{q} + k_v\dot{\tilde{q}} + \left[\ddot{q}_d + \lambda\dot{\tilde{q}}\right]\hat{\theta} + mgl\,\sin(q) \end{aligned}$$

and

$$\begin{aligned} \hat{\theta}(t) &= \gamma\int_0^t \Phi\left[\dot{\tilde{q}} + \lambda\tilde{q}\right]\,ds + \hat{\theta}(0) \\ &= \gamma\int_0^t \left[\ddot{q}_d + \lambda\dot{\tilde{q}}\right]\left[\dot{\tilde{q}} + \lambda\tilde{q}\right]\,ds + \hat{\theta}(0), \end{aligned}$$

where $k_p > 0$, $k_v > 0$, $\lambda = k_p/k_v$, $\gamma > 0$ and $\hat{\theta}(0) \in \mathbb{R}$. Figure 16.1 shows the block-diagram corresponding to this adaptive controller.

$\diamond$

We present next, the design of the PD control law with adaptive compensation, the Slotine and Li controller, for a planar 2-DOF robot. As the reader should notice, the resulting adaptive controller presents a higher degree of complexity than the previous example.

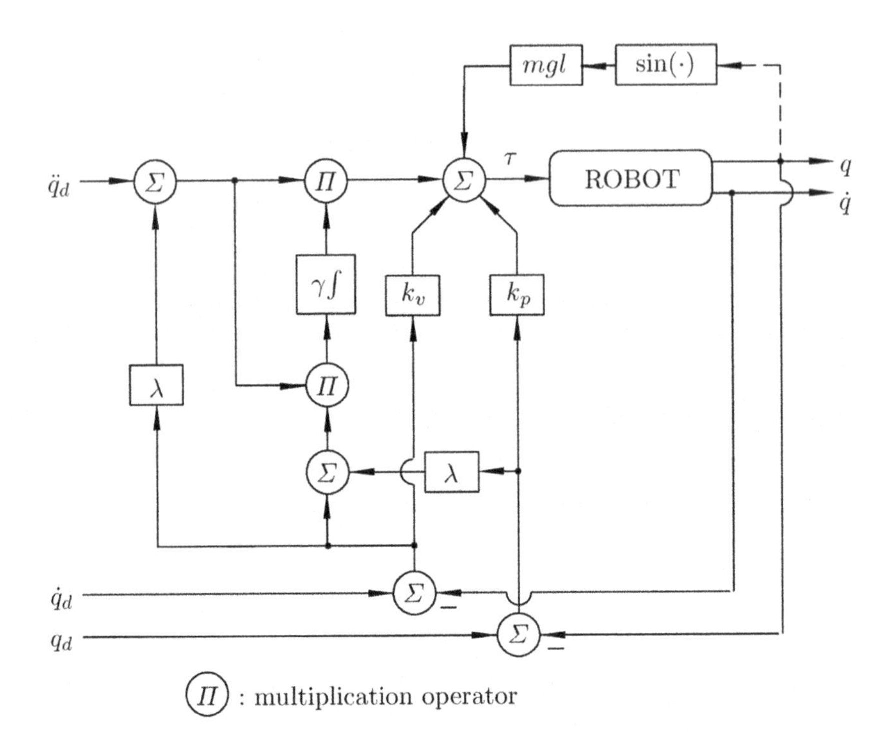

Figure 16.1. Block-diagram: pendulum under PD control with adaptive compensation

---

Example 16.2. Consider the 2-DOF planar manipulator shown in Figure 16.2 and whose dynamic model is described in Example 3.3.

The dynamic model of this planar manipulator is given by (3.8)–(3.9), and may be written as

$$\theta_1 \ddot{q}_1 + (\theta_3 C_{21} + \theta_4 S_{21})\, \ddot{q}_2 - \theta_3 S_{21} \dot{q}_2^2 + \theta_4 C_{21} \dot{q}_2^2 = \tau_1$$

$$(\theta_3 C_{21} + \theta_4 S_{21})\, \ddot{q}_1 + \theta_2 \ddot{q}_2 + \theta_3 S_{21} \dot{q}_1^2 - \theta_4 C_{21} \dot{q}_1^2 = \tau_2$$

where

$$\begin{aligned}
\theta_1 &= m_1 l_{c1}^2 + m_2 l_1^2 + I_1 \\
\theta_2 &= m_2 l_{c2}^2 + I_2 \\
\theta_3 &= m_2 l_1 l_{c2} \cos(\delta) \\
\theta_4 &= m_2 l_1 l_{c2} \sin(\delta)\,.
\end{aligned}$$

The four dynamic parameters θ_1, θ_2, θ_3 and θ_4 depend on the physical characteristics of the manipulator such as the masses and inertias

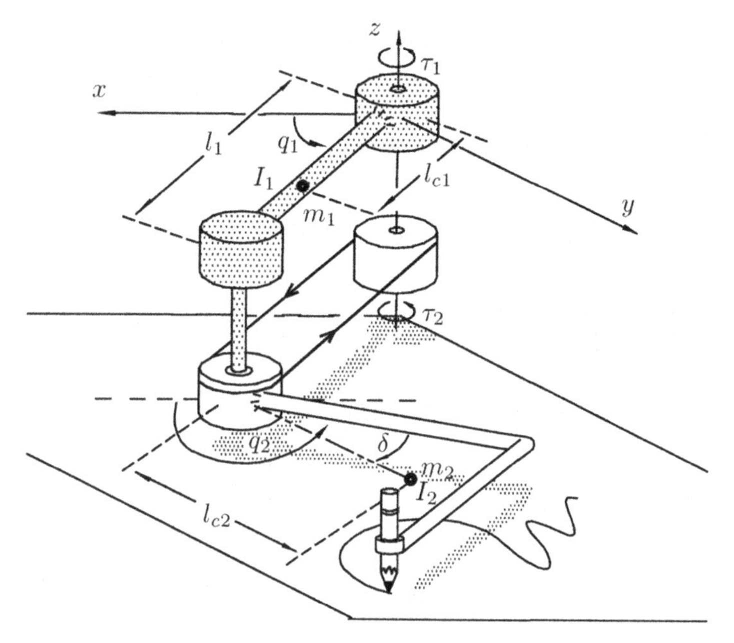

Figure 16.2. Planar 2-DOF manipulator

of its links. The vector of unknown constant dynamic parameters $\boldsymbol{\theta}$ is defined as

$$\boldsymbol{\theta} = \begin{bmatrix} \theta_1 \\ \theta_2 \\ \theta_3 \\ \theta_4 \end{bmatrix}.$$

In Example 14.6 we obtained the parameterization (14.9) of the dynamic model, *i.e.*

$$\begin{aligned} M(\boldsymbol{q},\boldsymbol{\theta})\boldsymbol{u} + C(\boldsymbol{q},\boldsymbol{w},\boldsymbol{\theta})\boldsymbol{v} + \boldsymbol{g}(\boldsymbol{q},\boldsymbol{\theta}) \\ = \begin{bmatrix} \Phi_{11} & \Phi_{12} & \Phi_{13} & \Phi_{14} \\ \Phi_{21} & \Phi_{22} & \Phi_{23} & \Phi_{24} \end{bmatrix} \begin{bmatrix} \theta_1 \\ \theta_2 \\ \theta_3 \\ \theta_4 \end{bmatrix} \\ + M_0(\boldsymbol{q})\boldsymbol{u} + C_0(\boldsymbol{q},\boldsymbol{w})\boldsymbol{v} + \boldsymbol{g}_0(\boldsymbol{q}) \end{aligned}$$

where

$$\begin{aligned} \Phi_{11} &= u_1 \\ \Phi_{12} &= 0 \\ \Phi_{13} &= C_{21}u_2 - S_{21}w_2v_2 \\ \Phi_{14} &= S_{21}u_2 + C_{21}w_2v_2 \end{aligned}$$

$$
\begin{aligned}
\Phi_{21} &= 0 \\
\Phi_{22} &= u_2 \\
\Phi_{23} &= C_{21} u_1 + S_{21} w_1 v_1 \\
\Phi_{24} &= S_{21} u_1 - C_{21} w_1 v_1
\end{aligned}
$$

$$
\begin{aligned}
M_0(\boldsymbol{q}) &= 0 \in \mathbb{R}^{2\times 2} \\
C_0(\boldsymbol{q}, \dot{\boldsymbol{q}}) &= 0 \in \mathbb{R}^{2\times 2} \\
\boldsymbol{g}_0(\boldsymbol{q}) &= \boldsymbol{0} \in \mathbb{R}^2 .
\end{aligned}
$$

Hence, the parameterization (16.2) yields

$$
\begin{aligned}
M(\boldsymbol{q}, \boldsymbol{\theta}) &\left[\ddot{\boldsymbol{q}}_d + \Lambda \dot{\tilde{\boldsymbol{q}}}\right] + C(\boldsymbol{q}, \dot{\boldsymbol{q}}, \boldsymbol{\theta}) \left[\dot{\boldsymbol{q}}_d + \Lambda \tilde{\boldsymbol{q}}\right] + \boldsymbol{g}(\boldsymbol{q}, \boldsymbol{\theta}) \\
&= \begin{bmatrix} \theta_1 & \theta_3 C_{21} + \theta_4 S_{21} \\ \theta_3 C_{21} + \theta_4 S_{21} & \theta_2 \end{bmatrix} \left[\ddot{\boldsymbol{q}}_d + \Lambda \dot{\tilde{\boldsymbol{q}}}\right] \\
&\quad + \begin{bmatrix} 0 & (\theta_4 C_{21} - \theta_3 S_{21})\, \dot{q}_2 \\ (\theta_3 S_{21} - \theta_4 C_{21})\, \dot{q}_1 & 0 \end{bmatrix} \left[\dot{\boldsymbol{q}}_d + \Lambda \tilde{\boldsymbol{q}}\right] \\
&= \Phi \boldsymbol{\theta}
\end{aligned}
$$

where this time,

$$
\boldsymbol{u} = \ddot{\boldsymbol{q}}_d + \Lambda \dot{\tilde{\boldsymbol{q}}} = \begin{bmatrix} u_1 \\ u_2 \end{bmatrix} = \begin{bmatrix} \ddot{q}_{d1} + \lambda_{11} \dot{\tilde{q}}_1 + \lambda_{12} \dot{\tilde{q}}_2 \\ \ddot{q}_{d2} + \lambda_{21} \dot{\tilde{q}}_1 + \lambda_{22} \dot{\tilde{q}}_2 \end{bmatrix}
$$

$$
\boldsymbol{v} = \dot{\boldsymbol{q}}_d + \Lambda \tilde{\boldsymbol{q}} = \begin{bmatrix} v_1 \\ v_2 \end{bmatrix} = \begin{bmatrix} \dot{q}_{d1} + \lambda_{11} \tilde{q}_1 + \lambda_{12} \tilde{q}_2 \\ \dot{q}_{d2} + \lambda_{21} \tilde{q}_1 + \lambda_{22} \tilde{q}_2 \end{bmatrix}
$$

$$
\boldsymbol{w} = \dot{\boldsymbol{q}} = \begin{bmatrix} w_1 \\ w_2 \end{bmatrix} = \begin{bmatrix} \dot{q}_1 \\ \dot{q}_2 \end{bmatrix}
$$

and

$$
\Lambda = \begin{bmatrix} \lambda_{11} & \lambda_{12} \\ \lambda_{21} & \lambda_{22} \end{bmatrix} \in \mathbb{R}^{2\times 2} .
$$

The adaptive control system is given by Equations (16.8) and (16.9). Notice that $M_0 = C_0 = \boldsymbol{g}_0 = 0$ therefore, the control law becomes

$$
\boldsymbol{\tau} = K_p \tilde{\boldsymbol{q}} + K_v \dot{\tilde{\boldsymbol{q}}} + \begin{bmatrix} \Phi_{11} & \Phi_{12} & \Phi_{13} & \Phi_{14} \\ \Phi_{21} & \Phi_{22} & \Phi_{23} & \Phi_{24} \end{bmatrix} \begin{bmatrix} \hat{\theta}_1 \\ \hat{\theta}_2 \\ \hat{\theta}_3 \\ \hat{\theta}_4 \end{bmatrix} ,
$$

while the adaptive law is

$$\begin{bmatrix} \hat{\theta}_1(t) \\ \hat{\theta}_2(t) \\ \hat{\theta}_3(t) \\ \hat{\theta}_4(t) \end{bmatrix} = \Gamma \int_0^t \begin{bmatrix} \Phi_{11}[v_1 - \dot{q}_1] \\ \Phi_{22}[v_2 - \dot{q}_2] \\ \Phi_{13}[v_1 - \dot{q}_1] + \Phi_{23}[v_2 - \dot{q}_2] \\ \Phi_{14}[v_1 - \dot{q}_1] + \Phi_{24}[v_2 - \dot{q}_2] \end{bmatrix} ds + \begin{bmatrix} \hat{\theta}_1(0) \\ \hat{\theta}_2(0) \\ \hat{\theta}_3(0) \\ \hat{\theta}_4(0) \end{bmatrix}$$

where $K_p = K_p^T > 0$, $K_v = K_v^T > 0$, $\Lambda = K_v^{-1} K_p$, $\Gamma = \Gamma^T > 0$ and $\hat{\boldsymbol{\theta}}(0) \in \mathbb{R}^m$. ◇

We end this section with an example that illustrates the performance of PD control with adaptive compensation on the Pelican robot prototype.

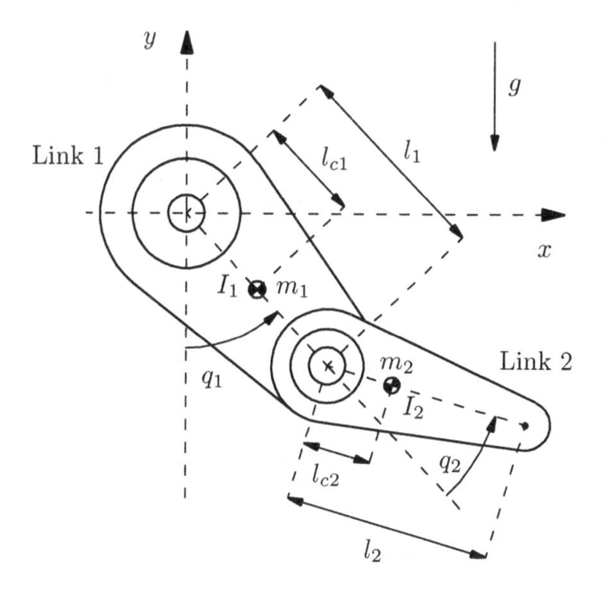

Figure 16.3. Diagram of the Pelican robot

Example 16.3. Consider the Pelican robot presented in Chapter 5 and shown in Figure 16.3. Its dynamic model is recalled below for ease of reference:

$$\underbrace{\begin{bmatrix} M_{11}(\boldsymbol{q}) & M_{12}(\boldsymbol{q}) \\ M_{21}(\boldsymbol{q}) & M_{22}(\boldsymbol{q}) \end{bmatrix}}_{M(\boldsymbol{q})} \ddot{\boldsymbol{q}} + \underbrace{\begin{bmatrix} C_{11}(\boldsymbol{q}, \dot{\boldsymbol{q}}) & C_{12}(\boldsymbol{q}, \dot{\boldsymbol{q}}) \\ C_{21}(\boldsymbol{q}, \dot{\boldsymbol{q}}) & C_{22}(\boldsymbol{q}, \dot{\boldsymbol{q}}) \end{bmatrix}}_{C(\boldsymbol{q}, \dot{\boldsymbol{q}})} \dot{\boldsymbol{q}} + \underbrace{\begin{bmatrix} g_1(\boldsymbol{q}) \\ g_2(\boldsymbol{q}) \end{bmatrix}}_{\boldsymbol{g}(\boldsymbol{q})} = \boldsymbol{\tau}$$

where

$$\begin{aligned}
M_{11}(\boldsymbol{q}) &= m_1 l_{c1}^2 + m_2 \left[l_1^2 + l_{c2}^2 + 2 l_1 l_{c2} \cos(q_2)\right] + I_1 + I_2 \\
M_{12}(\boldsymbol{q}) &= m_2 \left[l_{c2}^2 + l_1 l_{c2} \cos(q_2)\right] + I_2 \\
M_{21}(\boldsymbol{q}) &= m_2 \left[l_{c2}^2 + l_1 l_{c2} \cos(q_2)\right] + I_2 \\
M_{22}(\boldsymbol{q}) &= m_2 l_{c2}^2 + I_2 \\
C_{11}(\boldsymbol{q}, \dot{\boldsymbol{q}}) &= -m_2 l_1 l_{c2} \sin(q_2) \dot{q}_2 \\
C_{12}(\boldsymbol{q}, \dot{\boldsymbol{q}}) &= -m_2 l_1 l_{c2} \sin(q_2) \left[\dot{q}_1 + \dot{q}_2\right] \\
C_{21}(\boldsymbol{q}, \dot{\boldsymbol{q}}) &= m_2 l_1 l_{c2} \sin(q_2) \dot{q}_1 \\
C_{22}(\boldsymbol{q}, \dot{\boldsymbol{q}}) &= 0 \\
g_1(\boldsymbol{q}) &= \left[m_1 l_{c1} + m_2 l_1\right] g \sin(q_1) + m_2 l_{c2} g \sin(q_1 + q_2) \\
g_2(\boldsymbol{q}) &= m_2 l_{c2} g \sin(q_1 + q_2) \, .
\end{aligned}$$

For this example we selected as unknown parameters, the mass m_2, the inertia I_2 and the distance to the center of mass, l_{c2}.

We wish to design a controller that is capable of driving to zero the articular position error $\tilde{\boldsymbol{q}}$. It is desired that the robot tracks the trajectories $\boldsymbol{q}_d(t)$, $\dot{\boldsymbol{q}}_d(t)$ and $\ddot{\boldsymbol{q}}_d(t)$ represented by Equations (5.7)–(5.9). To that end, we use PD control with adaptive compensation.

In Example 14.6 we derived the parameterization (14.9) of the dynamic model, *i.e.*

$$\begin{aligned}
&M(\boldsymbol{q}, \boldsymbol{\theta})\boldsymbol{u} + C(\boldsymbol{q}, \boldsymbol{w}, \boldsymbol{\theta})\boldsymbol{v} + \boldsymbol{g}(\boldsymbol{q}, \boldsymbol{\theta}) \\
&\quad = \begin{bmatrix} \Phi_{11} & \Phi_{12} & \Phi_{13} \\ \Phi_{21} & \Phi_{22} & \Phi_{23} \end{bmatrix} \begin{bmatrix} \theta_1 \\ \theta_2 \\ \theta_3 \end{bmatrix} \\
&\quad\quad + M_0(\boldsymbol{q})\boldsymbol{u} + C_0(\boldsymbol{q}, \boldsymbol{w})\boldsymbol{v} + \boldsymbol{g}_0(\boldsymbol{q})
\end{aligned}$$

where

$$\begin{aligned}
\Phi_{11} &= l_1^2 u_1 + l_1 g \sin(q_1) \\
\Phi_{12} &= 2 l_1 \cos(q_2) u_1 + l_1 \cos(q_2) u_2 - l_1 \sin(q_2) w_2 v_1 \\
&\quad - l_1 \sin(q_2) [w_1 + w_2] v_2 + g \sin(q_1 + q_2) \\
\Phi_{13} &= u_1 + u_2 \\
\Phi_{21} &= 0 \\
\Phi_{22} &= l_1 \cos(q_2) u_1 + l_1 \sin(q_2) w_1 v_1 + g \sin(q_1 + q_2) \\
\Phi_{23} &= u_1 + u_2
\end{aligned}$$

$$\boldsymbol{\theta} = \begin{bmatrix} \theta_1 \\ \theta_2 \\ \theta_3 \end{bmatrix} = \begin{bmatrix} m_2 \\ m_2 l_{c2} \\ m_2 l_{c2}^2 + I_2 \end{bmatrix} = \begin{bmatrix} 2.0458 \\ 0.047 \\ 0.0126 \end{bmatrix}$$

$$M_0(\boldsymbol{q}) = \begin{bmatrix} m_1 l_{c1}^2 + I_1 & 0 \\ 0 & 0 \end{bmatrix}$$

$$C_0(\boldsymbol{q}, \boldsymbol{w}) = \begin{bmatrix} 0 & 0 \\ 0 & 0 \end{bmatrix}$$

$$\boldsymbol{g}_0(\boldsymbol{q}) = \begin{bmatrix} m_1 l_{c1} g \sin(q_1) \\ 0 \end{bmatrix}.$$

Since the numerical values of m_2, I_2 and l_{c2} are assumed unknown, the vector of dynamic parameters $\boldsymbol{\theta}$ is also unknown. This hypothesis obviously complicates our task of designing a controller that is capable of satisfying our control objective.

Let us now see the form that the PD control law with adaptive compensation takes for the Pelican prototype. For this, we first define

$$\Lambda = \begin{bmatrix} \lambda_{11} & \lambda_{12} \\ \lambda_{21} & \lambda_{22} \end{bmatrix} \in \mathbb{R}^{2\times 2}.$$

Then, the parameterization (16.2) is simply

$$\begin{aligned} M(\boldsymbol{q}, \boldsymbol{\theta}) \left[\ddot{\boldsymbol{q}}_d + \Lambda \dot{\tilde{\boldsymbol{q}}}\right] + C(\boldsymbol{q}, \dot{\boldsymbol{q}}, \boldsymbol{\theta}) \left[\dot{\boldsymbol{q}}_d + \Lambda \tilde{\boldsymbol{q}}\right] + \boldsymbol{g}(\boldsymbol{q}, \boldsymbol{\theta}) \\ = \Phi \boldsymbol{\theta} + M_0(\boldsymbol{q}) \left[\ddot{\boldsymbol{q}}_d + \Lambda \dot{\tilde{\boldsymbol{q}}}\right] + \boldsymbol{g}_0(\boldsymbol{q}), \end{aligned}$$

where

$$\boldsymbol{u} = \ddot{\boldsymbol{q}}_d + \Lambda \dot{\tilde{\boldsymbol{q}}} = \begin{bmatrix} u_1 \\ u_2 \end{bmatrix} = \begin{bmatrix} \ddot{q}_{d1} + \lambda_{11}\dot{\tilde{q}}_1 + \lambda_{12}\dot{\tilde{q}}_2 \\ \ddot{q}_{d2} + \lambda_{21}\dot{\tilde{q}}_1 + \lambda_{22}\dot{\tilde{q}}_2 \end{bmatrix}$$

$$\boldsymbol{v} = \dot{\boldsymbol{q}}_d + \Lambda \tilde{\boldsymbol{q}} = \begin{bmatrix} v_1 \\ v_2 \end{bmatrix} = \begin{bmatrix} \dot{q}_{d1} + \lambda_{11}\tilde{q}_1 + \lambda_{12}\tilde{q}_2 \\ \dot{q}_{d2} + \lambda_{21}\tilde{q}_1 + \lambda_{22}\tilde{q}_2 \end{bmatrix}$$

$$\boldsymbol{w} = \dot{\boldsymbol{q}} = \begin{bmatrix} w_1 \\ w_2 \end{bmatrix} = \begin{bmatrix} \dot{q}_1 \\ \dot{q}_2 \end{bmatrix}.$$

The adaptive control system is given by Equations (16.8) and (16.9). Therefore the control law becomes

$$\begin{aligned} \boldsymbol{\tau} = K_p \tilde{\boldsymbol{q}} + K_v \dot{\tilde{\boldsymbol{q}}} + \begin{bmatrix} \Phi_{11} & \Phi_{12} & \Phi_{13} \\ \Phi_{21} & \Phi_{22} & \Phi_{23} \end{bmatrix} \begin{bmatrix} \hat{\theta}_1 \\ \hat{\theta}_2 \\ \hat{\theta}_3 \end{bmatrix} \\ + \begin{bmatrix} (m_1 l_{c1}^2 + I_1) u_1 \\ 0 \end{bmatrix} + \begin{bmatrix} m_1 l_{c1} g \sin(q_1) \\ 0 \end{bmatrix}, \end{aligned}$$

while the adaptive law is

$$\begin{bmatrix} \hat{\theta}_1(t) \\ \hat{\theta}_2(t) \\ \hat{\theta}_3(t) \end{bmatrix} = \Gamma \int_0^t \begin{bmatrix} \Phi_{11}[v_1 - \dot{q}_1] \\ \Phi_{12}[v_1 - \dot{q}_1] + \Phi_{22}[v_2 - \dot{q}_2] \\ \Phi_{13}[v_1 - \dot{q}_1] + \Phi_{23}[v_2 - \dot{q}_2] \end{bmatrix} ds + \begin{bmatrix} \hat{\theta}_1(0) \\ \hat{\theta}_2(0) \\ \hat{\theta}_3(0) \end{bmatrix}.$$

In this experiment the symmetric positive definite design matrices were chosen as

$$
\begin{aligned}
K_p &= \operatorname{diag}\{200,\ 150\} \quad [\mathrm{N\,m/rad}]\,, \\
K_v &= \operatorname{diag}\{3\} \quad [\mathrm{N\,m\,s/rad}]\,, \\
\Gamma &= \operatorname{diag}\{1.6\ [\mathrm{kg\,s^2/m^2}],\ 0.004\ [\mathrm{kg\,s^2}],\ 0.004\ [\mathrm{kg\,m^2\,s^2}]\},
\end{aligned}
$$

and therefore $\Lambda = K_v^{-1} K_p = \operatorname{diag}\{66.6,\ 50\}\ [1/s]$.

The corresponding initial conditions for the positions, velocities and adaptive parameters are chosen as

$$
\begin{aligned}
q_1(0) &= 0, & q_2(0) &= 0 \\
\dot{q}_1(0) &= 0, & \dot{q}_2(0) &= 0 \\
\hat{\theta}_1(0) &= 0, & \hat{\theta}_2(0) &= 0 \\
\hat{\theta}_3(0) &= 0\,.
\end{aligned}
$$

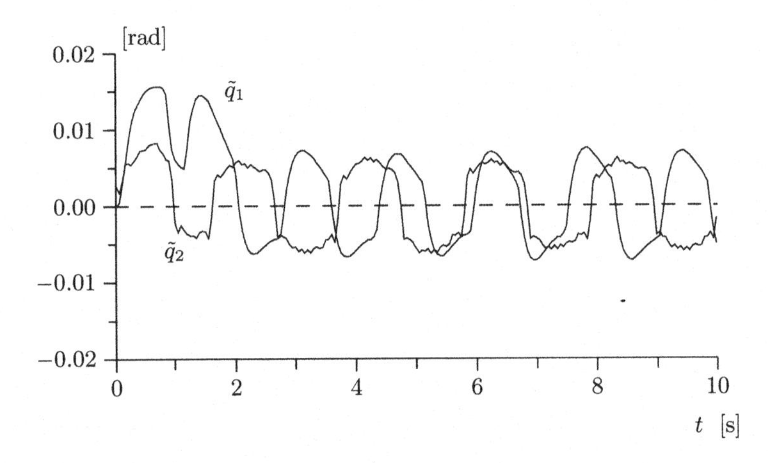

Figure 16.4. Graphs of position errors $\tilde{q}_1$ and $\tilde{q}_2$

Figures 16.4 and 16.5 show the experimental results. In particular, Figure 16.4 shows the steady state tracking position errors $\tilde{\boldsymbol{q}}(t)$, which, by virtue of friction phenomena in the actual robot, are not zero. It is remarkable, however, that if we take the upper-bound on the position errors as a measure of performance, we see that the latter is better than or similar to that of other nonadaptive control systems (compare with Figures 10.2, 10.4, 11.3, 11.5 and 12.5).

Finally, Figure 16.5 shows the evolution in time of the adaptive parameters. As mentioned before, these parameters were arbitrarily assumed to be zero at the initial instant. This has been done for no specific reason since we did not have any knowledge, *a priori*, about any of the dynamic parameters $\boldsymbol{\theta}$.

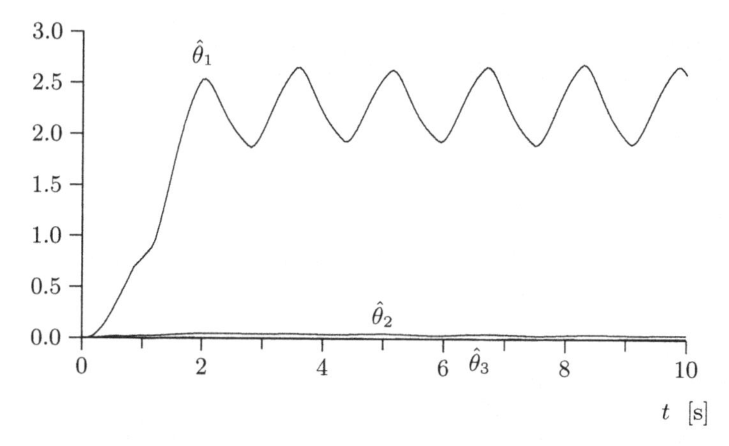

Figure 16.5. Graphs of adaptive parameters $\hat{\theta}_1$, $\hat{\theta}_2$, and $\hat{\theta}_3$

16.4 Conclusions

We conclude this chapter with the following remarks.

- For PD control with adaptive compensation the origin of the state space corresponding to the closed-loop equation, *i.e.* $\left[\tilde{\boldsymbol{q}}^T \ \dot{\tilde{\boldsymbol{q}}}^T \ \tilde{\boldsymbol{\theta}}^T\right]^T = \mathbf{0}$, is stable for any choice of symmetric positive definite matrices K_p, K_v and Γ. Moreover, the motion control objective is achieved globally. That is, for any initial position error $\tilde{\boldsymbol{q}}(0) \in \mathrm{IR}^n$ velocity error $\dot{\tilde{\boldsymbol{q}}}(0) \in \mathrm{IR}^n$, and arbitrary uncertainty over the dynamic parameters $\boldsymbol{\theta} \in \mathrm{IR}^m$ of the robot model, $\lim_{t\to\infty} \tilde{\boldsymbol{q}}(t) = \mathbf{0}$.

Bibliography

PD control with adaptive compensation was originally proposed in

- Slotine J. J., Li W., 1987, *"On the adaptive control of robot manipulators"*, The International Journal of Robotics Research, Vol. 6, No. 3, pp. 49–59.

This controller has also been the subject of study (among many others) in

- Slotine J. J., Li W., 1988. *"Adaptive manipulator control: A case study"*, IEEE Transactions on Automatic Control, Vol. AC-33, No. 11, November, pp. 995–1003.
- Spong M., Vidyasagar M., 1989, *"Robot dynamics and control"*, John Wiley and Sons.
- Slotine J. J., Li W., 1991, *"Applied nonlinear control"*, Prentice-Hall.
- Lewis F. L., Abdallah C. T., Dawson D. M., 1993, *"Control of robot manipulators"*, Macmillan.

The Lyapunov function (16.14) used in the stability analysis of PD control with adaptive compensation follows

- Spong M., Ortega R., Kelly R., 1990, *"Comments on "Adaptive manipulator control: A case study" "*, IEEE Transactions on Automatic Control, Vol. 35, No. 6, June, pp. 761–762.

Example 16.2 has been adapted from Section III.B of

- Slotine J. J., Li W., 1988. *"Adaptive manipulator control: A case study"*, IEEE Transactions on Automatic Control, Vol. AC-33, No. 11, November, pp. 995-1003.

Parameter convergence was shown in

- J.J. Slotine and W. Li, 1987, "Theoretical issues in adaptive manipulator control", *5th Yale Workshop on Applied Adaptive Systems Theory*, pp. 252–258.

Global uniform asymptotic stability for the closed-loop equation; in particular, *uniform* parameter convergence, for robots with revolute joints under PD control with adaptive compensation, was first shown in

- A. Loría, R. Kelly and A. Teel, 2003, "Uniform parametric convergence in the adaptive control of manipulators: a case restudied", in Proceedings of International Conference on Robotics and Automation, Taipei, Taiwan, pp. 106–1067.

Problems

1. Consider Example 16.1 in which we studied control of the pendulum

$$J\ddot{q} + mgl\,\sin(q) = \tau\,.$$

Supposing that the inertia J is unknown, the PD control law with adaptive compensation is given by

$$\tau = k_p\tilde{q} + k_v\dot{\tilde{q}} + \left[\ddot{q}_d + \lambda\dot{\tilde{q}}\right]\hat{\theta} + mgl\,\sin(q)$$

$$\hat{\theta} = \gamma\int_0^t \left[\ddot{q}_d + \lambda\dot{\tilde{q}}\right]\left[\dot{\tilde{q}} + \lambda\tilde{q}\right]\,ds + \hat{\theta}(0)$$

where $k_p > 0$, $k_v > 0$, $\lambda = k_p/k_v$, $\gamma > 0$ and $\hat{\theta}(0) \in \mathbb{R}$.

a) Obtain the closed-loop equation in terms of the state vector $\begin{bmatrix}\tilde{q} & \dot{\tilde{q}} & \tilde{\theta}\end{bmatrix}^T \in \mathbb{R}^3$ where $\tilde{\theta} = \hat{\theta} - J$.

b) Show that the origin of the closed-loop equation is a stable equilibrium, by using the following Lyapunov function candidate:

$$V(\tilde{q}, \dot{\tilde{q}}, \tilde{\theta}) = \frac{1}{2}\begin{bmatrix}\tilde{q}\\ \dot{\tilde{q}}\\ \tilde{\theta}\end{bmatrix}^T \begin{bmatrix} 2k_p + I\lambda^2 & \lambda I & 0\\ I\lambda & I & 0\\ 0 & 0 & \gamma^{-1}\end{bmatrix}\begin{bmatrix}\tilde{q}\\ \dot{\tilde{q}}\\ \tilde{\theta}\end{bmatrix}.$$

2. Consider again Example 16.1 in which we studied control of the pendulum

$$J\ddot{q} + mgl\,\sin(q) = \tau\,.$$

Assume now that, in addition, the inertia J and the mass m are unknown. Design a PD controller with adaptive compensation, *i.e.* give explicitly the control and adaptive laws.

3. On pages 366–368 we showed that $\tilde{\boldsymbol{q}} \in L_2^n$. Use similar arguments to prove also that $\dot{\tilde{\boldsymbol{q}}} \in L_2^n$. May we conclude that $\lim_{t\to\infty}\dot{\tilde{\boldsymbol{q}}}(t) = \boldsymbol{0}$?

4. Consider the 2-DOF Cartesian robot showed in Figure 16.6. The corresponding dynamic model is given by $M(\boldsymbol{q})\ddot{\boldsymbol{q}} + \boldsymbol{g}(\boldsymbol{q}) = \boldsymbol{\tau}$ where $\boldsymbol{q} = [q_1\ \ q_2]^T$ and

$$M(\boldsymbol{q}) = \begin{bmatrix} m_1 + m_2 & 0\\ 0 & m_2\end{bmatrix}$$

$$\boldsymbol{g}(\boldsymbol{q}) = \begin{bmatrix}(m_1 + m_2)g\\ 0\end{bmatrix}.$$

Assume that the masses m_1 and m_2 are constant but unknown.

a) Design a PD controller with adaptive compensation to achieve the motion control objective. Specifically, determine the matrix Φ for the control law and for the adaptive law

$$\boldsymbol{\tau} = K_p\tilde{\boldsymbol{q}} + K_v\dot{\tilde{\boldsymbol{q}}} + \Phi\hat{\boldsymbol{\theta}}$$

$$\hat{\boldsymbol{\theta}} = \int_0^t \Phi^T\left[\dot{\tilde{\boldsymbol{q}}} + \Lambda\tilde{\boldsymbol{q}}\right]\,ds + \hat{\boldsymbol{\theta}}(0),$$

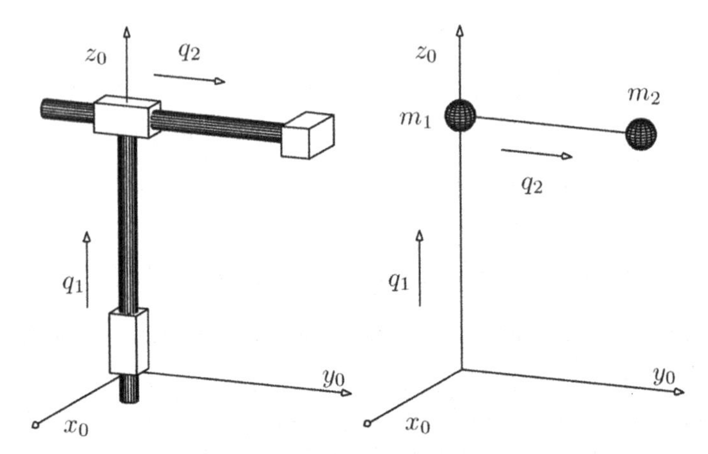

Figure 16.6. Problem 4. Cartesian 2-DOF robot.

where

$$\boldsymbol{\theta} = \begin{bmatrix} m_1 + m_2 \\ (m_1 + m_2)g \\ m_2 \end{bmatrix} .$$

Appendices

A

Mathematical Support

In this appendix we present additional mathematical tools that are employed in the textbook, mainly in the advanced topics of Part IV. It is recommended that the graduate student following these chapters read first this appendix, specifically the material from Section A.3 which is widely used in the text. As for other chapters and appendices, references are provided at the end.

A.1 Some Lemmas on Linear Algebra

The following lemmas, whose proofs may be found in textbooks on linear algebra, are used to prove certain properties of the dynamic model of the robot stated in Chapter 4.

Lemma A.1. *Consider a vector $\boldsymbol{x} \in I\!R^n$. Its Euclidean norm, $\|\boldsymbol{x}\|$, satisfies*

$$\|\boldsymbol{x}\| \leq n \left[\max_i \{|x_i|\} \right] .$$

Lemma A.2. *Consider a symmetric matrix $A \in I\!R^{n\times n}$ and denote by a_{ij} its ijth element. Let $\lambda_1\{A\}, \cdots, \lambda_n\{A\}$ be its eigenvalues. Then, it holds that*

$$|\lambda_k\{A\}| \leq n \left[\max_{i,j} \{|a_{ij}|\} \right] \qquad \text{for all } k = 1, \cdots, n .$$

Lemma A.3. *Consider a symmetric matrix $A = A^T \in I\!R^{n\times n}$ and denote by a_{ij} its ijth element. The spectral norm of the matrix A, $\|A\|$, induced by the vectorial Euclidean norm satisfies*

$$\|A\| = \sqrt{\lambda_{\text{Max}}\{A^T A\}} \leq n \left[\max_{i,j} \{|a_{ij}|\} \right] .$$

We present here a useful theorem on partitioned matrices which is taken from the literature.

Theorem A.1. *Assume that a symmetric matrix is partitioned as*

$$\begin{bmatrix} A & B \\ B^T & C \end{bmatrix} \tag{A.1}$$

where A and C are square matrices. The matrix is positive definite if and only if

$$\begin{aligned} A &> 0 \\ C - B^T A^{-1} B &> 0 \,. \end{aligned}$$

A.2 Vector Calculus

Theorem A.2. Mean-value
Consider the continuous function $f : \mathbb{R}^n \to \mathbb{R}$. If moreover $f(z_1, z_2, \cdots, z_n)$ has continuous partial derivatives then, for any two constant vectors $\boldsymbol{x}, \boldsymbol{y} \in \mathbb{R}^n$ we have

$$f(\boldsymbol{x}) - f(\boldsymbol{y}) = \begin{bmatrix} \left.\dfrac{\partial f(\boldsymbol{z})}{\partial z_1}\right|_{\boldsymbol{z}=\boldsymbol{\xi}} \\ \left.\dfrac{\partial f(\boldsymbol{z})}{\partial z_2}\right|_{\boldsymbol{z}=\boldsymbol{\xi}} \\ \vdots \\ \left.\dfrac{\partial f(\boldsymbol{z})}{\partial z_n}\right|_{\boldsymbol{z}=\boldsymbol{\xi}} \end{bmatrix}^T [\boldsymbol{x} - \boldsymbol{y}]$$

where $\boldsymbol{\xi} \in \mathbb{R}^n$ is a vector suitably chosen on the line segment which joins the vectors $\boldsymbol{x}$ and $\boldsymbol{y}$, i.e. which satisfies

$$\begin{aligned} \boldsymbol{\xi} &= \boldsymbol{y} + \alpha[\boldsymbol{x} - \boldsymbol{y}] \\ &= \alpha \boldsymbol{x} + (1-\alpha)\boldsymbol{y} \end{aligned}$$

for some real α in the interval $(0,1)$. Notice moreover, that the norm of $\boldsymbol{\xi}$ verifies

$$\|\boldsymbol{\xi}\| \leq \|\boldsymbol{y}\| + \|\boldsymbol{x} - \boldsymbol{y}\|$$

and also

$$\|\boldsymbol{\xi}\| \leq \|\boldsymbol{x}\| + \|\boldsymbol{y}\| \,.$$

An extension of the mean-value theorem for vectorial functions is presented next

Theorem A.3. Mean-value theorem for vectorial functions *Consider the continuous vectorial function* $\boldsymbol{f} : \mathbb{R}^n \to \mathbb{R}^m$. *If* $f_i(z_1, z_2, \cdots, z_n)$ *has continuous partial derivatives for* $i = 1, \cdots, m$, *then for each pair of vectors* $\boldsymbol{x}, \boldsymbol{y} \in \mathbb{R}^n$ *and each* $\boldsymbol{w} \in \mathbb{R}^m$ *there exists* $\boldsymbol{\xi} \in \mathbb{R}^n$ *such that*

$$[\boldsymbol{f}(\boldsymbol{x}) - \boldsymbol{f}(\boldsymbol{y})]^T \boldsymbol{w} = \boldsymbol{w}^T \underbrace{\begin{bmatrix} \left.\frac{\partial f_1(\boldsymbol{z})}{\partial z_1}\right|_{\boldsymbol{z}=\boldsymbol{\xi}} & \left.\frac{\partial f_1(\boldsymbol{z})}{\partial z_2}\right|_{\boldsymbol{z}=\boldsymbol{\xi}} & \cdots & \left.\frac{\partial f_1(\boldsymbol{z})}{\partial z_n}\right|_{\boldsymbol{z}=\boldsymbol{\xi}} \\ \left.\frac{\partial f_2(\boldsymbol{z})}{\partial z_1}\right|_{\boldsymbol{z}=\boldsymbol{\xi}} & \left.\frac{\partial f_2(\boldsymbol{z})}{\partial z_2}\right|_{\boldsymbol{z}=\boldsymbol{\xi}} & \cdots & \left.\frac{\partial f_2(\boldsymbol{z})}{\partial z_n}\right|_{\boldsymbol{z}=\boldsymbol{\xi}} \\ \vdots & \vdots & \ddots & \vdots \\ \left.\frac{\partial f_m(\boldsymbol{z})}{\partial z_1}\right|_{\boldsymbol{z}=\boldsymbol{\xi}} & \left.\frac{\partial f_m(\boldsymbol{z})}{\partial z_2}\right|_{\boldsymbol{z}=\boldsymbol{\xi}} & \cdots & \left.\frac{\partial f_m(\boldsymbol{z})}{\partial z_n}\right|_{\boldsymbol{z}=\boldsymbol{\xi}} \end{bmatrix}}_{\text{Jacobian of } \boldsymbol{f} \text{ evaluated in } \boldsymbol{z}=\boldsymbol{\xi}} [\boldsymbol{x} - \boldsymbol{y}]$$

$$= \boldsymbol{w}^T \left.\frac{\partial \boldsymbol{f}(\boldsymbol{z})}{\partial \boldsymbol{z}}\right|_{\boldsymbol{z}=\boldsymbol{\xi}} [\boldsymbol{x} - \boldsymbol{y}]$$

where $\boldsymbol{\xi}$ *is a vector on the line segment that joins the vectors* $\boldsymbol{x}$ *and* $\boldsymbol{y}$, *and consequently satisfies*

$$\boldsymbol{\xi} = \boldsymbol{y} + \alpha[\boldsymbol{x} - \boldsymbol{y}]$$

for some real α *in the interval* $(0, 1)$.

We present next a useful corollary, which follows from the statements of Theorems A.2 and A.3.

Corollary A.1. *Consider the smooth matrix-function* $A : \mathbb{R}^n \to \mathbb{R}^{n \times n}$. *Assume that the partial derivatives of the elements of the matrix* A *are bounded functions, that is, that there exists a finite constant* δ *such that*

$$\left| \left.\frac{\partial a_{ij}(\boldsymbol{z})}{\partial z_k}\right|_{\boldsymbol{z}=\boldsymbol{z}_0} \right| \leq \delta$$

for $i, j, k = 1, 2, \cdots, n$ *and all vectors* $\boldsymbol{z}_0 \in \mathbb{R}^n$.

Define now the vectorial function

$$[A(\boldsymbol{x}) - A(\boldsymbol{y})]\, \boldsymbol{w},$$

with $\boldsymbol{x}, \boldsymbol{y}, \boldsymbol{w} \in \mathbb{R}^n$. *Then, the norm of this function satisfies*

$$\|[A(\boldsymbol{x}) - A(\boldsymbol{y})]\, \boldsymbol{w}\| \leq n^2 \max_{i,j,k,\boldsymbol{z}_0} \left\{ \left| \left.\frac{\partial a_{ij}(\boldsymbol{z})}{\partial z_k}\right|_{\boldsymbol{z}=\boldsymbol{z}_0} \right| \right\} \|\boldsymbol{x} - \boldsymbol{y}\| \; \|\boldsymbol{w}\|, \quad \text{(A.2)}$$

where $a_{ij}(\boldsymbol{z})$ *denotes the* ij*th element of the matrix* $A(\boldsymbol{z})$ *while* z_k *denotes the* k*th element of the vector* $\boldsymbol{z} \in \mathbb{R}^n$.

Proof. The proof of the corollary may be carried out by the use of Theorems A.2 or A.3. Here we use Theorem A.2.

The norm of the vector $A(\boldsymbol{x})\boldsymbol{w} - A(\boldsymbol{y})\boldsymbol{w}$ satisfies

$$\|A(\boldsymbol{x})\boldsymbol{w} - A(\boldsymbol{y})\boldsymbol{w}\| \leq \|A(\boldsymbol{x}) - A(\boldsymbol{y})\| \, \|\boldsymbol{w}\| \ .$$

Considering Lemma A.3, we get

$$\|A(\boldsymbol{x})\boldsymbol{w} - A(\boldsymbol{y})\boldsymbol{w}\| \leq n \left[\max_{i,j} \left\{|a_{ij}(\boldsymbol{x}) - a_{ij}(\boldsymbol{y})|\right\}\right] \|\boldsymbol{w}\| \ . \qquad \text{(A.3)}$$

On the other hand, since by hypothesis the matrix $A(\boldsymbol{z})$ is a smooth function of its argument, its elements have continuous partial derivatives. Consequently, given two constant vectors $\boldsymbol{x}, \boldsymbol{y} \in \mathbb{R}^n$, according to the mean-value Theorem (*cf.* Theorem A.2), there exists a real number α_{ij} in the interval $[0,1]$ such that

$$a_{ij}(\boldsymbol{x}) - a_{ij}(\boldsymbol{y}) = \begin{bmatrix} \left.\dfrac{\partial a_{ij}(\boldsymbol{z})}{\partial z_1}\right|_{\boldsymbol{z}=\boldsymbol{y}+\alpha_{ij}[\boldsymbol{x}-\boldsymbol{y}]} \\ \left.\dfrac{\partial a_{ij}(\boldsymbol{z})}{\partial z_2}\right|_{\boldsymbol{z}=\boldsymbol{y}+\alpha_{ij}[\boldsymbol{x}-\boldsymbol{y}]} \\ \vdots \\ \left.\dfrac{\partial a_{ij}(\boldsymbol{z})}{\partial z_n}\right|_{\boldsymbol{z}=\boldsymbol{y}+\alpha_{ij}[\boldsymbol{x}-\boldsymbol{y}]} \end{bmatrix}^T [\boldsymbol{x} - \boldsymbol{y}].$$

Therefore, taking the absolute value on both sides of the previous equation and using the triangle inequality, $|\boldsymbol{a}^T\boldsymbol{b}| \leq \|\boldsymbol{a}\|\|\boldsymbol{b}\|$, we obtain the inequality

$$\begin{aligned} |a_{ij}(\boldsymbol{x}) - a_{ij}(\boldsymbol{y})| &\leq \left\| \begin{matrix} \left.\dfrac{\partial a_{ij}(\boldsymbol{z})}{\partial z_1}\right|_{\boldsymbol{z}=\boldsymbol{y}+\alpha_{ij}[\boldsymbol{x}-\boldsymbol{y}]} \\ \left.\dfrac{\partial a_{ij}(\boldsymbol{z})}{\partial z_2}\right|_{\boldsymbol{z}=\boldsymbol{y}+\alpha_{ij}[\boldsymbol{x}-\boldsymbol{y}]} \\ \vdots \\ \left.\dfrac{\partial a_{ij}(\boldsymbol{z})}{\partial z_n}\right|_{\boldsymbol{z}=\boldsymbol{y}+\alpha_{ij}[\boldsymbol{x}-\boldsymbol{y}]} \end{matrix} \right\| \|\boldsymbol{x} - \boldsymbol{y}\| \\ &\leq n \left[\max_k \left\{ \left| \left.\frac{\partial a_{ij}(\boldsymbol{z})}{\partial z_k}\right|_{\boldsymbol{z}=\boldsymbol{y}+\alpha_{ij}[\boldsymbol{x}-\boldsymbol{y}]} \right| \right\}\right] \|\boldsymbol{x} - \boldsymbol{y}\| , \end{aligned}$$

where for the last step we used Lemma A.1 ($\|\boldsymbol{x}\| \leq n \, [\max_i \{|x_i|\}]$).

Moreover, since it has been assumed that the partial derivatives of the elements of A are bounded functions then, we may claim that

$$|a_{ij}(\boldsymbol{x}) - a_{ij}(\boldsymbol{y})| \leq n \left[\max_{k,z_0} \left\{ \left| \frac{\partial a_{ij}(\boldsymbol{z})}{\partial z_k} \bigg|_{\boldsymbol{z}=\boldsymbol{z}_0} \right| \right\} \right] \|\boldsymbol{x} - \boldsymbol{y}\| \,.$$

From the latter expression and from (A.3) we conclude the statement contained in (A.2).

Truncated Taylor Representation of a Function

We present now a result well known from calculus and optimization. In the first case, it comes from the 'theorem of Taylor' and in the second, it comes from what is known as 'Lagrange's residual formula'. Given the importance of this lemma in the study of positive definite functions in Appendix B the proof is presented in its complete form.

Lemma A.4. *Let $f : \mathbb{R}^n \to \mathbb{R}$ be a continuous function with continuous partial derivatives up to at least the second one. Then, for each $\boldsymbol{x} \in \mathbb{R}^n$, there exists a real number α $(1 \geq \alpha \geq 0)$ such that*

$$f(\boldsymbol{x}) = f(\boldsymbol{0}) + \frac{\partial f}{\partial \boldsymbol{x}}(\boldsymbol{0})^T \boldsymbol{x} + \frac{1}{2}\boldsymbol{x}^T H(\alpha \boldsymbol{x})\boldsymbol{x}$$

where $H(\alpha \boldsymbol{x})$ is the Hessian matrix (that is, its second partial derivative) of $f(\boldsymbol{x})$ evaluated at $\alpha \boldsymbol{x}$.

Proof. Let $\boldsymbol{x} \in \mathbb{R}^n$ be a constant vector. Consider the time derivative of $f(t\boldsymbol{x})$

$$\begin{aligned} \frac{d}{dt} f(t\boldsymbol{x}) &= \left[\frac{\partial f(\boldsymbol{s})}{\partial \boldsymbol{s}} \bigg|_{\boldsymbol{s}=t\boldsymbol{x}} \right]^T \boldsymbol{x} \\ &= \frac{\partial f}{\partial \boldsymbol{x}}(t\boldsymbol{x})^T \boldsymbol{x} \,. \end{aligned}$$

Integrating from $t = 0$ to $t = 1$,

$$\begin{aligned} \int_{f(0\cdot \boldsymbol{x})}^{f(1\cdot \boldsymbol{x})} df(t\boldsymbol{x}) &= \int_0^1 \frac{\partial f}{\partial \boldsymbol{x}}(t\boldsymbol{x})^T \boldsymbol{x}\; dt \\ f(\boldsymbol{x}) - f(\boldsymbol{0}) &= \int_0^1 \frac{\partial f}{\partial \boldsymbol{x}}(t\boldsymbol{x})^T \boldsymbol{x}\; dt \,. \end{aligned} \tag{A.4}$$

The integral on the right-hand side above may be written as

$$\int_0^1 \boldsymbol{y}(t)^T \boldsymbol{x}\; dt \tag{A.5}$$

where

$$\boldsymbol{y}(t) = \frac{\partial f}{\partial \boldsymbol{x}}(t\boldsymbol{x})\,. \tag{A.6}$$

Defining

$$u = \boldsymbol{y}(t)^T \boldsymbol{x}$$
$$v = t - 1$$

and consequently

$$\frac{du}{dt} = \dot{\boldsymbol{y}}(t)^T \boldsymbol{x}$$
$$\frac{dv}{dt} = 1,$$

the integral (A.5) may be solved by parts[1]

$$\int_0^1 \boldsymbol{y}(t)^T \boldsymbol{x}\ dt = -\int_0^1 [t-1]\dot{\boldsymbol{y}}(t)^T \boldsymbol{x}\ dt + \boldsymbol{y}(t)^T \boldsymbol{x}[t-1]\Big|_0^1$$
$$= \int_0^1 [1-t)]\dot{\boldsymbol{y}}(t)^T \boldsymbol{x}\ dt + \boldsymbol{y}(0)^T \boldsymbol{x}\,. \tag{A.7}$$

Now, using the mean-value theorem for integrals[2], and noting that $(1-t) \geq 0$ for all t between 0 and 1, the integral on the right-hand side of Equation (A.7) may be written as

$$\int_0^1 (1-t)\dot{\boldsymbol{y}}(t)^T \boldsymbol{x}\ dt = \dot{\boldsymbol{y}}(\alpha)^T \boldsymbol{x} \int_0^1 (1-t)\ dt$$
$$= \frac{1}{2}\dot{\boldsymbol{y}}(\alpha)^T \boldsymbol{x}$$

for some α $(1 \geq \alpha \geq 0)$.

Incorporating this in (A.7) we get

[1] We recall here the formula:

$$\int_0^1 u\frac{dv}{dt}dt = -\int_0^1 v\frac{du}{dt}dt + uv|_0^1\,.$$

[2] Recall that for functions $h(t)$ and $g(t)$, continuous on the closed interval $a \leq t \leq b$, and where $g(t) \geq 0$ for each t from the interval, there always exists a number c such that $a \leq c \leq b$ and

$$\int_a^b h(t)g(t)\ dt = h(c)\int_a^b g(t)\ dt\,.$$

$$\int_0^1 \boldsymbol{y}(t)^T \boldsymbol{x}\, dt = \frac{1}{2}\dot{\boldsymbol{y}}(\alpha)^T \boldsymbol{x} + \boldsymbol{y}(0)^T \boldsymbol{x}$$

and therefore, (A.4) may be written as

$$f(\boldsymbol{x}) - f(\boldsymbol{0}) = \frac{1}{2}\dot{\boldsymbol{y}}(\alpha)^T \boldsymbol{x} + \boldsymbol{y}(0)^T \boldsymbol{x}. \tag{A.8}$$

On the other hand, using the definition of $\boldsymbol{y}(t)$ given in (A.6), we get

$$\dot{\boldsymbol{y}}(t) = H(t\boldsymbol{x})\boldsymbol{x},$$

and therefore $\dot{\boldsymbol{y}}(\alpha) = H(\alpha\boldsymbol{x})\boldsymbol{x}$. Incorporating this and (A.6) in (A.8), we obtain

$$f(\boldsymbol{x}) - f(\boldsymbol{0}) = \frac{1}{2}\boldsymbol{x}^T H(\alpha\boldsymbol{x})^T \boldsymbol{x} + \frac{\partial f}{\partial \boldsymbol{x}}(\boldsymbol{0})^T \boldsymbol{x}$$

which is what we wanted to prove.

We present next a simple example with the aim of illustrating the use of the statement of Lemma A.4.

Example A.1. Consider the function $f : \mathbb{R} \to \mathbb{R}$ defined by

$$f(x) = e^x .$$

According to Lemma A.4, the function $f(x)$ may be written as

$$f(x) = e^x = 1 + x + \frac{1}{2}e^{\alpha x}x^2$$

where for each $x \in \mathbb{R}$ there exists an α $(1 \geq \alpha \geq 0)$. Specifically, for $x = 0 \in \mathbb{R}$ any $\alpha \in [0,1]$ applies (indeed, any $\alpha \in \mathbb{R}$). In the case that $x \neq 0 \in \mathbb{R}$ then α is explicitly given by

$$\alpha = \frac{\ln\left(\dfrac{2(e^x - 1 - x)}{x^2}\right)}{x} .$$

Figure A.1 shows the corresponding graph of α versus x.

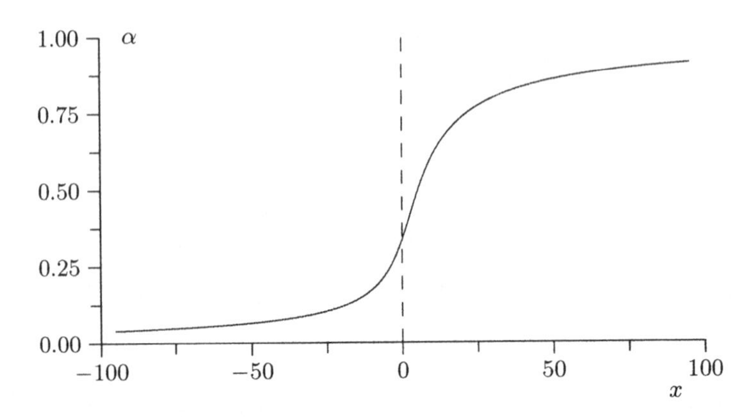

Figure A.1. Example A.1: graph of α

A.3 Functional Spaces

A special class of vectorial spaces are the so-called L_p^n (pronounce "el/pi:/en") where n is a positive integer and $p \in (0, \infty]$. The elements of the L_p^n spaces are functions with particular properties.

The linear spaces denoted by L_2^n and L_∞^n, which are defined below, are often employed in the analysis of interconnected dynamical systems in the theory of input–output stability. Formally, this methodology involves the use of operators that characterize the behavior of the distinct parts of the interconnected dynamic systems.

We present next a set of definitions and properties of spaces of functions that are useful in establishing certain convergence properties of solutions of differential equations.

For the purposes of this book, we say that a function $\boldsymbol{f} : \mathbb{R}^n \to \mathbb{R}^m$ is said to be *continuous* if

$$\lim_{\boldsymbol{x} \to \boldsymbol{x}_0} \boldsymbol{f}(\boldsymbol{x}) = \boldsymbol{f}(\boldsymbol{x}_0) \qquad \forall\, \boldsymbol{x}_0 \in \mathbb{R}^n.$$

A necessary condition for a function to be continuous is that it is defined at every point $\boldsymbol{x} \in \mathbb{R}^n$. It is also apparent that it is not necessary for a function to be continuous that the function's derivative be defined everywhere. For instance the derivative of the continuous function $f(x) = |x|$ is not defined at the origin, *i.e.* at $x = 0$. However, if a function's derivative is defined everywhere then the function is continuous.

The space L_2^n consists in the set of all the continuous functions $\boldsymbol{f} : \mathbb{R}_+ \to \mathbb{R}^n$ such that

$$\int_0^\infty \boldsymbol{f}(t)^T \boldsymbol{f}(t)\, dt = \int_0^\infty \|\boldsymbol{f}(t)\|^2\, dt < \infty\,.$$

In words, a function $\boldsymbol{f}$ belongs to the L_2^n space ($\boldsymbol{f} \in L_2^n$) if the integral of its Euclidean norm squared, is bounded from above. We also say that f is square-integrable.

The L_∞^n space consists of the set of all continuous functions $\boldsymbol{f} : \mathbb{R}_+ \to \mathbb{R}^n$ such that their Euclidean norms are upperbounded as[3],

$$\sup_{t \geq 0} \|\boldsymbol{f}(t)\| < \infty .$$

The symbols L_2 and L_∞ denote the spaces L_2^1 and L_∞^1 respectively.

We present next an example to illustrate the above-mentioned definitions.

Example A.2. Consider the continuous functions $f(t) = e^{-\alpha t}$ and $g(t) = \alpha \sin(t)$ where $\alpha > 0$. We want to determine whether f and g belong to the spaces of L_2 and L_∞.

Consider first the function $f(t)$:

$$\begin{aligned}\int_0^\infty |f(t)|^2 \, dt &= \int_0^\infty f^2(t) \, dt \\ &= \int_0^\infty e^{-2\alpha t} \, dt \\ &= \frac{1}{2\alpha} < \infty\end{aligned}$$

hence, $f \in L_2$. On the other hand, $|f(t)| = |e^{-\alpha t}| \leq 1 < \infty$ for all $t \geq 0$, hence $f \in L_\infty$. We conclude that $f(t)$ is *bounded* and *square-integrable, i.e.* $f \in L_\infty \cap L_2$ respectively.

Consider next the function $g(t)$. Notice that the integral

$$\int_0^\infty |g(t)|^2 \, dt = \alpha^2 \int_0^\infty \sin^2(t) \, dt$$

does not converge; consequently $g \notin L_2$. Nevertheless $|g(t)| = |\alpha \sin(t)| \leq \alpha < \infty$ for all $t \geq 0$, and therefore $g \in L_\infty$. ◇

A useful observation for analysis of convergence of solutions of differential equations is that if we consider a function $\boldsymbol{x} : \mathbb{R}_+ \to \mathbb{R}^n$ and a radially unbounded positive definite function $W : \mathbb{R}^n \to \mathbb{R}_+$ then, since $W(\boldsymbol{x})$ is continuous in $\boldsymbol{x}$ the composition $w(t) := W(\boldsymbol{x}(t))$ satisfies $w \in L_\infty$ if and only if $\boldsymbol{x} \in L_\infty^n$.

[3] For those readers not familiar with the sup of a function $f(t)$, it corresponds to the smallest possible number which is larger than $f(t)$ for all $t \geq 0$. For instance $\sup |\tanh(t)| = 1$ but $|\tanh(t)|$ has no maximal value since $\tanh(t)$ is ever increasing and *tends* to 1 as $t \to \infty$.

We remark that a continuous function $\boldsymbol{f}$ belonging to the space L_2^n may not have a limit. We present next a result from the functional analysis literature which provides sufficient conditions for functions belonging to the L_2^n space to have a limit at zero. This result is very often used in the literature of motion control of robot manipulators and in general, in the adaptive control literature.

Lemma A.5. *Consider a once continuously differentiable function* $\boldsymbol{f} : \mathbb{R}_+ \to \mathbb{R}^n$. *Suppose that* $\boldsymbol{f}$ *and its time derivative satisfy the following*

- $\boldsymbol{f}, \dot{\boldsymbol{f}} = \frac{d}{dt}\boldsymbol{f} \in L_\infty^n$,
- $\boldsymbol{f} \in L_2^n$.

Then, necessarily $\lim_{t\to\infty} \boldsymbol{f}(t) = \mathbf{0} \in \mathbb{R}^n$.

Proof. It follows by contradiction[4]. Specifically we show that if the conclusion of the lemma does not hold then the hypothesis that $\boldsymbol{f} \in L_2^n$ is violated.

To that end we first need to establish a convenient bound for the function $\|\boldsymbol{f}(t)\|^2 = \boldsymbol{f}(t)^T\boldsymbol{f}(t)$. Its total time derivative is $2\boldsymbol{f}(t)^T\dot{\boldsymbol{f}}(t)$ and is continuous by assumption so we may invoke the mean value theorem (see Theorem A.2) to conclude that for any pair of numbers $t, t_1 \in \mathbb{R}_+$ there exists a number s laying on the line segment that joins t and t_1, such that

$$\left| \|\boldsymbol{f}(t)\|^2 - \|\boldsymbol{f}(t_1)\|^2 \right| \leq 2\boldsymbol{f}(s)^T\dot{\boldsymbol{f}}(s)\,|t - t_1|\ .$$

On the other hand, since $\boldsymbol{f}, \dot{\boldsymbol{f}} \in L_\infty^n$ it follows that there exists $k > 0$ such that

$$\left| \|\boldsymbol{f}(t)\|^2 - \|\boldsymbol{f}(t_1)\|^2 \right| \leq k\,|t - t_1| \qquad \forall\, t, t_1 \in \mathbb{R}_+\,. \tag{A.9}$$

Next, notice that

$$\|\boldsymbol{f}(t)\|^2 = \|\boldsymbol{f}(t)\|^2 - \|\boldsymbol{f}(t_1)\|^2 + \|\boldsymbol{f}(t_1)\|^2$$

for all $t, t_1 \in \mathbb{R}_+$. Now we use the inequality $|a+b| \geq |a| - |b|$ which holds for all $a,\, b \in \mathbb{R}$, with $a = \|\boldsymbol{f}(t_1)\|^2$ and $b = \left[\, \|\boldsymbol{f}(t)\|^2 - \|\boldsymbol{f}(t_1)\|^2 \,\right]$ to see that

[4] Proof "by contradiction" or, "by *reductio ad absurdum*", is a technique widely used in mathematics to prove theorems and other *truths.* To illustrate the method consider a series of logical statements denoted A, B, C, *etc.* and their negations, denoted $\overline{A}$, $\overline{B}$, $\overline{C}$, *etc.* Then, to prove by contradiction the claim, "A AND $B \Longrightarrow C$", we proceed as follows. *Assume that* A AND B *hold but not* C. *Then, we seek for a series of implications that lead to a negation of* A AND B, *i.e. we look for other statements* D, E, *etc. such that* $\overline{C} \Longrightarrow D \Longrightarrow E \Longrightarrow \overline{A \text{ AND } B}$. *So we conclude that* $\overline{C} \Longrightarrow \overline{A \text{ AND } B}$. *However, in view of the fact that* A AND B *must hold, this contradicts the initial hypothesis of the proof that* C *does not hold (i.e.* $\overline{C}$*). Notice that* $\overline{A \text{ AND } B} = \overline{A}$ OR $\overline{B}$.

$$\|f(t)\|^2 \geq \|f(t_1)\|^2 - \left| \| f(t)\|^2 - \|f(t_1)\|^2 \right|$$

for all $t, t_1 \in \mathbb{R}_+$. Then, we use (A.9) to obtain

$$\|f(t)\|^2 \geq \|f(t_1)\|^2 - k\,|t - t_1| \,. \tag{A.10}$$

Assume now that the conclusion of the lemma does not hold i.e, either $\lim_{t\to\infty} f(t) \neq \mathbf{0}$ or this limit does not exist. In either case, it follows that for each $T \geq 0$ there exists an infinite unbounded sequence $\{t_1\,, t_2\,, \ldots\}$, denoted $\{t_n\} \in \mathbb{R}_+$ with $t_n \to \infty$ as $n \to \infty$, and a constant $\varepsilon > 0$ such that

$$\|f(t_i)\|^2 > \varepsilon \qquad \forall\, t_i \geq T\,. \tag{A.11}$$

To better see this, we recall that if $\lim_{t\to\infty} f(t)$ exists and is zero then, for any ε there exists $T(\varepsilon)$ such that for all $t \geq T$ we have $\|f(t)\|^2 \leq \varepsilon$. Furthermore, without loss of generality, defining $\delta := \dfrac{\varepsilon}{2k}$, we may assume that for all $i \leq n$, $t_{i+1} - t_i \geq \delta$ —indeed, if this does not hold, we may always extract another infinite unbounded subsequence $\{t_i'\}$ such that $t_{i+1}' - t_i' \geq \delta$ for all i.

Now, since Inequality (A.10) holds for any t and t_1 it also holds for any element of $\{t_n\}$. Then, in view of (A.11) we have, for each t_i belonging to $\{t_n\}$ and for all $t \in \mathbb{R}_+$,

$$\|f(t)\|^2 > \varepsilon - k\,|t - t_i| \,. \tag{A.12}$$

Integrating Inequality (A.12) from t_i to $t_i + \delta$ we obtain

$$\int_{t_i}^{t_i+\delta} \|f(t)\|^2 \; dt > \int_{t_i}^{t_i+\delta} \varepsilon \; dt - \int_{t_i}^{t_i+\delta} k\,|t - t_i| \; dt\,. \tag{A.13}$$

Notice that in the integrals above, $t \in [t_i,\, t_i + \delta]$ therefore, $-k|t - t_i| \geq -k\delta$. From this and (A.13) it follows that

$$\int_{t_i}^{t_i+\delta} \|f(t)\|^2 \; dt > \varepsilon\delta - k\delta^2$$

and since by definition $\dfrac{\varepsilon}{2k} = \delta$ we finally obtain

$$\int_{t_i}^{t_i+\delta} \|f(t)\|^2 \; dt > \frac{\varepsilon\delta}{2} > 0\,. \tag{A.14}$$

On the other hand, since $t_{i+1} \geq t_i + \delta$ for each t_i, it also holds that

$$\lim_{t\to\infty} \int_0^t \|f(\tau)\|^2 \; d\tau \geq \sum_{\{t_i\}} \int_{t_i}^{t_{i+1}} \|f(\tau)\|^2 \; d\tau \tag{A.15}$$

$$\geq \sum_{\{t_i\}} \int_{t_i}^{t_i+\delta} \|f(\tau)\|^2 \; d\tau\,. \tag{A.16}$$

We see that on one hand, the term on the left-hand side of Inequality (A.15) is bounded by assumption (since $\boldsymbol{f} \in L_2^n$) and on the other hand, since $\{t_n\}$ is infinite and (A.14) holds for each t_i the term on the right-hand side of Inequality (A.16) is unbounded. From this contradiction we conclude that it must hold that $\lim_{t\to\infty} \boldsymbol{f}(t) = \mathbf{0}$ which completes the proof.

◊◊◊

As an application of Lemma A.5 we present below the proof of Lemma 2.2 used extensively in Parts II and III of this text.

Proof of Lemma 2.2. Since $V(t, \boldsymbol{x}, \boldsymbol{z}, h) \geq 0$ and $\dot{V}(t, \boldsymbol{x}, \boldsymbol{z}, h) \leq 0$ for all $\boldsymbol{x}$, $\boldsymbol{z}$ and h then these inequalities also hold for $\boldsymbol{x}(\tau)$, $\boldsymbol{z}(\tau)$ and $h(\tau)$ and all $\tau \geq 0$. Integrating on both sides of $\dot{V}(\tau, \boldsymbol{x}(\tau), \boldsymbol{z}(\tau), h(\tau)) \leq 0$ from 0 to t we obtain[5]

$$V(0, \boldsymbol{x}(0), \boldsymbol{z}(0), h(0)) \geq V(t, \boldsymbol{x}(t), \boldsymbol{z}(t), h(t)) \geq 0 \quad \forall\, t \geq 0\,.$$

Now, since $P(t)$ is positive definite for all $t \geq 0$ we may invoke the theorem of Rayleigh–Ritz which establishes that $\boldsymbol{x}^T K \boldsymbol{x} \geq \lambda_{\min}\{K\} \boldsymbol{x}^T \boldsymbol{x}$ where K is any symmetric matrix and $\lambda_{\min}\{K\}$ denotes the smallest eigenvalue of K, to conclude that there exists[6] $p_m > 0$ such that $\boldsymbol{y}^T P(t) \boldsymbol{y} \geq p_m\{P\} \|\boldsymbol{y}\|^2$ for all $\boldsymbol{y} \in \mathbb{R}^{n+m}$ and all $t \in \mathbb{R}_+$. Furthermore, with an abuse of notation, we will denote such constant by $\lambda_{min}\{P\}$. It follows that

$$\begin{bmatrix} \boldsymbol{x}(0) \\ \\ \boldsymbol{z}(0) \end{bmatrix}^T P(0) \begin{bmatrix} \boldsymbol{x}(0) \\ \\ \boldsymbol{z}(0) \end{bmatrix} + h(0) \;\geq\; \lambda_{\min}\{P\} \left\| \begin{matrix} \boldsymbol{x}(t) \\ \boldsymbol{z}(t) \end{matrix} \right\|^2 + h(t) \;\geq\; 0 \qquad \forall t \geq 0$$

hence, the functions $\boldsymbol{x}(t)$, $\boldsymbol{z}(t)$ and $h(t)$ are bounded for all $t \geq 0$. This proves item 1.

To prove item 2 consider the expression

$$\dot{V}(t, \boldsymbol{x}(t), \boldsymbol{z}(t), h(t)) = -\boldsymbol{x}(t)^T Q(t) \boldsymbol{x}(t)\,.$$

Integrating between 0 and $T \in \mathbb{R}_+$ we get

$$V(T, \boldsymbol{x}(T), \boldsymbol{z}(T), h(T)) - V(0, \boldsymbol{x}(0), \boldsymbol{z}(0), h(0)) = -\int_0^T \boldsymbol{x}(\tau)^T Q(\tau) \boldsymbol{x}(\tau)\, d\tau$$

[5] One should not confuse $V(t, \boldsymbol{x}, \boldsymbol{z}, h)$ with $V(t, \boldsymbol{x}(t), \boldsymbol{z}(t), h(t))$ as often happens in the literature. The first denotes a function of four variables while the second is a functional. In other words, the second corresponds to the function $V(t, \boldsymbol{x}, \boldsymbol{z}, h)$ evaluated on certain trajectories which depend on time. Therefore, $V(t, \boldsymbol{x}(t), \boldsymbol{z}(t), h(t))$ is a function of time.

[6] In general, for such a bound to exist it may not be sufficient that P is positive definite for each t but we shall not deal with such issues here and rather, we assume that P *is* such that the bound exists. See also Remark 2.1 on page 25.

which, using the fact that $V(0, \boldsymbol{x}(0), \boldsymbol{z}(0), h(0)) \geq V(T, \boldsymbol{x}(T), \boldsymbol{z}(T), h(T)) \geq 0$ yields the inequality

$$V(0, \boldsymbol{x}(0), \boldsymbol{z}(0), h(0)) \geq \int_0^T \boldsymbol{x}(\tau)^T Q(\tau) \boldsymbol{x}(\tau)\, d\tau \quad \forall T \in \mathbb{R}_+ .$$

Notice that this inequality continues to hold as $T \to \infty$ hence,

$$V(0, \boldsymbol{x}(0), \boldsymbol{z}(0), h(0)) \geq \int_0^\infty \boldsymbol{x}(\tau)^T Q(\tau) \boldsymbol{x}(\tau)\, d\tau$$

so using that Q is positive definite we obtain[7] $\boldsymbol{x}^T Q(t) \boldsymbol{x} \geq \lambda_{\min}\{Q\} \|\boldsymbol{x}\|^2$ for all $\boldsymbol{x} \in \mathbb{R}^n$ and $t \in \mathbb{R}_+$ therefore

$$\frac{V(0, \boldsymbol{x}(0), \boldsymbol{z}(0), h(0))}{\lambda_{\min}\{Q\}} \geq \int_0^\infty \boldsymbol{x}(\tau)^T \boldsymbol{x}(\tau)\, d\tau .$$

The term on the left-hand side of this inequality is finite, which means that $\boldsymbol{x} \in L_2^n$.

Finally, since by assumption $\dot{\boldsymbol{x}} \in L_\infty^n$, invoking Lemma A.5 we may conclude that $\lim_{t\to\infty} \boldsymbol{x}(t) = \boldsymbol{0}$.

The following result is stated without proof. It can be established using the so-called Barbălat's lemma (see the Bibliography at the end of the appendix).

Lemma A.6. *Let $\boldsymbol{f} : \mathbb{R}_+ \to \mathbb{R}^n$ be a continuously differentiable function satisfying*

- $\lim_{t\to\infty} \boldsymbol{f}(t) = \boldsymbol{0}$
- $\boldsymbol{f}, \dot{\boldsymbol{f}}, \ddot{\boldsymbol{f}} \in L_\infty^n$.

Then,

- $\lim_{t\to\infty} \dot{\boldsymbol{f}}(t) = \boldsymbol{0}$.

Another useful observation is the following.

Lemma A.7. *Consider the two functions $\boldsymbol{f} : \mathbb{R}_+ \to \mathbb{R}^n$ and $h : \mathbb{R}_+ \to \mathbb{R}$ with the following characteristics:*

- $\boldsymbol{f} \in L_2^n$
- $h \in L_\infty$.

Then, the product $h\boldsymbol{f}$ satisfies

- $h\boldsymbol{f} \in L_2^n$.

[7] See footnote 6 on page 394.

Proof. According to the hypothesis made, there exist finite constants $k_f > 0$ and $k_h > 0$ such that

$$\int_0^\infty \boldsymbol{f}(t)^T \boldsymbol{f}(t)\, dt \le k_f$$

$$\sup_{t\ge 0} |h(t)| \le k_h\,.$$

Therefore

$$\int_0^\infty [h(t)\boldsymbol{f}(t)]^T [h(t)\boldsymbol{f}(t)]\, dt = \int_0^\infty h(t)^2 \boldsymbol{f}(t)^T \boldsymbol{f}(t)\, dt$$

$$\le k_h^2 \int_0^\infty \boldsymbol{f}(t)^T \boldsymbol{f}(t)\, dt \le k_h^2 k_f,$$

which means that $h\boldsymbol{f} \in L_2^n$.

Consider a dynamic linear system described by the following equations

$$\dot{\boldsymbol{x}} = A\boldsymbol{x} + B\boldsymbol{u}$$

$$\boldsymbol{y} = C\boldsymbol{x}$$

where $\boldsymbol{x} \in \mathbb{R}^m$ is the system's state $\boldsymbol{u} \in \mathbb{R}^n$, stands for the input, $\boldsymbol{y} \in \mathbb{R}^n$ for the output and $A \in \mathbb{R}^{m\times m}$, $B \in \mathbb{R}^{m\times n}$ and $C \in \mathbb{R}^{n\times m}$ are matrices having constant real coefficients. The transfer matrix function $H(s)$ of the system is then defined as $H(s) = C(sI - A)^{-1}B$ where s is a complex number ($s \in \mathbf{C}$).

The following result allows one to draw conclusions on whether $\boldsymbol{y}$ and $\dot{\boldsymbol{y}}$ belong to L_2^n or L_∞^n depending on whether $\boldsymbol{u}$ belongs to L_2^n or L_∞^n.

Lemma A.8. *Consider the square matrix function of dimension n, $H(s) \in \mathbb{R}^{n\times n}(s)$ whose elements are rational strictly proper[8] functions of the complex variable s. Assume that the denominators of all its elements have all their roots on the left half of the complex plane (i.e. they have negative real parts).*

1. *If $\boldsymbol{u} \in L_2^n$ then $\boldsymbol{y} \in L_2^n \cap L_\infty^n$, $\dot{\boldsymbol{y}} \in L_2^n$ and $\boldsymbol{y}(t) \to \mathbf{0}$ as $t \to \infty$.*
2. *If $\boldsymbol{u} \in L_\infty^n$ then $\boldsymbol{y} \in L_\infty^n$, $\dot{\boldsymbol{y}} \in L_\infty^n$.*

To illustrate the utility of the lemma above consider the differential equation

$$\dot{\boldsymbol{x}} + A\boldsymbol{x} = \boldsymbol{u}$$

where $\boldsymbol{x} \in \mathbb{R}^n$ and $A \in \mathbb{R}^{n\times n}$ is a constant positive definite matrix. If $\boldsymbol{u} \in L_2^n$, then we have from Lemma A.8 that $\boldsymbol{x} \in L_2^n \cap L_\infty^n$, $\dot{\boldsymbol{x}} \in L_2^n$ and $\boldsymbol{x}(t) \to \mathbf{0}$ when $t \to \infty$.

Finally, we present the following corollary whose proof follows immediately from Lemma A.8.

[8] That is, the degree of the denominator is strictly larger than that of the numerator.

Corollary A.2. *For the transfer matrix function $H(s) \in \mathbb{R}^{n\times n}(s)$, let $\boldsymbol{u}$ and $\boldsymbol{y}$ denote its inputs and outputs respectively and let the assumptions of Lemma A.8 hold. If $\boldsymbol{u} \in L_2^n \cap L_\infty^n$, then*

- $\boldsymbol{y} \in L_2^n \cap L_\infty^n$
- $\dot{\boldsymbol{y}} \in L_2^n \cap L_\infty^n$
- $\boldsymbol{y}(t) \to \boldsymbol{0}$ *when* $t \to \infty$.

The following interesting result may be proved without much effort from the definitions of positive definite function and decrescent function.

Lemma A.9. *Consider a continuous function $\boldsymbol{x} : \mathbb{R}_+ \to \mathbb{R}^n$ and a radially unbounded, positive definite, decrescent continuous function $V : \mathbb{R}_+ \times \mathbb{R}^n \to \mathbb{R}_+$. The composition $v(t) := V(t, \boldsymbol{x}(t))$ satisfies $v \in L_\infty$ if and only if $\boldsymbol{x} \in L_\infty^n$.*

Bibliography

Lemma A.2 appears in

- Marcus M., Minc H., 1965, *"Introduction to linear algebra"*, Dover Publications, p. 207.
- Horn R. A., Johnson C. R., 1985, *"Matrix analysis"*, Cambridge University Press, p. 346.

Theorem A.1 on partitioned matrices is taken from

- Horn R. A., Johnson C. R., 1985, *"Matrix analysis"*, Cambridge University Press.

The statement of the mean-value theorem for vectorial functions may be consulted in

- Taylor A. E., Mann W. R., 1983, *"Advanced calculus"*, John Wiley and Sons.

The definition of L_p spaces are clearly exposed in Chapter 6 of

- Vidyasagar M., 1993, *"Nonlinear systems analysis"*, Prentice-Hall, New Jersey.

The proof of Lemma A.5 is based on the proof of the so-called Barbălat's lemma originally reported in

- Barbălat B., 1959, "Systèmes d'équations différentielles d'oscillations non-linéaires", *Revue de mathématiques pures et appliquées*, Vol. 4, No. 2, pp. 267–270.

See also Lemma 2.12 in

- Narendra K., Annaswamy A., 1989, *Stable adaptive systems*, Prentice-Hall, p. 85.

Lemma A.8 is taken from

- Desoer C., Vidyasagar M., 1975, *"Feedback systems: Input–output properties"*, Academic Press, New York, p. 59.

Problems

1. Consider the continuous function

$$f(t) = \begin{cases} 2^{n+2}[t-n] & \text{if } n < t < n + \dfrac{1}{2^{n+2}} \\ 1 - 2^{n+2}\left[t - \left[n + \dfrac{1}{2^{n+2}}\right]\right] & \text{if } n + \dfrac{1}{2^{n+2}} \leq t < n + \dfrac{1}{2^{n+1}} \\ 0 & \text{if } n + \dfrac{1}{2^{n+1}} \leq t \leq n+1 \end{cases}$$

with $n = 0, 1, 2, \cdots$. The limit when $t \to \infty$ of $f(t)$ does not exist (see the Figure A.2). Show that $f(t)$ belongs to L_2.

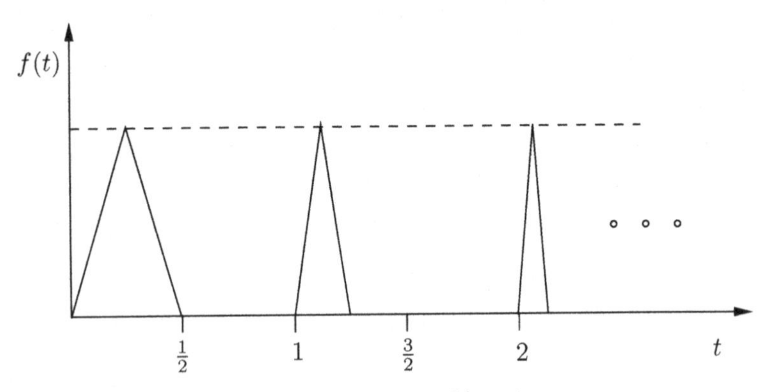

Figure A.2. Problem 1

Hint: Notice that $f^2(t) \leq h^2(t)$ where

$$h^2(t) = \begin{cases} 1 \text{ if } n < t < n + \dfrac{1}{2^{n+2}} \\ 1 \text{ if } n + \dfrac{1}{2^{n+2}} \leq t < n + \dfrac{1}{2^{n+1}} \\ 0 \text{ if } n + \dfrac{1}{2^{n+1}} \leq t \leq n + 1 \end{cases}$$

and $\int_0^\infty \mid h(t) \mid^2 dt = \sum_{i=1}^{\infty} (1/2^i)$.

B

Support to Lyapunov Theory

B.1 Conditions for Positive Definiteness of Functions

The interest of Lemma A.4 in this textbook resides in that it may be used to derive sufficient conditions for a function to be positive definite (locally or globally). We present such conditions in the statement of the following lemma.

Lemma B.1. *Let $f : \mathbb{R}^n \to \mathbb{R}$ be a continuously differentiable function with continuous partial derivatives up to at least second order. Assume that*

- $f(\mathbf{0}) = 0 \in \mathbb{R}$
- $\dfrac{\partial f}{\partial \boldsymbol{x}}(\mathbf{0}) = \mathbf{0} \in \mathbb{R}^n$.

Furthermore,

- *if the Hessian matrix satisfies $H(\mathbf{0}) > 0$, then $f(\boldsymbol{x})$ is a positive definite function (at least locally).*
- *If the Hessian matrix $H(\boldsymbol{x}) > 0$ for all $\boldsymbol{x} \in \mathbb{R}^n$, then $f(\boldsymbol{x})$ is a globally positive definite function.*

Proof. Considering Lemma A.4 and the hypothesis made on the function $f(\boldsymbol{x})$ we see that for each $\boldsymbol{x} \in \mathbb{R}^n$ there exists an α $(1 \geq \alpha \geq 0)$ such that

$$f(\boldsymbol{x}) = \frac{1}{2}\boldsymbol{x}^T H(\alpha \boldsymbol{x})\boldsymbol{x}\,.$$

Under the hypothesis of continuity up to the second partial derivative, if the Hessian matrix evaluated at $\boldsymbol{x} = \mathbf{0}$ is positive definite $(H(\mathbf{0}) > 0)$, then the Hessian matrix is also positive definite in a neighborhood of $\boldsymbol{x} = \mathbf{0} \in \mathbb{R}^n$, *e.g.* for all $\boldsymbol{x} \in \mathbb{R}^n$ such that $\|\boldsymbol{x}\| \leq \varepsilon$ and for some $\varepsilon > 0$, *i.e.*

$$H(\boldsymbol{x}) > 0 \qquad \forall\, \boldsymbol{x} \in \mathbb{R}^n : \|\boldsymbol{x}\| \leq \varepsilon\,.$$

Of course, $H(\alpha \boldsymbol{x}) > 0$ for all $\boldsymbol{x} \in \mathbb{R}^n$ such that $\|\boldsymbol{x}\| \leq \varepsilon$ and for any α $(1 \geq \alpha \geq 0)$. Since for all $\boldsymbol{x} \in \mathbb{R}^n$ there exists an α $(1 \geq \alpha \geq 0)$ and

$$f(\boldsymbol{x}) = \frac{1}{2}\boldsymbol{x}^T H(\alpha \boldsymbol{x})\boldsymbol{x}\,,$$

then $f(\boldsymbol{x}) > 0$ for all $\boldsymbol{x} \neq \boldsymbol{0} \in \mathbb{R}^n$ such that $\|\boldsymbol{x}\| \leq \varepsilon$. Furthermore, since by hypothesis $f(\boldsymbol{0}) = 0$, it follows that $f(\boldsymbol{x})$ is positive definite at least locally. On the other hand, if the Hessian matrix $H(\boldsymbol{x})$ is positive definite for all $\boldsymbol{x} \in \mathbb{R}^n$, it follows that so is $H(\alpha \boldsymbol{x})$ and this, not only for $1 \geq \alpha \geq 0$ but for any real α. Therefore, $f(\boldsymbol{x}) > 0$ for all $\boldsymbol{x} \neq \boldsymbol{0} \in \mathbb{R}^n$ and, since we assumed that $f(\boldsymbol{0}) = 0$ we conclude that $f(\boldsymbol{x})$ is globally positive definite.

Next, we present some examples to illustrate the application of the previous lemma.

Example B.1. Consider the following function $f : \mathbb{R}^2 \to \mathbb{R}$ used in the study of stability of the origin of the differential equation that models the behavior of an ideal pendulum, that is,

$$f(x_1, x_2) = mgl[1 - \cos(x_1)] + J\frac{x_2^2}{2}\,.$$

Clearly, we have $f(0,0) = 0$ that is, the origin is an equilibrium point. The gradient of $f(x_1, x_2)$ is given by

$$\frac{\partial f}{\partial \boldsymbol{x}}(\boldsymbol{x}) = \begin{bmatrix} mgl\,\sin(x_1) \\ Jx_2 \end{bmatrix}$$

which, evaluated at $\boldsymbol{x} = \boldsymbol{0} \in \mathbb{R}^2$ is zero. Next, the Hessian matrix is given by

$$H(\boldsymbol{x}) = \begin{bmatrix} mgl\,\cos(x_1) & 0 \\ 0 & J \end{bmatrix}$$

and is positive definite at $\boldsymbol{x} = \boldsymbol{0} \in \mathbb{R}^2$. Hence, according to Lemma B.1 the function $f(x_1, x_2)$ is positive definite at least locally. Notice that this function is not globally positive definite since $\cos(x_1) = 0$ for all $x_1 = \dfrac{n\pi}{2}$ with $n = 1, 2, 3 \ldots$ and $\cos(x_1) < 0$ for all $x_1 \in \left(\dfrac{n\pi}{2}, \dfrac{(n+2)\pi}{2}\right)$ for all $n = 1, 5, 7, 9, \ldots$ ◊

The following example, less trivial than the previous one, presents a function that is used as part of Lyapunov functions in the study of stability of various control schemes of robots.

Example B.2. Consider the function $f : \mathbb{R}^n \to \mathbb{R}$ defined as

$$f(\tilde{\boldsymbol{q}}) = \mathcal{U}(\boldsymbol{q}_d - \tilde{\boldsymbol{q}}) - \mathcal{U}(\boldsymbol{q}_d) + \boldsymbol{g}(\boldsymbol{q}_d)^T \tilde{\boldsymbol{q}} + \frac{1}{\varepsilon} \tilde{\boldsymbol{q}}^T K_p \tilde{\boldsymbol{q}}$$

where $K_p = K_p^T > 0$, $\boldsymbol{q}_d \in \mathbb{R}^n$ is a constant vector, ε is a real positive constant and $\mathcal{U}(\boldsymbol{q})$ stands for the potential energy of the robot. Here, we assume that all the joints of the robot are revolute.

The objective of this example is to show that if K_p is selected so that [1]

$$\lambda_{\min}\{K_p\} > \frac{\varepsilon}{2} k_g$$

then $f(\tilde{\boldsymbol{q}})$ is a globally positive definite function.

To prove the latter we use Lemma B.1. Notice first that $f(\boldsymbol{0}) = 0$.

The gradient of $f(\tilde{\boldsymbol{q}})$ with respect to $\tilde{\boldsymbol{q}}$ is

$$\frac{\partial}{\partial \tilde{\boldsymbol{q}}} f(\tilde{\boldsymbol{q}}) = \frac{\partial \mathcal{U}(\boldsymbol{q}_d - \tilde{\boldsymbol{q}})}{\partial \tilde{\boldsymbol{q}}} + \boldsymbol{g}(\boldsymbol{q}_d) + \frac{2}{\varepsilon} K_p \tilde{\boldsymbol{q}} .$$

Recalling from (3.20) that $\boldsymbol{g}(\boldsymbol{q}) = \partial \mathcal{U}(\boldsymbol{q}) / \partial \boldsymbol{q}$ and[2]

$$\frac{\partial}{\partial \tilde{\boldsymbol{q}}} \mathcal{U}(\boldsymbol{q}_d - \tilde{\boldsymbol{q}}) = \frac{\partial (\boldsymbol{q}_d - \tilde{\boldsymbol{q}})}{\partial \tilde{\boldsymbol{q}}}^T \frac{\partial \mathcal{U}(\boldsymbol{q}_d - \tilde{\boldsymbol{q}})}{\partial (\boldsymbol{q}_d - \tilde{\boldsymbol{q}})}$$

we finally obtain the expression:

$$\frac{\partial}{\partial \tilde{\boldsymbol{q}}} f(\tilde{\boldsymbol{q}}) = -\boldsymbol{g}(\boldsymbol{q}_d - \tilde{\boldsymbol{q}}) + \boldsymbol{g}(\boldsymbol{q}_d) + \frac{2}{\varepsilon} K_p \tilde{\boldsymbol{q}} .$$

Clearly the gradient of $f(\tilde{\boldsymbol{q}})$ is zero at $\tilde{\boldsymbol{q}} = \boldsymbol{0} \in \mathbb{R}^n$.

On the other hand, the (symmetric) Hessian matrix $H(\tilde{\boldsymbol{q}})$ of $f(\tilde{\boldsymbol{q}})$, defined as

---

[1] The constant k_g has been defined in Property 4.3 and satisfies

$$k_g \geq \left\| \frac{\partial \boldsymbol{g}(\boldsymbol{q})}{\partial \boldsymbol{q}} \right\| .$$

[2] Let $f : \mathbb{R}^n \to \mathbb{R}$, $\boldsymbol{g} : \mathbb{R}^n \to \mathbb{R}^n$, $\boldsymbol{x}, \boldsymbol{y} \in \mathbb{R}^n$ and $\boldsymbol{x} = \boldsymbol{g}(\boldsymbol{y})$. Then,

$$\frac{\partial f(\boldsymbol{x})}{\partial \boldsymbol{y}} = \left[\frac{\partial \boldsymbol{g}(\boldsymbol{y})}{\partial \boldsymbol{y}} \right]^T \frac{\partial f(\boldsymbol{x})}{\partial \boldsymbol{x}} .$$

$$H(\tilde{\boldsymbol{q}}) = \frac{\partial}{\partial \tilde{\boldsymbol{q}}}\left[\frac{\partial f(\tilde{\boldsymbol{q}})}{\partial \tilde{\boldsymbol{q}}}\right] = \begin{bmatrix} \frac{\partial^2 f(\tilde{\boldsymbol{q}})}{\partial \tilde{q}_1 \partial \tilde{q}_1} & \frac{\partial^2 f(\tilde{\boldsymbol{q}})}{\partial \tilde{q}_1 \partial \tilde{q}_2} & \cdots & \frac{\partial^2 f(\tilde{\boldsymbol{q}})}{\partial \tilde{q}_1 \partial \tilde{q}_n} \\ \frac{\partial^2 f(\tilde{\boldsymbol{q}})}{\partial \tilde{q}_2 \partial \tilde{q}_1} & \frac{\partial^2 f(\tilde{\boldsymbol{q}})}{\partial \tilde{q}_2 \partial \tilde{q}_2} & \cdots & \frac{\partial^2 f(\tilde{\boldsymbol{q}})}{\partial \tilde{q}_2 \partial \tilde{q}_n} \\ \vdots & \vdots & \ddots & \vdots \\ \frac{\partial^2 f(\tilde{\boldsymbol{q}})}{\partial \tilde{q}_n \partial \tilde{q}_1} & \frac{\partial^2 f(\tilde{\boldsymbol{q}})}{\partial \tilde{q}_n \partial \tilde{q}_2} & \cdots & \frac{\partial^2 f(\tilde{\boldsymbol{q}})}{\partial \tilde{q}_n \partial \tilde{q}_n} \end{bmatrix},$$

actually corresponds to[3]

$$H(\tilde{\boldsymbol{q}}) = \frac{\partial \boldsymbol{g}(\boldsymbol{q}_d - \tilde{\boldsymbol{q}})}{\partial(\boldsymbol{q}_d - \tilde{\boldsymbol{q}})} + \frac{2}{\varepsilon} K_p .$$

According to Lemma B.1, if $H(\tilde{\boldsymbol{q}}) > 0$ for all $\tilde{\boldsymbol{q}} \in \mathbb{R}^n$, then the function $f(\tilde{\boldsymbol{q}})$ is globally positive definite.

To show that $H(\tilde{\boldsymbol{q}}) > 0$ for all $\tilde{\boldsymbol{q}} \in \mathbb{R}^n$, we appeal to the following result. Let $A, B \in \mathbb{R}^{n\times n}$ be symmetric matrices. Assume moreover that the matrix A is positive definite, but B may not be so. If $\lambda_{\min}\{A\} > \|B\|$, then the matrix $A + B$ is positive definite[4]. Defining

---

[3] Let $\boldsymbol{f}, \boldsymbol{g} : \mathbb{R}^n \to \mathbb{R}^n$, $\boldsymbol{x}, \boldsymbol{y} \in \mathbb{R}^n$ and $\boldsymbol{x} = \boldsymbol{g}(\boldsymbol{y})$. Then,

$$\frac{\partial \boldsymbol{f}(\boldsymbol{x})}{\partial \boldsymbol{y}} = \frac{\partial \boldsymbol{f}(\boldsymbol{x})}{\partial \boldsymbol{x}} \frac{\partial \boldsymbol{g}(\boldsymbol{y})}{\partial \boldsymbol{y}} .$$

[4] *Proof.* Since by hypothesis $\lambda_{\min}\{A\} > \|B\|$, then

$$\lambda_{\min}\{A\} \|\boldsymbol{x}\|^2 > \|B\| \|\boldsymbol{x}\|^2$$

for all $\boldsymbol{x} \neq \boldsymbol{0}$.

Observe that the left-hand side of the inequality above satisfies

$$\boldsymbol{x}^T A \boldsymbol{x} \geq \lambda_{\min}\{A\} \|\boldsymbol{x}\|^2$$

while the right-hand side satisfies

$$\begin{aligned} \|B\| \|\boldsymbol{x}\|^2 &= \|B\| \|\boldsymbol{x}\| \|\boldsymbol{x}\| \\ &\geq \|B\boldsymbol{x}\| \|\boldsymbol{x}\| \\ &\geq \left|\boldsymbol{x}^T B \boldsymbol{x}\right| \\ &\geq -\boldsymbol{x}^T B \boldsymbol{x} . \end{aligned}$$

Therefore,

$$\boldsymbol{x}^T A \boldsymbol{x} > -\boldsymbol{x}^T B \boldsymbol{x}$$

for all $\boldsymbol{x} \neq \boldsymbol{0}$, that is

$$\boldsymbol{x}^T [A + B] \boldsymbol{x} > \boldsymbol{0} ,$$

which is equivalent to matrix $A + B$ being positive definite.

$A = \frac{2}{\varepsilon}K_p$, $B = \dfrac{\partial \boldsymbol{g}(\boldsymbol{q})}{\partial \boldsymbol{q}}$, and using the latter result, we conclude that the Hessian matrix is positive definite provided that

$$\lambda_{\min}\{K_p\} > \frac{\varepsilon}{2}\left\|\frac{\partial \boldsymbol{g}(\boldsymbol{q})}{\partial \boldsymbol{q}}\right\| . \tag{B.1}$$

Since the constant k_g satisfies $k_g \geq \left\|\frac{\partial \boldsymbol{g}(\boldsymbol{q})}{\partial \boldsymbol{q}}\right\|$, the condition (B.1) is implied by

$$\lambda_{\min}\{K_p\} > \frac{\varepsilon}{2}k_g .$$

♢

The following example may be considered as a corollary of the previous example.

Example B.3. This example shows that the function $f(\tilde{\boldsymbol{q}})$ defined in the previous example is lower-bounded by a quadratic function of $\tilde{\boldsymbol{q}}$ and therefore it is positive definite.

Specifically we show that

$$\mathcal{U}(\boldsymbol{q}_d - \tilde{\boldsymbol{q}}) - \mathcal{U}(\boldsymbol{q}_d) + \boldsymbol{g}(\boldsymbol{q}_d)^T\tilde{\boldsymbol{q}} + \frac{1}{2}\tilde{\boldsymbol{q}}^T K_p \tilde{\boldsymbol{q}} \geq \frac{1}{2}\left[\lambda_{\min}\{K_p\} - k_g\right]\|\tilde{\boldsymbol{q}}\|^2$$

is valid for all $\tilde{\boldsymbol{q}} \in \mathbb{R}^n$, with $K_p = K_p^T$ such that $\lambda_{\min}\{K_p\} > k_g$, where $\boldsymbol{q}_d \in \mathbb{R}^n$ is a constant vector and $\mathcal{U}(\boldsymbol{q})$ corresponds to the potential energy of the robot. As usual, we assume that all the joints of the robot are revolute.

To carry out the proof, we appeal to the argument of showing that the function

$$f(\tilde{\boldsymbol{q}}) = \mathcal{U}(\boldsymbol{q}_d - \tilde{\boldsymbol{q}}) - \mathcal{U}(\boldsymbol{q}_d) + \boldsymbol{g}(\boldsymbol{q}_d)^T\tilde{\boldsymbol{q}} + \frac{1}{2}\tilde{\boldsymbol{q}}^T K_p \tilde{\boldsymbol{q}} - \frac{1}{2}\left[\lambda_{\min}\{K_p\} - k_g\right]\|\tilde{\boldsymbol{q}}\|^2$$

is globally positive definite. With this objective in mind we appeal to Lemma B.1. Notice first that $f(\boldsymbol{0}) = 0$.

The gradient of $f(\tilde{\boldsymbol{q}})$ with respect to $\tilde{\boldsymbol{q}}$ is

$$\frac{\partial}{\partial \tilde{\boldsymbol{q}}} f(\tilde{\boldsymbol{q}}) = -\boldsymbol{g}(\boldsymbol{q}_d - \tilde{\boldsymbol{q}}) + \boldsymbol{g}(\boldsymbol{q}_d) + K_p\tilde{\boldsymbol{q}} - \left[\lambda_{\min}\{K_p\} - k_g\right]\tilde{\boldsymbol{q}} .$$

Clearly the gradient of $f(\tilde{\boldsymbol{q}})$ is zero at $\tilde{\boldsymbol{q}} = \boldsymbol{0} \in \mathbb{R}^n$.

The Hessian matrix $H(\tilde{\boldsymbol{q}})$ of $f(\tilde{\boldsymbol{q}})$ becomes

$$H(\tilde{\boldsymbol{q}}) = \frac{\partial \boldsymbol{g}(\boldsymbol{q}_d - \tilde{\boldsymbol{q}})}{\partial(\boldsymbol{q}_d - \tilde{\boldsymbol{q}})} + K_p - \left[\lambda_{\min}\{K_p\} - k_g\right] I .$$

We show next that the latter is positive definite. For this, we start from the fact that the constant k_g satisfies

$$k_g > \left\| \frac{\partial \boldsymbol{g}(\boldsymbol{q})}{\partial \boldsymbol{q}} \right\|$$

for all $\boldsymbol{q} \in \mathbb{R}^n$. Therefore, it holds that

$$\lambda_{\min}\{K_p\} - \lambda_{\min}\{K_p\} + k_g > \left\| \frac{\partial \boldsymbol{g}(\boldsymbol{q})}{\partial \boldsymbol{q}} \right\|$$

or equivalently,

$$\lambda_{\min}\{K_p\} - \lambda_{\mathrm{Max}}\left\{[\lambda_{\min}\{K_p\} - k_g]\, I\right\} > \left\| \frac{\partial \boldsymbol{g}(\boldsymbol{q})}{\partial \boldsymbol{q}} \right\| .$$

By virtue of the fact that for two symmetric matrices A and B we have that $\lambda_{\min}\{A - B\} \geq \lambda_{\min}\{A\} - \lambda_{\mathrm{Max}}\{B\}$, it follows that

$$\lambda_{\min}\left\{K_p - [\lambda_{\min}\{K_p\} - k_g]\, I\right\} > \left\| \frac{\partial \boldsymbol{g}(\boldsymbol{q})}{\partial \boldsymbol{q}} \right\| .$$

Finally, invoking the fact that for any given symmetric positive definite matrix A, and a symmetric matrix B it holds that $A + B > 0$ provided that $\lambda_{\min}\{A\} > \|B\|$, we conclude that

$$K_p - [\lambda_{\min}\{K_p\} - k_g]\, I + \frac{\partial \boldsymbol{g}(\boldsymbol{q})}{\partial \boldsymbol{q}} > 0$$

which corresponds precisely to the expression for the Hessian. Therefore, the latter is positive definite and according to Lemma B.1, we conclude that the function $f(\tilde{\boldsymbol{q}})$ is globally positive definite. ♢

C

Proofs of Some Properties of the Dynamic Model

Proof of Property 4.1.3

The proof of the inequality (4.2) follows straightforward invoking Corollary A.1. This is possible due to the fact that the inertia matrix $M(\boldsymbol{q})$ is continuous in $\boldsymbol{q}$ as well as the partial derivative of each of its elements $M_{ij}(\boldsymbol{q})$. Since moreover we considered the case of robots whose joints are all revolute, we obtain the additional characteristic that

$$\left| \left. \frac{\partial M_{ij}(\boldsymbol{q})}{\partial q_k} \right|_{\boldsymbol{q}=\boldsymbol{q}_0} \right|$$

is a function of $\boldsymbol{q}_0$ bounded from above.

Therefore, given any two vectors $\boldsymbol{x}, \boldsymbol{y} \in \mathbb{R}^n$, according to Corollary A.1, the norm of the vector $M(\boldsymbol{x})\boldsymbol{z} - M(\boldsymbol{y})\boldsymbol{z}$ satisfies

$$\|M(\boldsymbol{x})\boldsymbol{z} - M(\boldsymbol{y})\boldsymbol{z}\| \ \le \ n^2 \max_{i,j,k,\boldsymbol{q}_0} \left\{ \left| \left. \frac{\partial M_{ij}(\boldsymbol{q})}{\partial q_k} \right|_{\boldsymbol{q}=\boldsymbol{q}_0} \right| \right\} \|\boldsymbol{x} - \boldsymbol{y}\| \ \|\boldsymbol{z}\| \ .$$

Now, choosing the constant k_M in accordance with (4.3), *i.e.*

$$k_M = n^2 \max_{i,j,k,\boldsymbol{q}_0} \left\{ \left| \left. \frac{\partial M_{ij}(\boldsymbol{q})}{\partial q_k} \right|_{\boldsymbol{q}=\boldsymbol{q}_0} \right| \right\},$$

we obtain

$$\|M(\boldsymbol{x})\boldsymbol{z} - M(\boldsymbol{y})\boldsymbol{z}\| \ \le \ k_M \, \|\boldsymbol{x} - \boldsymbol{y}\| \ \|\boldsymbol{z}\|$$

which corresponds to the inequality stated in (4.2). 

Proof of Property 4.2.6

To carry out the proof of inequality (4.5) we start by considering (4.4) which allows one to express the vector $C(\boldsymbol{x}, \boldsymbol{z})\boldsymbol{w} - C(\boldsymbol{y}, \boldsymbol{v})\boldsymbol{w}$ as

$$C(\boldsymbol{x}, \boldsymbol{z})\boldsymbol{w} - C(\boldsymbol{y}, \boldsymbol{v})\boldsymbol{w} = \begin{bmatrix} \boldsymbol{w}^T C_1(\boldsymbol{x})\boldsymbol{z} \\ \boldsymbol{w}^T C_2(\boldsymbol{x})\boldsymbol{z} \\ \vdots \\ \boldsymbol{w}^T C_n(\boldsymbol{x})\boldsymbol{z} - \end{bmatrix} \begin{bmatrix} \boldsymbol{w}^T C_1(\boldsymbol{y})\boldsymbol{v} \\ \boldsymbol{w}^T C_2(\boldsymbol{y})\boldsymbol{v} \\ \vdots \\ \boldsymbol{w}^T C_n(\boldsymbol{y})\boldsymbol{v} \end{bmatrix} \tag{C.1}$$

where $C_k(\boldsymbol{q})$ is a symmetric matrix of dimension n, continuous in $\boldsymbol{q}$ and with the characteristic of that all of its elements $C_{k_{ij}}(\boldsymbol{q})$ are bounded for all $\boldsymbol{q} \in \mathbb{R}^n$ and moreover, so are its partial derivatives $(C_{k_{ij}}(\boldsymbol{q}) \in \mathcal{C}^\infty)$.

According to Equation (C.1), the vector $C(\boldsymbol{x}, \boldsymbol{z})\boldsymbol{w} - C(\boldsymbol{y}, \boldsymbol{v})\boldsymbol{w}$ may also be written as

$$C(\boldsymbol{x}, \boldsymbol{z})\boldsymbol{w} - C(\boldsymbol{y}, \boldsymbol{v})\boldsymbol{w} = \begin{bmatrix} \boldsymbol{w}^T \left[C_1(\boldsymbol{x}) - C_1(\boldsymbol{y})\right] \boldsymbol{z} - \boldsymbol{w}^T C_1(\boldsymbol{y})[\boldsymbol{v} - \boldsymbol{z}] \\ \boldsymbol{w}^T \left[C_2(\boldsymbol{x}) - C_2(\boldsymbol{y})\right] \boldsymbol{z} - \boldsymbol{w}^T C_2(\boldsymbol{y})[\boldsymbol{v} - \boldsymbol{z}] \\ \vdots \\ \boldsymbol{w}^T \left[C_n(\boldsymbol{x}) - C_n(\boldsymbol{y})\right] \boldsymbol{z} - \boldsymbol{w}^T C_n(\boldsymbol{y})[\boldsymbol{v} - \boldsymbol{z}] \end{bmatrix}$$

$$= \begin{bmatrix} \boldsymbol{w}^T \left[C_1(\boldsymbol{x}) - C_1(\boldsymbol{y})\right] \boldsymbol{z} \\ \boldsymbol{w}^T \left[C_2(\boldsymbol{x}) - C_2(\boldsymbol{y})\right] \boldsymbol{z} \\ \vdots \\ \boldsymbol{w}^T \left[C_n(\boldsymbol{x}) - C_n(\boldsymbol{y})\right] \boldsymbol{z} \end{bmatrix} - C(\boldsymbol{y}, \boldsymbol{v} - \boldsymbol{z})\boldsymbol{w}.$$

Evaluating the norms of the terms on each side of the equality above we immediately obtain

$$\|C(\boldsymbol{x}, \boldsymbol{z})\boldsymbol{w} - C(\boldsymbol{y}, \boldsymbol{v})\boldsymbol{w}\| \leq \left\| \begin{matrix} \boldsymbol{w}^T \left[C_1(\boldsymbol{x}) - C_1(\boldsymbol{y})\right] \boldsymbol{z} \\ \boldsymbol{w}^T \left[C_2(\boldsymbol{x}) - C_2(\boldsymbol{y})\right] \boldsymbol{z} \\ \vdots \\ \boldsymbol{w}^T \left[C_n(\boldsymbol{x}) - C_n(\boldsymbol{y})\right] \boldsymbol{z} \end{matrix} \right\| + \|C(\boldsymbol{y}, \boldsymbol{v} - \boldsymbol{z})\boldsymbol{w}\|. \tag{C.2}$$

We proceed next to determine upper-bounds on the two normed terms on the right-hand side of this inequality. First, using Lemma A.1 we get

$$\left\| \begin{matrix} \boldsymbol{w}^T \left[C_1(\boldsymbol{x}) - C_1(\boldsymbol{y})\right] \boldsymbol{z} \\ \boldsymbol{w}^T \left[C_2(\boldsymbol{x}) - C_2(\boldsymbol{y})\right] \boldsymbol{z} \\ \vdots \\ \boldsymbol{w}^T \left[C_n(\boldsymbol{x}) - C_n(\boldsymbol{y})\right] \boldsymbol{z} \end{matrix} \right\| \leq n \max_k \left\{ \left| \boldsymbol{w}^T \left[C_k(\boldsymbol{x}) - C_k(\boldsymbol{y})\right] \boldsymbol{z} \right| \right\}. \tag{C.3}$$

Furthermore, since the partial derivatives of the elements of the matrices $C_k(\boldsymbol{q})$ are bounded functions, Corollary A.1 leads to

$$\begin{aligned}\left|\boldsymbol{w}^T\left[C_k(\boldsymbol{x})-C_k(\boldsymbol{y})\right]\boldsymbol{z}\right| &\leq \left\|\left[C_k(\boldsymbol{x})-C_k(\boldsymbol{y})\right]\boldsymbol{z}\right\|\left\|\boldsymbol{w}\right\| \\ &\leq n^2 \max_{i,j,l,q_0}\left\{\left|\left.\frac{\partial C_{k_{ij}}(\boldsymbol{q})}{\partial q_l}\right|_{\boldsymbol{q}=\boldsymbol{q}_0}\right|\right\}\left\|\boldsymbol{x}-\boldsymbol{y}\right\| \\ &\quad \left\|\boldsymbol{z}\right\|\left\|\boldsymbol{w}\right\|,\end{aligned}$$

and therefore, it follows that

$$\begin{aligned}n\max_k\left\{\left|\boldsymbol{w}^T\left[C_k(\boldsymbol{x})-C_k(\boldsymbol{y})\right]\boldsymbol{z}\right|\right\} &\leq n^3 \max_{i,j,k,l,q_0}\left\{\left|\left.\frac{\partial C_{k_{ij}}(\boldsymbol{q})}{\partial q_l}\right|_{\boldsymbol{q}=\boldsymbol{q}_0}\right|\right\} \\ &\quad \left\|\boldsymbol{x}-\boldsymbol{y}\right\|\left\|\boldsymbol{z}\right\|\left\|\boldsymbol{w}\right\|.\end{aligned}$$

Incorporating this last inequality in (C.3) we finally obtain

$$\begin{aligned}&\left\|\begin{matrix}\boldsymbol{w}^T\left[C_1(\boldsymbol{x})-C_1(\boldsymbol{y})\right]\boldsymbol{z}\\ \boldsymbol{w}^T\left[C_2(\boldsymbol{x})-C_2(\boldsymbol{y})\right]\boldsymbol{z}\\ \vdots \\ \boldsymbol{w}^T\left[C_n(\boldsymbol{x})-C_n(\boldsymbol{y})\right]\boldsymbol{z}\end{matrix}\right\| \leq \\ &n^3 \max_{i,j,k,l,q_0}\left\{\left|\left.\frac{\partial C_{k_{ij}}(\boldsymbol{q})}{\partial q_l}\right|_{\boldsymbol{q}=\boldsymbol{q}_0}\right|\right\}\left\|\boldsymbol{x}-\boldsymbol{y}\right\|\left\|\boldsymbol{z}\right\|\left\|\boldsymbol{w}\right\|.\end{aligned} \tag{C.4}$$

On the other hand, using (4.8) it follows that the second normed term on the right-hand side of Inequality (C.2) may be bounded as

$$\left\|C(\boldsymbol{y},\boldsymbol{v}-\boldsymbol{z})\boldsymbol{w}\right\| \leq n^2\left(\max_{k,i,j,q}\left|C_{k_{ij}}(\boldsymbol{q})\right|\right)\left\|\boldsymbol{v}-\boldsymbol{z}\right\|\;\left\|\boldsymbol{w}\right\|. \tag{C.5}$$

Defining the constant k_{C_1} and k_{C_2} in accordance with table 4.1, *i.e.*

$$\begin{aligned}k_{C_1} &= n^2\left(\max_{i,j,k,q}\left|C_{k_{ij}}(\boldsymbol{q})\right|\right)\\ k_{C_2} &= n^3\left(\max_{i,j,k,l,q}\left|\frac{\partial C_{k_{ij}}(\boldsymbol{q})}{\partial q_l}\right|\right),\end{aligned}$$

and using (C.4) and (C.5) in the Inequality (C.2), we finally get

$$\left\|C(\boldsymbol{x},\boldsymbol{z})\boldsymbol{w}-C(\boldsymbol{y},\boldsymbol{v})\boldsymbol{w}\right\| \leq k_{C_1}\left\|\boldsymbol{v}-\boldsymbol{z}\right\|\;\left\|\boldsymbol{w}\right\| + k_{C_2}\left\|\boldsymbol{x}-\boldsymbol{y}\right\|\left\|\boldsymbol{z}\right\|\left\|\boldsymbol{w}\right\|,$$

which is what we wanted to demonstrate. 

Proof of Property 4.3.3

The proof of inequality (4.10) follows invoking Theorem A.3. Since the vector of gravitational torques $\boldsymbol{g}(\boldsymbol{q})$ is a vectorial continuous function, then for any two vectors $\boldsymbol{x}, \boldsymbol{y} \in \mathbb{R}^n$, we have

$$\boldsymbol{g}(\boldsymbol{x}) - \boldsymbol{g}(\boldsymbol{y}) = \left.\frac{\partial \boldsymbol{g}(\boldsymbol{q})}{\partial \boldsymbol{q}}\right|_{\boldsymbol{q}=\boldsymbol{\xi}} [\boldsymbol{x} - \boldsymbol{y}]$$

where $\boldsymbol{\xi} = \boldsymbol{y} + \alpha[\boldsymbol{x} - \boldsymbol{y}]$ and α is a number suitably chosen within the interval $[0, 1]$. Evaluating the norms of the terms on both sides of the equation above we obtain

$$\|\boldsymbol{g}(\boldsymbol{x}) - \boldsymbol{g}(\boldsymbol{y})\| \leq \left\| \left.\frac{\partial \boldsymbol{g}(\boldsymbol{q})}{\partial \boldsymbol{q}}\right|_{\boldsymbol{q}=\boldsymbol{\xi}} \right\| \|\boldsymbol{x} - \boldsymbol{y}\| . \tag{C.6}$$

On the other hand, using Lemma A.3, we get

$$\left\| \left.\frac{\partial \boldsymbol{g}(\boldsymbol{q})}{\partial \boldsymbol{q}}\right|_{\boldsymbol{q}=\boldsymbol{\xi}} \right\| \leq n \max_{i,j} \left\{ \left| \left.\frac{\partial g_i(\boldsymbol{q})}{\partial q_j}\right|_{\boldsymbol{q}=\boldsymbol{\xi}} \right| \right\} .$$

Furthermore, since we considered the case of robots with only revolute joints, the function

$$\left| \frac{\partial g_i(\boldsymbol{q})}{\partial q_j} \right|$$

is bounded. Therefore, it is also true that

$$\left\| \left.\frac{\partial \boldsymbol{g}(\boldsymbol{q})}{\partial \boldsymbol{q}}\right|_{\boldsymbol{q}=\boldsymbol{\xi}} \right\| \leq n \max_{i,j,q} \left\{ \left| \frac{\partial g_i(\boldsymbol{q})}{\partial q_j} \right| \right\} .$$

Incorporating this inequality in (C.6), we obtain

$$\|\boldsymbol{g}(\boldsymbol{x}) - \boldsymbol{g}(\boldsymbol{y})\| \leq n \max_{i,j,q} \left\{ \left| \frac{\partial g_i(\boldsymbol{q})}{\partial q_j} \right| \right\} \|\boldsymbol{x} - \boldsymbol{y}\| .$$

Choosing next the constant k_g as in (4.11), *i.e.*

$$k_g = n \left(\max_{i,j,q} \left| \frac{\partial g_i(\boldsymbol{q})}{\partial q_j} \right| \right)$$

which by the way implies, from Lemma A.3, that

$$k_g \geq \left\| \frac{\partial \boldsymbol{g}(\boldsymbol{q})}{\partial \boldsymbol{q}} \right\| ,$$

we finally get the expression

$$\|\boldsymbol{g}(\boldsymbol{x}) - \boldsymbol{g}(\boldsymbol{y})\| \leq k_g \|\boldsymbol{x} - \boldsymbol{y}\|$$

which is what we were seeking. 

D

Dynamics of Direct-current Motors

The actuators of robot manipulators may be electrical, hydraulic or pneumatic. The simplest electrical actuators used in robotics applications are permanent-magnet direct-current motors (DC).

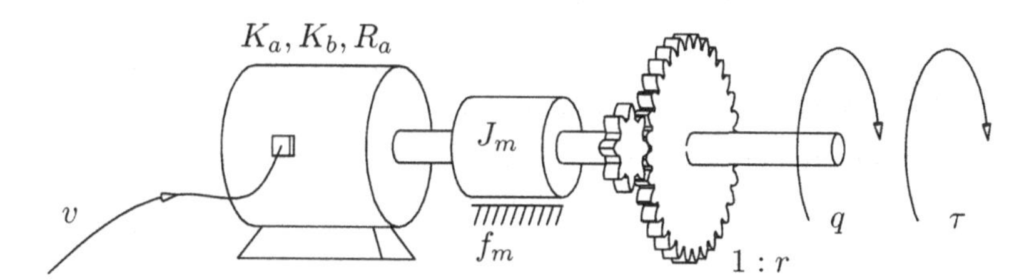

Figure D.1. DC motor

An idealized mathematical model that characterizes the behavior of a permanent-magnet DC motor controlled by the armature voltage is typically described by the set of equations (see Figure D.1)

$$\tau_m = K_a i_a \tag{D.1}$$

$$v = R_a i_a + L_a \frac{di_a}{dt} + e_b \tag{D.2}$$

$$e_b = K_b \dot{q}_m \tag{D.3}$$

$$q_m = rq,$$

where

- K_a : motor-torque constant (N m /A)
- R_a : armature resistance (Ω)
- L_a : armature inductance (H)
- K_b : back emf constant (V s/rad)

- τ_m : torque at the axis of the motor (N m)
- i_a : armature current (A)
- e_b : back emf (V)
- q_m : angular position of the axis of the motor (rad)
- q : angular position of the axis of the mechanical load[1] (rad)
- r : gears reduction ratio (in general $r \gg 1$)
- v : armature voltage (V).

The equation of motion for this system is

$$J_m \ddot{q}_m = \tau_m - f_m(\dot{q}_m) - \frac{\tau}{r} \tag{D.4}$$

where τ is the torque applied after the gear box at the axis of the load, J_m is the rotor inertia of the rotor, and $f_m(\dot{q}_m)$ is the torque due to friction between the rotor and its bearings, which in general, is a nonlinear function of its argument.

From a dynamic systems viewpoint, the DC motor may be regarded as a device whose input is the voltage v and output is the torque τ, which is applied after the gear box. Eventually, the time derivative $\dot{\tau}$ of the torque τ may also be considered as an output.

The dynamic model that relates the voltage v to the torque τ is obtained in the following manner. First, we proceed to replace i_a from (D.1) and e_b from (D.3) in (D.2) to get

$$v = \frac{R_a}{K_a}\tau_m + L_a \frac{di_a}{dt} + K_b \dot{q}_m \,. \tag{D.5}$$

Next, evaluating the time derivative on both sides of Equation (D.1) we obtain $\frac{di_a}{dt} = \dot{\tau}_m / K_a$ which, when replaced in (D.5) yields

$$v = \frac{R_a}{K_a}\tau_m + \frac{L_a}{K_a}\dot{\tau}_m + K_b \dot{q}_m \,. \tag{D.6}$$

On the other hand, from (D.4) we get τ_m as

$$\tau_m = J_m \ddot{q}_m + f_m(\dot{q}_m) + \frac{\tau}{r}$$

and whose time derivative is

$$\dot{\tau}_m = J_m \frac{d}{dt}\ddot{q}_m + \frac{\partial f_m(\dot{q}_m)}{\partial \dot{q}_m}\ddot{q}_m + \frac{\dot{\tau}}{r}$$

which, substituted in (D.6) yields

[1] For instance, a link of a robot.

$$v = \frac{R_a}{K_a}\left[J_m\ddot{q}_m + f_m(\dot{q}_m) + \frac{\tau}{r}\right] + \frac{L_a}{K_a}\left[J_m\frac{d}{dt}\ddot{q}_m + \frac{\partial f_m(\dot{q}_m)}{\partial \dot{q}_m}\ddot{q}_m + \frac{\dot{\tau}}{r}\right] + K_b\dot{q}_m\,.$$

Finally, using the relation $q_m = rq$, the previous equation may be written as

$$\frac{K_a}{rR_a}v = \frac{L_aJ_m}{R_a}\frac{d}{dt}\ddot{q} + \left(J_m + \frac{L_a}{R_a}\frac{\partial f_m(r\dot{q})}{\partial(r\dot{q})}\right)\ddot{q} + \frac{1}{r}f_m(r\dot{q}) + \frac{K_aK_b}{R_a}\dot{q} + \frac{\tau}{r^2} + \frac{L_a}{r^2R_a}\dot{\tau}\,, \tag{D.7}$$

which may also be expressed in terms of the state vector $[q \;\; \dot{q} \;\; \ddot{q}\,]$, as

$$\frac{d}{dt}\begin{bmatrix} q \\ \dot{q} \\ L_a\ddot{q}\end{bmatrix} = \begin{bmatrix} \dot{q} \\ \ddot{q} \\ \frac{1}{J_m}\left[\frac{K_a}{r}v - \left(R_aJ_m + L_a\frac{\partial f_m(r\dot{q})}{\partial(r\dot{q})}\right)\ddot{q} + g(\dot{q},\tau,\dot{\tau})\right]\end{bmatrix} \tag{D.8}$$

where

$$g(\dot{q},\tau,\dot{\tau}) = -\frac{R_a}{r}f_m(r\dot{q}) - K_aK_b\dot{q} - R_a\frac{\tau}{r^2} - L_a\frac{\dot{\tau}}{r^2}\,.$$

Equation (D.8) constitutes a differential equation of third order. In addition, this equation is nonlinear if the friction term $f_m(\cdot)$ is a nonlinear function of its argument. The presence of the armature inductance L_a multiplying $\frac{d}{dt}\ddot{q}$, causes the equation to be a 'singularly-perturbed'[2] differential equation for "small" inductance values.

Negligible Armature Inductance ($L_a \approx 0$)

In several applications, the armature inductance L_a is negligible ($L_a \approx 0$). In the rest of the present appendix we assume that this is the case. Thus, considering $L_a = 0$, Equation (D.7) becomes

$$J_m\ddot{q} + \frac{1}{r}f_m(r\dot{q}) + \frac{K_aK_b}{R_a}\dot{q} + \frac{\tau}{r^2} = \frac{K_a}{rR_a}v \tag{D.9}$$

or equivalently, in terms of the state vector $[q \;\; \dot{q}]$

$$\frac{d}{dt}\begin{bmatrix} q \\ \dot{q}\end{bmatrix} = \begin{bmatrix} \dot{q} \\ \frac{1}{J_m}\left[\frac{K_a}{rR_a}v - \frac{f_m(r\dot{q})}{r} - \frac{K_aK_b}{R_a}\dot{q} - \frac{\tau}{r^2}\right]\end{bmatrix}.$$

[2] See for instance H. Khalil, *Nonlinear Systems*, Prentice-Hall, 1996.

This important equation relates the voltage v applied to the armature of the motor, to the torque τ applied to the mechanical load, in terms of the angular position, velocity and acceleration of the latter.

Example D.1. Model of a motor with a mechanical load whose center of mass is located at the axis of rotation.
In the particular case when the load is modeled by a single inertia J_L with friction torques $f_L(\dot{q})$ as illustrated in Figure D.2, the torque τ is obtained from the equation of motion associated with the load as

$$J_L\ddot{q} = \tau - f_L(\dot{q}) \,. \tag{D.10}$$

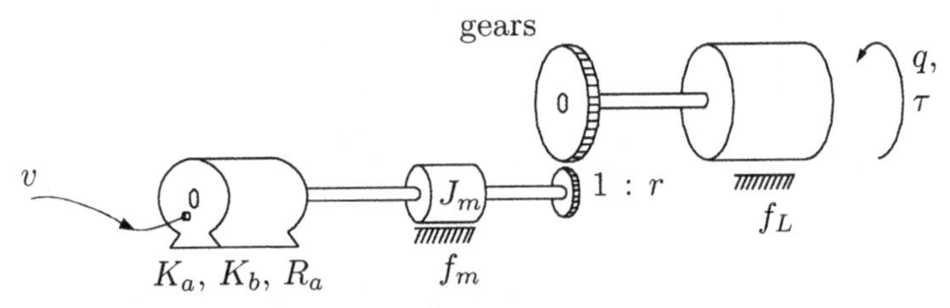

Figure D.2. DC motor with cylindrical inertia

The model of the motor-with-load for this case is obtained by substituting τ from (D.10) in (D.9), that is,

$$\left(\frac{J_L}{r^2} + J_m\right)\ddot{q} + \frac{1}{r}f_m(r\dot{q}) + \frac{1}{r^2}f_L(\dot{q}) + \frac{K_aK_b}{R_a}\dot{q} = \frac{K_a}{rR_a}v \,. \tag{D.11}$$

Example D.2. Model of a pendular device.
Consider the pendular device depicted in Figure D.3. The joint consists of a DC motor connected through gears to the pendular arm.

The equation of motion for the arm and its load is given by

$$\left(J + ml^2\right)\ddot{q} + f_L(\dot{q}) + (m_bl_b + ml)\,g\,\sin(q) = \tau \tag{D.12}$$

where

- J : arm inertia (without load)
- m_b : arm mass (without load)
- l_b : distance from the axis of rotation to the center of mass of the arm (without load)

- m : load mass (assumed concentrated at the center of mass)
- l : distance from the axis of rotation to the load m
- g : gravity acceleration
- τ : applied torque at the axis of rotation
- $f_L(\dot{q})$: friction torque between the arm and its bearings.

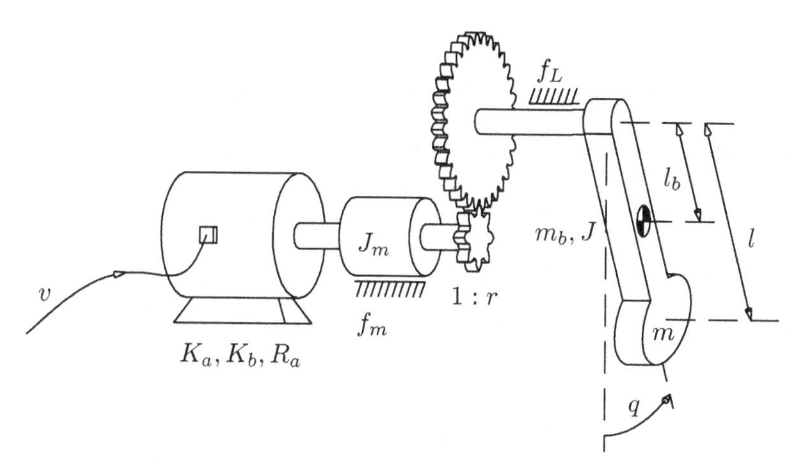

Figure D.3. Pendular device

Equation (D.12) may rewritten in compact form

$$J_L\ddot{q} + f_L(\dot{q}) + k_L\ \sin(q) = \tau \tag{D.13}$$

where

$$\begin{aligned} J_L &= J + ml^2 \\ k_L &= (m_b l_b + ml)\, g\,. \end{aligned}$$

Thus, the complete model of the pendular device that we obtain by substituting τ from (D.13) in the model of the DC motor (D.9) is

$$\left(\frac{J_L}{r^2} + J_m\right)\ddot{q} + \frac{1}{r}f_m(r\dot{q}) + \frac{1}{r^2}f_L(\dot{q}) + \frac{K_a K_b}{R_a}\dot{q} + \frac{k_L}{r^2}\ \sin(q) = \frac{K_a}{rR_a}v\,. \tag{D.14}$$

Notice that the previous model constitutes a nonlinear differential equation due not only to the friction torques $f_m(r\dot{q})$ and $f_L(\dot{q})$ but also due to the term $(k_L/r^2)\ \sin(q)$.

In the particular case where the gear reduction ratio r is high ($r \gg 1$) then, neglecting terms in $1/r^2$, the model (D.14) may be approximated by the model of the DC motor (D.9) with zero torque ($\tau = 0$). Also, observe that if the center of mass of the arm and that of the load are both located on the axis of rotation (*i.e.* $l_b = l = 0 \quad \Rightarrow$

$k_L = 0$) then, Equation (D.11) corresponds to the model of a motor with load whose center of mass is located on the axis of rotation.

D.1 Motor Model with Linear Friction

In spite of the complexity of friction phenomena, typically linear models are used to characterize their behavior, that is,

$$f_m(\dot{q}_m) = f_m \dot{q}_m \tag{D.15}$$

where f_m is a positive constant.

Considering the linear model above for the friction torques, Equation (D.9), that relates the voltage v applied at the armature of the motor to the torque τ applied on the load, takes the form

$$J_m \ddot{q} + \left(f_m + \frac{K_a K_b}{R_a} \right) \dot{q} + \frac{\tau}{r^2} = \frac{K_a}{r R_a} v \, . \tag{D.16}$$

Example D.3. Model of the motor with a load whose center of mass is located on the axis of rotation (linear friction).

Consider the linear model (D.15) for the friction torque corresponding to the rotor with respect to its bearings as well as the following linear equation for the friction torque between the load and its bearings:

$$f_L(\dot{q}) = f_L \dot{q}$$

where f_L is a positive constant. The model of the motor-with-load (D.11) reduces to

$$\left(\frac{J_L}{r^2} + J_m \right) \ddot{q} + \left(f_m + \frac{f_L}{r^2} + \frac{K_a K_b}{R_a} \right) \dot{q} = \frac{K_a}{r R_a} v$$

or, in compact form

$$\left(\frac{J_L}{r^2} + J_m \right) \ddot{q} + \frac{f_L}{r^2} \dot{q} + b \dot{q} = k v$$

where

$$b = f_m + \frac{K_a K_b}{R_a}$$

$$k = \frac{K_a}{r R_a} \, .$$

D.2 Motor Model with Nonlinear Friction

A more realistic model that characterizes the friction torques is given by the nonlinear expression

$$f_m(\dot{q}_m) = f_m \dot{q}_m + c_1 \, \text{sign}(\dot{q}_m) \tag{D.17}$$

where f_m and c_1 are positive constants (see Figure D.4) and $\text{sign}(\cdot)$ is the sign function which is defined as $\text{sign}(s) = -1$ if $s < 0$, $\text{sign}(s) = +1$ if $s > 0$ and at zero this function is discontinuous since clearly, its limits from the left and from the right are different.

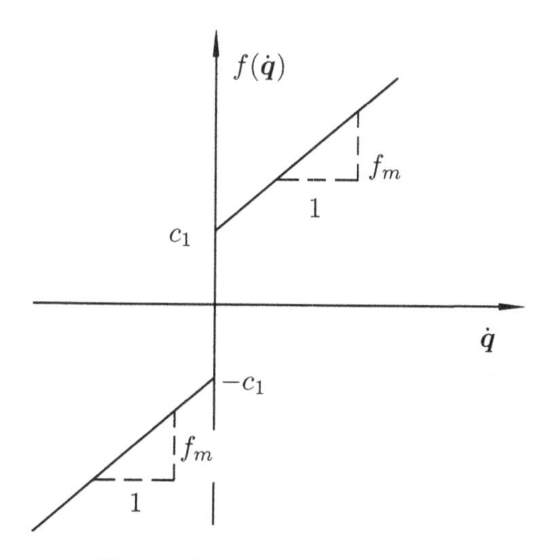

Figure D.4. Nonlinear friction

Strictly speaking one must be very careful in studying the stability of a system described by a differential equation which contain discontinuous functions. None of the theorems of stability presented here apply to this case. In general one cannot even guarantee the basic assumption that we made on page 28 that the solutions are unique for each initial condition. All this makes the study of Lyapunov stability for these systems a highly complex problem that is beyond the scope of this textbook.

Nevertheless, for the sake of completeness we present below the dynamic model of the motor with a discontinuous friction term. Considering the nonlinear model above for the friction torques, Equation (D.9) relating the voltage v applied to the armature of the motor together with the torque τ applied to the load, becomes

$$J_m \ddot{q} + \left(f_m + \frac{K_a K_b}{R_a} \right) \dot{q} + \frac{c_1}{r} \, \text{sign}(r\dot{q}) + \frac{\tau}{r^2} = \frac{K_a}{r R_a} v \, .$$

Example D.4. Model of the motor with load whose center of mass is located on the axis of rotation (nonlinear friction).
Consider the model of nonlinear friction (D.17) for the friction torque between the axis of the rotor and its bearings, and the corresponding load's friction,

$$f_L(\dot{q}) = f_L\dot{q} + c_2 \operatorname{sign}(\dot{q}) \tag{D.18}$$

where f_L and c_2 are positive constants.

Taking into account the functions (D.17) and (D.18), the motor-with-load model (D.11) becomes

$$(J_L + J_m)\ddot{q} + \left(f_m + \frac{K_a K_b}{R_a} + f_L\right)\dot{q} + (c_1 + c_2)\ \operatorname{sign}(\dot{q}) = \frac{K_a}{R_a}v$$

where for simplicity, we took $r = 1$. ♢

Bibliography

Derivation of the dynamic model of DC motors may be found in many texts, among which we suggest the reader to consult the following on control and robotics, respectively:

- Ogata K., 1970, *"Modern control engineering"*, Prentice-Hall.
- Spong M., Vidyasagar M., 1989, *"Robot dynamics and control"*, John Wiley and Sons, Inc.

Various nonlinear models of friction for DC motors are presented in

- Canudas C., Åström K. J., Braun K., 1987, *" Adaptive friction compensation in DC-motor drives"*, IEEE Journal of Robotics and Automation, Vol. RA-3, No. 6, December.
- Canudas C., 1988, *"Adaptive control for partially known systems—Theory and applications"*, Elsevier Science Publishers.
- Canudas C., Olsson H., Åström K. J., Lischinsky P., 1995, *"A new model for control of systems with friction"*, IEEE Transactions on Automatic Control, Vol. 40, No. 3, March, pp. 419–425.

An interesting paper dealing with the problem of definition of solutions for mechanical systems with discontinuous friction is

- Seung-Jean K., In-Joong Ha, 1999, *"On the existence of Carathéodory solutions in mechanical systems with friction"*, IEEE Transactions on Automatic Control, Vol. 44, No. 11, pp. 2086–2089.

Index